Methods in Molecular Biology

Series editor:
John M.Walker
School of Life and Medical Sciences
University of Hertfordshire
Hatfield, Hertfordshire, AL10 9AB, UK

For further volumes:
http://www.springer.com/series/7651

Transposons and Retrotransposons

Methods and Protocols

Edited by

Jose L. Garcia-Pérez

GENYO (Center for Genomics and Oncological Research), Pfizer/Universidad de Granada/Junta de Andalucia, PTS Granada, Spain;
Institute of Genetics and Molecular Medicine (IGMM), University of Edinburgh, United Kingdom

Editor
Jose L. Garcia-Pérez
GENYO (Center for Genomics and Oncological Research)
Pfizer/Universidad de Granada/Junta de Andalucia
PTS Granada, Spain

Institute of Genetics and Molecular Medicine (IGMM)
University of Edinburgh
United Kingdom

ISSN 1064-3745 ISSN 1940-6029 (electronic)
Methods in Molecular Biology
ISBN 978-1-4939-3370-9 ISBN 978-1-4939-3372-3 (eBook)
DOI 10.1007/978-1-4939-3372-3

Library of Congress Control Number: 2015960958

Springer New York Heidelberg Dordrecht London

Printed on acid-free paper

Humana Press is a brand of Springer
Springer Science+Business Media LLC New York is part of Springer Science+Business Media (www.springer.com)

Preface

Upon completion of the human genome and the ENCODE projects, together with "the genomics revolution" that has (and is providing) provided an enormous amount of DNA sequencing information of virtually any living creature in the world, today it is undeniable that transposable elements (TEs) have been main drivers during genome evolution. Indeed, and although often classified as "selfish DNA" or termed "junk DNA," it is obvious that TEs manifest a tremendous impact on genome biology and its regulation.

More than 25 years have passed since the seminal discovery that revealed that some TEs are still active in humans, acting sporadically as insertional mutagens. In the past years, the field of TE biology has dramatically expanded the horizons of research associated with "junk DNA," and the impact of TEs in genomes has provided exciting and provocative hypothesis like their putative role in brain biology and cancer biology. Coupled with new sequencing approaches, it is clear that the field of TE biology has entered a new and exciting era in which more and more scientists are challenging the concept of "junk DNA" using very different approaches and experimental settings. Thus, it is my hope that this collection of detailed protocols will serve as a driver to inspire new research in this fascinating area of research.

In this book, I intent to cover common but often laborious protocols developed by very talented scientists and used by laboratories located in all parts of the world. Because of their repeated nature, the identification of new TE insertions has proven an extremely complicated approach. This Method book covers some of the latest protocols designed to identify and characterize TEs in genomes, ancient or recently inserted. Additionally, this book includes a series of protocols designed to understand how host genomes act to regulate the activity of TEs, from elegant genetic mobilization assays to key biochemical methods that have proven very important to understand the complex interplay between TEs and genomes. Because of their high prevalence in genomes, understanding TE genome wide regulation and its impact is an area of intensive research and a number of chapters will cover these topics. Finally, this book also includes chapters that describe how TEs can be used for useful biotechnological applications.

I am really indebted to a group of friends and very talented scientific colleagues that have put together an outstanding compilation of Methods that are very easy to follow. Generous, fun, and hard working scientists that have been brave to, despite their "junk" nature, dedicate their careers to understand the impact of TEs in genomes form this field of research. And while doing that, they have developed sophisticated tools that I hope might be of interest to many other researchers. I would also like to dedicate this book to previous talented researchers that, despite not having a chapter in this book, have acted as main drivers in the field. Thank you to all contributors and past researchers in this field! It is a great honor to work with you in the preparation of this book and I hope it becomes a reference book in the future that will be used by contributors but also will help others to expand their horizons of research in the near term.

Granada, Spain *Jose L. Garcia-Pérez Ph.D,*

Contents

Contributors

CATHERINE ADE • *Department of Epidemiology, Tulane Cancer Center, Tulane University Health Sciences Center, New Orleans, LA, USA*

SUYAPA AMADOR • *Department of Human DNA Variability, Pfizer/University of Granada and Andalusian Regional Government Center for Genomics and Oncology (GENYO), Granada, Spain*

ANGELA ATWOOD-MOORE • *Section on Eukaryotic Transposable Elements, Program in Cellular Regulation and Metabolism, The Eunice Kennedy Shriver National Institute of Child Health and Human Development, National Institutes of Health, Bethesda, MD, USA*

RICHARD M. BADGE • *Department of Genetics, University of Leicester, Leicester, UK*

EUGÉNIA BASYUK • *Institut de Génétique Moléculaire de Montpellier, CNRS (UMR5535), Montpellier cedex 5, France*

VICTORIA P. BELANCIO • *Department of Structural and Cellular biology, Tulane School of Public Health, New Orleans, LA, USA; Tulane Cancer Center and Center for Aging, New Orleans, LA, USA*

ANJA BOCK • *Division of Medical Biotechnology, Paul-Ehrlich-Institut, Langen, Germany*

MAXIME BODAK • *Department of Biology, Swiss Federal Institute of Technology, Zurich, Switzerland*

JEF D. BOEKE • *Department of Biochemistry and Molecular Pharmacology, Institute for Systems Genetics, New York University Langone School of Medicine, New York, NY, USA*

GUILLAUME BOURQUE • *Department of Human Genetics, McGill University, Montreal, QC, Canada; McGill University and Génome Québec Innovation Center, Montréal, OC, Canada*

MIGUEL R. BRANCO • *Blizard Institute, School of Medicine and Dentistry, QMUL, London, UK*

KATHLEEN H. BURNS • *Department of Pathology, Johns Hopkins University School of Medicine, Baltimore, MD, USA*

DAVID CANO • *Department of Human DNA Variability, Pfizer/University of Granada and Andalusian Regional Government Center for Genomics and Oncology (GENYO), Granada, Spain*

CONSTANCE CIAUDO • *Department of Biology, Swiss Federal Institute of Technology, Zurich, Switzerland*

GAEL CRISTOFARI • *INSERM/CNRS (U1081 and UMR 7284) Institute for Research on Cancer and Aging of Nice (IRCAN), Nice, France; Faculty of Medicine, University of Nice-Sophia-Antipolis, Nice, France*

LIXIN DAI • *Department of Molecular Biology and Genetics, High Throughput Biology Center, Johns Hopkins University School of Medicine, Baltimore, MD, USA; Modern Meadow Inc., Brooklyn, USA*

Prescott Deininger • *Department of Epidemiology, Tulane School of Public Health, New Orleans, LA, USA; Tulane Cancer Center and Center for Aging, New Orleans, LA, USA*
Aurélien J. Doucet • *Department of Human Genetics, University of Michigan Medical School, Ann Arbor, MI, USA; INSERM/CNRS (U1081 and UMR 7284) Institute for Research on Cancer and Aging of Nice (IRCAN), Nice cedex 2, France; Faculty of Medicine, University of Nice-Sophia-Antipolis, Nice cedex 2, France*
Tara T. Doucet • *McKusick-Nathans Institute of Genetic Medicine, Johns Hopkins University School of Medicine, Baltimore, MD, USA*
Gabriela Ecco • *School of Life Sciences, École Polytechnique Fédérale de Lausanne (EPFL), Lausanne, Switzerland*
Geoffrey J. Faulkner • *Mater Research Institute, University of Queensland, Woolloongabba, QLD, Australia; Queensland Brain Institute, University of Queensland, Brisbane, QLD, Australia*
Pedro M. Fernández-Salguero • *Departamento de Bioquimica y Biologia Molecular, Facultad de Ciencias, Universidad de Extremadura, Badajoz, Spain*
Diane A. Flasch • *Department of Human Genetics, University of Michigan Medical School, Ann Arbor, MI, USA*
Jose L. Garcia-Pérez • *GENYO (Center for Genomics and Oncological Research), Pfizer/Universidad de Granada/Junta de Andalucia, PTS Granada, Spain; Institute of Genetics and Molecular Medicine (IGMM), University of Edinburgh, United Kingdom*
Fernando Manuel García-Rodríguez • *Grupo de Ecología Genética de la Rizosfera, Estación Experimental del Zaidín, Consejo Superior de Investigaciones Científicas, Granada, Spain*
Nicolas Gilbert • *Institute for Regenerative Medicine and Biotherapy, INSERM(U1183), Montpellier, France*
Jeffrey S. Han • *Department of Biochemistry and Molecular Biology, Tulane University School of Medicine, New Orleans, LA, USA*
Axel V. Horn • *Department of Biochemistry and Molecular Biology, Tulane University School of Medicine, New Orleans, LA, USA*
Zoltan Ivics • *Division of Medical Biotechnology, Paul Ehrlich Institute, Langen, Germany*
Haig H. Kazazian Jr. • *McKusick-Nathans Institute of Genetic Medicine, Johns Hopkins University School of Medicine, Baltimore, MD, USA; Department of Pediatrics, Johns Hopkins University School of Medicine, Baltimore, MD, USA*
Huira C. Kopera • *Department of Human Genetics, University of Michigan Medical School, Ann Arbor, MI, USA*
John LaCava • *Laboratory of Cellular and Structural Biology, The Rockefeller University, New York, NY, USA; Department of Biochemistry and Molecular Pharmacology, Institute for Systems Genetics, New York University Langone School of Medicine, New York, NY, USA*
Peter A. Larson • *Department of Human Genetics, University of Michigan Medical School, Ann Arbor, MI, USA*
Henry L. Levin • *Section on Eukaryotic Transposable Elements, Program in Cellular Regulation and Metabolism, The Eunice Kennedy Shriver National Institute of Child Health and Human Development, National Institutes of Health, Bethesda, MD, USA*
Ying Liu • *Howard Hughes Medical Institute, Chevy Chase, MD, USA*

CESAR LOPEZ-RUIZ • *Department of Human DNA Variability, Pfizer/University of Granada and Andalusian Regional Government Center for Genomics and Oncology (GENYO), Granada, Spain*
DIXIE L. MAGER • *Terry Fox Laboratory, British Columbia Cancer Agency, Vancouver, BC, Canada; Department of Medical Genetics, University of British Columbia, Vancouver, BC, Canada*
PRABHAT K. MANDAL • *Department of Biotechnology, Indian Institute of Technology Roorkee, Roorkee, Uttarakhand, India*
FRANCISCO MARTÍNEZ-ABARCA • *Grupo de Ecología Genética de la Rizosfera, Estación Experimental del Zaidín, Consejo Superior de Investigaciones Científicas, Granada, Spain*
PAOLO MITA • *Department of Biochemistry and Molecular Pharmacology, Institute for Systems Genetics, New York University Langone School of Medicine, New York, NY, USA*
TOMOICHIRO MIYOSHI • *Department of Human Genetics, University of Michigan Medical School, Ann Arbor, MI, USA*
JOHN B. MOLDOVAN • *Cellular and Molecular Biology Program, University of Michigan Medical School, Ann Arbor, MI, USA*
MARÍA DOLORES MOLINA-SÁNCHEZ • *Grupo de Ecología Genética de la Rizosfera, Estación Experimental del Zaidín, Consejo Superior de Investigaciones Científicas, Granada, Spain*
ANTONIO MORALES-HERNÁNDEZ • *Departamento de Bioquimica y Biologia Molecular, Facultad de Ciencias, Universidad de Extremadura, Badajoz, Spain*
JOHN V. MORAN • *Department of Human Genetics, University of Michigan Medical School, Ann Arbor, MI, USA; Cellular and Molecular Biology Program, University of Michigan Medical School, Ann Arbor, MI, USA; Department of Internal Medicine, University of Michigan Medical School, Ann Arbor, MI, USA; Howard Hughes Medical Institute, Chevy Chase, MD, USA*
SANTIAGO MORELL • *Department of Human DNA Variability, Pfizer/University of Granada and Andalusian Regional Government Center for Genomics and Oncology (GENYO), Granada, Spain*
MARTIN MUÑOZ-LOPEZ • *Department of Human DNA Variability, Pfizer/University of Granada and Andalusian Regional Government Center for Genomics and Oncology (GENYO), Granada, Spain*
ALYSSON R. MUOTRI • *Stem Cell Program, Department of Cellular and Molecular Medicine, University of California San Diego, School of Medicine, Pediatrics/Rady Children's Hospital San Diego, La Jolla, CA, USA*
MITSUHIRO NAKAMURA • *Department of Human Genetics, University of Michigan Medical School, Ann Arbor, MI, USA*
RAFAEL NISA-MARTÍNEZ • *Grupo de Ecología Genética de la Rizosfera, Estación Experimental del Zaidín, Consejo Superior de Investigaciones Científicas, Granada, Spain*
ANDRES J. PULGARIN • *Department of Human DNA Variability, Pfizer/University of Granada and Andalusian Regional Government Center for Genomics and Oncology (GENYO), Granada, Spain*
RAHELEH RAHBARI • *The Wellcome Trust Sanger Institute, Wellcome Trust Genome Campus Hinxton, Cambridge, UK*
SUDHIR K. RAI • *Section on Eukaryotic Transposable Elements, Program in Cellular Regulation and Metabolism, The Eunice Kennedy Shriver National Institute of Child Health and Human Development, National Institutes of Health, Bethesda, MD, USA*

LEEANN RAMSAY • *Department of Human Genetics, McGill University, Montreal, Canada*
RITA REBOLLO • *Terry Fox Laboratory, British Columbia Cancer Agency, Vancouver, BC, Canada; Department of Medical Genetics, University of British Columbia, Vancouver, BC, Canada*
LORENZO DE LA RICA • *Blizard Institute, School of Medicine and Dentistry, London, UK*
SANDRA R. RICHARDSON • *Mater Research Institute, University of Queensland, Woolloongabba, Australia; Department of Human Genetics, University of Michigan Medical School, Ann Arbor, MI, USA*
NEMANJA RODIĆ • *Department of Pathology, Johns Hopkins University School of Medicine, Baltimore, MD, USA*
ÁNGEL CARLOS ROMÁN • *Champalimaud Neuroscience Programme, Champalimaud Center for the Unknown, Lisbon, Portugal*
MICHAEL P. ROUT • *Laboratory of Cellular and Structural Biology, The Rockefeller University, New York, NY, USA*
HELEN M. ROWE • *School of Life Sciences, École Polytechnique Fédérale de Lausanne, Lausanne (EPFL), Switzerland; Division of Infection and Immunity, Centre for Medical Molecular Virology, University College London, London, UK*
ASTRID M. ROY-ENGEL • *Department of Epidemiology, Tulane Cancer Center, Tulane University Health Sciences Center, New Orleans, LA, USA*
FRANCISCO J. SANCHEZ-LUQUE • *Mater Research Institute, University of Queensland, Woolloongabba, QLD, Australia; Pfizer-University of Granada-Andalusian Goverment Centre for Genomics and Oncological Research, Granada, Spain*
MAYA SANGESLAND • *Section on Eukaryotic Transposable Elements, Program in Cellular Regulation and Metabolism, The Eunice Kennedy Shriver National Institute of Child Health and Human Development, National Institutes of Health, Bethesda, MD, USA*
GERALD G. SCHUMANN • *Division of Medical Biotechnology, Paul-Ehrlich-Institut, Langen, Germany*
ATTILA SEBE • *Division of Medical Biotechnology, Paul Ehrlich Institute, Langen, Germany*
REEMA SHARMA • *Department of Pathology, Johns Hopkins University School of Medicine, Baltimore, MD, USA*
JATINDER S. STANLEY • *Blizard Institute, School of Medicine and Dentistry, QMUL, London, UK*
MARTIN S. TAYLOR • *Department of Pathology, Johns Hopkins University School of Medicine, Baltimore, MD, USA; Department of Pathology, Massachusetts General Hospital, Boston, MA, USA*
NICOLAS TORO • *Grupo de Ecología Genética de la Rizosfera, Estación Experimental del Zaidín, Consejo Superior de Investigaciones Científicas, Granada, Spain*
PABLO TRISTAN-RAMOS • *Department of Human DNA Variability, Pfizer/University of Granada and Andalusian Regional Government Center for Genomics and Oncology (GENYO), Granada, Spain*
DIDIER TRONO • *School of Life Sciences, École Polytechnique Fédérale de Lausanne (EPFL), Lausanne, Switzerland*
RAQUEL VILAR-ASTASIO • *Department of Human DNA Variability, Pfizer/University of Granada and Andalusian Regional Government Center for Genomics and Oncology (GENYO), Granada, Spain*
SÉBASTIEN VIOLLET • *INSERM/CNRS (U1081 and UMR 7284) Institute for Research on Cancer and Aging of Nice (IRCAN), Nice, France; Faculty of Medicine, University of Nice-Sophia-Antipolis, Nice, France*

Chapter 1

Study of Transposable Elements and Their Genomic Impact

Martin Muñoz-Lopez, Raquel Vilar-Astasio, Pablo Tristan-Ramos, Cesar Lopez-Ruiz, and Jose L. Garcia-Pérez

Abstract

Transposable elements (TEs) have been considered traditionally as *junk DNA*, i.e., DNA sequences that despite representing a high proportion of genomes had no evident cellular functions. However, over the last decades, it has become undeniable that not only TE-derived DNA sequences have (and had) a fundamental role during genome evolution, but also TEs have important implications in the origin and evolution of many genomic disorders. This concise review provides a brief overview of the different types of TEs that can be found in genomes, as well as a list of techniques and methods used to study their impact and mobilization. Some of these techniques will be covered in detail in this Method Book.

Key words Transposable element, Retrotransposon, DNA transposon, LINE-1, SINE, Retrotransposition, Transposition

1 Types of Transposable Elements

We start with the description of the major types of TEs that can be identified in genomes.

(a) *DNA Transposons.* This group of TEs is widely represented in different taxa of organisms, from bacteria, ciliates, rotifers, fungi, and plants to nematodes, insects, and vertebrates [1–6]. The genomic structure of DNA transposons is constituted by a gene, which encodes a *transposase* protein, flanked by two *terminal inverted repeats* (TIRs) (Fig. 1). DNA transposons move by a "*cut-and-paste*" mechanism, according to which the *transposase* is able to recognize the TIRs, performs the transposon DNA excision by cleaving both ends, then cleaves in a specific target site, and performs the integration of the transposon DNA in a new genomic location [7] (Fig. 1). Upon insertion, a sequence in the target site is duplicated generating a *target site duplication* (TSD) [7]. In general, no active DNA transposons have been described in mammals, with some

Jose L. Garcia-Pérez (ed.), *Transposons and Retrotransposons: Methods and Protocols*, Methods in Molecular Biology, vol. 1400, DOI 10.1007/978-1-4939-3372-3_1, © Springer Science+Business Media New York 2016

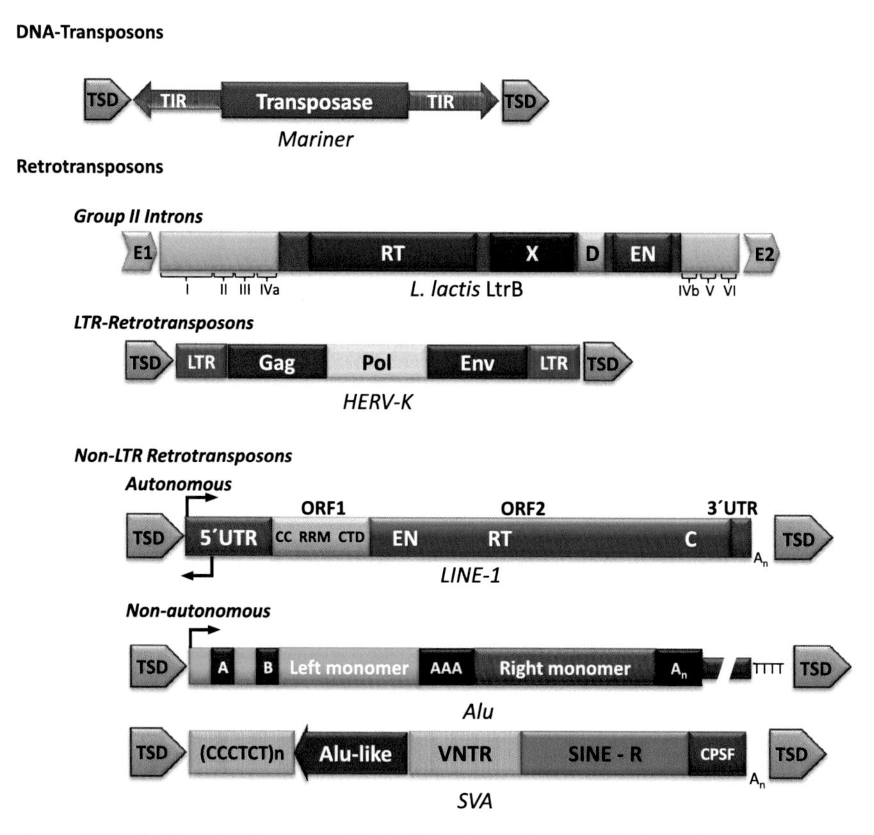

Fig. 1 Structure of TEs. Further details are provided within the main text and cited references. *DNA transposons*, TSD (target site duplication), TIR (terminal inverted repeat); *Group II introns*, I-VI (RNA domains), E (Exon), RT (reverse transcriptase domain), X (X domain), D (D motif), EN (endonuclease domain); *LTR retrotransposons*, LTR (long terminal repeats), Gag (group-specific antigen), Pol (polymerase), Env (dysfunctional envelope protein); *LINE-1*, UTR (untranslated region), ORF (open reading frame), CC (coiled coil), RRM (RNA recognition motif), CTD (carboxyl-terminal domain), C (cysteine-rich domain); *Alu*, A and B (component sequences of the RNA polymerase III promoter), left and right 7SL monomers, AAA (adenosine-rich region); *SVA*, VNTR (variable number of tandem repeats), SINE-R (domain derived from an HERV-K), CPSF (cleavage polyadenylation-specific factor)

exceptions in some bat species [8–10]. Nonetheless, DNA transposons are becoming very useful as *genomic tools* (in *transgenesis* and as *insertional mutagens*) in the genome of different mammals, including humans, due to their specific features [11–13]. One of the most popular DNA transposons used as a genomic tool is *Sleeping Beauty*, a DNA transposon fossil reanimated from salmonid-type fishes [14], and used widely to discover genes implicated in cancer by *insertional mutagenesis* [15–17], *transgenesis* [18, 19], and *gene therapy* [20, 21]. *PiggyBac*, from the moth *Trichoplusia ni*, and *Tol2*, from Medaka fish (*Oryzias latipes*) [22], have also been used to knock out genes [23–25] and in transgenesis experiments [26–29].

(b) *Retrotransposons*. Within this group, retrotransposons can be subclassified into different subgroups as follows:

- *Group II introns*. Group II introns are bacterial TEs that move to new genomic locations by a *reverse splicing* mechanism from a precursor intron RNA. These elements are indeed mobile ribozymes constituted by a catalytically active intron RNA that encode a reverse transcriptase-containing protein (IEP, intron-encoded protein) (Fig. 1). When group II intron elements are transcribed, the resultant RNA catalyzes its own *splicing* giving rise to spliced 5′ and 3′ exons and the excised intron (*forward splicing*) [30]. IEP participates in the mobilization by stabilizing the RNA to perform the *splicing*, forming ribonucleoprotein particles. Next, the intron RNA is integrated in a different genomic site and reverse transcribed by the IEP to DNA (*reverse splicing*) [30]. Thus, by this cycle of *forward* and *reverse splicing*, group II introns proliferate in the bacterial and organellar genomes without relevant alteration of gene expression. Interestingly, characteristics associated with group II introns suggest that they might represent the origin of eukaryotic retrotransposons and spliceosomal introns.
- *LTR Retrotransposons. Long terminal repeat* (LTR) retroelements share features with retrovirus regarding their structure and mobilization mechanisms. However, LTR retrotransposons may have lost the extracellular mobility and thus they are also known as endogenous retroviruses (ERVs). In general, LTR retrotransposons lack a functional envelope gene (Env) but contains functional Gag and Pol genes organized in an open reading frame (ORF) flanked by the LTRs (Fig. 1). In the human genome, human endogenous retroviruses (HERVs) comprise almost a tenth of our genome [31] although they are thought to be no longer active. However, some HERV-K might retain some mobilization capability [32], at least some rare alleles. On the other hand, in the mouse genome, mouse ERVs move actively by a "*copy-and-paste*" mechanism with a considerable mutagenic capacity [33]. Furthermore, ERVs in the mouse are widely represented by several subfamilies as intracisternal A particles (IAP), Mus D elements, mammalian apparent LTR retrotransposon (MALRs), and non-autonomous early transposons (ETns) [33].
- *Non-LTR Retrotransposons*. Long interspersed element class 1 (LINE-1 or L1) is the only active autonomous transposable element in our genome. L1s comprise more than 17 % of the human genome, although only 80–100

L1 copies are currently active [34]. In addition, two non-autonomous retrotransposons known as short interspersed elements (SINEs), including *Alu* and *SVA* retrotransposons (and comprising approximately 10 and 1 % of our genome, respectively), can mobilize using the L1 enzymatic machinery [34] (Fig. 1). L1s and SINEs move by a "*copy-and-paste*" mechanism using an intermediate RNA. An active L1 sequence is 6 kb in length and includes a 5′UTR (untranslated region), two non-overlapping ORFs, and ends in a 3′UTR containing a poly(A) tail ([35] reviewed in [34]) (Fig. 1). The L1-5′UTR contains a sense promoter that drives the L1 expression [36], and a CpG island that regulates its expression [37, 38]. In addition, the L1-5′UTR harbors a conserved antisense promoter [39] whose function has been related to L1 self-regulation by the generation of interference RNAs [40–42]. ORF1 and ORF2 encode an RNA-binding protein and a protein with *endonuclease* (EN) and *reverse transcriptase* (RT) activities, respectively (reviewed in [34, 43, 44]). Both proteins are strictly required for L1 retrotransposition [34, 45]. Retrotransposition begins with the transcription from an active L1; the L1-mRNA is transported to the cytoplasm and after translation of several ORF1p proteins and only one molecule of ORF2p [46, 47], these proteins bind *in cis* to the L1-mRNA [48] forming a ribonucleoprotein particle (RNP). The RNP gets into the nucleus and performs retrotransposition by a process termed *target prime reverse transcription* (TPRT) [49, 50]. During TPRT, the EN activity of ORF2p recognizes and cleaves a loose consensus sequence [51, 52]. Next, using the free 3′OH released, the RT activity starts first-strand cDNA synthesis (reviewed in [34, 43, 44]). Then, second-strand cDNA synthesis occurs by an analogous mechanism, although this process is not totally understood. This mechanism results in the generation of a *de novo* L1 insertion, usually flanked by TSDs. Notably, most new L1 insertions are dead on arrival due to frequent 5′-truncation processes [49, 52, 53].

2 Mobilization of LINE-1 in the Human Genome

As a general rule, TEs are considered as *selfish DNA* or *molecular parasites*. Thus, these elements would have only one purpose: to be perpetuated in genomes over generations. If this is really their goal, the mobilization of L1 would only make sense in cellular types where new L1 insertions could be transmitted to a newborn. Therefore, L1 activity must be prominent in the germ line or during early embryonic development. Indeed, the activity of L1 has been detected during early embryonic development [54–56], although germ cell L1 insertions seem to be less frequent [55, 57]. Nevertheless, more recently L1 retrotransposition has been associated with viability of fetal oocytes in mice [58]. L1 activity during early embryogenesis has been analyzed in human *embryonic stem cells* (hESCs) [56, 59] and in human *induced pluripotent stem cell* (iPSCs) [38], which are considered models of early embryonic development. In hESCs and human iPSCs, high levels of L1 expression have been reported (with respect to somatic cells [38, 56]). Furthermore, engineered L1 retrotransposition assays have demonstrated that L1 can move in both cell types [38, 56], suggesting that the minimal host factors required for retrotransposition are present in these cells. Work in iPSCs has revealed that, upon reprogramming, L1 expression is reactivated with respect to the parental cells, by methylation changes in the L1 promoter [38]. In sum, different studies suggest that heritable L1 insertions can accumulate during human early embryonic development.

Surprisingly, L1 activity is not restricted to early human embryogenesis, and L1 activity has been described in neural progenitor cells (NPCs) [60–62], tumor cells [63–68], and more recently in neurons of the human brain [69, 70]. There is not a clear explanation to support a role for L1 activity in the human brain; indeed, this somatic-in-nature activity of L1 could be just a consequence derived from the nature of this cellular type. However, it is also possible that L1 is performing a "function" in the brain, and ongoing retrotransposition might represent a real-time *domestication* phenomenon. On the other hand, studies in cancer cells have revealed that there is a good correlation between DNA methylation levels and the load of L1 insertions in tumors, which in some cases may implicate L1 in the origin and evolution of cancer by the mutation of genes involved in tumor genesis [63–68, 71].

3 Impact of LINE-1 Retrotransposition on the Human Genome

L1 constitutes a source of mutations in our genome. Therefore its control must be important to maintain human genome integrity. The source of genomic variability represented by L1 has been very significant during our evolution, as high variability might imply a higher evolving capacity. However, at an individual level, this source of mutation can be problematic because it can result in the generation of genomic disorders (reviewed in [72]), including cancer [63–68, 71]. Indeed, in the same way as L1 has participated in human genome evolution, it could also participate in the origin and evolution of cancer. It is known from 20 years ago that the ongoing activity of L1 can likely be a driver in cancer. In fact, a seminal study demonstrated that a *de novo* L1 insertion in the tumor-suppressor gene APC was probably the cause of a sporadic case of colorectal cancer [71]. More recently, the genomics revolution allowed *Iskow and colleagues* (and others [63]) to demonstrate that endogenous L1s move with high frequency in lung cancer genomes. Similar results have been reported by others in different epithelial tumors (colorectal cancer, prostate cancer, ovarian cancer, etc.) [64–66]. Indeed, it is worth noting that many de novo L1 insertions were located in genes implicated in tumorigenesis. These observations suggest that this "abnormal somatic L1 activity" could be involved in the etiology of cancer. Furthermore, *Shukla and colleagues* reported that in hepatocellular cancer, L1 insertions in a tumor suppressor (MCC) and in an oncogene (ST18) resulted in decreased MCC expression and ST18 over-expression, giving rise to aberrant oncogenic signaling [67]. Thus, it is becoming more and more evident that LINE-1 retrotransposition might be involved in the progression of cancer [73]. Therefore, we could consider LINE-1 regulation as an additional target in cancer biology study.

4 TE Regulation

This is an active area of research in the field, and below we provide an overview of host factors and pathways that have been associated with the regulation of TE mobilization.

(a) *DNA Methylation*. DNA methylation is considered an important mechanism used for the epigenetic control of L1 expression. Pioneer studies by the *Bestor lab* clearly established a main role for DNA methylation in controlling TE expression [37, 74]. More recently, *Wissing and colleagues* [38] exploited cellular reprogramming to reveal how DNA methylation controls LINE-1 expression during embryonic development. LINE-1 DNA hypomethylation, L1 expression, and engineered L1 activity are similar in iPSCs and hESCs, and significantly higher to that observed generally in somatic cells (excluding cancer cells [63–68] and NPCs [60–62]). More recently, work from the *Trono lab* has demonstrated the complex epigenetic regulation of LINE-1s during early embryogenesis [75]. In mice, Dnmt3L is expressed in the precursor of spermatogonial stem cells. Dnmt3L depletion avoids DNA methylation of retrotransposon promoters, resulting in high level of expression of these elements in spermatogonia and spermatocytes [37]. The reactivation of retrotransposon expression and subsequent severely deregulated retrotransposition upon Dnmt3L deletion would explain the meiotic failure observed in spermatocytes in KO animals [37]. Therefore,

DNA methylation by Dnmt3L keeps retroelements under control at this stage, allowing normal meiosis in germ cells. A similar effect has been observed in the absence of Tex19.1 (and other testis expression-specific factors), causing defective spermatogenesis by reactivation of endogenous retroviruses [76], and growth retardation in mouse placenta by activation of IAP and L1s [77].

(b) *Histone Modifications.* Histone modifications and associated factors are key regulators of L1 expression, especially in tissues characterized by severe DNA hypomethylation of genomes, as hESCs. Indeed, L1 regulation during early embryogenesis is a very dynamic process, where multiple factors might act to control the rate of retrotransposition in humans [75, 78].

(c) *TE Control by RNAs.* Several TE control mechanisms that involve small RNA have been described. Piwi-interacting RNAs (piRNAs) have been associated with the regulation of TE in *Drosophila* and mammals [79, 80]. piRNAs are generated by transcription from TE clusters, exported to the cytoplasm, and loaded on MILI-containing complexes to be used as guides to degrade the complementary TE RNA by an endonucleolytic processing [79, 80]. Moreover, the activity of piRNAs has also been related with DNA methylation of TEs by Dnmt3L; thus L1 silencing in mouse germ cells might be more complex than simply DNA methylation [80]. Recently, the *microprocessor complex,* which is involved in the generation of microRNAs (miRNAs), has also been described as a post-transcriptional mechanism to control TE retrotransposition in mammals [81]. The *microprocessor* is composed by DGCR8 (which recognizes the primary miRNA, pri-miRNAs, with hairpin RNA structures) and Drosha (which cleaves and processes pri-miRNA to pre-miRNA). This pre-miRNA is exported to the cytoplasm where Dicer generates mature miRNAs. Notably, the *microprocessor* is able to recognize presumably structured RNA domains present in L1, Alu, and SVA RNAs, and likely control the expression levels of these elements by processing their transcripts [81]. Furthermore, it has been described that Dicer can also process sense/antisense pairing RNA from L1 promoters, generating *repeat-associated small interfering RNA* (rasiRNAs). rasiRNAs can be used by the RISC complex to degrade L1 mRNAs [42]. Notably, Dicer has also been related with the methylation of the L1 promoter [42], by a non-defined mechanism. In sum, there is a complex regulation of TE by small RNAs (reviewed in [82]).

(d) *Post-transcriptional Control of LINE-1.* Several host factors have been described as restriction factors for L1 retrotransposition. These factors are implicated in the restriction of retroviral replication such as RNaseL [83], APOBEC proteins [59,

84, 85], Trex1 [86], SAMHD1 [87], MOV10 [88], or hnRNPL [89]. These are host factors that can reduce L1 reverse transcription (SAMHD1 [87]), perform editing of L1 sequences presumably during insertion by deamination (APOBEC3A [84]), degrade single-strand L1 RNA (RNaseL [83], MOV10 [88]), or reduce reverse-transcribed L1 DNAs (Trex1 [86]). The use of epitope-tagged L1s has allowed to discover a list of host factors that interact with L1 and might be involved in the regulation of retrotransposition [90–93], but the mechanism of action for most of these factors remains to be described.

5 Methods Used to Study TE Regulation and Its Genomic Impact

In the following sections, we provide a brief overview of techniques commonly used to study TEs, and some of them will be further covered in this book.

5.1 TE Expression Analysis

(a) *Level of Transcription.* The traditional way to quantify the levels of any TE is Northern blot. This technique is still the gold standard to infer the expression level of full-length L1 mRNAs [94]. Because of their high prevalence in genomes and their presence in transcripts (coding or not [95, 96]), it is clear that Northern blot is the most accurate method to distinguish *bona fide* full-length L1 transcripts from other transcripts that may contain portions of TEs [95, 96]. More recently, other indirect methods as RT-qPCR have emerged as a simpler method to quantify expression of L1s. One primer, or a combination of different sets of primers, can be used to amplify different regions of the L1 sequence (5′-UTR, ORF1, or ORF2) and these values are used to quantify L1 expression in any RNA sample [97]. It is particularly interesting that these primers tend to amplify the youngest subfamily of LINE-1, L1Hs, which includes the set of currently active L1s. L1-5′UTR primers will mostly quantify the amount of full-length transcripts [97], while primers located on ORF1p or ORF2p represent the expression from full-length and truncated LINEs present within long non-coding RNAs and other cellular mRNAs. More recently *next-generation sequencing* (NGS) has provided a large source of information that has also allowed us to analyze L1 expression. To do this, the number of reads corresponding to the TE sequence must be determined. A widely accepted estimation of expression level is reads per kilobase per million (RPKM) [66, 98], which express the level of transcription as the number of gene-mapped reads, normalizing regarding transcript length and the total number of genome-mapped reads [98]. In addition, RNA-seq data can also be used to dissect the effect of TE insertions in the genome, analyzing the expression from the targeted genomic region [66]. Some of these methods will be covered with additional details in this book.

(b) *Level of Translation*. The level of translation or the amount of L1-encoded proteins can be determined using antibodies directed against L1-ORF1p and/or L1-ORF2p, either in L1 RNP preparations, in total lysates, or in tissue sections. The L1-RNP consists of the L1-mRNA, several ORF1p proteins and maybe only one molecule of ORF2p. These particles can be isolated by ultracentrifugation on a sucrose cushion [99, 100]. Thus, Western blot can be used to quantify the amount of L1-RNPs using L1-ORF1p and L1-ORF2p antibodies. On the other hand, it is possible to determine the presence, distribution, features, and functionality of L1-RNPs in live cells by exploiting epitope tag engineered L1 constructs [92, 101, 102]. Similarly, L1-RNPs from transfected L1s have been used to detect enzymatic activities associated with the L1-encoded proteins [103, 104]. These methods will be fully described in this book.

5.2 TE Regulation Analyses

(a) *Epigenetic Regulation Analyses*. Different protocols have been described to analyze L1 methylation [97]. Similarly, other TEs like LTR retrotransposons are regulated by DNA methylation and by the interaction with other epigenetic modifiers [105]. However, it is becoming more and more evident that genome-wide regulation studies might be required to fully understand TE regulation. In this book, several methods that determine the genome-wide regulation of TE by DNA methylation and other modifications will be discussed.

(b) *L1-Interacting Protein Analyses*. Protein co-immunoprecipitation is a technique that allows to identify proteins that interact with L1-RNPs, which therefore could be modulating L1 expr ession/retrotransposition. To identify these proteins, the use of epitope tag engineered L1 constructs (together with mass spectrometry) has allowed the identification of L1-RNP interactors with confidence [90, 92, 93]. Different proteins have been reported to interact with L1-RNPs, such as helicases, RNA transporters, splicing factors, post-transcriptional modifiers, and regulators of gene expression [90–93]. Some of these methods will be covered in detail in this book.

5.3 In Vitro Retrotransposition Assays

The development of in vitro retrotransposition assays has clearly augmented our knowledge of TE domains involved in TE mobilization and has also allowed to discover how host factors regulate this mobility. Below, we briefly cover some of the most common used in vitro assays that will be fully described in this Method Book.

Thirty years ago, *Boeke and colleagues* developed the first system to study the mobilization of an LTR retrotransposon termed Ty1 in yeast [106]. The assay is based on the development of a reporter gene whose expression can only be activated after a round of retrotransposition, exploiting a backward intron cloned into a reporter gene. The same trick was later adapted by the *Heidmann and Kazazian laboratories* to demonstrate that other retrotransposons are active in the human and mouse genomes (reviewed in [107, 108]). Within the L1 field, this assay has been instrumental to reveal how domains within L1-encoded proteins are required for retrotransposition [45], to determine which L1s are active in the human and mouse genome [109–112], to identify active LINEs in other species [113] and a large etc. (reviewed in [34, 43, 44]). Additionally, modifications of this assay have been determinant to decipher the impact of L1 retrotransposition in genomes, by the characterization of engineered L1 insertions [52, 114–117].

More recently, modifications of this assay have been developed to study in vitro the retrotransposition of SINEs, including Alu and SVA [118–120]. In essence, the assay (and modifications) is based on the mechanism of retrotransposition, exploiting the fact that retrotransposition occurs using an intermediate RNA. The fundament of this assay is to detect retrotransposition events from an exogenous engineered transfected construct, and is based on the activation of a reporter gene as a result of the retrotransposition process (reviewed in [107, 108]). Additionally, as the assay uses reporter-tagged retrotransposons containing unique sequences, de novo insertions can be isolated and characterized by different methodologies as "inverse PCR" [38, 48, 56, 115] or the "recovery" assay [115, 117, 121]. Finally, the same assay has allowed to generate mouse and rat models of L1 retrotransposition in vivo [55, 122, 123].

5.4 Methods Used to Detect In Vivo Retrotransposition Events

The identification of de novo endogenous retrotransposition events is analogous to finding a needle in a haystack, due to their repeated nature in genomes. With the genomics revolution, the tools available to study these processes have grown exponentially, and some of the current methodologies will be covered in this book.

Currently, *next-generation DNA sequencing* allows searching for polymorphic mobile element insertions (pMEIs) or even detecting de novo MEIs in an individual [124, 125], a specific tissue [62, 64, 67], or even single cells isolated from an individual [69, 70, 126]. Different methods have been developed for the identification of MEIs using *high-throughput sequencing*. Indeed, MEIs can be analyzed directly from *whole-genome sequencing* data [64, 66, 69], although several methodologies have been developed to maximize MEI searching efficiency by MEI enrichment [127].

MEI enrichment from genomic DNA can be performed by different strategies such as PCR amplification, hybridization, or a combination of both [67, 125, 127–129].

DNA fragmentation, either by enzymatic digestion or mechanical shearing, is required for the generation of sequencing libraries and MEI enrichment [67, 125, 127, 128]. The goal of MEI enrichment is to increase the representation of DNA fragments consisting of a TE end and its flanking genomic DNA, which will allow the location of the insertion in the genome. Each MEI enrichment procedure depends on the specific methodology; for example *ATLAS* method [128] enriches by primer extension from the L1 region in single-strand DNA loop structures [128]. In *Transposon-Seq* [63], the option is a hemispecific PCR by a primer that anneals on the TE end and random primers that anneal on the flanking DNA [63]. Moreover, *L1-Display* [125] combines both techniques, primer extension and a hemispecific PCR [125]. Otherwise, ME-Scan [129] performs the enrichment by recovering biotinylated DNA amplicons generated from a prior hemispecific PCR [129]. The main issue derived from enrichment by PCR is the bias that can arise from the stochastic nature of this technique. In addition, the use of random primers can entail differential capacity to amplify different regions of the genome. To avoid this bias, an alternative method is based on enrichment by hybridization, like in the Retrotransposon Capture-sequencing (RC-seq) protocol [62, 67, 70]. In this protocol, upon DNA shearing, TE-containing fragments are enriched/captured using specific biotinylated probes [62, 67, 70]. Thus, this methodology does not rely on PCR amplification, except for a few number of PCR cycles required during the sequencing library generation.

Upon *deep sequencing*, the goal is always to detect polymorphic or de novo MEIs by a computational analysis. Thus, different algorithms have been developed to find MEIs, and two are the principal strategies to do it: "*paired-end-reads mapping*" and "*split-read mapping*." In the first case, the algorithm needs paired-end reads, which do not usually cover fully the insertion junction, and middle regions are not represented. In the presence of an MEI, a read will match the genomic position ("*anchor read*") and the other pair read will match to the TE end. Therefore, after genome mapping, clustering, and distribution analysis of the reads, there will be a discordant span and orientation of the reads in this point, indicating the putative presence of an MEI [130, 131].

"*Split-read mapping*" methods search for reads that contain the MEI-genome junction [132, 133], i.e., that contain the MEI end and the flanking DNA. Thus, using this algorithm it is possible to obtain the exact MEI position, and even TSDs and poly-A tails. Therefore, this type of reads has the highest confidence for MEI resolution. However, to apply this algorithm longer reads are needed to increase the probability to find reads that overlap accurately the MEI end and unique flanking DNA sequences [132]. Currently, practically all the high-throughput sequencing platforms generate reads long enough for a *split-read mapping*. Thus, the most accurate method to find MEIs is to combine both algorithms, "*anchored read-pair mapping*" to infer putative MEIs and "*split-read mapping*" to confirm the MEI presence by detecting *soft-clipped* reads. Some examples of programs that integrate both methodologies are *Tea* [64, 126], *RetroSeq* [130], and *Tangram* [133], which are able to detect MEI breakpoint with a single-nucleotide resolution as well as other characteristics such as TSDs and poly-A tails. However, it is also highly recommended to confirm MEI presence and hallmarks associated with retrotransposition by conventional PCR validation/Sanger DNA sequencing from the original genomic DNA, in order to avoid annotation of artifacts [134].

Acknowledgements

We acknowledge members of the J.L.G.-P. lab for critically reading this succinct review. M.M.-L. research is partially funded by a Consejeria de Salud/Junta de Andalucia Research Grant (PI-0224-2011) and by an ERC-Consolidator (ERC-STG-2012-233764) grant to J.L.G.-P. J.L.G.P's lab is supported by CICE-FEDER-P09-CTS-4980, CICE-FEDER-P12-CTS-2256, Plan Nacional de I+D+I 2008-2011 and 2013-2016 (FIS-FEDER-PI11/01489 and FIS-FEDER-PI14/02152), PCIN-2014-115-ERA-NET NEURON II, the European Research Council (ERC-Consolidator ERC-STG-2012-233764), and an International Early Career Scientist grant from the Howard Hughes Medical Institute (IECS-55007420). R.V.-A. and C.L.-R. are fully supported by an International Early Career Scientist grant (Howard Hughes Medical Institute).

References

1. Lazarow K, Doll ML, Kunze R (2013) Molecular biology of maize Ac/Ds elements: an overview. Methods Mol Biol 1057:59–82. doi:10.1007/978-1-62703-568-2_5
2. Plasterk RH, Izsvak Z, Ivics Z (1999) Resident aliens: the Tc1/mariner superfamily of transposable elements. Trends Genet 15(8):326–332
3. Kleckner N (1990) Regulation of transposition in bacteria. Annu Rev Cell Biol 6:297–327. doi:10.1146/annurev.cb.06.110190.001501
4. Ding S, Wu X, Li G, Han M, Zhuang Y, Xu T (2005) Efficient transposition of the piggyBac (PB) transposon in mammalian cells and mice. Cell 122(3):473–483. doi:10.1016/j.cell.2005.07.013
5. Arkhipova IR, Meselson M (2005) Diverse DNA transposons in rotifers of the class Bdelloidea. Proc Natl Acad Sci U S A 102(33):11781–11786. doi:10.1073/pnas.0505333102
6. Munoz-Lopez M, Siddique A, Bischerour J, Lorite P, Chalmers R, Palomeque T (2008) Transposition of Mboumar-9: identification of a new naturally active mariner-family transposon. J Mol Biol 382(3):567–572. doi:10.1016/j.jmb.2008.07.044

7. Richardson JM, Dawson A, O'Hagan N, Taylor P, Finnegan DJ, Walkinshaw MD (2006) Mechanism of Mos1 transposition: insights from structural analysis. EMBO J 25(6):1324–1334. doi:10.1038/sj.emboj.7601018
8. Ray DA, Feschotte C, Pagan HJ, Smith JD, Pritham EJ, Arensburger P, Atkinson PW, Craig NL (2008) Multiple waves of recent DNA transposon activity in the bat, Myotis lucifugus. Genome Res 18(5):717–728. doi:10.1101/gr.071886.107
9. Ray DA, Pagan HJ, Thompson ML, Stevens RD (2007) Bats with hATs: evidence for recent DNA transposon activity in genus Myotis. Mol Biol Evol 24(3):632–639. doi:10.1093/molbev/msl192
10. Mitra R, Li X, Kapusta A, Mayhew D, Mitra RD, Feschotte C, Craig NL (2013) Functional characterization of piggyBat from the bat Myotis lucifugus unveils an active mammalian DNA transposon. Proc Natl Acad Sci U S A 110(1):234–239. doi:10.1073/pnas.1217548110
11. Ivics Z, Garrels W, Mates L, Yau TY, Bashir S, Zidek V, Landa V, Geurts A, Pravenec M, Rulicke T, Kues WA, Izsvak Z (2014) Germline transgenesis in pigs by cytoplasmic microinjection of Sleeping Beauty transposons. Nat Protoc 9(4):810–827. doi:10.1038/nprot.2014.010
12. Collier LS, Largaespada DA (2007) Transposons for cancer gene discovery: sleeping beauty and beyond. Genome Biol 8(Suppl 1):S15. doi:10.1186/gb-2007-8-s1-s15
13. Izsvak Z, Ivics Z (2004) Sleeping beauty transposition: biology and applications for molecular therapy. Mol Ther 9(2):147–156. doi:10.1016/j.ymthe.2003.11.009
14. Radice AD, Bugaj B, Fitch DH, Emmons SW (1994) Widespread occurrence of the Tc1 transposon family: Tc1-like transposons from teleost fish. Mol Gen Genet 244(6):606–612
15. Dupuy AJ, Akagi K, Largaespada DA, Copeland NG, Jenkins NA (2005) Mammalian mutagenesis using a highly mobile somatic Sleeping Beauty transposon system. Nature 436(7048):221–226
16. Dupuy AJ, Jenkins NA, Copeland NG (2006) Sleeping beauty: a novel cancer gene discovery tool. Hum Mol Genet 15(Spec No 1):R75–R79. doi:10.1093/hmg/ddl061
17. Howell VM, Colvin EK (2014) Genetically engineered insertional mutagenesis in mice to model cancer: sleeping beauty. Methods Mol Biol 1194:367–383. doi:10.1007/978-1-4939-1215-5_21
18. Mates L, Chuah MK, Belay E, Jerchow B, Manoj N, Acosta-Sanchez A, Grzela DP, Schmitt A, Becker K, Matrai J, Ma L, Samara-Kuko E, Gysemans C, Pryputniewicz D, Miskey C, Fletcher B, VandenDriessche T, Ivics Z, Izsvak Z (2009) Molecular evolution of a novel hyperactive Sleeping Beauty transposase enables robust stable gene transfer in vertebrates. Nat Genet 41(6):753–761. doi:10.1038/ng.343
19. Kowarz E, Loescher D, Marschalek R (2015) Optimized Sleeping Beauty transposons rapidly generate stable transgenic cell lines. Biotechnol J. doi:10.1002/biot.201400821
20. Ohlfest JR, Frandsen JL, Fritz S, Lobitz PD, Perkinson SG, Clark KJ, Nelsestuen G, Key NS, McIvor RS, Hackett PB, Largaespada DA (2005) Phenotypic correction and long-term expression of factor VIII in hemophilic mice by immunotolerization and nonviral gene transfer using the Sleeping Beauty transposon system. Blood 105(7):2691–2698. doi:10.1182/blood-2004-09-3496
21. Boehme P, Doerner J, Solanki M, Zhang W, Ehrhardt A (2015) The sleeping beauty transposon vector system for treatment of rare genetic diseases: an unrealized hope? Curr Gene Ther 15:255–265
22. Kawakami K, Shima A, Kawakami N (2000) Identification of a functional transposase of the Tol2 element, an Ac-like element from the Japanese medaka fish, and its transposition in the zebrafish germ lineage. Proc Natl Acad Sci U S A 97(21):11403–11408. doi:10.1073/pnas.97.21.11403
23. Pettitt SJ, Tan EP, Yusa K (2015) piggyBac transposon-based insertional mutagenesis in mouse haploid embryonic stem cells. Methods Mol Biol 1239:15–28. doi:10.1007/978-1-4939-1862-1_2
24. Abe G, Suster ML, Kawakami K (2011) Tol2-mediated transgenesis, gene trapping, enhancer trapping, and the Gal4-UAS system. Methods Cell Biol 104:23–49. doi:10.1016/B978-0-12-374814-0.00002-1
25. Rad R, Rad L, Wang W, Strong A, Ponstingl H, Bronner IF, Mayho M, Steiger K, Weber J, Hieber M, Veltkamp C, Eser S, Geumann U, Ollinger R, Zukowska M, Barenboim M, Maresch R, Cadinanos J, Friedrich M, Varela I, Constantino-Casas F, Sarver A, Ten Hoeve J, Prosser H, Seidler B, Bauer J, Heikenwalder M, Metzakopian E, Krug A, Ehmer U, Schneider G, Knosel T, Rummele P, Aust D, Grutzmann R, Pilarsky C, Ning Z, Wessels L, Schmid RM, Quail MA, Vassiliou G, Esposito I, Liu P, Saur D, Bradley A (2015) A conditional piggyBac transposition system for

genetic screening in mice identifies oncogenic networks in pancreatic cancer. Nat Genet 47(1):47–56. doi:10.1038/ng.3164

26. Yusa K, Rad R, Takeda J, Bradley A (2009) Generation of transgene-free induced pluripotent mouse stem cells by the piggyBac transposon. Nat Methods 6(5):363–369. doi:10.1038/nmeth.1323
27. Kawakami K (2007) Tol2: a versatile gene transfer vector in vertebrates. Genome Biol 8(Suppl 1):S7. doi:10.1186/gb-2007-8-s1-s7
28. Mosimann C, Zon LI (2011) Advanced zebrafish transgenesis with Tol2 and application for Cre/lox recombination experiments. Methods Cell Biol 104:173–194. doi: 10.1016/B978-0-12-374814-0.00010-0
29. Ley D, Van Zwieten R, Puttini S, Iyer P, Cochard A, Mermod N (2014) A PiggyBac-mediated approach for muscle gene transfer or cell therapy. Stem Cell Res 13(3 Pt A):390–403. doi:10.1016/j.scr.2014.08.007
30. Lambowitz AM, Zimmerly S (2011) Group II introns: mobile ribozymes that invade DNA. Cold Spring Harb Perspect Biol 3(8):a003616. doi:10.1101/cshperspect.a003616
31. Lander ES, Linton LM, Birren B, Nusbaum C, Zody MC, Baldwin J, Devon K, Dewar K, Doyle M, FitzHugh W, Funke R, Gage D, Harris K, Heaford A, Howland J, Kann L, Lehoczky J, LeVine R, McEwan P, McKernan K, Meldrim J, Mesirov JP, Miranda C, Morris W, Naylor J, Raymond C, Rosetti M, Santos R, Sheridan A, Sougnez C, Stange-Thomann N, Stojanovic N, Subramanian A, Wyman D, Rogers J, Sulston J, Ainscough R, Beck S, Bentley D, Burton J, Clee C, Carter N, Coulson A, Deadman R, Deloukas P, Dunham A, Dunham I, Durbin R, French L, Grafham D, Gregory S, Hubbard T, Humphray S, Hunt A, Jones M, Lloyd C, McMurray A, Matthews L, Mercer S, Milne S, Mullikin JC, Mungall A, Plumb R, Ross M, Shownkeen R, Sims S, Waterston RH, Wilson RK, Hillier LW, McPherson JD, Marra MA, Mardis ER, Fulton LA, Chinwalla AT, Pepin KH, Gish WR, Chissoe SL, Wendl MC, Delehaunty KD, Miner TL, Delehaunty A, Kramer JB, Cook LL, Fulton RS, Johnson DL, Minx PJ, Clifton SW, Hawkins T, Branscomb E, Predki P, Richardson P, Wenning S, Slezak T, Doggett N, Cheng JF, Olsen A, Lucas S, Elkin C, Uberbacher E, Frazier M, Gibbs RA, Muzny DM, Scherer SE, Bouck JB, Sodergren EJ, Worley KC, Rives CM, Gorrell JH, Metzker ML, Naylor SL, Kucherlapati RS, Nelson DL, Weinstock GM, Sakaki Y, Fujiyama A, Hattori M, Yada T, Toyoda A, Itoh T, Kawagoe C, Watanabe H, Totoki Y, Taylor T, Weissenbach J, Heilig R, Saurin W, Artiguenave F, Brottier P, Bruls T, Pelletier E, Robert C, Wincker P, Smith DR, Doucette-Stamm L, Rubenfield M, Weinstock K, Lee HM, Dubois J, Rosenthal A, Platzer M, Nyakatura G, Taudien S, Rump A, Yang H, Yu J, Wang J, Huang G, Gu J, Hood L, Rowen L, Madan A, Qin S, Davis RW, Federspiel NA, Abola AP, Proctor MJ, Myers RM, Schmutz J, Dickson M, Grimwood J, Cox DR, Olson MV, Kaul R, Shimizu N, Kawasaki K, Minoshima S, Evans GA, Athanasiou M, Schultz R, Roe BA, Chen F, Pan H, Ramser J, Lehrach H, Reinhardt R, McCombie WR, de la Bastide M, Dedhia N, Blocker H, Hornischer K, Nordsiek G, Agarwala R, Aravind L, Bailey JA, Bateman A, Batzoglou S, Birney E, Bork P, Brown DG, Burge CB, Cerutti L, Chen HC, Church D, Clamp M, Copley RR, Doerks T, Eddy SR, Eichler EE, Furey TS, Galagan J, Gilbert JG, Harmon C, Hayashizaki Y, Haussler D, Hermjakob H, Hokamp K, Jang W, Johnson LS, Jones TA, Kasif S, Kaspryzk A, Kennedy S, Kent WJ, Kitts P, Koonin EV, Korf I, Kulp D, Lancet D, Lowe TM, McLysaght A, Mikkelsen T, Moran JV, Mulder N, Pollara VJ, Ponting CP, Schuler G, Schultz J, Slater G, Smit AF, Stupka E, Szustakowski J, Thierry-Mieg D, Thierry-Mieg J, Wagner L, Wallis J, Wheeler R, Williams A, Wolf YI, Wolfe KH, Yang SP, Yeh RF, Collins F, Guyer MS, Peterson J, Felsenfeld A, Wetterstrand KA, Patrinos A, Morgan MJ, de Jong P, Catanese JJ, Osoegawa K, Shizuya H, Choi S, Chen YJ (2001) Initial sequencing and analysis of the human genome. Nature 409(6822):860–921
32. Lee YN, Bieniasz PD (2007) Reconstitution of an infectious human endogenous retrovirus. PLoS Pathog 3(1):e10. doi:10.1371/journal.ppat.0030010
33. Maksakova IA, Romanish MT, Gagnier L, Dunn CA, van de Lagemaat LN, Mager DL (2006) Retroviral elements and their hosts: insertional mutagenesis in the mouse germ line. PLoS Genet 2(1):e2. doi:10.1371/journal.pgen.0020002
34. Beck CR, Garcia-Perez JL, Badge RM, Moran JV (2011) LINE-1 elements in structural variation and disease. Annu Rev Genomics Hum Genet 12:187–215. doi:10.1146/annurev-genom-082509-141802
35. Scott AF, Schmeckpeper BJ, Abdelrazik M, Comey CT, O'Hara B, Rossiter JP, Cooley T, Heath P, Smith KD, Margolet L (1987)

Origin of the human L1 elements: proposed progenitor genes deduced from a consensus DNA sequence. Genomics 1(2):113–125
36. Swergold GD (1990) Identification, characterization, and cell specificity of a human LINE-1 promoter. Mol Cell Biol 10(12): 6718–6729
37. Bourc'his D, Bestor TH (2004) Meiotic catastrophe and retrotransposon reactivation in male germ cells lacking Dnmt3L. Nature 431(7004):96–99. doi:10.1038/nature02886, nature02886 [pii]
38. Wissing S, Munoz-Lopez M, Macia A, Yang Z, Montano M, Collins W, Garcia-Perez JL, Moran JV, Greene WC (2012) Reprogramming somatic cells into iPS cells activates LINE-1 retroelement mobility. Hum Mol Genet 21(1):208–218
39. Macia A, Munoz-Lopez M, Cortes JL, Hastings RK, Morell S, Lucena-Aguilar G, Marchal JA, Badge RM, Garcia-Perez JL (2011) Epigenetic control of retrotransposon expression in human embryonic stem cells. Mol Cell Biol 31(2):300–316
40. Li J, Kannan M, Trivett AL, Liao H, Wu X, Akagi K, Symer DE (2014) An antisense promoter in mouse L1 retrotransposon open reading frame-1 initiates expression of diverse fusion transcripts and limits retrotransposition. Nucleic Acids Res 42(7):4546–4562. doi:10.1093/nar/gku091
41. Yang N, Kazazian HH Jr (2006) L1 retrotransposition is suppressed by endogenously encoded small interfering RNAs in human cultured cells. Nat Struct Mol Biol 13(9):763–771
42. Ciaudo C, Jay F, Okamoto I, Chen CJ, Sarazin A, Servant N, Barillot E, Heard E, Voinnet O (2013) RNAi-dependent and independent control of LINE1 accumulation and mobility in mouse embryonic stem cells. PLoS Genet 9(11):e1003791. doi:10.1371/journal.pgen.1003791
43. Levin HL, Moran JV (2011) Dynamic interactions between transposable elements and their hosts. Nat Rev Genet 12(9):615–627. doi:10.1038/nrg3030
44. Macia A, Blanco-Jimenez E, Garcia-Perez JL (2015) Retrotransposons in pluripotent cells: impact and new roles in cellular plasticity. Biochim Biophys Acta 1849(4):417–426. doi:10.1016/j.bbagrm.2014.07.007
45. Moran JV, Holmes SE, Naas TP, DeBerardinis RJ, Boeke JD, Kazazian HH Jr (1996) High frequency retrotransposition in cultured mammalian cells. Cell 87(5):917–927, doi:S0092-8674(00)81998-4 [pii]
46. Alisch RS, Garcia-Perez JL, Muotri AR, Gage FH, Moran JV (2006) Unconventional translation of mammalian LINE-1 retrotransposons. Genes Dev 20(2):210–224
47. Dmitriev SE, Andreev DE, Terenin IM, Olovnikov IA, Prassolov VS, Merrick WC, Shatsky IN (2007) Efficient translation initiation directed by the 900-nucleotide-long and GC-rich 5' untranslated region of the human retrotransposon LINE-1 mRNA is strictly cap dependent rather than internal ribosome entry site mediated. Mol Cell Biol 27(13):4685–4697
48. Wei W, Gilbert N, Ooi SL, Lawler JF, Ostertag EM, Kazazian HH, Boeke JD, Moran JV (2001) Human L1 retrotransposition: cis preference versus trans complementation. Mol Cell Biol 21(4):1429–1439
49. Goodier JL, Kazazian HH Jr (2008) Retrotransposons revisited: the restraint and rehabilitation of parasites. Cell 135(1):23–35. doi:10.1016/j.cell.2008.09.022
50. Luan DD, Korman MH, Jakubczak JL, Eickbush TH (1993) Reverse transcription of R2Bm RNA is primed by a nick at the chromosomal target site: a mechanism for non-LTR retrotransposition. Cell 72(4):595–605
51. Jurka J (1997) Sequence patterns indicate an enzymatic involvement in integration of mammalian retroposons. Proc Natl Acad Sci U S A 94(5):1872–1877
52. Gilbert N, Lutz S, Morrish TA, Moran JV (2005) Multiple fates of L1 retrotransposition intermediates in cultured human cells. Mol Cell Biol 25(17):7780–7795. doi:10.1128/MCB.25.17.7780-7795.2005
53. Grimaldi G, Skowronski J, Singer MF (1984) Defining the beginning and end of KpnI family segments. EMBO J 3(8):1753–1759
54. van den Hurk JA, Meij IC, Seleme MC, Kano H, Nikopoulos K, Hoefsloot LH, Sistermans EA, de Wijs IJ, Mukhopadhyay A, Plomp AS, de Jong PT, Kazazian HH, Cremers FP (2007) L1 retrotransposition can occur early in human embryonic development. Hum Mol Genet 16(13):1587–1592. doi:10.1093/hmg/ddm108, ddm108 [pii]
55. Kano H, Godoy I, Courtney C, Vetter MR, Gerton GL, Ostertag EM, Kazazian HH Jr (2009) L1 retrotransposition occurs mainly in embryogenesis and creates somatic mosaicism. Genes Dev 23(11):1303–1312. doi:10.1101/gad.1803909
56. Garcia-Perez JL, Marchetto MC, Muotri AR, Coufal NG, Gage FH, O'Shea KS, Moran JV (2007) LINE-1 retrotransposition in human

embryonic stem cells. Hum Mol Genet 16(13):1569–1577. doi:10.1093/hmg/ddm105, ddm105 [pii]

57. Freeman P, Macfarlane C, Collier P, Jeffreys AJ, Badge RM (2011) L1 hybridization enrichment: a method for directly accessing de novo L1 insertions in the human germline. Hum Mutat 32(8):978–988. doi:10.1002/humu.21533
58. Malki S, van der Heijden GW, O'Donnell KA, Martin SL, Bortvin A (2014) A role for retrotransposon LINE-1 in fetal oocyte attrition in mice. Dev Cell 29(5):521–533. doi:10.1016/j.devcel.2014.04.027
59. Wissing S, Montano M, Garcia-Perez JL, Moran JV, Greene WC (2011) Endogenous APOBEC3B restricts LINE-1 retrotransposition in transformed cells and human embryonic stem cells. J Biol Chem 286(42):36427–36437. doi:10.1074/jbc.M111.251058
60. Coufal NG, Garcia-Perez JL, Peng GE, Yeo GW, Mu Y, Lovci MT, Morell M, O'Shea KS, Moran JV, Gage FH (2009) L1 retrotransposition in human neural progenitor cells. Nature 460(7259):1127–1131
61. Muotri AR, Chu VT, Marchetto MC, Deng W, Moran JV, Gage FH (2005) Somatic mosaicism in neuronal precursor cells mediated by L1 retrotransposition. Nature 435(7044):903–910
62. Baillie JK, Barnett MW, Upton KR, Gerhardt DJ, Richmond TA, De Sapio F, Brennan PM, Rizzu P, Smith S, Fell M, Talbot RT, Gustincich S, Freeman TC, Mattick JS, Hume DA, Heutink P, Carninci P, Jeddeloh JA, Faulkner GJ (2011) Somatic retrotransposition alters the genetic landscape of the human brain. Nature 479(7374):534–537. doi:10.1038/nature10531
63. Iskow RC, McCabe MT, Mills RE, Torene S, Pittard WS, Neuwald AF, Van Meir EG, Vertino PM, Devine SE (2010) Natural mutagenesis of human genomes by endogenous retrotransposons. Cell 141(7):1253–1261
64. Lee E, Iskow R, Yang L, Gokcumen O, Haseley P, Luquette LJ III, Lohr JG, Harris CC, Ding L, Wilson RK, Wheeler DA, Gibbs RA, Kucherlapati R, Lee C, Kharchenko PV, Park PJ (2012) Landscape of somatic retrotransposition in human cancers. Science 337(6097):967–971. doi:10.1126/science.1222077, science.1222077 [pii]
65. Solyom S, Ewing AD, Rahrmann EP, Doucet T, Nelson HH, Burns MB, Harris RS, Sigmon DF, Casella A, Erlanger B, Wheelan S, Upton KR, Shukla R, Faulkner GJ, Largaespada DA, Kazazian HH Jr (2012) Extensive somatic L1 retrotransposition in colorectal tumors. Genome Res 22(12):2328–2338. doi:10.1101/gr.145235.112
66. Helman E, Lawrence MS, Stewart C, Sougnez C, Getz G, Meyerson M (2014) Somatic retrotransposition in human cancer revealed by whole-genome and exome sequencing. Genome Res 24(7):1053–1063. doi:10.1101/gr.163659.113
67. Shukla R, Upton KR, Munoz-Lopez M, Gerhardt DJ, Fisher ME, Nguyen T, Brennan PM, Baillie JK, Collino A, Ghisletti S, Sinha S, Iannelli F, Radaelli E, Dos Santos A, Rapoud D, Guettier C, Samuel D, Natoli G, Carninci P, Ciccarelli FD, Garcia-Perez JL, Faivre J, Faulkner GJ (2013) Endogenous retrotransposition activates oncogenic pathways in hepatocellular carcinoma. Cell 153(1):101–111. doi:10.1016/j.cell.2013.02.032
68. Tubio JM, Li Y, Ju YS, Martincorena I, Cooke SL, Tojo M, Gundem G, Pipinikas CP, Zamora J, Raine K, Menzies A, Roman-Garcia P, Fullam A, Gerstung M, Shlien A, Tarpey PS, Papaemmanuil E, Knappskog S, Van Loo P, Ramakrishna M, Davies HR, Marshall J, Wedge DC, Teague JW, Butler AP, Nik-Zainal S, Alexandrov L, Behjati S, Yates LR, Bolli N, Mudie L, Hardy C, Martin S, McLaren S, O'Meara S, Anderson E, Maddison M, Gamble S, Group IBC, Group IBC, Group IPC, Foster C, Warren AY, Whitaker H, Brewer D, Eeles R, Cooper C, Neal D, Lynch AG, Visakorpi T, Isaacs WB, van't Veer L, Caldas C, Desmedt C, Sotiriou C, Aparicio S, Foekens JA, Eyfjord JE, Lakhani SR, Thomas G, Myklebost O, Span PN, Borresen-Dale AL, Richardson AL, Van de Vijver M, Vincent-Salomon A, Van den Eynden GG, Flanagan AM, Futreal PA, Janes SM, Bova GS, Stratton MR, McDermott U, Campbell PJ (2014) Mobile DNA in cancer. Extensive transduction of nonrepetitive DNA mediated by L1 retrotransposition in cancer genomes. Science 345(6196):1251343. doi:10.1126/science.1251343
69. Evrony GD, Cai X, Lee E, Hills LB, Elhosary PC, Lehmann HS, Parker JJ, Atabay KD, Gilmore EC, Poduri A, Park PJ, Walsh CA (2012) Single-neuron sequencing analysis of L1 retrotransposition and somatic mutation in the human brain. Cell 151(3):483–496. doi:10.1016/j.cell.2012.09.035
70. Upton KR, Gerhardt DJ, Jesuadian JS, Richardson SR, Sanchez-Luque FJ, Bodea GO, Ewing AD, Salvador-Palomeque C, van der Knaap MS, Brennan PM, Vanderver A, Faulkner GJ (2015) Ubiquitous L1 mosaicism in hippocampal neurons. Cell 161(2):228–239. doi:10.1016/j.cell.2015.03.026

71. Miki Y, Nishisho I, Horii A, Miyoshi Y, Utsunomiya J, Kinzler KW, Vogelstein B, Nakamura Y (1992) Disruption of the APC gene by a retrotransposal insertion of L1 sequence in a colon cancer. Cancer Res 52(3):643–645
72. Hancks DC, Kazazian HH Jr (2012) Active human retrotransposons: variation and disease. Curr Opin Genet Dev 22(3):191–203. doi:10.1016/j.gde.2012.02.006
73. Munoz-Lopez M, Medina PP, Garcia-Perez JL (2013) Wiping DNA methylation: Wip1 regulates genomic fluidity on cancer. Cancer Cell 24(4):405–407. doi:10.1016/j.ccr.2013.10.002
74. Bestor TH (2003) Cytosine methylation mediates sexual conflict. Trends Genet 19(4):185–190, doi:S0168952503000490 [pii]
75. Castro-Diaz N, Ecco G, Coluccio A, Kapopoulou A, Yazdanpanah B, Friedli M, Duc J, Jang SM, Turelli P, Trono D (2014) Evolutionally dynamic L1 regulation in embryonic stem cells. Genes Dev 28(13):1397–1409. doi:10.1101/gad.241661.114
76. Ollinger R, Childs AJ, Burgess HM, Speed RM, Lundegaard PR, Reynolds N, Gray NK, Cooke HJ, Adams IR (2008) Deletion of the pluripotency-associated Tex19.1 gene causes activation of endogenous retroviruses and defective spermatogenesis in mice. PLoS Genet 4(9):e1000199. doi:10.1371/journal.pgen.1000199
77. Reichmann J, Reddington JP, Best D, Read D, Ollinger R, Meehan RR, Adams IR (2013) The genome-defence gene Tex19.1 suppresses LINE-1 retrotransposons in the placenta and prevents intra-uterine growth retardation in mice. Hum Mol Genet 22(9):1791–1806. doi:10.1093/hmg/ddt029
78. Garcia-Perez JL, Morell M, Scheys JO, Kulpa DA, Morell S, Carter CC, Hammer GD, Collins KL, O'Shea KS, Menendez P, Moran JV (2010) Epigenetic silencing of engineered L1 retrotransposition events in human embryonic carcinoma cells. Nature 466(7307):769–773. doi:10.1038/nature09209
79. Siomi MC, Sato K, Pezic D, Aravin AA (2011) PIWI-interacting small RNAs: the vanguard of genome defence. Nat Rev Mol Cell Biol 12(4):246–258. doi:10.1038/nrm3089
80. Aravin AA, Sachidanandam R, Bourc'his D, Schaefer C, Pezic D, Toth KF, Bestor T, Hannon GJ (2008) A piRNA pathway primed by individual transposons is linked to de novo DNA methylation in mice. Mol Cell 31(6):785–799. doi:10.1016/j.molcel.2008.09.003
81. Heras SR, Macias S, Plass M, Fernandez N, Cano D, Eyras E, Garcia-Perez JL, Caceres JF (2013) The Microprocessor controls the activity of mammalian retrotransposons. Nat Struct Mol Biol 20(10):1173–1181. doi:10.1038/nsmb.2658
82. Heras SR, Macias S, Caceres JF, Garcia-Perez JL (2014) Control of mammalian retrotransposons by cellular RNA processing activities. Mob Genet Elements 4:e28439
83. Zhang A, Dong B, Doucet AJ, Moldovan JB, Moran JV, Silverman RH (2014) RNase L restricts the mobility of engineered retrotransposons in cultured human cells. Nucleic Acids Res 42(6):3803–3820. doi:10.1093/nar/gkt1308
84. Richardson SR, Narvaiza I, Planegger RA, Weitzman MD, Moran JV (2014) APOBEC3A deaminates transiently exposed single-strand DNA during LINE-1 retrotransposition. Elife 3:e02008. doi:10.7554/eLife.02008
85. Schumann GG (2007) APOBEC3 proteins: major players in intracellular defence against LINE-1-mediated retrotransposition. Biochem Soc Trans 35(Pt 3):637–642. doi:10.1042/BST0350637
86. Stetson DB, Ko JS, Heidmann T, Medzhitov R (2008) Trex1 prevents cell-intrinsic initiation of autoimmunity. Cell 134(4):587–598. doi:10.1016/j.cell.2008.06.032
87. Zhao K, Du J, Han X, Goodier JL, Li P, Zhou X, Wei W, Evans SL, Li L, Zhang W, Cheung LE, Wang G, Kazazian HH Jr, Yu XF (2013) Modulation of LINE-1 and Alu/SVA retrotransposition by Aicardi-Goutieres syndrome-related SAMHD1. Cell Rep 4(6):1108–1115. doi:10.1016/j.celrep.2013.08.019
88. Goodier JL, Cheung LE, Kazazian HH Jr (2012) MOV10 RNA helicase is a potent inhibitor of retrotransposition in cells. PLoS Genet 8(10):e1002941. doi:10.1371/journal.pgen.1002941
89. Peddigari S, Li PW, Rabe JL, Martin SL (2013) hnRNPL and nucleolin bind LINE-1 RNA and function as host factors to modulate retrotransposition. Nucleic Acids Res 41(1):575–585. doi:10.1093/nar/gks1075
90. Goodier JL, Cheung LE, Kazazian HH Jr (2013) Mapping the LINE1 ORF1 protein interactome reveals associated inhibitors of human retrotransposition. Nucleic Acids Res 41(15):7401–7419. doi:10.1093/nar/gkt512
91. Dai L, Taylor MS, O'Donnell KA, Boeke JD (2012) Poly(A) binding protein C1 is essential

for efficient L1 retrotransposition and affects L1 RNP formation. Mol Cell Biol 32(21):4323–4336. doi:10.1128/MCB.06785-11

92. Taylor MS, Lacava J, Mita P, Molloy KR, Huang CR, Li D, Adney EM, Jiang H, Burns KH, Chait BT, Rout MP, Boeke JD, Dai L (2013) Affinity proteomics reveals human host factors implicated in discrete stages of LINE-1 retrotransposition. Cell 155(5):1034–1048. doi:10.1016/j.cell.2013.10.021
93. Moldovan JB, Moran JV (2015) The zinc-finger antiviral protein ZAP inhibits LINE and Alu retrotransposition. PLoS Genet 11(5):e1005121. doi:10.1371/journal.pgen.1005121
94. Perepelitsa-Belancio V, Deininger P (2003) RNA truncation by premature polyadenylation attenuates human mobile element activity. Nat Genet 35(4):363–366
95. Faulkner GJ, Kimura Y, Daub CO, Wani S, Plessy C, Irvine KM, Schroder K, Cloonan N, Steptoe AL, Lassmann T, Waki K, Hornig N, Arakawa T, Takahashi H, Kawai J, Forrest AR, Suzuki H, Hayashizaki Y, Hume DA, Orlando V, Grimmond SM, Carninci P (2009) The regulated retrotransposon transcriptome of mammalian cells. Nat Genet 41(5):563–571
96. Kapusta A, Kronenberg Z, Lynch VJ, Zhuo X, Ramsay L, Bourque G, Yandell M, Feschotte C (2013) Transposable elements are major contributors to the origin, diversification, and regulation of vertebrate long non-coding RNAs. PLoS Genet 9(4):e1003470. doi:10.1371/journal.pgen.1003470
97. Munoz-Lopez M, Garcia-Canadas M, Macia A, Morell S, Garcia-Perez JL (2012) Analysis of LINE-1 expression in human pluripotent cells. Methods Mol Biol 873:113–125. doi:10.1007/978-1-61779-794-1_7
98. Mortazavi A, Williams BA, McCue K, Schaeffer L, Wold B (2008) Mapping and quantifying mammalian transcriptomes by RNA-Seq. Nat Methods 5(7):621–628. doi:10.1038/nmeth.1226
99. Kulpa DA, Moran JV (2005) Ribonucleoprotein particle formation is necessary but not sufficient for LINE-1 retrotransposition. Hum Mol Genet 14(21):3237–3248. doi:10.1093/hmg/ddi354, ddi354 [pii]
100. Hohjoh H, Singer MF (1996) Cytoplasmic ribonucleoprotein complexes containing human LINE-1 protein and RNA. EMBO J 15(3):630–639
101. Goodier JL, Zhang L, Vetter MR, Kazazian HH Jr (2007) LINE-1 ORF1 protein localizes in stress granules with other RNA-binding proteins, including components of RNA interference RNA-induced silencing complex. Mol Cell Biol 27(18):6469–6483
102. Doucet AJ, Hulme AE, Sahinovic E, Kulpa DA, Moldovan JB, Kopera HC, Athanikar JN, Hasnaoui M, Bucheton A, Moran JV, Gilbert N (2010) Characterization of LINE-1 ribonucleoprotein particles. PLoS Genet 6(10). doi: 10.1371/journal.pgen.1001150, e1001150 [pii]
103. Kulpa DA, Moran JV (2006) Cis-preferential LINE-1 reverse transcriptase activity in ribonucleoprotein particles. Nat Struct Mol Biol 13(7):655–660
104. Monot C, Kuciak M, Viollet S, Mir AA, Gabus C, Darlix JL, Cristofari G (2013) The specificity and flexibility of l1 reverse transcription priming at imperfect T-tracts. PLoS Genet 9(5):e1003499. doi:10.1371/journal.pgen.1003499
105. Imbeault M, Trono D (2014) As time goes by: KRABs evolve to KAP endogenous retroelements. Dev Cell 31(3):257–258. doi:10.1016/j.devcel.2014.10.019
106. Boeke JD, Garfinkel DJ, Styles CA, Fink GR (1985) Ty elements transpose through an RNA intermediate. Cell 40(3):491–500, doi:0092-8674(85)90197-7 [pii]
107. Moran JV (1999) Human L1 retrotransposition: insights and peculiarities learned from a cultured cell retrotransposition assay. Genetica 107(1–3):39–51
108. Rangwala SH, Kazazian HH (2009) The L1 retrotransposition assay: a retrospective and toolkit. Methods 49(3):219–226
109. DeBerardinis RJ, Goodier JL, Ostertag EM, Kazazian HH Jr (1998) Rapid amplification of a retrotransposon subfamily is evolving the mouse genome. Nat Genet 20(3):288–290
110. Goodier JL, Ostertag EM, Du K, Kazazian HH Jr (2001) A novel active L1 retrotransposon subfamily in the mouse. Genome Res 11(10):1677–1685
111. Naas TP, DeBerardinis RJ, Moran JV, Ostertag EM, Kingsmore SF, Seldin MF, Hayashizaki Y, Martin SL, Kazazian HH (1998) An actively retrotransposing, novel subfamily of mouse L1 elements. EMBO J 17(2):590–597
112. Brouha B, Schustak J, Badge RM, Lutz-Prigge S, Farley AH, Moran JV, Kazazian HH Jr (2003) Hot L1s account for the bulk of retrotransposition in the human population. Proc Natl Acad Sci U S A 100(9):5280–5285

113. Sugano T, Kajikawa M, Okada N (2006) Isolation and characterization of retrotransposition-competent LINEs from zebrafish. Gene 365:74–82
114. Moran JV, DeBerardinis RJ, Kazazian HH Jr (1999) Exon shuffling by L1 retrotransposition. Science 283(5407):1530–1534
115. Gilbert N, Lutz-Prigge S, Moran JV (2002) Genomic deletions created upon LINE-1 retrotransposition. Cell 110(3):315–325
116. Morrish TA, Garcia-Perez JL, Stamato TD, Taccioli GE, Sekiguchi J, Moran JV (2007) Endonuclease-independent LINE-1 retrotransposition at mammalian telomeres. Nature 446(7132):208–212. doi:10.1038/nature05560
117. Symer DE, Connelly C, Szak ST, Caputo EM, Cost GJ, Parmigiani G, Boeke JD (2002) Human l1 retrotransposition is associated with genetic instability in vivo. Cell 110(3):327–338
118. Dewannieux M, Esnault C, Heidmann T (2003) LINE-mediated retrotransposition of marked Alu sequences. Nat Genet 35(1): 41–48
119. Hancks DC, Goodier JL, Mandal PK, Cheung LE, Kazazian HH (2011) Retrotransposition of marked SVA elements by human L1s in cultured cells. Hum Mol Genet 20(17): 3386–3400
120. Raiz J, Damert A, Chira S, Held U, Klawitter S, Hamdorf M, Lower J, Stratling WH, Lower R, Schumann GG (2012) The non-autonomous retrotransposon SVA is trans-mobilized by the human LINE-1 protein machinery. Nucleic Acids Res 40: 1666–1683
121. Wagstaff BJ, Hedges DJ, Derbes RS, Campos Sanchez R, Chiaromonte F, Makova KD, Roy-Engel AM (2012) Rescuing Alu: recovery of new inserts shows LINE-1 preserves Alu activity through A-tail expansion. PLoS Genet 8(8):e1002842. doi:10.1371/journal.pgen.1002842
122. Babushok DV, Ostertag EM, Courtney CE, Choi JM, Kazazian HH Jr (2006) L1 integration in a transgenic mouse model. Genome Res 16(2):240–250
123. Ostertag EM, DeBerardinis RJ, Goodier JL, Zhang Y, Yang N, Gerton GL, Kazazian HH Jr (2002) A mouse model of human L1 retrotransposition. Nat Genet 32(4):655–660
124. Ewing AD, Kazazian HH Jr (2011) Whole-genome resequencing allows detection of many rare LINE-1 insertion alleles in humans. Genome Res. doi:10.1101/gr.114777.110, gr.114777.110 [pii]
125. Ewing AD, Kazazian HH Jr (2010) High-throughput sequencing reveals extensive variation in human-specific L1 content in individual human genomes. Genome Res 20(9):1262–1270. doi:10.1101/gr.106419.110, gr.106419.110 [pii]
126. Evrony GD, Lee E, Mehta BK, Benjamini Y, Johnson RM, Cai X, Yang L, Haseley P, Lehmann HS, Park PJ, Walsh CA (2015) Cell lineage analysis in human brain using endogenous retroelements. Neuron 85(1):49–59. doi:10.1016/j.neuron.2014.12.028
127. Xing J, Witherspoon DJ, Jorde LB (2013) Mobile element biology: new possibilities with high-throughput sequencing. Trends Genet 29(5):280–289. doi:10.1016/j.tig.2012.12.002
128. Badge RM, Alisch RS, Moran JV (2003) ATLAS: a system to selectively identify human-specific L1 insertions. Am J Hum Genet 72(4):823–838
129. Witherspoon DJ, Xing J, Zhang Y, Watkins WS, Batzer MA, Jorde LB (2010) Mobile element scanning (ME-Scan) by targeted high-throughput sequencing. BMC Genomics 11:410. doi:10.1186/1471-2164-11-410, 1471-2164-11-410 [pii]
130. Keane TM, Wong K, Adams DJ (2013) RetroSeq: transposable element discovery from next-generation sequencing data. Bioinformatics 29(3):389–390. doi:10.1093/bioinformatics/bts697
131. Sveinbjornsson JI, Halldorsson BV (2012) PAIR: polymorphic Alu insertion recognition. BMC Bioinformatics 13(Suppl 6):S7. doi:10.1186/1471-2105-13-S6-S7
132. Stewart C, Kural D, Stromberg MP, Walker JA, Konkel MK, Stutz AM, Urban AE, Grubert F, Lam HY, Lee WP, Busby M, Indap AR, Garrison E, Huff C, Xing J, Snyder MP, Jorde LB, Batzer MA, Korbel JO, Marth GT, Genomes P (2011) A comprehensive map of mobile element insertion polymorphisms in humans. PLoS Genet 7(8):e1002236. doi:10.1371/journal.pgen.1002236
133. Wu J, Lee WP, Ward A, Walker JA, Konkel MK, Batzer MA, Marth GT (2014) Tangram: a comprehensive toolbox for mobile element insertion detection. BMC Genomics 15:795. doi:10.1186/1471-2164-15-795
134. Goodier JL (2014) Retrotransposition in tumors and brains. Mob DNA 5:11. doi:10.1186/1759-8753-5-11

Chapter 2

Bacterial Group II Introns: Identification and Mobility Assay

Nicolás Toro, María Dolores Molina-Sánchez, Rafael Nisa-Martínez, Francisco Martínez-Abarca, and Fernando Manuel García-Rodríguez

Abstract

Group II introns are large catalytic RNAs and mobile retroelements that encode a reverse transcriptase. Here, we provide methods for their identification in bacterial genomes and further analysis of their splicing and mobility capacities.

Key words Group II introns, Catalytic RNAs, Maturase, Mobile DNA, Retroelements, Reverse transcriptase, RNA, Ribozymes

1 Introduction

Group II introns are mobile metalloribozymes that self-splice from precursor RNA to generate excised intron lariat RNA that invade new DNA genomic locations by reverse splicing [1–3]. These retroelements also encode a reverse transcriptase that stabilizes the RNA structure for forward and reverse splicing and finally converts the inserted intron RNA back to DNA. Although not numerous, group II introns are found widely across the domains of life, being present in eubacteria, archaebacteria, and eukaryotic organelles.

Typically, a complete group II intron consists of a conserved RNA structure of 500–800 nt organized in six domains and an intron-encoded protein (IEP; 400–700 amino acids) enclosed in one of the RNA domains (domain IV). The IEP is a multifunctional protein that has two conserved domains: an N-terminal RT domain and a Maturase/thumb domain (also named X-domain). Some IEPs additionally exhibit a C-terminal DNA-binding (D) region followed by a DNA endonuclease (En) domain (Fig. 1).

The retromobility mechanism of group II introns has been well studied biochemically and genetically elsewhere ([2], and references in). For identification of group II introns, it is relevant to

Jose L. Garcia-Pérez (ed.), *Transposons and Retrotransposons: Methods and Protocols*, Methods in Molecular Biology, vol. 1400, DOI 10.1007/978-1-4939-3372-3_2, © Springer Science+Business Media New York 2016

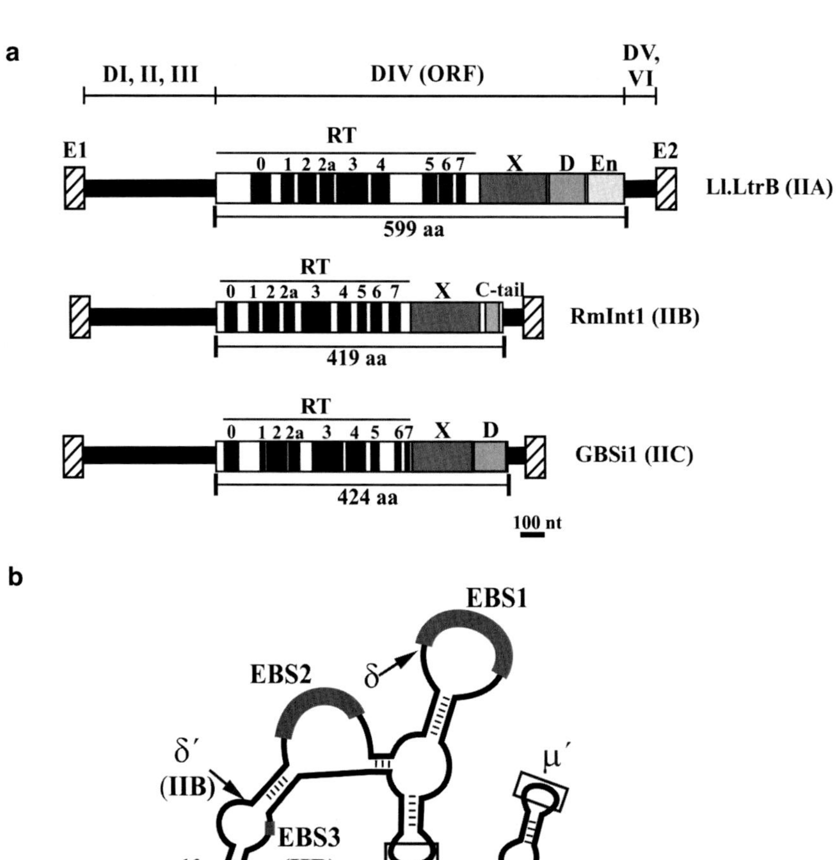

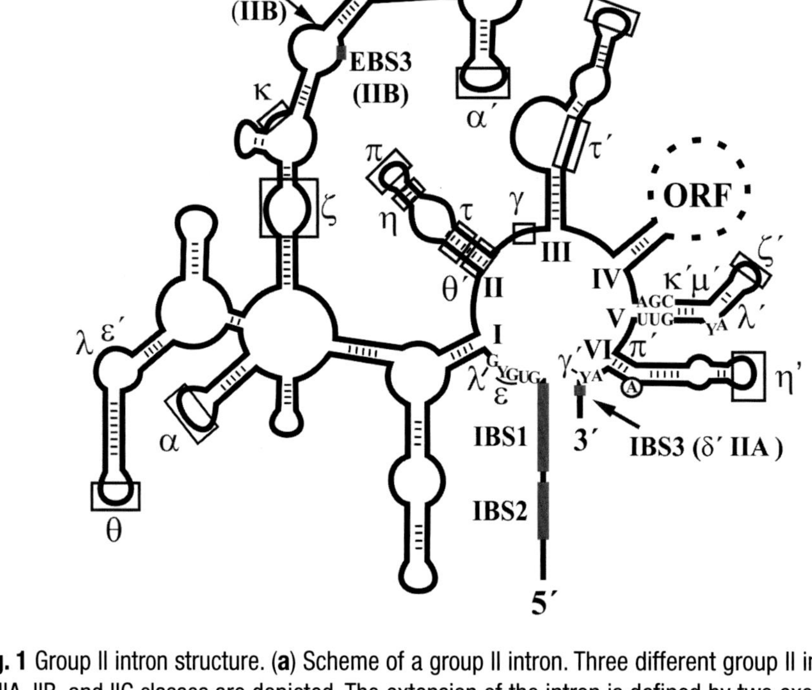

Fig. 1 Group II intron structure. (**a**) Scheme of a group II intron. Three different group II introns representative of IIA, IIB, and IIC classes are depicted. The extension of the intron is defined by two exon edges (E1, E2) and contains six domains (I-VI) spanning a region of approximately 2 kb, which include an ORF region (IEP). In the IEP can be identified a reverse transcriptase domain (RT), maturase (X), and eventually DNA-binding motifs (D) followed by a canonical DNA endonuclease domain (En). Schematics of introns and ORFs are to scale (**b**) group II intron RNA secondary structure. Structure of the representative bacterial IIB1 intron RmInt1 (not to scale). *Boxes* indicate sequences involved in tertiary interactions (Greek letters, EBS, IBS). The loop of DIV (ORF), which encodes the IEP, is depicted by *dashed lines*. Specific nucleotides common in the intron boundaries as well as the conserved domain V are indicated

know that most group II introns have very high sequence selectivity for a long DNA target insertion site known as a homing site (20–35 bp). Often, the homing site is in a conserved gene and the intron boundaries can be defined and confirmed based on a non-interrupted ORF. However, group II introns tend to evolve toward an inactive form by fragmentation, mainly, with loss of the 3′ terminus making it difficult to identify them [4, 5].

2 Materials

2.1 Primer Extension

1. Total RNA preparations from at least three individual colonies harboring a plasmid which constitutively expresses the intron to be characterized (donor plasmid): This donor plasmid must contain the full-length group II intron flanked by at least 50 bp, which includes the intron-binding sites (IBSs). RNA extraction can be performed by a commercial kit or a suitable protocol described for your bacteria. A high yield protocol for bacterial RNA extraction is detailed in [6].
2. 20-mer oligonucleotide complementary to a sequence within the first 100 nt of the intron RNA sequence, that is, the extension has to proceed towards the 5′ end of the intron.
3. T4 polynucleotide kinase supplied with its own concentrated buffer 10×.
4. γ-[^{32}P] ATP 3000 Ci/mmol; 10 mCi/μl.
5. Illustra™ microspin G-25 columns (GE Healthcare).
6. Annealing buffer: 10 mM PIPES pH 7.5, 400 mM NaCl.
7. Extension mixture: 50 mM Tris pH 8.0, 60 mM NaCl, 10 mM DTT, 6 mM magnesium acetate, 1 mM each of all four dNTPs, 60 μg/ml actinomycin D, 2 U of RNase Inhibitor (Invitrogen), and 7 U AMV reverse transcriptase.
8. 3 M sodium acetate pH 4.8.
9. 100 % ethanol and 70 % ethanol (kept at −20 °C).
10. TE 1×: 10 mM Tris pH 8.0, 1 mM EDTA.
11. STOP solution: 0.3 % (w/v) each: Bromophenol blue and xylene cyanol FF, 10 mM EDTA, 97.5 % deionized formamide.
12. 8 M Urea/polyacrylamide gel in TBE 1× (89 mM Tris pH 8.0, 89 mM boric acid, 2 mM EDTA).
13. Imaging Plate 2040 or X-ray film.

2.2 RT-qPCR

1. Total RNA preparations from at least three individual colonies containing the intron donor plasmid (*see* Subheading 2.1.).
2. A primer pair spanning the splicing junction (Tm 60 °C): For instance, a 20-mer oligonucleotide which sequence corresponds

to the 5′ end of the insertion site (designed into the 50 nt upstream of the 5′ intron boundary) and the other 20-mer oligonucleotide complementary to the 3′ end of the intron insertion site (designed into the 50 nt downstream of the intron end).

3. qPCR mix: 20 mM Tris pH 8.4, 50 mM KCl, 3 mM $MgCl_2$, 0.2 mM each of all four dNTPs, Sybr Green I mix, and 0.5 U Taq DNA polymerase. To prepare the Sybr Green I mix, combine 1 μl Sybr Green I 10,000×, 1 μl fluorescein, and 18 μl DMSO, and then dilute this mixture 1:250 (add 2.5 μl of the dilution to the 25 μl final volume reaction).
4. The consumable material will depend on the technology used for amplification, but generally consist of see-through 96-well plates and cover films.

2.3 Plasmid to Plasmid Mobility Assay

1. A donor plasmid (see above, Subheading 2.1).
2. A receptor plasmid that contains the intron DNA target sequence. This DNA target must contain the sequences corresponding to the IBSs that interact with the exon-binding sites (EBSs) of the intron. Generally a region of 100 nucleotides spanning from −50 to +50 of the intron insertion site (+1) should include all the requirements as a homing site. Donor and receptor plasmids must be compatible.
3. Plasmid DNA extraction kit adapted to your bacteria.
4. Restriction enzymes and 10× buffers.

2.4 Southern Blot

1. Agarose gel electrophoresis system: Generally 1 % agarose is used in TAE 1× (40 mM Tris-acetate, 2 mM EDTA; pH 8.0).
2. DIG-labeled DNA molecular weight marker.
3. Fluorescent nucleic acid gel stain (Gel Red, EtBr, …).
4. Ultraviolet transilluminator.
5. 20× SSC: 3 M NaCl, 300 mM trisodium citrate (adjusted to pH 7.0 with HCl).
6. VacuGene™ XL Vacuum Blotting System (GE Healthcare Amersham™).
7. 1 N NaOH solution.
8. Positively charged nylon membranes.
9. Whatman 3MM paper.
10. Heated vacuum desiccator.
11. Hybridization oven and tubes.
12. Prehybridization solution: 50 % v/v formamide, 5× SSC, 2 % w/v blocking reagent (Roche), 0.1 % w/v *N*-lauroylsarcosine, 0.02 % w/v SDS, and deionized water. Heat until just about to boil to dissolve the blocking reagent. Store at −20 °C.

13. DIG-labeled probe: 300–500 nt PCR DIG-labeled product which hybridizes at the 5′ or 3′ region of the insertion site. PCR amplification must be performed using the following dNTP mixture: 0.1 mM dATP, 0.1 mM dCTP, 0.1 mM dGTP, 0.065 mM dTTP, 0.035 mM DIG-dUTP.
14. Digoxigenin-11-dUTP, alkali-stable.
15. Thermocycler.
16. Probe solution (5–10 ml): 60 ng of DIG-labeled probe/ml prehybridization solution.
17. Solution 1: 2× SSC, 0.1 % (w/v) SDS and deionized water.
18. Solution 2: 0.1× SSC, 0.1 % (w/v) SDS and deionized water.
19. Solution 3: Buffer 1 (0.1 M malic acid, 0.15 M NaCl, deionized water; adjusted at pH 7.5 with NaOH) and 0.3 % (w/v) Tween 20.
20. Solution 4: 1 % of Blocking Reagent (Roche) in Buffer 1. Heat the solution to dissolve the blocking reagent, but do not let it boil. Prepare 100 ml fresh.
21. Solution 5: Prepare a 1:10,000 dilution of anti-digoxigenin-AP, Fab fragments (Roche) in 20 ml of Solution 4.
22. Solution 6: 0.1 M Tris, 0.1 M NaCl, and deionized water. Adjust at pH 9.5 with HCl.
23. Solution 7: Prepare a 1:100 dilution of CSPD (Roche) in Solution 6.
24. X-ray film and exposure holder.
25. X-ray developer and fixer.

3 Methods

3.1 How to Identify a Group II Intron

3.1.1 Identification of the Intron-Encoded Protein

Generally, a first indication of the presence of a group II intron is the identification of its reverse transcriptase domain (RT). Among others, group II intron RTs are characterized by a high conservation degree in all RT domains (0–7 domains). Examples of ORFs associated to group II introns can be identified (annotated) as (1) reverse transcriptase, (2) RNA-directed DNA polymerase, (3) retron-type reverse transcriptase, or even only (4) DNA polymerase. Alternatively, they can be searched by using representative group II intron-encoded proteins obtained from intron database (http://webapps2.ucalgary.ca/~groupii/) by BlastP searches on particular genomes or other sequence data. Finally, the query RT sequence could be aligned to the reported RT data set [7, 8], and performed phylogenetic analysis using FastTree so that it could be identified as an RT potentially associated to a group II intron RNA.

3.1.2 Identification of Intron RNA Component

In annotated genomes, frequently the RNA component of a group II intron has not been identified. In several genomes, ORFs overlapping the ribozyme sequence have been automatically annotated making it difficult to interpret if a particular genome sequence contains a canonical group II intron. In order to identify if a full-length group II intron is associated to a particular reverse transcriptase ORF, nucleotide sequences (4–5 kb) containing 2 kb upstream and 2 kb downstream of the ORF can be analyzed for the presence of characteristic group II intron RNA domains. In these cases, a conserved catalytic domain V (34–36 nt) can be easily identified, generally just downstream from the ORF sequence. As example, a local BlastN using the conserved domain V of RmInt1 (GAGCGGTGTGAATCGAGAGGTTTACGCACCGTTC) on these 6–8 kb can identify the potential domain V of the RNA structure. Additionally, the bioinformatics tool "intron boundaries" from the intron database (http://webapps2.ucalgary.ca/~groupii/primes.html) can be used to delimit the borders of the putative group II intron [9, 10]. In a canonical group II intron a "GUGYG" and "AY" as boundaries of 5′ and 3′ end, respectively, can be confirmed using this tool. A further study using Mfold (http://mfold.rna.albany.edu/?q=mfold;) for the detection of potential group II intron RNA structures can be performed in order to identify the remaining RNA domains (I, II, III, and VI). At this stage, the domain VI defining the 3′ end of the group II intron can be easily confirmed. In some cases, the identification of the domains I, II, and III (domains upstream of the ORF sequence) should require the uses of nucleotide-level comparisons with group II introns already described [10].

3.1.3 Identification of a Putative Exon Junction

The confirmation of the presence "in silico" of a full-length intron can be finalized with an examination of putative exon junctions. In some cases the nucleotide corresponding with the IBSs (generally between −15 and +1) of the intron insertion site can be deduced as complementary stretches to some loop structures proposed for domain I of the RNA. Sometimes it is useful to investigate if this group II intron is interrupting a particular gene. A BlastX using the corresponding DNA fragment lacking the proposed group II intron sequence can reveal a new copy of the interrupted ORF but in this case without the group II intron. This fact is also indicative that our sequence contains a full-length mobile group II intron.

3.2 Group II Intron Splicing

The excision of group II introns can be detected by primer extension [11], and more accurately by RT-qPCR [12]. This protocol has been adapted to be carried out using a microfuge.

3.2.1 Primer Extension

1. Prepare the (5′-^{32}P)-labeled oligonucleotide. For that, mix 10 pmol of oligonucleotide with 10 μCi of γ-[^{32}P] ATP, 8 U of

T4 polynucleotide kinase, and the corresponding buffer. Incubate for 1 h at 37 °C. Remove the non-incorporate radioactive nucleotides using G-25 columns and measure 1 μl in a scintillation counter.

2. For the annealing mixture, combine 10 μg of total RNA with 300,000 cpm (~0.2 pmol) of (5′-^{32}P)-labeled primer in annealing buffer (10 μl final volume). Heat the mixture at 85 °C for 5 min in a heater block, then cool the block fast to 60 °C in water bath, and, finally, leave it at room temperature for slowly cooling until 44–45 °C.
3. Prepare the extension mixture and preincubate it at 42 °C for 2–5 min. When the annealing is completed, add 40 μl of preheated extension mixture and incubate at 42 °C for 60 min.
4. Stop the reverse transcription reaction by adding 15 μl 3 M sodium acetate pH 4.8 and 150 μl of cold 100 % ethanol. Keep for at least 60 min at −80 °C. Alternatively, sample precipitation could be placed at −20 °C overnight.
5. Centrifuge at 16,000 *g* for 15 min, and remove the supernatant. Wash samples with 200 μl of cold 70 % ethanol. Centrifuge at 16,000 *g* for 5–10 min, remove supernatant with a tip, and let the pellets to dry at room temperature.
6. Resuspend reactions in 4 μl of TE1x + 4 μl of STOP solution.
7. Load 4 μl of the sample from **step 6** on a denaturing, polyacrylamide gel. Pre-run the gel for 30 min at 45 W. Denature the samples by heating for 3 min at 95 °C, and keep them on ice until loading. Run the gel at 50 W until the bromophenol blue reaches the bottom of the gel. Dry the gel and expose to an imaging plate or an X-ray film.
8. Acquire images after 24–48 h of exposure, with a laser scanning system (for the imaging plates), or by conventional developing (developer/fixer, for the X-ray films).

3.2.2 RT-qPCR

1. Prepare cDNA from total RNA samples extracted from cells harboring the intron whose splicing efficiency will be quantified. For this purpose, we reverse transcribe 10 μg of the RNA using random hexamers (5 ng/μl final concentration) and AMV RT (see above in the primer extension section, but omitting actinomycin D in the extension mixture).
2. The qPCR experiments can be conducted with a real-time PCR detection system. Each reaction must contain 1 μg of cDNA template and 5 pmol of each primer in 25 μl of qPCR mix final volume.
3. The PCR cycling conditions are as follows: 5-min hot start at 95 °C, followed by 45 cycles of denaturation at 95 °C for 10 s, annealing at 63–65 °C for 2 s, and extension at 72 °C for 2 s.

4. Relative or absolute quantification can be performed. Relative quantification requires a parallel reaction with a primer pair amplifying a reference gene; any housekeeping gene with constitutive expression is suitable. Thus, the mRNA levels will be calculated according to the 2-ΔΔCT method [13], where ΔΔCT = ΔCT (calibrator or WT) − ΔCT (test or mutant). For absolute quantification, a standard curve is needed preferably from a plasmid containing the exon junction.

3.3 Group II Intron Mobility Assay

An outline of the double-plasmid retrohoming assay [14] is shown in Fig. 2.

1. Transform bacteria with both donor and recipient plasmids.
2. Grow bacteria in liquid media supplemented with the corresponding antibiotics and extract the plasmid DNA according to a suitable protocol or kit.

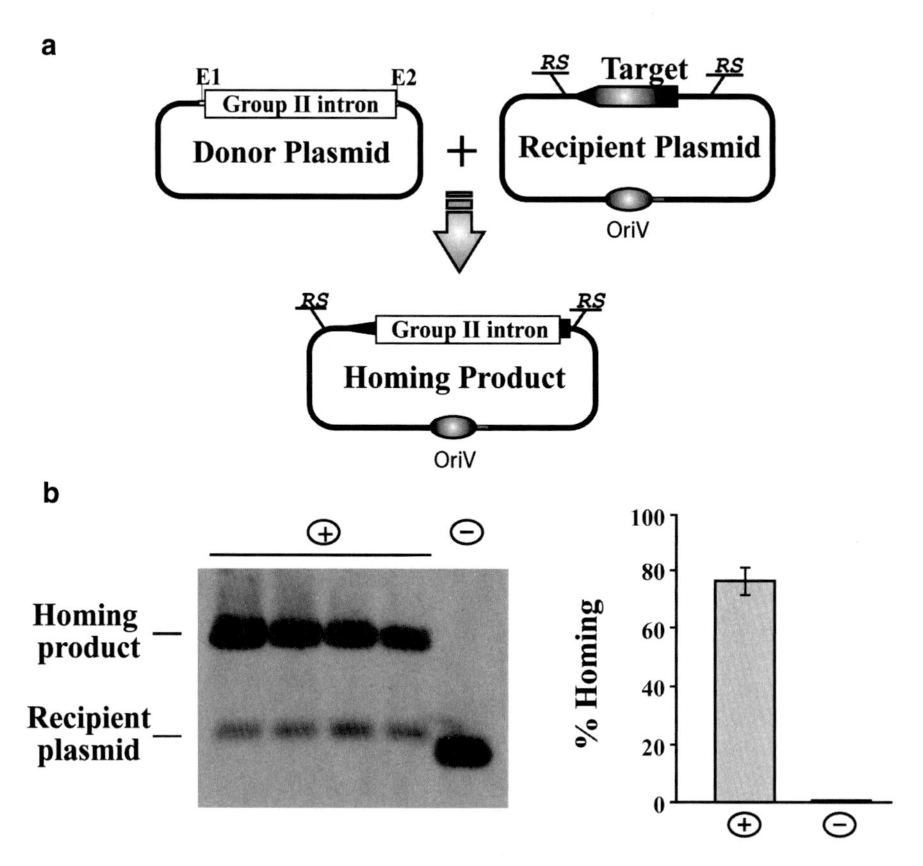

Fig. 2 Plasmid-to-plasmid group II intron mobility assay. (**a**) Outline of the double-plasmid retrohoming assay: the target on the recipient plasmid could be invaded by an intron from the donor, resulting in the homing product. (**b**) Southern hybridization with a DNA probe specific to the target. Negative control is indicated as a *minus sign in a circle* above the blot. Target invasion rate was calculated as described in Subheading 3 and is plotted in the histogram shown beside the blot

3. Digest DNA samples with one or a combination of restriction enzymes which release the DNA fragment containing the invaded or non-invaded target site from the recipient plasmid.
4. Run equal amount of digested DNA samples in a preparative agarose gel using a DIG-labeled DNA molecular weight marker.
5. Reveal the bands using a fluorescent nucleic acid gel dye. Expose the gel to the UV during 15–20 min to facilitate the DNA break and transfer to the membrane.
6. Cut the nylon membrane and wet it with distilled water. Note that the membrane dimensions must exceed 1–1.5 cm from the gel size. Equilibrate by immersion in 20× SSC for 5 min with occasional swinging.
7. Assemble the transfer system. Place first the nylon membrane on the foam bed followed by the perforated, plastic mask in such a way that the membrane overlaps with the open gap boundaries in the mask. Close the tab to fix the components and finally put the gel on the top covering the mask window to ensure the sealing of the system.
8. Connect the vacuum pump and regulate the power to 50–60 mbar.
9. Add 1 N NaOH until the gel gets covered. Let the transfer during 2 h.
10. Disassemble the transfer system and incubate the nylon membrane on a tray containing 2× SSC during 15 min. Swing occasionally.
11. Let the membrane dry on Whatman 3MM paper. Introduce the membrane in a Whatman 3MM paper envelope and fix the DNA at 120 °C on a heated vacuum desiccator during 35 min.
12. Introduce the nylon membrane in a hybridization tube in such a way that the DNA side remains opposite to the glass walls. Add 50 ml of prehybridization solution. Incubate in a preheated hybridization oven at 42 °C for at least 2 h.
13. Denaturalize the probe solution in boiling water during 10 min and cool quickly on ice for 10 min.
14. Remove the prehybridization solution from the hybridization tube. Add the denatured probe solution in the membrane container. Incubate at 42 °C for 12–16 h (overnight).
15. Collect the probe solution from the hybridization tube and store it at −20 °C. The probe solution can be reused at least for four to five additional hybridizations.
16. Wash the membrane for 5 min twice with 100 ml of Solution 1.
17. Wash with 100 ml of Solution 2 for 15 min at 68 °C twice. These two washing steps will remove all non-hybridized probe.

18. Chill the hybridization tube at room temperature (RT). Incubate the membrane with 100 ml of Solution 3 for 5 min at RT. Hereafter, all the incubation steps will be performed at RT.
19. Block the membrane by adding 80 ml of Solution 4 to the tube and incubate for 30 min.
20. Drain the Solution 4 and add 20 ml of Solution 5. Incubate for 30 min to allow the antibody to recognize the DIG molecules.
21. Wash with 100 ml of Solution 3 for 15 min. Repeat this washing step to remove the unbounded antibody.
22. Equilibrate the membrane by incubating in 50 ml of Solution 6 for 5 min at room temperature.
23. Finally, incubate in 5 ml of Solution 7 for 5 min.
24. Place the membrane on Whatman 3MM paper and let it dry for at least 5 min.
25. Wrap the membrane in plastic wrap, avoiding wrinkles. Incubate at 37 °C for 15 min in darkness to activate the alkaline phosphatase.
26. Expose the wrapped membrane to X-ray film during 4–12 h (overnight) in an X-ray exposure holder.
27. Reveal the X-ray film using developer and fixer as indicated by the manufacturer.
28. Digitalize the image obtained from the exposure and quantify the band intensities using proper software.
29. The homing efficiency is calculated as the percentage of the ratio of homing product (H) to the addition of homing product and non-invaded recipient plasmid (H + R), (H/(H + R) × 100) [14].

4 Notes

4.1 Group II Intron Splicing

1. For RNA handling precautions are essential: keep all solutions and material free of RNase contamination, wear gloves during all the steps, and preserve the RNA and cDNA samples at 4 °C during all the processes.
2. Radioactivity protocols require supervising and properly conducting.
3. Do not dry the primer extension samples using high temperature because it should be difficult to resuspend them.
4. The size and the percentage of the polyacrylamide gel to run the primer extension samples depend on the cDNA length. For instance, a 100 nt cDNA could be separated in a denaturing, 6 % polyacrylamide gel of 40 cm length/0.2 mm thick.

5. In the qPCR, the same cDNA has to be amplified in triplicates, and three cDNAs are recommended to be synthesized from each RNA sample. Use fine calibrated pipettes.
6. To obtain a better reproducibility in the RT-qPCR experiments use the master mixes available from several commercial suppliers either for first-strand cDNA synthesis and/or for qPCR amplification.
7. Depending on the size of the PCR product, it would be necessary to suppress the extension step in order to minimize the plausible amplification of the unspliced RNA precursor.

4.2 Group II Intron Mobility

1. Since it is required to release the fragment containing the target DNA by restriction enzymes, design your recipient plasmid suitably. It is also recommended that the enzyme linearize the donor plasmid.
2. Plasmid-to-plasmid assay must be performed in an intronless strain to obtain an accurate quantification of the homing rates.
3. It is important to design a negative control of invasion, like a receptor plasmid lacking the insertion site.
4. In order to seal the transfer system, you can cover the edges of the gel with 2 % agarose.
5. Use gloves to manipulate both the Whatman paper and the nylon membrane.
6. Before using the anti-digoxigenin-AP antibody, vortex and centrifuge the stock for 3 min at full speed. Take the necessary volume near the surface, not from the bottom of the tube.
7. If you have no vacuum transfer system, you can opt for the traditional capillary transfer protocol.
8. There are several devices to acquire chemiluminescent signal that can be used in these analyses.

Acknowledgments

This work was supported by research grants CSD 2009-0006 from the Consolider-Ingenio program, and BIO2011-24401 from the currently Ministerio de Economía y Competitividad, including ERDF (European Regional Development Funds).

References

1. Toro N, Jiménez-Zurdo JI, García-Rodríguez FM (2007) Bacterial group II introns: not just splicing. FEMS Microbiol Rev 31:342–358
2. Lambowitz AM, Zimmerly S (2010) Group II introns: mobile ribozymes that invade DNA. Cold Spring Harb Perspect Biol. doi:10.1101/cshperspect.a003616
3. Enyeart PJ, Mohr G, Ellington AD, Lambowitz AM (2014) Biotechnological applications of mobile group II introns and their reverse

transcriptases: gene targeting, RNA-seq, and non-coding RNA analysis. Mob DNA. doi:10.1186/1759-8753-5-2

4. Dai L, Zimmerly S (2002) Compilation and analysis of group II intron insertions in bacterial genomes: evidence for retroelement behavior. Nucleic Acids Res 30(5):1091–1102
5. Toro N, Martínez-Rodríguez L, Martínez-Abarca F (2014) Insights into the history of a bacterial group II intron remnant from the genomes of the nitrogen-fixing symbionts *Sinorhizobium meliloti* and *Sinorhizobium medicae*. Heredity (Edinb). doi:10.1038/hdy.2014.32
6. Molina-Sánchez MD, Martínez-Abarca F, Toro N (2006) Excision of the *Sinorhizobium meliloti* group II intron RmInt1 as circles *in vivo*. J Biol Chem 281:28737–28744
7. Toro N, Martínez-Abarca F (2013) Comprehensive phylogenetic analysis of bacterial group II intron-encoded ORFs lacking the DNA endonuclease domain reveals new varieties. PLoS One. doi:10.1371/journal.pone.0055102
8. Toro N, Nisa-Martínez R (2014) Compre hensive phylogenetic analysis of bacterial reverse transcriptases. PLoS One. doi:10.1371/journal.pone.0114083
9. Candales MA, Duong A, Hood KS et al (2012) Database for bacterial group II introns. Nucleic Acids Res. doi:10.1186/1759-8753-4-28
10. Abebe M, Candales MA, Duong A et al (2013) A pipeline of programs for collecting and analyzing group II intron retroelement sequences from GenBank. Mob DNA. doi:10.1186/1759-8753-4-28
11. Muñoz-Adelantado E, San Filippo J, Martínez-Abarca F et al (2003) Mobility of the *Sinorhizobium meliloti* group II intron RmInt1 occurs by reverse splicing into DNA, but requires an unknown reverse transcriptase priming mechanism. J Mol Biol 327:931–943
12. Chillón I, Martínez-Abarca F, Toro N (2011) Splicing of the *Sinorhizobium meliloti* RmInt1 group II intron provides evidence of retroelement behavior. Nucleic Acids Res 39:1095–1104
13. Livak KJ, Schmittgen TD (2001) Analysis of relative gene expression data using real-time quantitative PCR and the 2(-Delta Delta C (T)) Method. Methods 25:402–408
14. Nisa-Martinez R, Jiménez-Zurdo JI, Martínez-Abarca F et al (2007) Dispersion of the RmIntI group II intron in the *Sinorhizobium meliloti* genome upon acquisition by conjugative transfer. Nucleic Acids Res 35:214–222

Chapter 3

In Silico Methods to Identify Exapted Transposable Element Families

LeeAnn Ramsay and Guillaume Bourque

Abstract

Transposable elements (TEs) have recently been shown to have many regulatory roles within the genome. In this chapter, we will examine two in silico methods for analyzing TEs and identifying families that may have acquired such functions. The first method will look at how the overrepresentation of a repeat family in a set of genomic features can be discovered. The example situation of OCT4 binding sites originating from LTR7 TE sequences will be used to show how this method could be applied. The second method will describe how to determine if a TE family exhibits a cell type-specific expression pattern. As an example, we will look at the expression of HERV-H, an endogenous retrovirus known to act as an lncRNA in embryonic stem cells. We will use this example to demonstrate how RNA-seq data can be used to compare cell type expression of repeats.

Key words Transposable elements, Repeats, Bioinformatics, RNA-seq, Transcriptional regulation, Endogenous retrovirus, lncRNAs

1 Introduction

Transposable elements (TEs) make up about 50 % of the human genome. Most of these repetitive elements fall into one of four classes: long interspersed elements (LINEs), short interspersed elements (SINEs), long terminal repeats (LTR) retrotransposons, and DNA transposons. These different classes indicate which method the TE utilizes to propagate through the genome [1]. Repeats were once thought to be "selfish" sequences which did not play any role in normal genome function [2, 3], but recently evidence has been found to the contrary. TEs that play a role in the host genome are also known as "exapted" repeats [4].

A number of recent studies have identified TEs with various regulatory functions. DNA sequences that have, or had, a function in transposing the element can allow the genome to modify regulation more quickly than random mutation would allow. In this way, TEs provide the genome with a rich source of novel genetic

Jose L. Garcia-Pérez (ed.), *Transposons and Retrotransposons: Methods and Protocols*, Methods in Molecular Biology, vol. 1400, DOI 10.1007/978-1-4939-3372-3_3, © Springer Science+Business Media New York 2016

material. Confirming that this does occur, a number of studies have found specific retrotransposons that have contributed transcription factor binding sites. The long terminal repeats (LTRs) of endogenous retroviruses (ERVs) naturally have transcriptional regulatory elements [5], which are important in the transposition of ERVs. Several studies have discovered LTRs that have been exapted for use by the host regulatory system. For example, a study by Wang et al. identified specific LTR transposons which act as p53 transcription factor binding sites [6]. Additionally, from a comparison of mouse and human binding profiles researchers found that OCT4 (a transcription factor essential to embryonic stem cell development) has a very different binding profile in humans [7]. This may be because many OCT4 binding sites are co-opted LTR7 transposable elements [8]. Transcription factor binding sites are not the only regulatory contribution of the TEs. A 2014 study identified RNA binding proteins which bind TEs to effect changes on transcript abundance and splicing [9]. A specific example of this is when the protein HuR binds antisense Alu elements. Normally HuR stabilizes the target transcript, but in the case where it binds an Alu element the target is downregulated.

The majority of the human genome is transcribed, however most of the resulting RNA products do not code for proteins [10]. TEs are rarely found in protein coding genes since generally an insertion into a gene is detrimental, although non-lethal insertions have been observed [11]. TEs do however make up a large portion of lncRNA transcripts. Long non-coding RNAs (lncRNAs) are an important genomic regulatory element for many species [12]. A recent study showed that significantly more TEs are found in lncRNAs than any other type of RNA, including tRNAs and small ncRNAs. Seventy-five percent of human lncRNAs contain an exon which is at least partly made up of a transposable element [13]. A number of studies characterizing lncRNAs highlight the prevalence of TEs in these sequences. For example, lnc-RoR is implicated in the modulation of reprogramming of human iPS cells [14]. lnc-RoR transcription initiates in the LTR7 region of HERV-H and its exons consist mainly of 6 different TEs. HERV-H itself plays an important role in human embryonic stem cells (hESCs). These sequences act as lncRNAs almost exclusively in hESCs and are essential to maintaining pluripotency in this cell type [15].

In this chapter, we will present two in silico methods for identifying some of the repeats described above. The first method will describe how one can determine that the LTR7 repeat is overrepresented in OCT4 binding regions. This technique can be generalized to find other TEs which are overrepresented in a set of genomic features. The second method will look at the expression of HERV-H sequences in hESC cells versus their expression in GM12878 cells. This method describes standard RNA-seq alignment procedures and gives a simplified look at how to compare repeat expression of two cell types. This method is intended to give a conceptual understanding of the procedure, which would typically be expanded to include biological replicates and several cell types.

2 Materials

1. OCT4 binding site coordinates. Download for BED format at http://www.ncbi.nlm.nih.gov/geo/query/acc.cgi?acc=GSM518373
   (a) Rename oct4_hg18.bed
2. hg18 to hg19 LiftOver chain file: http://hgdownload.soe.ucsc.edu/goldenPath/hg18/liftOver/hg18ToHg19.over.chain.gz
3. hg19 chromosome sizes. UCSC download link: http://genome.ucsc.edu/goldenpath/help/hg19.chrom.sizes
4. RepeatMasker track from UCSC:
   (a) http://hgdownload.soe.ucsc.edu/goldenPath/hg19/database/rmsk.txt.gz
5. nestedRepeats track UCSC: http://hgdownload.soe.ucsc.edu/goldenPath/hg19/database/nestedRepeats.txt.gz
6. ENCODE RNA-seq data [16]. UCSC fastq download links:
   (a) http://hgdownload.cse.ucsc.edu/goldenPath/hg19/encodeDCC/wgEncodeCshlLongRnaSeq/wgEncodeCshlLongRnaSeqH1hescCellLongnonpolyaFastqRd1Rep1.fastq.gz
   (b) http://hgdownload.cse.ucsc.edu/goldenPath/hg19/encodeDCC/wgEncodeCshlLongRnaSeq/wgEncodeCshlLongRnaSeqH1hescCellLongnonpolyaFastqRd2Rep1.fastq.gz
   (c) http://hgdownload.cse.ucsc.edu/goldenPath/hg19/encodeDCC/wgEncodeCshlLongRnaSeq/wgEncodeCshlLongRnaSeqGm12878CellLongnonpolyaFastqRd1Rep1.fastq.gz
   (d) http://hgdownload.cse.ucsc.edu/goldenPath/hg19/encodeDCC/wgEncodeCshlLongRnaSeq/wgEncodeCshlLongRnaSeqGm12878CellLongnonpolyaFastqRd2Rep1.fastq.gz
   (e) More details about these data can be found at http://genome.ucsc.edu/cgi-bin/hgTrackUi?hgsid=409936327_wq5BDu3jPBE5D0gEEmUyFUA6tlTi&c=chr5&g=wgEncodeCshlLongRnaSeq

7. hg19 genome assembly file. UCSC link for 2bit file: http://hgdownload.soe.ucsc.edu/goldenPath/hg19/bigZips/hg19.2bit
8. LiftOver utility. UCSC Linux download link: http://hgdownload.soe.ucsc.edu/admin/exe/linux.x86_64/liftOver
   (a) Usage: http://genome.sph.umich.edu/wiki/LiftOver
9. twoBitToFa file conversion tool. UCSC Linux download link: http://hgdownload.cse.ucsc.edu/admin/exe/linux.x86_64/twoBitToFa
   (a) Usage: http://genome.ucsc.edu/goldenPath/help/blatSpec.html#twoBitToFaUsage
10. BEDTools [17]. Software download on Github: https://github.com/arq5x/bedtools2/releases
    (a) Use version 2.17.0 or later.
11. SAMtools [18]. Software download on SourceForge: http://sourceforge.net/projects/samtools/files/
    (a) Use version 0.1.19 or later.
12. R or some other statistical software.
    (a) Command to install on Linux systems:
    ```
    sudo apt-get update
    sudo apt-get install r-base
    ```
    (b) Install directions available at http://cran.r-project.org/
    (c) Use version 3.1.1 or later.
13. Install DESeq. In an R session use the command:
    (a)
    ```
    source("http://www.bioconductor.org/biocLite.R")
    biocLite("DESeq")
    ```
    (b) http://www.bioconductor.org/install/
    (c) Use version 3.1.2_3.0 or later.
14. STAR alignment software [19]. Software download on GitHub: https://github.com/alexdobin/STAR/releases
    (a) I use version 2.4.0e
    (b) Installation directions available in the STAR manual: https://rna-star.googlecode.com/files/STARmanual_2.3.0.1.pdf

3 Methods

3.1 Method for Identifying Features Overrepresented by Transposable Elements

OCT4 is a transcription factor important for embryonic development. Previous studies [7, 8] have found that a particular repetitive element, LTR7, is overrepresented in OCT4 binding regions and likely contributed OCT4 binding sites to the human genome. This section describes a method for identifying repeats like LTR7, which are overrepresented in a set of genomic intervals. We will use OCT4 as an example set of binding regions as a starting point.

1. The OCT4 binding sites are mapped to hg18. First convert the coordinates to hg19 using LiftOver.
   (a) `./LiftOver oct4_hg18.bed hg18ToHg19.over.chain.gz oct4.bed oct4_unlifted.bed`
2. Extract repeat files.
   (a) `gunzip rmsk.txt.gz`
3. Format repeat data into a BED file. See http://bedtools.readthedocs.org/en/latest/content/general-usage.html for more information on this format.
   (a) One way to format is using an `awk` command. This example assumes this is the full RepeatMasker file from UCSC.

      awk '{print $6"\t"$7"\t"$8"\t"$10"\t"$11}' rmsk.txt > rmsk.bed.
   (b) `Identify which repeats overlap OCT4 binding sites: intersectBed -f 0.1 -u -a rmsk.bed -b oct4.bed > rmsk_oct4_intersect.bed`
   (c) -f 0.1 requires that a minimum of 10 % of each repeat overlaps a binding site.
   (d) -u writes each repeat instance once if any overlaps are found.
4. Use the BEDTools program `shuffleBed` to randomly permute the OCT4 binding regions in the genome for use as a control (*see* **Note 1**).
   (a) `shuffleBed -i oct4.bed -g hg19.chrom.sizes -seed 300000 > oct4_shuffled.bed`
   (b) Where the `–seed` value is some integer value used to generate the pseudo-random numbers.
5. Repeat **step 3b** using the file generated by shuffleBed in place of the original repeat bed file.
   (a) `intersectBed -f 0.1 -u -a rmsk.bed -b oct4_shuffled.bed > rmsk_oct4_shuffled_intersect.bed`
6. Filter the two intersect files for LTR7 repeat elements (*see* **Note 2**).
   (a) `sed -n '/LTR7$/p' rmsk_oct4_intersect.bed > rmsk_oct4_intersect.ltr7.bed`
   (b) `sed -n '/LTR7$/p' rmsk_oct4_shuffled_intersect.bed > rmsk_oct4_shuffled_intersect.ltr7.bed`
7. Count the number of repeats which overlap their respective features.
   (a) For each file: wc -l *intersect*.bed > intersect_counts.txt
   (b) Results are shown in Table 1 and Fig. 1.

Table 1
The number of features and intersections we observed from the analysis

File	Counts
rmsk_oct4_intersect.bed	11,615
rmsk_oct4_intersect.ltr7.bed (number of LTR7 which overlap OCT4 binding sites)	318
rmsk_oct4_shuffed_intersect.bed	1597
rmsk_oct4_shuffled_intersect.ltr7.bed (expected value)	0
Total number of LTR7	2344
Total number of repeats in rmsk.txt	5,298,130

8. Using the values from Table 1 perform a one-sided binomial test to determine if LTR7 repeat elements overlap OCT4 binding sites more than expected by chance.
   (a) R code:
   ```
   >binom.test(318,2344,p=0/2344,alternative="greater",
   conf.level=0.95)
   ```
   (b) First command determines the statistical significance of LTR7 overlapping OCT4 binding sites.
   (c) See https://stat.ethz.ch/R-manual/R-patched/library/stats/html/binom.test.html for more information about R's binomial test function.
   (d) The result of this step is shown in Fig. 1.

This method can be expanded to include multiple tissue types or other genomic feature data. You may also want to determine if a different type of TE is overrepresented, which can be done by filtering the RepeatMasker data differently.

3.2 Method for Evaluating Transposable Element Expression Levels

Generally transposable elements are not located within protein coding genes, even so some are still expressed as non-coding RNAs. For example, HERV-H is a known lncRNA expressed almost exclusively in embryonic stem cells[2]. This method gives a simplified look at how this can be discovered by calculating the expression level of repeats using RNA-seq data. The initial steps will be standard RNA-seq analysis for aligning reads, which is followed by an expression analysis of repetitive elements.

1. For this analysis we will want to use a different repeat annotation: nestedRepeats.txt (*see* **Note 4**). Create a bed file from the UCSC text file.
   (a) `gunzip nestedRepeat.txt.gz`
   (b)
   ```
   awk '{print $2"\t"$3"\t"$4"\t"$7"\t"$5}' neste-
   dRepeats.txt > repeats.bed
   ```

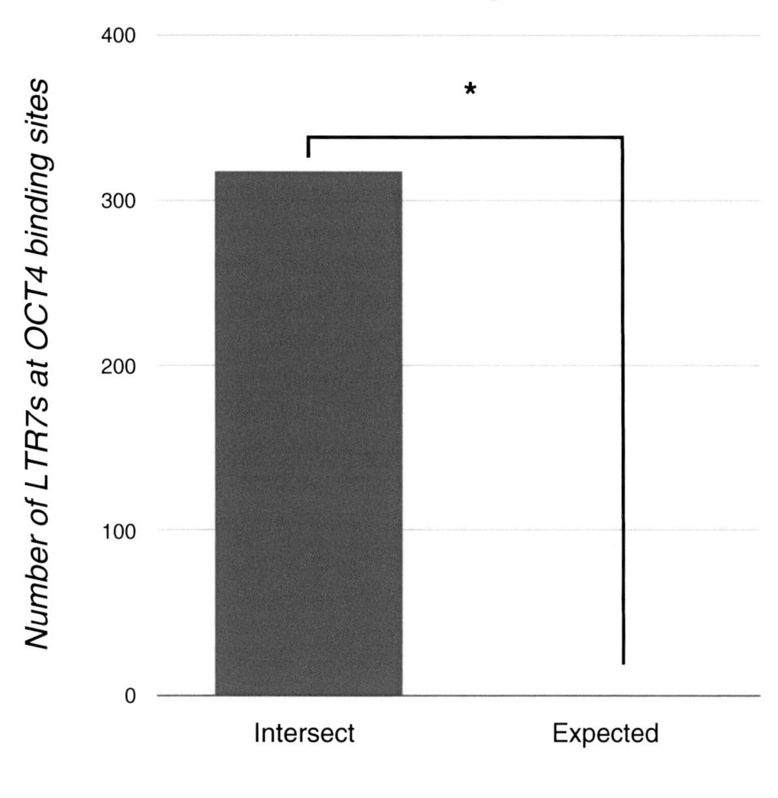

Fig. 1 The number of LTR7 that intersect with OCT4 binding regions and the expected number of intersections

2. Convert UCSC hg19 genome 2bit file to fasta format.
   (a) `./twoBitToFa hg19.2bit hg19.fa`
3. Gunzip all fastqfiles.
   (a) `gunzip` *`filename`*`.fastq.gz`
4. Here files names are shortened for readability.
   (a) `mv wgEncodeCshlLongRnaSeqH1hescCellLongnonpolyaFastqRd1Rep1.fastq CshlH1hescRd1Rep1.fastq`
   (b) `mv wgEncodeCshlLongRnaSeqH1hescCellLongnonpolyaFastqRd2Rep1.fastq CshlH1hescRd2Rep1.fastq`
   (c) `mv wgEncodeCshlLongRnaSeqGm12878CellLongnonpolyaFastqRd1Rep1.fastq CshlGm12878Rd1Rep1.fastq`
   (d) `mv wgEncodeCshlLongRnaSeqGm12878CellLongnonpolyaFastqRd2Rep1.fastq CshlGm12878Rd2Rep1.fastq`

5. Build STAR genome assembly library using the genome assembly fasta file.
   (a) `mkdir hg19_star_index`
   (b) `/pathToStarDir/STAR --runMode genome Generate --genomeDir hg19_star_index --genomeFastaFiles hg19.fa --runThreadN 12`
   (c) See other available options in the STAR manual https://code.google.com/p/rna-star/downloads/detail?name=STARmanual_2.3.0.1.pdf
   (d) This process is quite time and memory intensive. Allocate around 30 GB of RAM for this command.
6. Run STAR mapping job.
   (a) `mkdir star_out`
   (b) `STAR --genomeDir hg19_star_index --readFilesIn CshlH1hescRd1Rep1.fastq CshlH1hescRd2Rep1.fastq --runThreadN 15 --outFileNamePrefix star_out/CshlH1hescRep1`
   (c) `STAR --genomeDir hg19_star_index --readFilesIn CshlGm12878Rd1Rep1.fastq CshlGm12878Rd2Rep1.fastq --runThreadN 15 --outFileNamePrefix star_out/CshlGm12878Rep1`
   (d) See other available options in the STAR manual https://code.google.com/p/rna-star/downloads/detail?name=STARmanual_2.3.0.1.pdf
   (e) Check the `Log.final.out` file for the percent of uniquely mapped reads.
7. Convert the aligned SAM file output from each STAR run to BAM format and sort using SAMtools.
   (a) `cd star_out/`
   (b) `samtools view -bS` *`filename`*`Aligned.out.sam -o` *`filename`*`Rep1Aligned.out.bam`
   (c) `samtools sort` *`filename`*`Aligned.out.bam` *`filename`*`Rep1Aligned.out.sorted`
8. For this analysis we will look at the repeat HERVH-int and ERVL-int (*see* **Note 5**). Here, we diverge from the standard RNA-seq analysis by using repeats to evaluate expression instead of genes.
   (a) `awk '{if($5=="HERVH-int" || $5=="ERVL-int") print}' ../repeats.bed > testrepeats.bed`

   In subsequent commands italicized words *celltype* and *repeat* mean that the command should be run on each cell type and/or repeat being analyzed.

9. Count the number of reads that overlap each repeat using `coverageBed` from the BEDTools suite. Preform the following command on each of the files created in **step 7**.

   (a) `coverageBed -split -abam filenameAligned.out.sorted.bam -b testrepeats.bed > coverage_celltype.txt`

   (b) For each coverage file normalize read counts based on repeat length and library size; reads per kilobase per million mapped reads (RPKM) (*see* **Note 3**).

```
> libcount_celltype=$(samtools view -c filename-Aligned.out.sorted.bam)
```

```
> awk -v t=$libcount_celltype '{x=(($6*10^9)/(t*($3-$2))); print $0"\t"x}' coverage_celltype.txt > coverage_celltype_rpkm.txt
```

10. Filter for expressed HERV-H instances. Here, we define expressed as greater than or equal to 1 RPKM. Do this for each coverage_rpkm file.

    (a) `awk '{if ($5=="HERVH-int") print}' coverage_celltype_rpkm.txt > coverage_celltype_rpkm_hervh.txt`

    (b) `awk '{if ($5=="ERVL-int") print}' coverage_celltype_rpkm.txt > coverage_celltype_rpkm_ervl.txt`

    (c) `awk '{if($10>=1) print}' coverage_celltype_rpkm_repeat.txt | wc -l`

    (d) Results of running this step are shown in Table 2 and Fig. 2.

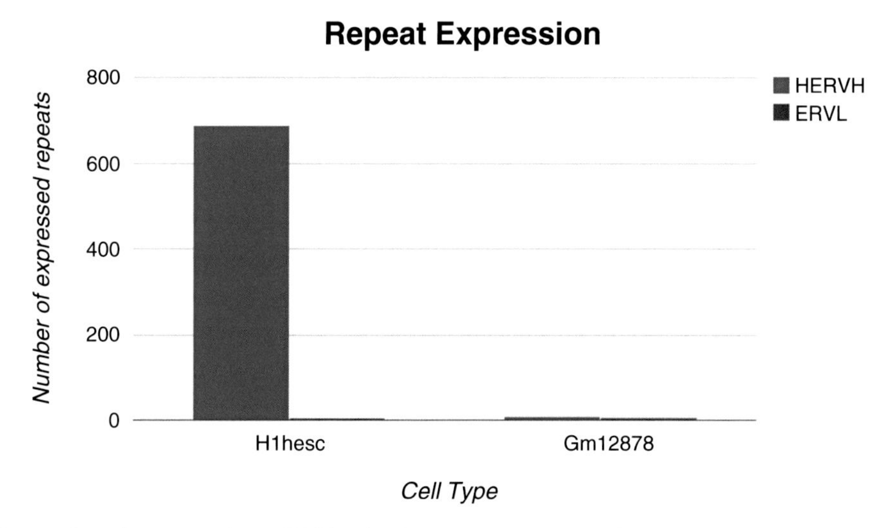

Fig. 2 The number of repeats whose RPKM value is greater than or equal to 1

Table 2
The number of repeats whose RPKM value is greater than or equal to 1 in each cell type

	HERVH in H1hesc	ERVL in H1hesc	HERVH in Gm12878	ERVL in Gm12878	Total # HERVH	Total # ERVL
Number expressed	688	5	9	6	1271	680

Table 3
The number of repeats whose RPKM value is more than two times in higher in H1hesc compared to Gm12878

	≥2× RPKM in H1hesc
HERVH	513
ERVL	18

11. Finally, calculate the ratios of the RPKM values for each repeat to compare the cell types. To do this first merge the coverage_rpkm_repeat files.
    (a) `awk '{print $1":"$2"-"$3"\t"$1"\t"$2"\t"$3"\t"$4"\t"$5"\t"$10}' coverage_celltype_rpkm_repeat.txt | sort > coverage_celltype_repeat.sorted`
    (b) `join -t$'\t' coverage_H1hesc_repeat.sorted coverage_Gm12878_repeat.sorted > coverage_joined_repeat.txt`
    (c) `awk '{if($13==0) ratio=$7; else rastio=$7/$13; print $2"\t"$3"\t"$4"\t"$5"\t"$6"\t"ratio}' coverage_joined_repeat.txt > coverage_ratio_repeat.txt`
12. Extract instances where H1hesc has at least two times higher RPKM than Gm12878.
    (a) `awk '{if($6>=2) print}' coverage_ratio_repeat.txt | wc -l`
    (b) Results of this step are shown in Table 3.

4 Notes

1. Shuffling the repeat intervals is used to determine how many repeats are expect to overlap OCT4 if they are randomly distributed through the genome. This value can be used later to determine whether the number of overlaps we see are more or less than expected by chance.

Usually one would want to shuffle the binding regions while keeping the distribution relative to genes [7, 8]. This makes for a more realistic, conservative shuffle since transcription factor binding sites are usually near genes whose expression they control.

Also, here only one shuffled set is used, but usually this analysis should be performed with many shuffled datasets. This will increase the confidence of your statistical values.

2. For this example analysis we examine LTR7 specifically because it is known OCT4 binding site. This analysis gives an example of how this could have been discovered.
3. Equation:

   N = number of reads mapped to each feature
   L = length of the feature
   T = total number of reads in the library

$$\text{RPKM} = \left(N \times 10^9\right) / \left(L \times \text{T}\right)$$

   This analysis was performed using only one replicate, but if more than one is used average the RPKM values: sort RPKM files, number lines, join files, and calculate average of each files RPKM value.
4. For the second analysis we use the nestedRepeats track from UCSC. This is because we are investigating the endogenous retrovirus HERVH. This repeat is frequently fragmented because of the age of the insertion. The nestedRepeats track joins repeats based on their RepeatMasker ID. This is to increase the accuracy of the expression normalization.
5. In this analysis we filter for the repeat HERVH because it is a known lncRNA expressed specifically in human embryonic stem cells. This analysis gives of a sense of how this lncRNA could have been discovered.
6. For this analysis we only used two cell types and one replicate to simplify the concept. Normally for this kind of analysis one should use as many replicates and cell types as possible. One tool that is useful for comparing multiple samples in this manner is Cuffdiff http://cole-trapnell-lab.github.io/cufflinks/cuffdiff/.

Acknowledgements

We would like to thank Patricia Goerner-Potvin for help reviewing the manuscript. This work was supported by a grant from the Canadian Institute of Health Research (CIHR MOP-115090).

References

1. Lander ES, Linton LM, Birren B, Nusbaum C, Zody MC, Baldwin J, Devon K, Dewar K, Doyle M, FitzHugh W, Funke R, Gage D, Harris K, Heaford A, Howland J, Kann L, Lehoczky J, LeVine R, McEwan P, McKernan K, Meldrim J, Mesirov JP, Miranda C, Morris W, Naylor J, Raymond C, Rosetti M, Santos R, Sheridan A, Sougnez C, Stange-Thomann N, Stojanovic N, Subramanian A, Wyman D, Rogers J, Sulston J, Ainscough R, Beck S, Bentley D, Burton J, Clee C, Carter N, Coulson A, Deadman R, Deloukas P, Dunham A, Dunham I, Durbin R, French L, Grafham D, Gregory S, Hubbard T, Humphray S, Hunt A, Jones M, Lloyd C, McMurray A, Matthews L, Mercer S, Milne S, Mullikin JC, Mungall A, Plumb R, Ross M, Shownkeen R, Sims S, Waterston RH, Wilson RK, Hillier LW, McPherson JD, Marra MA, Mardis ER, Fulton LA, Chinwalla AT, Pepin KH, Gish WR, Chissoe SL, Wendl MC, Delehaunty KD, Miner TL, Delehaunty A, Kramer JB, Cook LL, Fulton RS, Johnson DL, Minx PJ, Clifton SW, Hawkins T, Branscomb E, Predki P, Richardson P, Wenning S, Slezak T, Doggett N, Cheng JF, Olsen A, Lucas S, Elkin C, Uberbacher E, Frazier M, Gibbs RA, Muzny DM, Scherer SE, Bouck JB, Sodergren EJ, Worley KC, Rives CM, Gorrell JH, Metzker ML, Naylor SL, Kucherlapati RS, Nelson DL, Weinstock GM, Sakaki Y, Fujiyama A, Hattori M, Yada T, Toyoda A, Itoh T, Kawagoe C, Watanabe H, Totoki Y, Taylor T, Weissenbach J, Heilig R, Saurin W, Artiguenave F, Brottier P, Bruls T, Pelletier E, Robert C, Wincker P, Smith DR, Doucette-Stamm L, Rubenfield M, Weinstock K, Lee HM, Dubois J, Rosenthal A, Platzer M, Nyakatura G, Taudien S, Rump A, Yang H, Yu J, Wang J, Huang G, Gu J, Hood L, Rowen L, Madan A, Qin S, Davis RW, Federspiel NA, Abola AP, Proctor MJ, Myers RM, Schmutz J, Dickson M, Grimwood J, Cox DR, Olson MV, Kaul R, Raymond C, Shimizu N, Kawasaki K, Minoshima S, Evans GA, Athanasiou M, Schultz R, Roe BA, Chen F, Pan H, Ramser J, Lehrach H, Reinhardt R, McCombie WR, de la Bastide M, Dedhia N, Blocker H, Hornischer K, Nordsiek G, Agarwala R, Aravind L, Bailey JA, Bateman A, Batzoglou S, Birney E, Bork P, Brown DG, Burge CB, Cerutti L, Chen HC, Church D, Clamp M, Copley RR, Doerks T, Eddy SR, Eichler EE, Furey TS, Galagan J, Gilbert JG, Harmon C, Hayashizaki Y, Haussler D, Hermjakob H, Hokamp K, Jang W, Johnson LS, Jones TA, Kasif S, Kaspryzk A, Kennedy S, Kent WJ, Kitts P, Koonin EV, Korf I, Kulp D, Lancet D, Lowe TM, McLysaght A, Mikkelsen T, Moran JV, Mulder N, Pollara VJ, Ponting CP, Schuler G, Schultz J, Slater G, Smit AF, Stupka E, Szustakowski J, Thierry-Mieg D, Thierry-Mieg J, Wagner L, Wallis J, Wheeler R, Williams A, Wolf YI, Wolfe KH, Yang SP, Yeh RF, Collins F, Guyer MS, Peterson J, Felsenfeld A, Wetterstrand KA, Patrinos A, Morgan MJ, de Jong P, Catanese JJ, Osoegawa K, Shizuya H, Choi S, Chen YJ (2001) Initial sequencing and analysis of the human genome. Nature 409(6822):860–921. doi:10.1038/35057062
2. Doolittle WF, Sapienza C (1980) Selfish genes, the phenotype paradigm and genome evolution. Nature 284(5757):601–603
3. Orgel LE, Crick FH (1980) Selfish DNA: the ultimate parasite. Nature 284(5757):604–607
4. Gould SJ, Vrba ES (1982) Exaptation – a missing term in the science of form. Paleobiology 8(1):4–15. doi:10.2307/2400563
5. Cohen CJ, Lock WM, Mager DL (2009) Endogenous retroviral LTRs as promoters for human genes: a critical assessment. Gene 448(2):105–114. doi:10.1016/j.gene.2009.06.020
6. Wang T, Zeng J, Lowe CB, Sellers RG, Salama SR, Yang M, Burgess SM, Brachmann RK, Haussler D (2007) Species-specific endogenous retroviruses shape the transcriptional network of the human tumor suppressor protein p53. Proc Natl Acad Sci U S A 104(47):18613–18618. doi:10.1073/pnas.0703637104
7. Kunarso G, Chia NY, Jeyakani J, Hwang C, Lu X, Chan YS, Ng HH, Bourque G (2010) Transposable elements have rewired the core regulatory network of human embryonic stem cells. Nat Genet 42(7):631–634. doi:10.1038/ng.600
8. Jacques PE, Jeyakani J, Bourque G (2013) The majority of primate-specific regulatory sequences are derived from transposable elements. PLoS Genet 9(5):e1003504. doi:10.1371/journal.pgen.1003504
9. Kelley D, Hendrickson D, Tenen D, Rinn J (2014) Transposable elements modulate human RNA abundance and splicing via specific RNA-protein interactions. Genome Biol 15(12):537
10. Dinger ME, Amaral PP, Mercer TR, Mattick JS (2009) Pervasive transcription of the eukaryotic genome: functional indices and conceptual implications. Brief Funct Genomic Proteomic 8(6):407–423. doi:10.1093/bfgp/elp038
11. Morgan HD, Sutherland HG, Martin DI, Whitelaw E (1999) Epigenetic inheritance at the agouti locus in the mouse. Nat Genet 23(3):314–318. doi:10.1038/15490

12. Kelley D, Rinn J (2012) Transposable elements reveal a stem cell-specific class of long noncoding RNAs. Genome Biol 13(11):R107. doi:10.1186/gb-2012-13-11-r107
13. Kapusta A, Kronenberg Z, Lynch VJ, Zhuo X, Ramsay L, Bourque G, Yandell M, Feschotte C (2013) Transposable elements are major contributors to the origin, diversification, and regulation of vertebrate long noncoding RNAs. PLoS Genet 9(4):e1003470. doi:10.1371/journal.pgen.1003470
14. Loewer S, Cabili MN, Guttman M, Loh YH, Thomas K, Park IH, Garber M, Curran M, Onder T, Agarwal S, Manos PD, Datta S, Lander ES, Schlaeger TM, Daley GQ, Rinn JL (2010) Large intergenic non-coding RNA-RoR modulates reprogramming of human induced pluripotent stem cells. Nat Genet 42(12):1113–1117. doi:10.1038/ng.710
15. Lu X, Sachs F, Ramsay L, Jacques PE, Goke J, Bourque G, Ng HH (2014) The retrovirus HERVH is a long noncoding RNA required for human embryonic stem cell identity. Nat Struct Mol Biol 21(4):423–425. doi:10.1038/nsmb.2799
16. Rosenbloom KR, Sloan CA, Malladi VS, Dreszer TR, Learned K, Kirkup VM, Wong MC, Maddren M, Fang R, Heitner SG, Lee BT, Barber GP, Harte RA, Diekhans M, Long JC, Wilder SP, Zweig AS, Karolchik D, Kuhn RM, Haussler D, Kent WJ (2013) ENCODE data in the UCSC genome browser: year 5 update. Nucleic Acids Res 41(Database issue):D56–D63
17. Quinlan AR, Hall IM (2010) BEDTools: a flexible suite of utilities for comparing genomic features. Bioinformatics (Oxford) 26(6):841–842. doi:10.1093/bioinformatics/btq033
18. Li H, Handsaker B, Wysoker A, Fennell T, Ruan J, Homer N, Marth G, Abecasis G, Durbin R (2009) The sequence alignment/map format and SAMtools. Bioinformatics (Oxford) 25(16):2078–2079. doi:10.1093/bioinformatics/btp352
19. Dobin A, Davis CA, Schlesinger F, Drenkow J, Zaleski C, Jha S, Batut P, Chaisson M, Gingeras TR (2013) STAR: ultrafast universal RNA-seq aligner. Bioinformatics (Oxford) 29(1):15–21. doi:10.1093/bioinformatics/bts635

Chapter 4

Retrotransposon Capture Sequencing (RC-Seq): A Targeted, High-Throughput Approach to Resolve Somatic L1 Retrotransposition in Humans

Francisco J. Sanchez-Luque, Sandra R. Richardson, and Geoffrey J. Faulkner

Abstract

Mobile genetic elements (MGEs) are of critical importance in genomics and developmental biology. Polymorphic and somatic MGE insertions have the potential to impact the phenotype of an individual, depending on their genomic locations and functional consequences. However, the identification of polymorphic and somatic insertions among the plethora of copies residing in the genome presents a formidable technical challenge. Whole genome sequencing has the potential to address this problem; however, its efficacy depends on the abundance of cells carrying the new insertion. Robust detection of somatic insertions present in only a subset of cells within a given sample can also be prohibitively expensive due to a requirement for high sequencing depth. Here, we describe retrotransposon capture sequencing (RC-seq), a sequence capture approach in which Illumina libraries are enriched for fragments containing the 5′ and 3′ termini of specific MGEs. RC-seq allows the detection of known polymorphic insertions present in an individual, as well as the identification of rare or private germline insertions not previously described. Furthermore, RC-seq can be used to detect and characterize somatic insertions, providing a valuable tool to elucidate the extent and characteristics of MGE activity in healthy tissues and in various disease states.

Key words Retrotransposition, Somatic mosaicism, Whole genome sequencing (WGS), LINE-1, *Alu*, Mobile genetic element, Neurogenesis, Oncogenesis

1 Introduction

The degree to which somatic cell genomes within a single individual differ from one another is an active area of research. Somatic genome alterations can be generated by aberrant mutational processes, but in some instances are the products of precise, programmed genomic rearrangements. Somatic mutations vary in their physiological impact, depending on the nature of the alteration, as well as the type and number of cells affected. Abnormalities arising during cell division and DNA replication [1, 2] range from point mutations to large genomic alterations that affect chromosome content and structure.

Jose L. Garcia-Pérez (ed.), *Transposons and Retrotransposons: Methods and Protocols*, Methods in Molecular Biology, vol. 1400, DOI 10.1007/978-1-4939-3372-3_4, © Springer Science+Business Media New York 2016

By contrast, as an example of controlled genomic mosaicism, V(D)J recombination and somatic hypermutation are confined to the lymphocyte antigen receptor *loci* and generate variability necessary for adaptive immunity [3, 4].

Recent work has led to an increased appreciation of insertional mutagenesis generated by retrotransposons, a class of replicative MGE, as an additional source of somatic genome mosaicism in mammals. Their activity is frequently associated with deletion or duplication of genomic DNA, but can also facilitate mobilization of other cellular transcripts [5, 6]. Studies involving engineered retrotransposon reporter constructs and copy number variation (CNV) assays have demonstrated that retrotransposons are active in embryonic stem cells and also in adult neuronal progenitor cells [7–9]. Retrotransposon activity in embryonic stem cells can be attributed to the impetus for a "selfish" element to reach the germline and generate a heritable insertion. In contrast, retrotransposition in adult neural progenitor cells has no obvious evolutionary benefit for the retrotransposon. Such activity may therefore represent the "domestication" of retrotransposons into a functional role in the brain, although the physiological impact, if any, of neuronal retrotransposition remains to be elucidated. Well-characterized examples of transposon domestication include the Transib transposon-derived V(D)J recombinase mentioned above [10], as well as the *Drosophila* retroelements HeT-A, TART and Tahre which function to support telomere stability [11].

RC-seq was first published in 2011 [12]. At this time, L1 had been established as the only active autonomous retrotransposon in humans, and was also known to be responsible for the mobilization of other non-autonomous elements including *Alu* and SVA [13–15]. The canonical mechanism of L1 mobilization is Target-Primed Reverse Transcription (TPRT) [16, 17], which occurs at an L1-encoded endonuclease cleavage motif (5′-TT/AAAA-3′), and is characterized by the generation of short direct target site duplications (TSDs) flanking the new insertion (Fig. 1). There is also evidence that L1 can integrate at sites of existing DNA damage, which generally results in insertions lacking TSDs [18, 19].

Nearly one-third of the human genome is comprised of L1 or L1-dependent MGEs [20]. In 1988, a de novo L1 insertion into the X-linked Factor VIII gene was isolated as the cause of a case of hemophilia A in a male patient. This discovery constituted unequivocal evidence that L1s were actively retrotransposing in the human population [21]. Indeed, ~100 cases of human disease have been demonstrated to arise from L1-mediated mutagenesis, and current estimates suggest that there are hundreds of polymorphic L1 insertions segregating in the human population [22–26]. However, the discovery of L1 promoter activity during neurogenesis [8, 27], as well as age- and region-dependent L1 transcription and CNV in

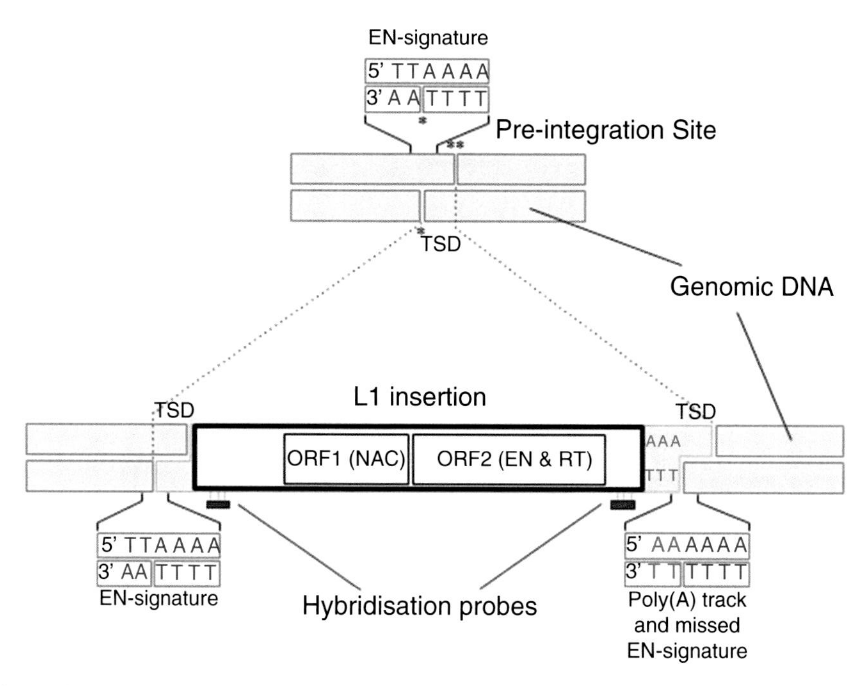

Fig. 1 Schematic representation of a pre-integration site and an inserted new L1 copy with TPRT signatures. The TPRT insertion depends on two single-strand cleavages of the target DNA. The *single asterisk* indicates the first cleavage that occurs in the EN consensus site, while the *double asterisk* indicates the second strand cleavage. The regions between the two cleavages becomes a target site duplication (TSD) flanking the new insertion. The 5′ TSD contains the EN consensus site in the 5′ end (*blue*), while the 3′ TSD starts just after the 3′ poly(A) tract at the end of the inserted copy (*green*). The L1 copy is represented by *thick-outlined white box* and the two open reading frames (ORF) are depicted by two *inside boxes* encoding nucleic acid chaperone (NAC), endonuclease (EN), and reverse transcriptase (RT) activities

the healthy human brain [27–30], and the mobilization of engineered L1s in vitro and in transgenic animals [8, 29] supported the existence of L1-driven somatic mosaicism in the human brain. Nonetheless, prior to the development of RC-seq, evidence for endogenous L1 retrotransposition in the brain lacked critical proof, namely genomic mapping and sequence characterization of the resulting L1 insertions. Indeed, such evidence is essential to elucidate a putative physiological impact of L1 activity during neurogenesis through some particular insertional pattern, perhaps mediated by differential euchromatinization [31]. RC-seq allowed the large-scale identification and genomic localization of somatic MGE insertions present in one or a few cells within a tissue, revealing the surprising extent of somatic mosaicism induced by L1 in the brain [12, 32].

RC-seq is typically performed on genomic DNA extracted from a tissue of interest and a matched control tissue from the same individual (i.e., brain and liver, Fig. 2a), in order to distinguish somatic L1 insertions from unannotated polymorphic

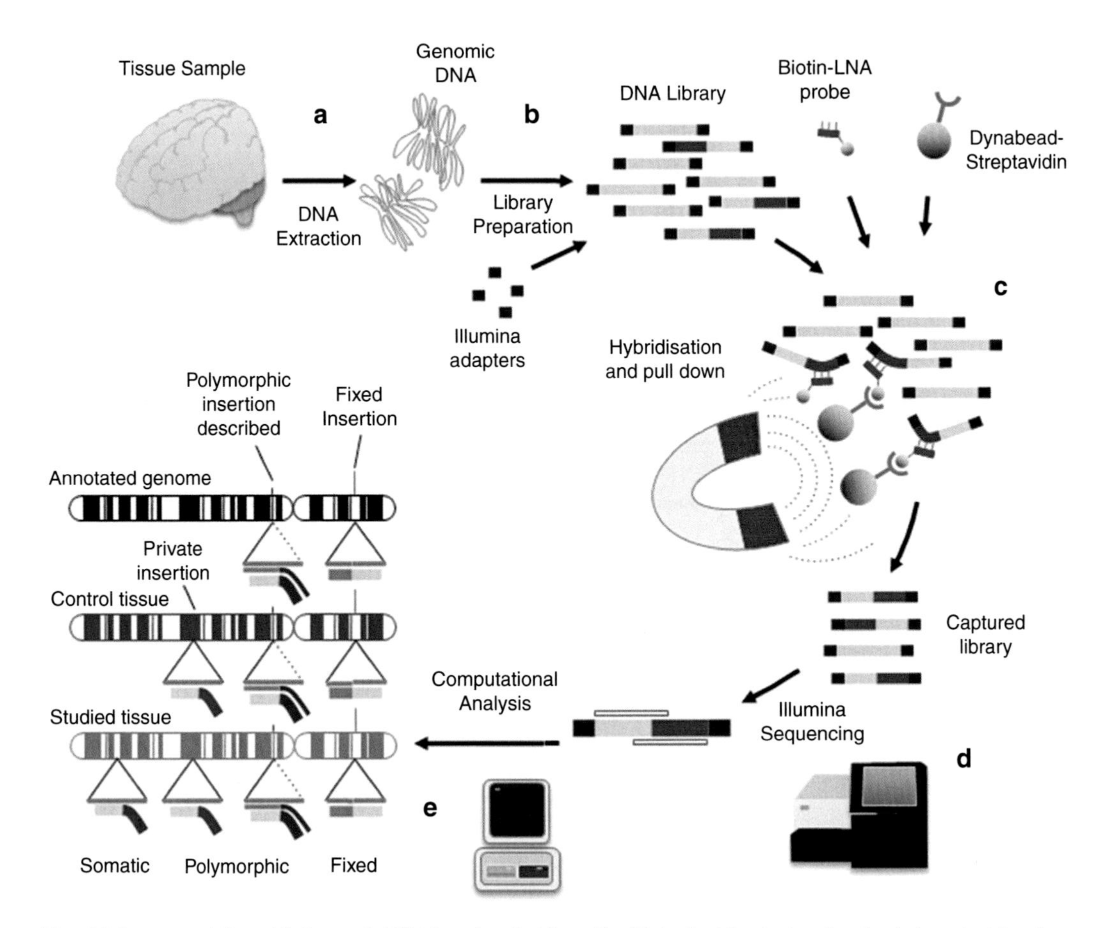

Fig. 2 RC-seq workflow. (*A*) Genomic DNA is extracted from the biological tissue by standard phenol-chloroform extraction. (*B*) Library preparation is performed using Illumina technology. (*C*) Hybridization and pull down of the subset of fragments containing the boundaries of L1 element using a L1-specific biotin-LNA probe and Dynabeads M-270 Streptavidin. (*D*) The captured library is sequenced using an Illumina platform. (*E*) Computational analysis maps new somatic insertions in the annotated genome, and also can provide information about the particular genotype concerning to the polymorphic insertions described in the population and even detect new unknown ones. *Note*: the *red sections* within the library fragments represent L1 sequences; the *grey sections* represent other genomic regions

insertions. Genomic DNA is sheared by sonication, and the sheared DNA serves as the substrate for Illumina sequencing library preparation (Fig. 2b). The library is then hybridized to two biotin-labeled locked nucleic acid (LNA) probes targeting the 5′ and 3′ termini of the L1-Ta consensus (Fig. 2c). Then, streptavidin-based pull down enriches the library for fragments containing the junctions between L1 sequence and flanking genomic DNA. The subsequent amplification and sequencing of the post-hybridization library by a high-throughput platform produces a collection of sequencing reads enriched for L1-genome junctions (Fig. 2d). The data are then analyzed in silico to identify previously annotated fixed or polymorphic insertions, as well as find any previously unknown polymorphic insertions present in the individual (Fig. 2e). The remaining L1-genome junctions represent putative somatic insertions unique to the tissue of interest.

The location of the L1-specific RC-seq probes at the extreme 5′ and 3′ termini of the L1-Ta consensus sequence lead to RC-seq generally only capturing the 5′ junctions of full-length or extremely 5′ truncated L1 insertions. Indeed, most insertions are only detected at their 3′ L1-genome junction, as the vast majority of L1 insertions are variably 5′ truncated but retain their 3′ end. For those insertions detected at only a 3′ L1-genome junction, PCR amplification and capillary sequencing of the matched 5′ L1-genome junction is the gold standard validation to confirm bona fide L1 insertions. In this scenario, a collection of primers tiled along, and oriented antisense to, the L1-Ta consensus are combined with a primer in the presumed 5′ genomic flank and sense oriented relative to the L1. Capillary sequencing the resulting amplicon(s) allows characterization of the 5′ L1-genome junction, typically revealing TPRT hallmark TSDs and an L1 EN motif and thereby distinguishing true insertions from putative artifacts.

RC-seq was first applied to elucidate L1-driven somatic mosaicism in the brain, but it has also been successfully applied to tumor samples, identifying insertional mutations that trigger oncogenic pathways [33]. In addition, RC-seq has the potential to profile the unique subset of annotated polymorphic insertions in a single individual. Thus it can be used as a genotyping tool to match different tissues or samples from the same donor, and also constitutes a source of potential traceable genetic markers applicable, for example, in genome wide association studies.

The most significant advantage of the sequence capture approach used in RC-seq is that the number of PCR cycles can be kept to a minimum. In contrast to other methods for identifying endogenous L1 insertions [23], minimization of PCR cycles preserves the integrity of library content by reducing PCR artifact generation and amplification, giving greater resolution in the detection of rare L1 variants in tissue samples. Note that RC-Seq was originally performed with custom sequence capture arrays consisting of a pool of DNA probes covering 1 and 0.2 Kb at either end of the L1-Ta consensus sequence in the first (V1) [12] and second (V2) [33] RC-seq designs. However, the technique described here, the third (V3) RC-seq design, utilizes two locked nucleic acid (LNA) probes targeting suitable regions at the 5′ and 3′ termini of the L1-Ta consensus sequence, and achieves more than 15-fold improvement in enrichment over previous iterations of RC-seq [34].

2 Materials

Solutions should be prepared with molecular grade water. Examples are water purified by filtration and deionization to achieve a resistivity of 18.2 MΩ cm at 25 °C (such as Mili-Q water produced by Millipore Corporation water filtration stations) or water distilled and filtered by 0.1 μm membrane filters (such as Ultrapure Water provided by Invitrogen—Life Technologies).

2.1 DNA Extraction

1. Benchtop centrifuge for 1.5 ml tubes.
2. Two thermoblocks.
3. Nanodrop spectrophotometer.
4. Scalpel blades, blade holder, forceps and spatula.
5. Disposable plastic Petri dishes.
6. 1.5 ml tubes.
7. Phenol (equilibrated with 10 mM TRIS pH 8).
8. Phenol:Chloroform:Isoamyl alcohol (12:12:1, TRIS saturated).
9. Chloroform:Isoamyl alcohol (24:1, TRIS saturated).
10. Isopropanol (molecular grade).
11. Absolute Ethanol (molecular grade).
12. TE buffer: 10 mM Tris–HCl, 1 mM EDTA, pH 8.
13. Lysis Buffer: TE, 2 % SDS, 100 μg/ml Proteinase K, *see* **Note 1**).
14. 3 M Sodium Acetate.
15. 10 mg/ml RNase A.
16. Dry ice (if the tissue sample is frozen).

2.2 DNA Shearing

1. Covaris M220 Focused-ultrasonicator electronically controlled by Sonolab 7 software.
2. Covaris MicroTube AFA Snap-Cap, 130 μl sample.
3. Buffer TE: 10 mM Tris–HCl, 1 mM EDTA, pH 8.
4. Molecular grade water.
5. Low lint paper wipes (such as Kimtech Science Kimwipes).

2.3 Library Preparation

1. Two thermocyclers.
2. Qubit® Fluorometric Quantification technology (Life Technologies) or similar. This includes Qubit® dsDNA HS Assay kit, Qubit® Assay Tubes and Qubit® Fluorometer.
3. Illumina® TruSeq® Nano DNA Sample Prep Kit. This kit provides End Repair Mix, A-Tailing Mix, Ligation Mix, Stop Ligation Mix, Resuspension Buffer and a set of Illumina

Barcoded Library Adapters that will be used in the following protocol.

4. DynaMag™-2 Side (Life Technologies) magnetic rack (*see* **Note 2**).
5. 0.2 and 1.5 ml tubes. 0.2 ml tubes must be PCR-grade.
6. Agentcourt® AMPure® XP beads (Beckman Coulter).
7. Absolute Ethanol (molecular grade).

2.4 Agarose Gel-Size Selection and PCR Enrichment of the DNA Library

1. Thermocycler.
2. Safe Blue Light Imager or UV trans-illuminator (Safe Blue Light source is preferable).
3. Gel tray, gel combs, electrophoresis tank and power pad.
4. Agilent Bioanalyzer technology or similar. This includes Agilent 2100 Bioanalyzer instrument, Agilent DNA 1000 Reagents and DNA Chips (Agilent Technologies).
5. UV source.
6. MiniElute® Gel Extraction Kit (Qiagen). This includes Buffer QG, Buffer PE, Buffer EB and MiniElute® Spin Columns.
7. Scalpel blades.
8. 0.2 and 1.5 ml tubes. 0.2 ml tubes must be PCR-grade.
9. UV or Safe Blue Light-transparent cling film.
10. High Resolution Agarose (Sigma-Aldrich).
11. Molecular grade water.
12. SYBR® Gold Nucleic Acid Gel Stain (Life Technologies).
13. TAE Buffer: 40 mM Tris–HCl, 20 mM Acetic acid, 1 mM EDTA.
14. Gel Loading Buffer: 20 % glycerol, 0.04 % Orange G stain.
15. Phusion® High-Fidelity PCR Master Mix 2× (New England Biolabs).
16. DNA ladder. Recommended: a ladder with several bands on the 200–400 bp rage, like GenRuler 1 Kb Plus DNA ladder 0.5 μg/μl (Thermo Scientific).
17. Isopropanol (molecular grade).
18. 100 μM each LM-PCR primers: TS-F Primer (5′AATGATACGGCGACCACCGAGA3′) and TS-F Primer (5′CAAGCAGAAGACGGCATACGAG3′).

2.5 Hybridization

1. Thermoblock with two tube removable holders, one of them able to host 1.5 ml tubes.
2. Two thermocyclers. One of them must be connected to a power source capable of holding uninterrupted supply for 3 days.

3. Roche NimbleGen Sequence Capture Kit. Components used in the hybridization step are 2× Hybridization Buffer and Hybridization Component A.
4. Roche Diagnostics Sequence Capture Developer Reagent.
5. 0.2 ml and 1.5 tubes. 0.2 ml tubes must be PCR-grade.
6. 100 μM Universal Blocking Oligo. Sequence: 5′AATGATACGGCGACCACCGAGATCTACACTCTTTCCCTACACGACGCTCTTCCGATC*/3ddC/3′.
7. 100 μM Index-specific Blocking Oligos. Sequence: 5′CAAGCAGAAGACGGCATACGAGATN_8GACTGGAGTTCAGACGTGTGCTCTTCCGATCT/3ddC/3′. N_8 fragment is an Illumina index-specific sequence indicated in **Note 3**.
8. 10 μM each Capture probes: LNA-5′ (/5Biosg/CTCCGGT+C+T+ACAGCTC+C+C+AGC) and LNA-3′ (/5Biosg/AG+A+TGAC+A+C+ATTAGTGGGTGC+A+GCG). Note that /5Biosg/ denotes the presence of a biotin moiety in the 5′ end and + denotes the LNA positions within each probe.
9. ToughTag stickers.

2.6 Capture Recovery and Amplification

1. Thermoblock.
2. Thermocycler (the same used for the Subheading 2.5).
3. Agilent Bioanalyzer technology or similar. This includes Agilent 2100 Bioanalyzer instrument, Agilent DNA 1000 Reagents and DNA Chips (Agilent Technologies).
4. Qubit® Fluorometric Quantification technology (Life Technologies) or similar. This includes Qubit® dsDNA HS Assay kit, Qubit® Assay Tubes and Qubit® Fluorometer.
5. Roche NimbleGen Capture Wash Kit. Components used in the capture recovery step are 10× Stringent Wash Buffer, 10× Wash Buffer I, II and III, and 2.5× Bead Wash Buffer.
6. MiniElute® Gel Extraction kit (Qiagen).
7. DynaMag™-2 Magnet rack (Life Technologies, *see* **Note 2**).
8. DynaMag™-96 Side Magnet plate (Life Technologies).
9. Agentcourt® AMPure® XP beads (Beckman Coulter).
10. Dynabeads® M-270 Streptavidin (Life Technologies).
11. Molecular grade water.
12. Ethanol.
13. Phusion® High-Fidelity PCR Master Mix 2× (New England Biolabs).
14. 100 μM each LM-PCR primers: TS-F Primer (5′AATGATACGGCGACCACCGAGA3′) and TS-F Primer (5′CAAGCAGAAGACGGCATACGAG3′).

2.7 Sequencing

RC-seq libraries can, theoretically, be sequenced on any Illumina platform able to sequence WGS libraries. Thus far, RC-seq has been performed on Illumina HiSeq2000, HiSeq2500 and MiSeq platforms. This choice is dependent upon how many libraries are to be sequenced, the desired depth of sequencing, and the maximum insert size that can be spanned by paired-end reads generated on a given platform.

2.8 Bioinformatic Analysis

The computational resources required for RC-seq library analysis scale with project size. A laptop or PC with 4 GB RAM, a Linux operating system, and 100 GB of hard disk space would likely be sufficient to run a single library over the course of 72 h, depending on library size and L1 enrichment. Most projects would involve multiple RC-seq libraries and therefore would require a larger server. For example, the Translational Research Institute server used by the Faulkner laboratory has 2200 CPUs, 8 TB RAM, and 3 PB hard disk space. This enables parallelization and accelerated analysis.

2.9 PCR Validation

The PCR validation is a highly variable procedure that requires researcher expertise in molecular biology, particularly in PCR amplification, cloning, and sequencing. The materials listed here provide a recommended starting point for the validation process. Additionally, more complex approaches might be necessary according to the inherent difficulty of amplification of some insertions.

For primer design:

1. Primer3 software.

For amplification:

2. Roche Expand Long Range dNTPack.
3. Platinum® Taq DNA Polymerase High Fidelity.

For DNA imaging and purification:

4. Agarose electrophoresis material and Safe Blue Light Imager or UV trans-illuminator (Safe Blue Light source is preferable).
5. Material for agarose gel-purification of DNA.

For cloning:

6. Promega pGEM®-T vector system.
7. Life Technologies TOPO® PCR cloning system.
8. Material for molecular cloning, bacteria transformation, culture, and DNA extraction.

3 Methods

As described in the Introduction, to effectively discern between previously unannotated polymorphic insertions and somatic insertions within a tissue of interest of a single donor, it is necessary to analyze a control tissue from the same donor. Those insertions not previously described as fixed or polymorphic insertions, but found in both tissues, very likely represent a germline insertion rather than a somatic one. Insertions present in only one of the tissues are putative somatic insertions. In the example described here, we describe the use of brain and liver tissues, with liver as the control tissue.

Due to limitations in the number of elongation cycles possible during sequencing, it is important to choose an appropriate fragment size for the library. The sequencing reaction described here is designed for a 300-cycle Illumina paired-end sequencing kit (150 cycles from each end). To produce a minimum overlap of the paired-reads initiated at both ends of a single molecule sufficient to reconstitute the fragment sequence, the preferable sequenced fragment size is 250 bp.

Unless otherwise indicated, it is recommended to bring thermocycles to reach the indicated block and lid temperatures before preparing the reaction. To do that, start the run of the thermocycler and pause it when it reaches the first step of the cycling protocol.

3.1 DNA Extraction

3.1.1 Preparation

1. Take an aliquot of TE 2%SDS with the appropriate volume necessary for the extraction (10 μl of Lysis Buffer per 1 mg of tissue) and add the Proteinase K to a final concentration of 100 μg/ml (*see* **Note 1**).
2. Place the forceps, spatula, one scalpel blade, one 1.5 ml tube and one Petri dish per sample on dry ice.
3. Set up one thermoblock at 37 °C and the other at 65 °C.
4. Prepare 500 μl of 70–80 % ethanol per sample with Ultrapure Water.

3.1.2 Procedure

1. Working on a bed of dry ice, place the tissue sample in a pre-cooled Petri dish using the forceps and shave it using a pre-cooled scalpel blade. The dissociation will be quicker if the shaving is thinner. Transfer the shaved sample to a pre-cooled 1.5 ml tube and keep it frozen until all the tissue samples are processed.
2. Add 10 μl of Lysis Buffer per 1 mg tissue to each tube and incubate at 65 °C in thermoblock until the tissue is completely dissolved. Shake the tubes every 10 min (*see* **Note 4**).

3. Once dissociated, allow the sample to cool briefly at room temperature and add RNase A to a final concentration of 20 μg/ml (*see* **Note 5**). Incubate the tubes in the thermoblock at 37 °C for 30 min.
4. Add an equal volume of phenol saturated solution and mix by inversion (not vigorously) until the sample is homogeneous.
5. Centrifuge at maximum speed for 10 min.
6. Transfer the aqueous phase to a fresh tube. Make the pipette tip end wider by cutting it. Repeat **steps 4–6** once if the sample is from a tissue highly rich in organic compounds (for example, liver. *See* **Note 6**).
7. Add an equal volume of Phenol:Chloroform:Isoamyl alcohol (12:12:1) and repeat **steps 5–6**. Repeat **steps** 7 and **5–6** if the sample is from a tissue highly rich in organic compounds (for example, liver).
8. Add an equal volume of Chloroform:Isoamyl alcohol (24:1) and repeat **steps 5–6**.
9. Add 0.1 volume of 3 M Sodium Acetate and 2 volumes of isopropanol and invert the tube several times to precipitate DNA until all DNA is dehydrated (when the viscous transparent goo turns white).
10. Spool DNA with a pipette tip and transfer to a new tube (*see* **Note** 7). Rinse DNA with 500 μl 70–80 % ethanol and briefly air dry. Stop drying samples before they turn clear. Overdrying can result in extremely difficult resuspension of the pellet.
11. Resuspend samples in 100–200 μl TE and incubate samples at 4 °C overnight for complete resuspension (*see* **Note 8**).
12. Quantify DNA concentration by Nanodrop.

SAFETY STOPPING POINT. You can store the samples at 4 °C for short periods, or freeze at –20/–80 °C for longer periods.

3.2 DNA Shearing

3.2.1 Preparation

1. Switch on the Covaris M220 Focused-Ultrasonicator (*see* **Note 9**) and the computer that manages the ultrasonicator. Open SonoLab 7.1 software and prepare the instrument for sonication. Ensure that the *water temperature* and the *water level* in the "Instrument Status" window are correct.
2. Select the appropriate sonication method. For a library with an ideal fragment size of 250 bp, the sonication method is: Peak power 50, Duty factor 20, Cycles per burst 200 and Timer 120 s (*see* **Note 9**).

3.2.2 Procedure

1. Dilute 1–5 μg of genomic DNA from each tissue sample into 130 μl of buffer TE (*see* **Note 10**). Do not use the whole

amount of DNA for the sonication, keep an aliquot and dilute it to 5 ng/μl to use for PCR validation reactions (*see* Subheading 3.9).

2. Transfer each DNA dilution to a MicroTube AFA Snap-Cap and sonicate the sample following manufacturer's instructions.
3. Wipe out any water remaining in the edge of the Cap of the MicroTube AFA Snap-Cap using a low lint paper wipe and transfer the sheared DNA solution to a fresh 1.5 ml tube.
4. Repeat the sonication with each sample. Check the water level every time (*see* **Note 11**).

SAFETY STOPPING POINT. You can store the samples at 4 °C for short periods, or freeze at −20/−80 °C for longer periods.

3.3 Library Preparation

3.3.1 Preparation

1. Take an aliquot of AMPure® XP beads, resuspend well by vortex and allow it to reach room temperature (~30 min).
2. Prepare 2–4 ml of 80 % ethanol per sample (*see* **Note 2**).
3. Set up a thermocycler with a 30 °C block, **40 °C lid** (**absolutely critical**).
4. Set up a thermocycler with A-tailing program: 37 °C for 30 min, 70 °C for 5 min, 4 °C hold; lid at 80 °C.

3.3.2 Procedure

To concentrate DNA.

1. Add 1.1 volumes of Ampure® XP beads to each tube, mix by pipetting ten times and incubate at room temperature for 15 min.
2. Place the tubes on a magnetic rack for 2 min. Aspirate and discard the supernatant.
3. Add 200–400 μl of 80 % ethanol to each tube without disturbing the beads and incubate at room temperature for 30 s (*see* **Note 2**).
4. Remove the ethanol and repeat **step 3**.
5. Remove the ethanol and leave the tubes air dry on the magnetic rack for 15 min.
6. Add 52 μl of resuspension buffer to each tube. Remove the tubes from the magnetic rack and flick it until beads are completely resuspended.
7. Incubate at room temperature for 2 min and place the tubes back in the magnetic rack. Incubate at room temperature for 2 min (the liquid must appear clear).
8. Transfer 50 μl of supernatant to a new fresh tube.
9. Quantify DNA concentration using Qubit® Fluorometric Technology. To check the efficiency of the sonication setting, use Agilent Bioanalyzer technology to check the DNA size distribution (Fig. 3a).

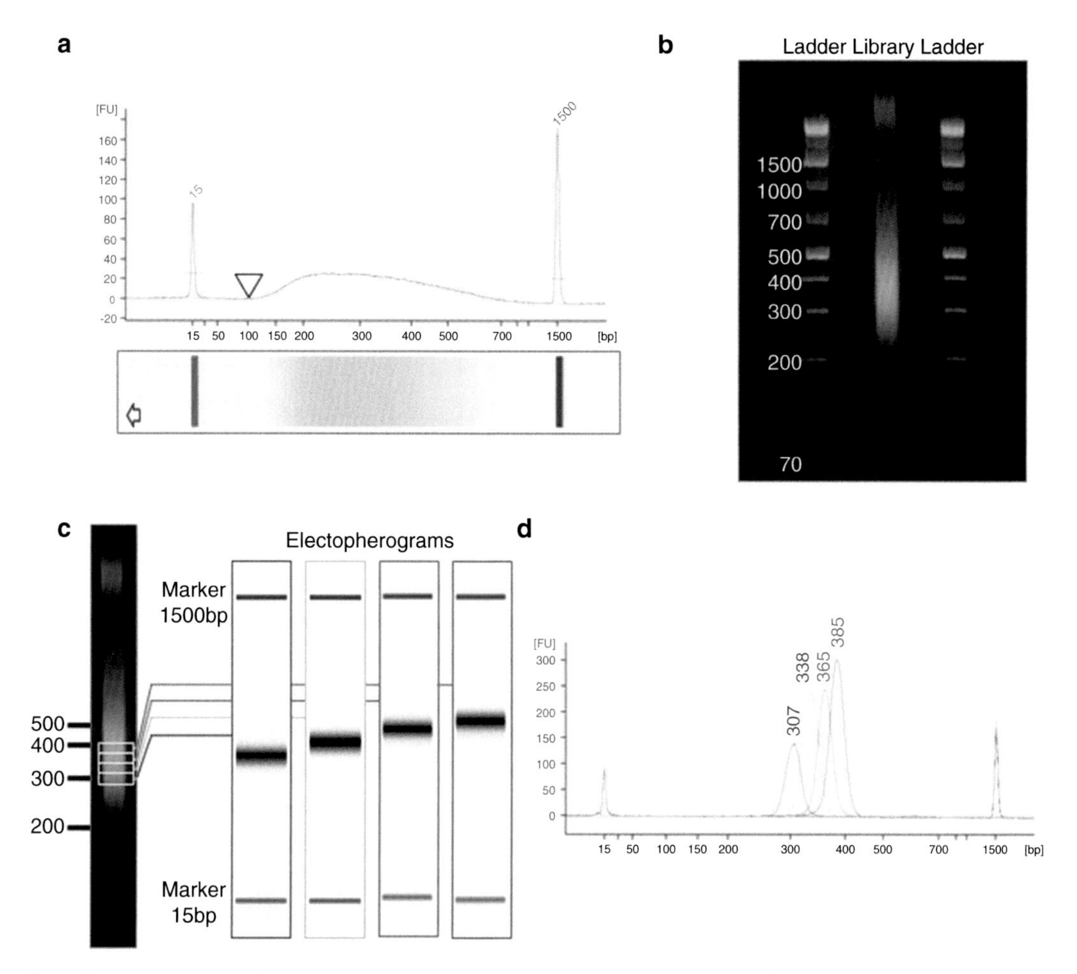

Fig. 3 Gel-size selection for library sequencing. (**a**) The figure shows an Agilent® 2100 Bioanalyzer distribution (graph and electropherogram) of a genomic DNA sample sheared by sonication after **step 9** of the Subheading 3.3.2. The post-shearing Agencourt® AMPure® beads size selection performed to concentrate DNA and remove fragments under 100 bp (**steps 1–8** in same Subheading). The *arrowhead* indicates the 100 bp cut-off in the size distribution. The *arrow* in the electropherogram indicates the electrophoresis direction. (**b**) A typical gel electrophoresis distribution of a pre-hybridization library is shown in the central lane. GenRuler 1 Kb Plus DNA Ladder is used as Marker at both sides. Note that there is an empty well at each side of the library to prevent cross-contamination. *White numbers* indicate the size (bp) of the corresponding fragment of the ladder. (**c**) An example of size selection by gel-cutting. *White boxes* represent a recreation of the gel-cuts performed on the gel shown in **b**, aiming for sizes between 290 and 400 bp. The electropherogram corresponding to four consecutive gel-cuts produced by the Agilent® 2100 Bioanalyzer instrument shows an effective size selection. (**d**) Size distribution corresponding to the electropherograms shown in **c**. The gel-cut shown in *green* has an approximate distribution between 330–400 bp and an average size of 365 bp

SAFETY STOPPING POINT. You can store the samples at 4 °C until the next day.

To prepare TruSeq Libraries:

10. Aliquot 1 μg of sheared DNA per sample into a 0.2 ml tube and bring it to a final volume of 60 μl with resuspension buffer.

Add 40 μl of End Repair Mix, mix by pipetting ten times, and place in the thermocycler at 30 °C for 30 min.

11. Transfer the whole volume of each tube to a 1.5 ml tube and add 100 μl of well resuspended room temperature AMPure® XP beads (ratio 1:1) and proceed as in **steps 1–5**.
12. Resuspend the beads in 19 μl of resuspension buffer by flicking and incubate at room temperature for 2 min.
13. Place the tube back in the magnetic rack and incubate at room temperature for 2 min. Transfer 17.5 μl of supernatant to a new 0.2 ml tube.
14. Add 12.5 μl of A-tailing mix to each sample and mix by pipetting ten times (do not vortex). Place the tubes in the thermocycler with the A-tailing program (37 °C for 30 min, 70 °C for 5 min, 4 °C hold) and resume the run.
15. Take the 0.2 ml tubes with the samples and add 2.5 μl of resuspension buffer, 2.5 μl of Ligase Mix and 2.5 μl of each adapter specific for each sample, and mix by pipetting up and down ten times (do not vortex, *see* **Note 12**).
16. Place the tubes in a thermocycler at 30 °C for 10 min.
17. Add 5 μl of Stop Ligase Mix and mix by pipetting ten times.
18. Transfer the whole volume (42.5 μl) to a 1.5 ml tube and add 42.5 μl of AMPure® XP beads. Mix by pipetting ten times and incubate at room temperature for 15 min.
19. Repeat **steps 2–8** (two ethanol washes and resuspension of the sample in 50 μl). *Pay attention to avoid confusing the elution step with a washing step* (*see* **Note 13**).
20. Add 50 μl of AMPure® XP beads to each tube. Mix by pipetting ten times and incubate at room temperature for 15 min.
21. Repeat **steps 2–5**.
22. Add 21 μl of resuspension buffer to each tube, resuspend by flicking and incubate at room temperature for 2 min.
23. Place the tubes in the magnetic rack and incubate them at room temperature for 2 min (the liquid must appear clear).
24. Transfer 20 μl of supernatant to a new fresh tube (avoid carry-over of beads).

SAFETY STOPPING POINT. You can store the samples at 4 °C until the next day.

3.4 Agarose Gel-Size Selection

3.4.1 Preparation

1. Wash the gel tray and comb with 70 % ethanol and wipe out. UV-irradiate the gel tray and comb to eliminate contaminant DNA.
2. Prepare a 2 % high-resolution agarose gel using the following proportions: 5 g Agarose in 250 ml of final volume 1× TAE

buffer. Partially cool the molten gel in a 65° water bath for ~20 to 30 min. Add 25 μl of SYBR® Gold Stain to the melted agarose solution. Leave the gel sit in a 4 °C room until it solidify (the gel can be prepared the previous day).

3. Set the agarose gel in the gel tank. Use TAE buffer 1× as electrophoresis buffer.
4. Set up a thermocycler with the LM-PCR cycling protocol: 98 °C 45 s; 98 °C 15 s, 60 °C 30 s and 72 °C 30 s for six times (*see* **Note 14**); 72 °C 5 min and 4 °C hold. Do not bring the thermocycler to the first step, wait until the reaction tubes are inside.
5. Wrap the trans-illuminator with cling film to protect the gel sample from cross-contamination during the gel cutting (this can be done during the **step 3** of the Subheading 3.4.2).

3.4.2 Procedure

1. Add 4 μl of Gel Loading Buffer 6× to each tube and load the 24 μl samples in the gel. Leave an empty space between each sample to prevent cross-contamination. Load also 5 μl of DNA ladder every two or three samples and make sure you have a DNA ladder line in both ends of the gel.
2. Run the gel electrophoresis at 120 mA until the separation of the fragmented DNA is enough to comfortably purify bands of Δ30–50 nt size fragments in the frame 250–400 bp (a typical run time for a 15 cm length gel is 3 h).
3. Place the gel at 4 °C for 5–20 min.
4. Perform cut out of bands ranging from 290–310 bp, 310–350 bp, 350–380 bp and 380–410 bp (*see* **Note 15** and Fig. 3b). Place in labeled tubes. Wrap the remaining pieces of gel in cling film and keep them as a backup at 4 °C in case of the need to do new gel cuts.
5. Proceed with at least the fragment of 350–380 bp size to DNA gel-purification using the MiniElute® Gel Extraction Kit by adding 6 μl of Buffer QG per 1 mg of gel band to each sample tube.
6. Dissolve the agarose at room temperature (avoid incubation at 50 °C as indicated in the manufacturer instructions, which can result in a GC-bias in the sequencing data).
7. Add 2 μl of isopropanol per 1 mg of gel band to each sample tube and mix by inverting.
8. Place the MiniElute® Spin Columns in provided 2 ml collection tubes and apply each sample to a single column. Centrifuge at maximum speed for 1 min.
9. Discard the flow-through and add 500 μl of Buffer QG. Centrifuge at maximum speed for 1 min.

10. Discard the flow-through and add 750 μl of Buffer PE. Incubate at room temperature for 2–5 min and centrifuge at maximum speed for 1 min.
11. Discard the flow-through and centrifuge again at maximum speed for 1 min.
12. Transfer the column to a fresh 1.5 ml tube and add 16 μl of Buffer EB pre-heated to 60 °C. Incubate at room temperature for 1 min and centrifuge at maximum speed for 1 min.
13. Repeat **step 12** in the same 1.5 ml tube, ending with slightly >30 μl final volume. Discard the column.
14. Transfer 30 μl of each sample to a 0.2 ml tube and add: 50 μl of Phusion® High-Fidelity PCR MasterMix (2×), 18 μl of Ultrapure Water, 1 μl of TS-F Primer (100 μM) and 1 μl of TS-R Primer (100 μM). Mix my pipetting ten times.
15. Put the tubes in the thermocycler with the LM-PCR program and start the run.
16. Transfer the whole volume (100 μl) to a 1.5 ml tube and add 110 μl of AMPure® XP beads to each tube. Mix by pipetting ten times and incubate at room temperature for 15 min.
17. Place the tubes on a magnetic rack for 2 min. Aspirate and discard the supernatant.
18. Add 200–400 μl of 80 % ethanol to each tube without disturbing the beads and incubate at room temperature for 30 s (*see* **Note 2**).
19. Remove the ethanol and repeat **step 3**.
20. Remove the ethanol and leave the tubes air dry on the magnetic rack for 15 min.
21. Add 32 μl of molecular grade water (do not use Resuspension Buffer, *see* **Note 16**), remove the tubes from the magnetic rack and flick them until completely resuspend the beads.
22. Incubate at room temperature for 2 min and place the tubes back in the magnetic rack. Incubate at room temperature for 2 min (the liquid must appear clear).
23. Transfer 30 μl of supernatant to a new fresh tube.
24. Quantify the concentration and size distribution of the DNA library by analyzing 1 μl of each sample in an Agilent DNA 1000 chip. The preferable fragment size distribution should be between 340 and 410 bp, with a median peak of 370 bp (*see* Fig. 3c, d, and **Note 17**).

SAFETY STOPPING POINT. You can store the samples at 4 °C until the next day.

3.5 Hybridization

3.5.1 Preparation

1. Pre-heat Speed-Vac at 70 °C.
2. Prepare two 0.2 ml tubes with 4.5 μl of each LNA capture probe (10 μM): LNA-5′ and LNA-3′.
3. Set up a thermocycler with 47 °C block, **57 °C lid** (**absolutely critical**). This one will host an incubation for 3 days.
4. Set up a thermocycler with 95 °C block, 105 °C lid.
5. Set up a thermoblock at 95 °C.

3.5.2 Procedure

1. Pool the same amount of brain and liver libraries in the same tube in a ratio 1:1 by molecular mass to reach 1 μg of amplified DNA (*see* **Note 18** about pooling several libraries). This will be used for both 5′ end and 3′ end captures. You can proceed with smaller amount of DNA (up to 100 ng) but keep the ratio 1:1 for the different libraries.
2. Add 10 μl of Sequence Capture Developer Reagent and 10 μl of Universal Blocking Oligo per 1 μg of total DNA. Add also Index-specific Blocker Oligos according to the libraries' indices at the same ratio (10 μl of Blocker Oligo per 1 μg of each specific library within the pool).
3. Mix the sample by pipetting up and down, spin down and take half of the sample to a fresh tube (each tube will be used for capturing the 5′ and the 3′ end, respectively).
4. Make a hole in the lid of the tubes using a needle.
5. Dry the whole sample by heat and vacuum in the Speed-Vac for 30–60 min.
6. When the sample is absolutely dry, cover the hole in the lid with a ToughTag sticker and add 7.5 μl of 2× Hybridization Buffer and 3 μl of Hybridization Component A. Be careful not to dislodge the dried sample when opening tube. Vortex the mixture and centrifuge the tubes at maximum speed for 10 s.
7. Incubate the tubes on the thermoblock at 95 °C for 5 min. Cover them with a 95 °C pre-heated heat block to prevent any evaporation of the sample through condensation in the tube lid.
8. Mix by flicking to ensure a complete resuspension of the sample and transfer the whole sample of each tube (10.5 μl) to the corresponding 0.2 ml tube with pre-aliquoted LNA probes.
9. Place the tubes in the thermocycler at 95 °C for 3 min.
10. Transfer the tubes to the thermocycler at 47 °C and incubate for 3 days (*see* **Note 19**).

3.6 Capture Recovery and Amplification

3.6.1 Preparation

1. Prepare 1× solutions of each wash buffer in the Roche NimbleGen Capture Wash Kit. The following amounts are indicated for a single RC-seq sample involving one 5′ and one 3′ capture.

	10× Stock (μl)	Water (μl)	Volume (μl)
Stringent Wash Buffer	100	900	1000
Wash Buffer I (first aliquot)	50	450	500
Wash Buffer I (second aliquot)	50	450	500
Wash Buffer II	50	450	500
Wash Buffer III	50	450	500
	2.5× Stock (μl)	Water (μl)	Volume (μl)
Beads Wash Buffer	440	660	1100

2. Set up a thermoblock at 47 °C and place one aliquot of 1× Wash Buffer I and the one of 1× Stringent Wash Buffer.
3. Set up a thermocycler with 47 °C block, **57 °C lid** (this one is the same that just held the 3 days incubation for the hybridization).
4. Set vortex on constant (not pressure-sensitive), at minimum speed.
5. Allow the streptavidin Dynabeads® M-270 Streptavidin and the AMPure® XP beads to warm to room temperature 30 min before use and resuspend them well by inverting or vortex for 1 min.
6. Set up a thermocycler with the LM-PCR cycling protocol: 98 °C 45 s; 98 °C 15 s, 60 °C 30 s and 72 °C 30 s for eight times (*see* **Note 14**); 72 °C 5 min and 4 °C hold. Do not bring the thermocycler to the first step, wait until the reaction tubes are inside.
7. Prepare 2–4 ml of 80 % ethanol per each pair of 5′ and 3′ captures of the same pooled sample (*see* **Note 2**).
8. Place an aliquot of 50 μl of MiniElute® Gel Extraction Kit buffer EB on a thermoblock at 60 °C (this thermoblock can be the same listed above at 47 °C, that must be set up after **step 18** in the Subheading 3.6.2).

3.6.2 Procedure

To pull down the captured library subset:

1. Dispense 200 μl of Dynabeads® M-270 Streptavidin per each pair of 5′ and 3′ captures in one 1.5 ml tube (up to a maximum of 400 μl per tube).
2. Place the tube in the magnetic rack. Aspirate and discard the supernatant once it becomes clear.
3. With the tube in the magnetic rack, add two volumes of 1× Bead Wash Buffer per initial volume of Dynabeads® M-270 Streptavidin.
4. Remove the tubes from the magnetic rack and vortex well.
5. Place it back in the magnetic rack until the liquid becomes clear. Aspirate and discard the supernatant.

6. Repeat **steps 3–5**.
7. Resuspend the Dynabeads® M-270 Streptavidin in the original volume of 1× Beads Wash Buffer and aliquot 100 μl of beads in 0.2 ml tubes, one tube for the 5′ capture and one tube for the 3′ capture.
8. Place the 0.2 ml tubes against the magnetic rack until the liquid becomes clear. Aspirate and discard the supernatant.
9. Place the 0.2 ml tubes for 30 s in the thermocycler at 47 °C where the hybridization reaction is taking place.
10. Transfer each 5′ and 3′ capture hybridization reactions to each 0.2 ml tube with Dynabeads® M-270 Streptavidin while keeping the tubes in the thermocycler block. Close the thermocycler lid and incubate for 30 s.
11. Take the tubes, secure the lid and flick them to resuspend the mixture. Spin down briefly (*see* **Note 20**).
12. Place the tube back in the thermocycler and incubate at 47 °C for 45 min. Resuspend the samples every 15 min by pipetting up and down ten times. During this incubation, warm up the magnetic rack for 0.2 ml tubes on the thermoblock at 47 °C.
13. Place the 0.2 ml tubes in the magnetic rack at 47 °C until the liquid is clear. Aspirate and discard the supernatant.
14. Place the tubes in the thermocycler at 47 °C, add 100 μl of 47 °C 1× Wash Buffer I to each tube and mix by pipetting for 10 s.
15. Repeat **step 13**.
16. Place the tubes in the thermocycler at 47 °C and add 200 μl of 47 °C 1× Stringent Wash Buffer to each tube. Mix by pipetting ten times and incubate for 5 min.
17. Repeat **step 13**.
18. Repeat **step 16–17**.
19. Remove the tubes from the magnetic rack, add 200 μl of room temperature 1× Wash Buffer I to each tube, mix by vortexing at minimum speed for 2 min.
20. Bring the magnetic rack to room temperature.
21. Place the tubes in the magnetic rack at room temperature until the liquid is clear. Aspirate and discard the supernatant.
22. Remove the tubes from the magnetic rack, add 200 μl of room temperature 1× Wash Buffer II to each tube, mix by vortexing at minimum speed for 1 min.
23. Repeat **step 21**.
24. Remove the tubes from the magnetic rack, add 200 μl of room temperature 1× Wash Buffer III to each tube, mix by vortexing at minimum speed for 30 s.

25. Repeat **step 21**.
26. Resuspend the Dynabeads® M-270 Streptavidin pellet of each tube in 50 μl of ultrapure water and transfer the whole mixture to a fresh 0.2 ml tube (including the beads).

SAFETY STOPPING POINT. You can store the samples at −20 °C or proceed immediately.

To amplify the post-hybridization library:

27. Add, *directly to the dynabeads bound to your sample*, 100 μl of Phusion® High-Fidelity PCR MasterMix (2×), 46 μl of Ultrapure Water, 2 μl of TS-F Primer (100 μM) and 2 μl of TS-R Primer (100 μM). Mix by pipetting ten times and split the sample in two tubes of 0.2 ml with 100 μl each (this is to ensure that the whole reaction volume is in the tube section contained within the block).
28. Place the tubes in the thermocycler with the LM-PCR program and start the run. Do not pre-heat cycler block to 95° prior to the run (*see* **step 6** in Subheading 3.6.1).
29. Combine the identical reactions (5′ + 5′, 3′ + 3′) into a 1.5 ml tube.
30. Proceed to a clean up reaction using the MiniElute® Gel Extraction Kit by adding 1 ml of Binding Buffer to each tube and mix well (including the beads).
31. Add 700 μl of each sample to a MiniElute® column and centrifuge at maximum speed for 1 min.
32. Discard the flow-through and add the remaining sample to the corresponding column. Centrifuge at maximum speed for 1 min and discard the flow-through.
33. Add 700 μl of Washing Buffer and incubate at room temperature for 2 min. Centrifuge at maximum speed for 1 min and discard the flow-through.
35. Rotate each column 180° in its well of the centrifuge rotor and centrifuge at maximum speed for 30 s.
36. Transfer each column to a fresh 1.5 ml tube, add 16 μl of 60 °C Elution Buffer in the center of the column and let it sit for 5 min.
37. Centrifuge at maximum speed for 1 min to elute the library. Discard the column.
38. Quantify each 5′ and 3′ capture analyzing 1 μl of the sample by a Qubit® dsDNA HS Assay. Usually, 3′ capture concentration is >1 ng/μl and 5′ capture concentration is >7 ng/μl (*see* **Note 21**).
39. Pool 5′ and 3′ capture samples in a ratio 3:7 by molecular mass.
40. Quantify the concentration and size distribution of the capture sample by analyzing 1 μl on an Agilent DNA 1000 Assay.

3.7 Sequencing

3.7.1 Preparation

1. Prepare the Sample Sheet for sequencing on Illumina Instruments using Illumina Experiment Manager software according to the following guidelines:
   - Choose the application "FASTQ Only" for the Sample Sheet, that appears in the "Other" category for MiSeq Instrument and is listed directly as application "HiSeq FASTQ Only" for HiSeq Instruments. This is designed to generate demultiplexed FASTQ files from any type of library.
   - Select the number of Index Reads according to the Illumina adapters used for the library preparation.
   - Select "Paired End" type for sequencing run. This sets the distribution of the Illumina reads after the sequencing reaction in two files according to their forward and reverse orientation in their fragment. The paired reads originated from the same fragment cluster will appear in the same position in both documents.
   - Select 151 cycles for each end. Illumina recommends this number of cycles for a sequencing reaction by a 300 elongation cycles kit, when preparing libraries of the fragment size described in this protocol. Cycle number will vary according to the fragment size and, thus, the sequencing kit.
   - Introduce Samples IDs and Indexes for each sample pooled in the hybridization mix.
2. Prepare an appropriate dilution of the capture library containing both 5′ and 3′ captures using molecular grade water (i.e., for MiSeq sequencing the protocol starts with a 4 nM dilution of the library).
3. Ensure that you have enough storage space in the hard disk of the instrument (i.e., a MiSeq run requires 100 GB).
4. Ensure that the Illumina Sequencing instrument is connected to an uninterrupted power supply or is equipped with a power supply unit able to support the instrument during the whole length of the run.

3.7.2 Procedure

1. Dilute the library and denature the DNA following Illumina instructions, according to the sequencing run you are performing.
2. Set up the sequencing components (flow cell, sequencing cartridge, and sequencing buffers) as described by Illumina instructions.
3. Start the sequencing run.
4. After the run, the Instrument performs an analysis of the primary data. Paired-end sequencing indicated here produces two different FASTQ files for each indexed library. These files contain the

reads from the forward and the reverse sequencing reaction of the same fragment, which are located in the same position of each file. Check that you transfer both files to the computer designated for the analysis.

3.8 Bioinformatic Analysis

The goal of RC-seq is to elucidate true positive L1 insertions, relying primarily on L1-genome junction sequences, rather than read count, to distinguish these events from artifacts. Although this can be computationally intensive, it is preferable to stringently exclude false positives (artifacts) bioinformatically rather than having to PCR validate every L1 insertion. Similarly, the analysis must not be so strict as to exclude any L1 insertion structure that is unusual, or cause false negatives. The procedure outlined below is intended to balance these considerations when searching for rare somatic L1 insertions in human brain tissue; a user should evaluate whether the algorithms and parameters in each step are appropriate to other retrotransposons, spatiotemporal contexts or types of L1 insertions (e.g. germline polymorphisms). A user will also need to be able to write scripts to process the output of each step. Custom scripts referred to below are written in Python, similar scripts may be written in other programming languages.

3.8.1 Preparation

1. Paired-end Illumina sequencing will have generated two fastq files, one for each read pair end. Prior to analyzing these files, it is recommended to establish a file name nomenclature that is consistent across projects, including the date of sequencing, the project identifier and a sample identifier. For example, 011214.GBM.1T.R1.fastq and 120114.GBM.1T.R2.fastq are two files for RC-seq sequenced on December 1, 2014 for a glioblastoma (GBM) project sample identifier 1T (individual #1 tumor). Using this system will simplify project management and data archiving prior to publication.

3.8.2 Procedure

1. Write a Python script to trim reads in each fastq file from their 5′ and 3′ ends to remove any bases with Illumina quality scores <10.
2. Input fastq files into FLASH [35], using default parameters, to assemble overlapping read pairs into read contigs in a single fasta file. These read contigs potentially span L1-genome junctions.
3. Align read contig fasta file to the latest human reference genome build (e.g. hg19, available from UCSC Genome Browser) hg19 using SOAP2 [36] (parameters –M 4 –v 2 –r 1 –p 8). These parameters will retain only read contigs aligning to one genomic location each. SOAP2 can also output unmapped read contigs (option –u). L1 insertions not present in the reference genome will be unmapped, and are what a user is seeking.

4. Align the unmapped read contigs to an active L1 consensus sequence (e.g. L1.4 from [37]) using LAST [38] (parameters –s 2 –l 11 –d 30 –q 3 –e 30).
5. Write a Python script to retain read contigs aligned at >95 % identity to L1.4 and spanning ≥33 nt of one contig end and arrange read contigs with a 5′ non-retrotransposon section (≥33 nt) followed by a 3′ retrotransposon section (≥33 nt).
6. Align read contigs to hg19 using LAST (s2 –l11 –d30 –q3 –e30), which excels in reporting split alignments found for translocations and, for RC-seq, where one end of the assembled read contig maps to one location on the genome and the other end maps elsewhere (Fig. 4).
7. Write a Python script to remove any contig read with an alignment of the non-retrotransposon section plus 10 nt (potential molecular chimera or genomic rearrangement). The remaining contig reads with a uniquely mapped non-retrotransposon section indicate the nucleotide position and strand of an L1-genome junction not present in the reference genome, and the nucleotide position and end of L1.4 detected.
8. Write a Python script to cluster aligned contig reads join opposing clusters separated by ≤100 nt and detecting different ends of a common L1 insertion, and annotate this list of clusters with existing databases of polymorphic L1 insertions. L1 insertions found only in one brain sample by ≥1 RC-seq reads, and absent from any matched control RC-seq library, or previous publications, can be annotated as somatic insertions. These can be further filtered by, for example, removing L1-genome junctions that indicate substantial L1 3′ truncations or requiring reported insertions to present TSDs. This process should generate a table of putative somatic L1 insertions in each sample.

3.9 PCR Validation

The validation process is highly dependent on the researcher's inventiveness and ability. To validate a new somatic insertion it is required that both ends of the insertion represent a genomic continuity interrupted by the new L1 copy. The validation will be more robust if, additionally, canonical TPRT signatures like TSDs, a polyA track in the 3′ end of the element and an EN-motif at the 5′ end of the 5′ TSD are found, otherwise it will be necessary to assume that the copy has been inserted by non-canonical pathways already characterized via in vitro approaches. However, many of the insertions detected by RC-Seq are only indentified at one junction [12, 33] (Prof. Faulkner group, unpublished results), so the basis of the PCR validation resides in the specific PCR amplification of the missing junction of each insertion. To amplify that junction sequence, it is necessary to "recreate" the structure of the missing junction, deduced by combining the annotated genome sequence following the genomic section of the known junction

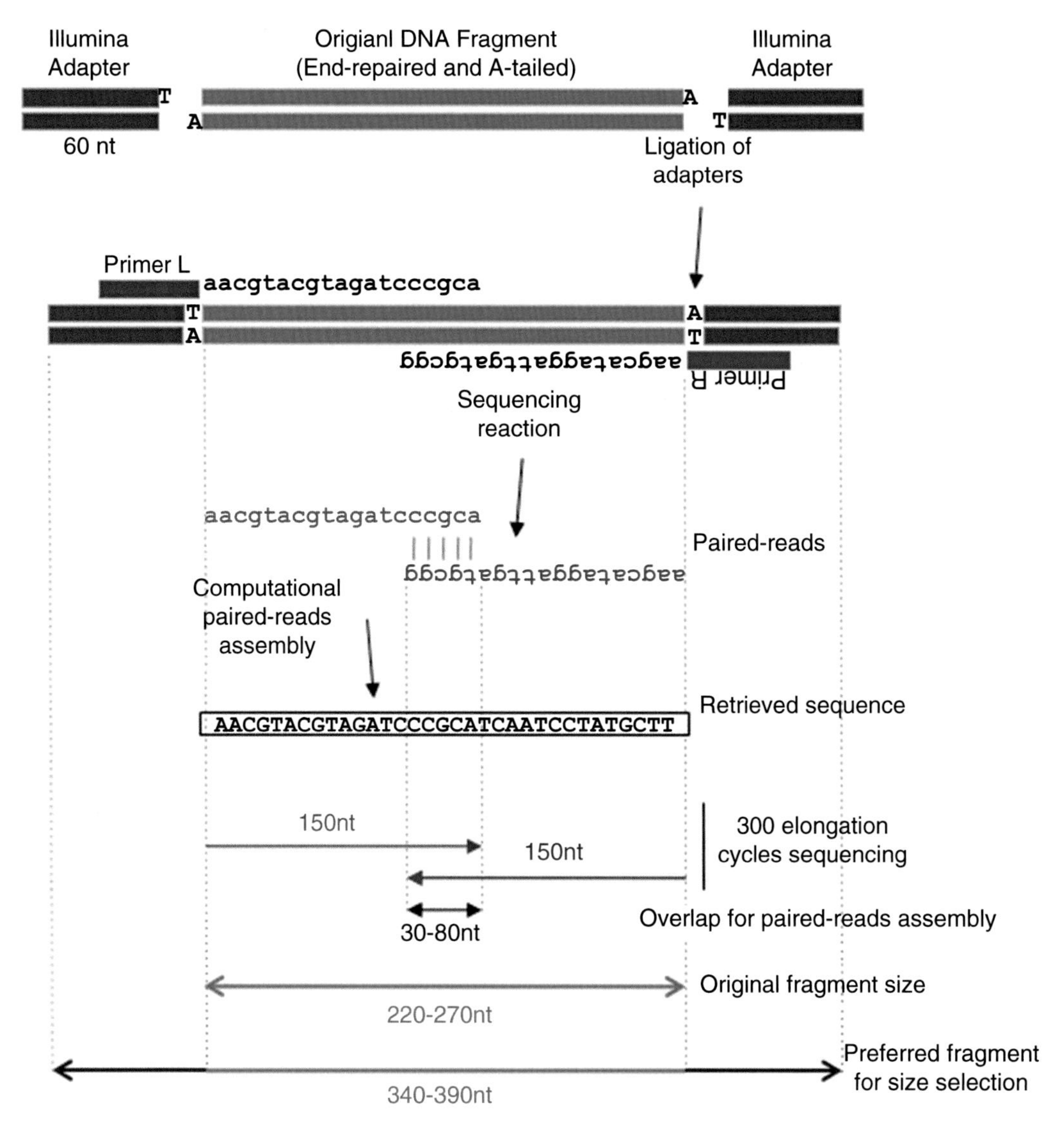

Fig. 4 Logic for the fragment size-selection by electrophoresis for a 300-cycle sequencing run. The sequencing reaction is primed by two oligonucleotides (Primer L in *red* and Primer R in *blue*), one from each end of the molecule. The 300-cycle run catalyzes 150 elongation cycles for the extension of each primer (*reads* and *arrows* in *red* and *blue*). Since 30–80 nt overlap between the paired reads is necessary for a successful alignment (*double-headed black arrow*) along the RC-seq pipeline, the overall size of the readable fragment is 220–290 bp. The agarose electrophoresis size-selection is performed after the ligation of Illumina sequencing adapters, increasing fragment size by 120 bp (60 bp per adapter). Therefore, for a library to be sequenced on a 300-cycle run, the preferred fragment for the size selection is 340–390 nt in length

and the consensus sequence of the L1 end opposite the one detected in the junction (Fig. 5). It will be necessary to design primers annealing within the L1 sequence and within the genomic region expected for the missing junction.

The PCR amplification of 3′ junctions consistently yields off-target amplicons [12]. Additionally, due to the location of the

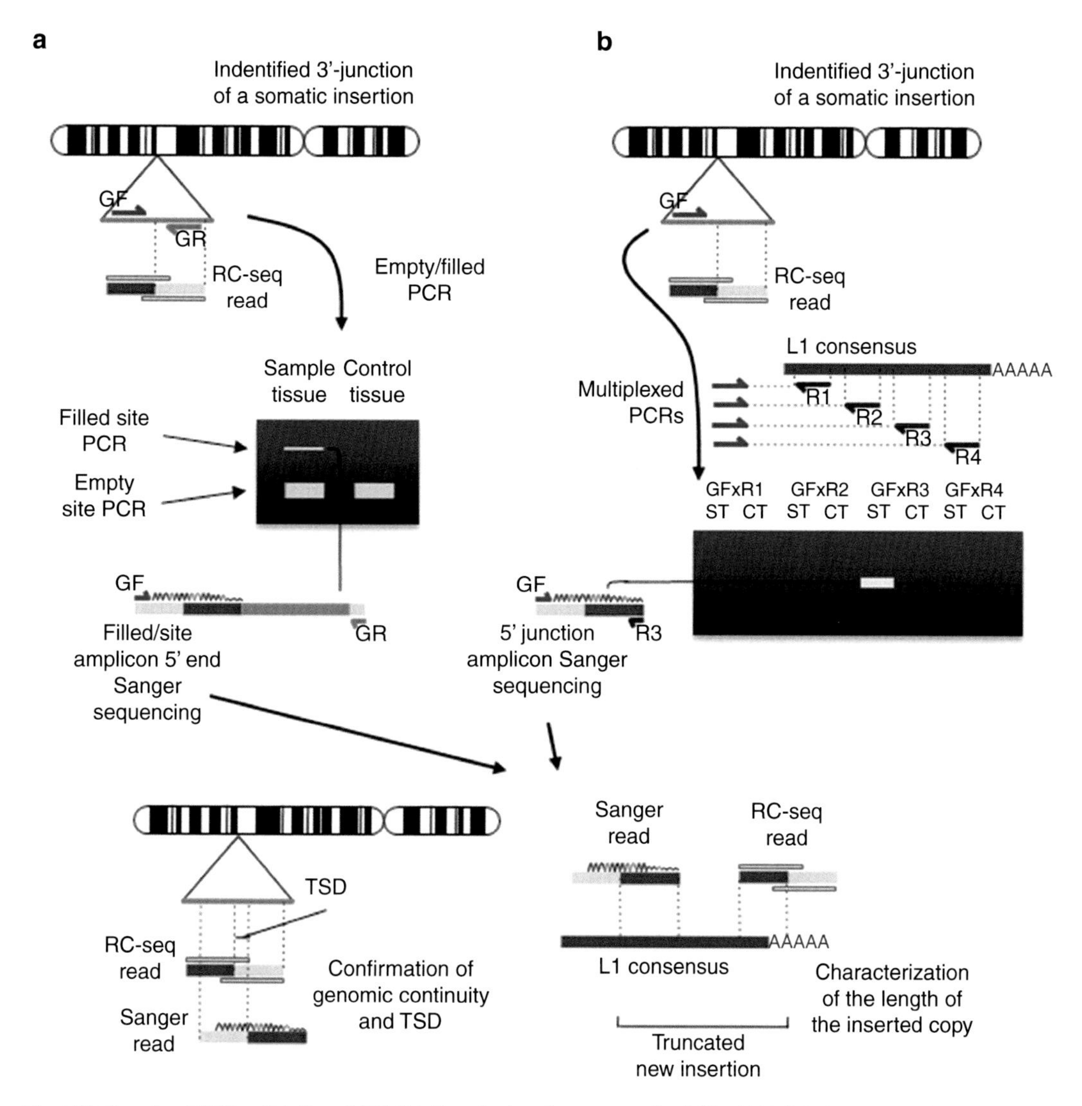

Fig. 5 Rationale of PCR validation. (**a**) Validation design for empty-site PCR. (**b**) Validation design for 5′ junction amplification

LNA-5′ probe and the fact that TPRT frequently generates 5′ truncated insertions, the 3′ junction is the most frequently captured one. The PCR validation described here is therefore based exclusively on PCR amplification of 5′ junctions.

The PCR validation of a 5′ junction has as a major handicap the fact that a priori the extent of 5′ truncation is not known, so it is impossible to predict the 5′ L1 sequence constituting the junction. As a guideline, there are two consecutive approaches: empty-filled site PCR and 5′ junction PCR.

The empty-filled site PCR consists of amplification by primers annealing in the 5′ and 3′ genomic flanks of the putative insertion

(Fig. 5a). For this PCR, we recommend the use of a high processive enzyme with PCR-cycling conditions able to amplify a ≤8 kb amplicon. As a starting point we suggest the Expand Long Range dNTPack (Roche) with the following conditions: 1× Buffer (with 2.5 mM $MgCl_2$ final concentration), 5 % DMSO, 2 mM dNTPs (0.5 mM each), 2 mM primers (1 mM each), 1.75 U of polymerase and 10 ng template DNA in a final volume of 25 μl with molecular grade water; and the suggested cycling conditions are: 92 °C 3 min; 92 °C 30 s, 60 °C 30 s and 68 °C 7.5 min for ten times; 92 °C 30 s, 58 °C 30 s and 68 °C 7 min plus Δ20 s increase per cycle for 30 times; 68 °C 10 min and 10 °C hold. The expected empty site is usually favored during the amplification and competes with the amplification of the filled site. It is important to screen faint bands in an electrophoresis gel in the size frame around 7 Kb. In any case, any band above the empty site amplicon is a candidate to be purified, cloned and sequenced in order to identify the 5′ junction. We recommend designing the genomic primers within 200 nt from the breaking point. Primer design can be initially attempted using a software like Primer3; if Primer3 fails to find good candidate primers then primers must be chosen by hand.

In the second approach, the 5′ junction amplification will require a primer annealing in the genomic section of the missing junction (designed as described above) and a primer annealing in the L1 sequence in reverse sense. If the empty-site PCR has provided positive results about the location of the junction (estimating by amplicon size or directly by sequencing), a specific primer can be designed in the known L1 region next to the truncation point and this PCR will confirm the result obtained with the empty-filled PCR. Otherwise, a set of L1 primers located along the whole L1 sequence in reverse sense can be used for multiple PCR reactions aiming to capture the 5′ L1-genome junction (Fig. 5b. *See* **Note 22** for a list of L1 primers to start with). Note that the length of full length insertions together with the small size of the empty site amplicon will strongly oppose the filled site amplification, so if the empty-filled PCR is not successful, proceed to the 5′ junction amplification using the battery of reverse L1 primers.

For this PCR reaction a high fidelity PCR enzyme is preferable, to avoid the formation of chimeras during the amplification due to the repetition of L1 sequences in the genome. The suggested enzyme is the Platinum® Taq DNA Polymerase High Fidelity. The starting conditions are: 1× High Fidelity PCR buffer, 2 mM $MgSO_4$, 0.8 mM dNTPs (0.2 mM each), 0.4 μM primers (0.2 μM each), 7.5 U polymerase, and 10 ng template DNA. The recommended cycling conditions are 94 °C 3 min; 94 °C 30 s, 57 °C 30 s and 68 °C 30 s for 35 cycles; 68 °C 10 min and 10 °C hold.

Sanger sequencing of the PCR products directly of via cloning will be required to fully characterize the paired junction of the insertion. According to the features of the amplification enzyme utilized, we recommend AT cloning by pGEM®-T vector system or

AT or blunt-ends cloning by TOPO® PCR system (Life Technologies) for regular size and large amplicons respectively. Due to the low yield of some amplification products such as full-length insertions, agarose gel-purification and re-amplification may be attempted to obtain enough amplified fragment for cloning. In some cases, phenol extraction from agarose and ethanol precipitation may increase the yield for faint bands of large size DNA fragments.

4 Notes

1. Prepare the Lysis Buffer without adding Proteinase K and store it at room temperature. Take an aliquot for each extraction and add Proteinase K to the aliquot at final concentration of 100 μg/ml just before use. If using Ambion® 20 mg/ml Proteinase K Solution (Life Technologies), add 5 μl of Proteinase K to each 995 μl of Lysis Buffer.
2. The whole protocol can be adapted to 0.2 ml tube-strips for multichannel adaptation. In this case, a DynaMag™-96 Side (Life Technologies) magnetic rack must be used and, consequently, the 80 % ethanol volume used for the washes must be 200 μl.
3. N_8 segment sequence of Illumina Indices-specific Blocking Oligos: Index 1, CGTGATGT; Index 2, ACATCGGT; Index 3, GCCTAAGT; Index 4, TGGTCAGT; Index 5, CACTGTGT; Index 6, ATTGGCGT; Index 7, GATCTGGT; Index 8, TCAAGTGT; Index 9, CTGATCGT; Index 10, AAGCTAGT; Index 11, GTAGCCGT; Index 12, TACAAGGT; Index 13, TGTTGACT; Index 14, ACGGAACT; Index 15, TCTGACAT; Index 16, CGGGACGG; Index 18, GTGCGGAC; Index 19, CGTTTCAC; Index 20, AAGGCCAC; Index 21, TCCGAAAC; Index 22, TACGTACG; Index 23, ATCCACTC; Index 25, ATATCAGT; Index 27, AAAGGAAT.
4. Additional Proteinase K can be added to the dissociation reaction if the tissues are not dissolving at a timely rate.
5. If using 10 mg/ml DNase-free, protease-free RNase A (Thermo Scientific), add 1 μl per 500 μl of sample.
6. The quality of the starting DNA solution is really important for the library preparation. It is worthwhile to take extra time to duplicate the phenol and phenol:chlorophorm:isoamyl alcohol extraction steps in order to ensure the success of the subsequent steps.
7. Alternatively the sample can be centrifuged at >12,000*g* in a benchtop centrifuge and supernatant thoroughly removed.

8. Short incubations (~30 min) at 65 °C can be used to aid resuspension, but high temperatures should be avoided where possible.
9. Alternatively, Covaris S220 Focused-Ultrasonicator can be used with the following parameters: Duty Cycle 10 %, Intensity 5, Pulses per Burst 200 and Duration 120 s.
10. DNA quantification by Nanodrop usually overestimates genomic DNA concentration. After shearing the genomic DNA, the overestimation is reduced but is still higher than results produced by analysis by Agilent Bioanalyzer Technology. In **step 10** from the Subheading 3.3.2, up to 1 μg of DNA is required, so if only a single library is being prepared, we recommend starting the whole process with 5 μg of genomic DNA.
11. If the sonication is interrupted by a sudden drop of the water level, immediately proceed to add more water with the Wash Bottle through the water sense aperture in the Tube Holder and click "resume" in the warning panel.
12. For two different libraries combine indices 6 and 12; or 5 and 19. For three libraries combine indices 2, 7 and 19; 5, 6 and 15; or any combination for two libraries plus any other index. For four different libraries combine indices 5, 6, 12 and 19; 2, 4, 7 and 16; or any combination for three libraries plus any other index. For 5–11 different libraries, use the combination for 4 different libraries plus any other adapter (for more information please visit http://supportres.illumina.com/documents/documentation/chemistry_documentation/samplepreps_truseq/truseqsampleprep/truseq_sample_prep_pooling_guide_15042173_a.pdf).
13. Pay special attention to these two consecutive AMPure XP beads purification steps. They are necessary to completely remove the adapter excess from the samples and avoid any interference in the following amplification steps. The procedure is highly repetitive and it is very easy to trash the library by mistaking the elution step for an ethanol washing step.
14. The number of cycles of PCR must be kept to a minimum. The amplification of certain repeated sequences is relatively unfavored, diminishing their relative abundance in the pool with increasing amplification cycles. In both the pre- and post-hybridization LM-PCR steps, start with the indicated number of cycles. The hybridization mix before completely drying the sample has no restrictions concerning sample volume, so it is preferable to include more pre-hybridization library input in case of low concentration instead of running more PCR cycles.
15. We recommend taking one or two initial cuts below and above the preferred one to increase the chances of obtaining a library with right size distribution. The gel cuts can be stored at 4 °C or can be processed together. Consider the fact that each

Agilent DNA chip has 12 wells, so it can be convenient to proceed with more than one gel cut depending on the number of libraries to maximize the chip yield.

16. Salt in the hybridization mix can dramatically affect to the reaction, so for this step is essential to use molecular grade water instead of resuspension buffer. This library will be substrate for the immediate hybridization and it is necessary to reduce the salt content to minimum.
17. After this analysis, two possible problems can come up: low amount of library (<20 ng/μl) or an inappropriate size. If the amount is low, one solution is to perform additional cycles of LM-PCR. To do that, go to **step 14** and proceed with a first attempt of 2–3 cycles depending on the concentration of the input library. If the fragment size is not appropriate, because it is too big or too small, then go to the backup gel-cuts of **step 4** and proceed with the immediately smaller or bigger one respectively. If two consecutive gel-cuts are obtained, one of them bigger and the other one smaller that the preferred size, it is possible to combine the two libraries and reanalyze the mixture. Usually the resulting peak will be the preferred size.
18. If analyzing tissues A and B from the same donor, pool 500 ng of each sample library; if analyzing tissues A, B and C, pool 333 ng of each sample library; if analyzing tissues A and B tissues from donor X and Y, then pool 250 ng of each library; and so.
19. It is very important the temperature of the incubation not to go below 47 °C. LNA probes have a high ability for hybridization and the reversibility is strongly disabled. The unspecific binding occurring below 47 °C will not dissociate upon bringing the temperature back to 47 °C. This is why it is very important that the thermocycler has an uninterrupted power supply during the 3 days the incubation requires. Any drop in the incubation temperature will result in a reduction in the library enrichment in L1-specific sequences.
20. Try to avoid holding the tubes out of the thermocycler more than 10 s. It is better to put them back in the thermocycler and do several resuspension re-attempts.
21. In case one or both libraries end with low concentration insufficient for sequencing, proceed as in **Note 17**. Briefly, bring the sample to 30 μl final volume with molecular grade water and go to **step 14–23** from the Subheading 3.4.2. Similarly, do a first attempt of reamplification by LM-PCR with 2–3 cycles. In **step 21**, add only 16.5 μl of resuspension buffer and in **step 23** recover 15 μl to a new tube.
22. This is a guide of suggested reverse primer mapping in the L1 consensus sequence with their approximate location in brackets. L1-1R(554) 5′CCAGAGGTGGAGCCTACAGA3′; L1-2R (1085) 5′ATGTCCTCCCGTAGCTCAGA3′; L1-3R (1426)

5′TGGTTCCATTCTCCACATCA3′;	L1-4R	(2080)
5′TCCAACTTGCCAGTCTGTGT3′;	L1-5R	(2591)
5′TAGGTGTGGTGTGGTGCTGA3′;	L1-6R	(3085)
5′ACCAGCTCCTGGATTCATTG3′;	L1-7R	(3550)
5′CCGGCTTTGGTATCAGAATG3′;	L1-8R	(4041)
5′TTCCTTCTCCTGCCTGATTG3′;	L1-9R	(4263)
5′TGGGAGTTCACCCATGATTT3′;	L1-10R	(5084)
5′TGCCTGTTCACTCTGATGGT3′;	L1-11R	(5627)
5′CATTTGGGTTGGTTCCAAGT3′;	L1-12R	(5799)
5′TGAGAATATGCGGTGTTTGG3′.		

Acknowledgments

G.J.F. acknowledges the support of an NHMRC Career Development Fellowship (GNT1045237). F.J.S-L. was supported by a postdoctoral fellowship from the Alfonso Martín Escudero Foundation (Spain). Work in the Faulkner laboratory was funded by Australian NHMRC Project grants GNT1042449, GNT1045991, GNT1067983 and GNT1068789, as well as the European Union's Seventh Framework Programme (FP7/2007–2013) under grant agreement No. 259743 underpinning the MODHEP consortium.

References

1. Taylor TH, Gitlin SA, Patrick JL et al (2014) The origin, mechanisms, incidence and clinical consequences of chromosomal mosaicism in humans. Hum Reprod Update 20:571–581
2. Stern C (1936) Somatic crossing over and segregation in Drosophila melanogaster. Genetics 21:625–730
3. Hozumi N, Tonegawa S (1976) Evidence for somatic rearrangement of immunoglobulin genes coding for variable and constant regions. Proc Natl Acad Sci U S A 73:3628–3632
4. Muramatsu M, Kinoshita K, Fagarasan S et al (2000) Class switch recombination and hypermutation require activation-induced cytidine deaminase (AID), a potential RNA editing enzyme. Cell 102:553–563
5. Gilbert N, Lutz S, Morrish TA et al (2005) Multiple fates of L1 retrotransposition intermediates in cultured human cells. Mol Cell Biol 25:7780–7795
6. Garcia-Perez JL, Doucet AJ, Bucheton A et al (2007) Distinct mechanisms for trans-mediated mobilization of cellular RNAs by the LINE-1 reverse transcriptase. Genome Res 17:602–611
7. Kano H, Godoy I, Courtney C et al (2009) L1 retrotransposition occurs mainly in embryogenesis and creates somatic mosaicism. Genes Dev 23:1303–1312
8. Muotri AR, Chu VT, Marchetto MC et al (2005) Somatic mosaicism in neuronal precursor cells mediated by L1 retrotransposition. Nature 435:903–910
9. McConnell MJ, Lindberg MR, Brennand KJ et al (2013) Mosaic copy number variation in human neurons. Science 342:632–637
10. Kapitonov VV, Jurka J (2005) RAG1 core and V(D)J recombination signal sequences were derived from Transib transposons. PLoS Biol 3, e181
11. Levis RW, Ganesan R, Houtchens K et al (1993) Transposons in place of telomeric repeats at a Drosophila telomere. Cell 75: 1083–1093
12. Baillie JK, Barnett MW, Upton KR et al (2011) Somatic retrotransposition alters the genetic landscape of the human brain. Nature 479: 534–537
13. Moran JV, Holmes SE, Naas TP, DeBerardinis RJ, Boeke JD, Kazazian HH Jr (1996) High frequency retrotransposition in cultured mammalian cells. Cell 87:917–927

14. Dewannieux M, Esnault C, Heidmann T (2003) LINE-mediated retrotransposition of marked Alu sequences. Nat Genet 35:41–48
15. Raiz J, Damert A, Chira S et al (2012) The non-autonomous retrotransposon SVA is trans-mobilized by the human LINE-1 protein machinery. Nucleic Acids Res 40:1666–1683
16. Jurka J (1997) Sequence patterns indicate an enzymatic involvement in integration of mammalian retroposons. Proc Natl Acad Sci U S A 94:1872–1877
17. Luan DD, Korman MH, Jakubczak JL et al (1993) Reverse transcription of R2Bm RNA is primed by a nick at the chromosomal target site: a mechanism for non-LTR retrotransposition. Cell 72:595–605
18. Ostertag EM, Kazazian HH Jr (2001) Twin priming: a proposed mechanism for the creation of inversions in L1 retrotransposition. Genome Res 11:2059–2065
19. Morrish TA, Gilbert N, Myers JS et al (2002) DNA repair mediated by endonuclease-independent LINE-1 retrotransposition. Nat Genet 31:159–165
20. Lander ES, Linton LM, Birren B et al (2001) Initial sequencing and analysis of the human genome. Nature 409:860–921
21. Kazazian HH Jr, Wong C, Youssoufian H et al (1988) Haemophilia A resulting from de novo insertion of L1 sequences represents a novel mechanism for mutation in man. Nature 332:164–166
22. Huang CR, Schneider AM, Lu Y et al (2010) Mobile interspersed repeats are major structural variants in the human genome. Cell 141:1171–1182
23. Ewing AD, Kazazian HH Jr (2010) High-throughput sequencing reveals extensive variation in human-specific L1 content in individual human genomes. Genome Res 20:1262–1270
24. Stewart C, Kural D, Stromberg MP et al (2011) A comprehensive map of mobile element insertion polymorphisms in humans. PLoS Genet 7, e1002236
25. Faulkner GJ (2011) Retrotransposons: mobile and mutagenic from conception to death. FEBS Lett 585:1589–1594
26. Hancks DC, Kazazian HH Jr (2013) Active human retrotransposons: variation and disease. Curr Opin Genet Dev 22:191–203
27. Muotri AR, Marchetto MC, Coufal NG et al (2010) L1 retrotransposition in neurons is modulated by MeCP2. Nature 468:443–446
28. Faulkner GJ, Kimura Y, Daub CO et al (2009) The regulated retrotransposon transcriptome of mammalian cells. Nat Genet 41:563–571
29. Coufal NG, Garcia-Perez JL, Peng GE et al (2009) L1 retrotransposition in human neural progenitor cells. Nature 460:1127–1131
30. Belancio VP, Roy-Engel AM, Pochampally RR et al (2010) Somatic expression of LINE-1 elements in human tissues. Nucleic Acids Res 38:3909–3922
31. Cost GJ, Golding A, Schlissel MS et al (2001) Target DNA chromatinization modulates nicking by L1 endonuclease. Nucleic Acids Res 29:573–577
32. Richardson SR, Morell S, Faulkner GJ (2014) L1 retrotransposons and somatic mosaicism in the brain. Annu Rev Genet 48:1–27
33. Shukla R, Upton KR, Munoz-Lopez M et al (2013) Endogenous retrotransposition activates oncogenic pathways in hepatocellular carcinoma. Cell 153:101–111
34. Upton KR, Gerhardt DJ, Jesuadian JS et al (2015) Ubiquitous L1 mosaicism in hippocampal neurons. Cell 161:228-239
35. Magoc T, Salzberg SL (2011) FLASH: fast length adjustment of short reads to improve genome assemblies. Bioinformatics (Oxford) 27:2957–2963
36. Li R, Yu C, Li Y et al (2009) SOAP2: an improved ultrafast tool for short read alignment. Bioinformatics (Oxford) 25:1966–1967
37. Dombroski BA, Scott AF, Kazazian HH Jr (1993) Two additional potential retrotransposons isolated from a human L1 subfamily that contains an active retrotransposable element. Proc Natl Acad Sci U S A 90:6513–6517
38. Kielbasa SM, Wan R, Sato K et al (2011) Adaptive seeds tame genomic sequence comparison. Genome Res 21:487–493

Chapter 5

Long Interspersed Element Sequencing (L1-Seq): A Method to Identify Somatic LINE-1 Insertions in the Human Genome

Tara T. Doucet and Haig H. Kazazian Jr.

Abstract

L1-seq is a high-throughput sequencing technique which is utilized to identify novel L1 insertions in genomic DNA samples of interest. Using special diagnostic nucleotides unique to the youngest and most active L1 sequence, we can amplify new somatic insertions. This technique has helped to establish the number of L1 insertions present in the general population as well as the variation among individuals with regard to their complement of active L1 elements. More recently, this technique has been employed to assess the level of retrotransposition occurring in various diseases such as cancer. These efforts try to establish a connection between the process of retrotransposition and disease development and/or progression.

Key words Non-LTR retrotransposon, Retroelement, LINE-1, L1, Next-Generation DNA sequencing

1 Introduction

Retrotransposons are nearly ubiquitous in eukaryotes from slime molds [1] to humans [2] and have contributed greatly to genome composition of these organisms. Retrotransposons make up 45 % of the human genome [2]. In particular, the LINE-1 (L1) element has contributed to approximately 17 % of the human genome and continues to add to it via a copy and paste mechanism with an RNA intermediate [2]. L1 is the only autonomous retrotransposon in the human genome because it encodes two proteins necessary for mobilization and reinsertion into the genome; however, these two proteins, once expressed can mobilize other types of retrotransposons as well as processed pseudogenes [3–5]. Each individual has a different complement of potentially active L1 elements, although the majority of the L1s in each individual's genome are truncated and therefore inactive. L1-seq [6] was developed to help characterize L1 variation among individuals because L1s have

Jose L. Garcia-Pérez (ed.), *Transposons and Retrotransposons: Methods and Protocols*, Methods in Molecular Biology, vol. 1400, DOI 10.1007/978-1-4939-3372-3_5, © Springer Science+Business Media New York 2016

contributed to a substantial fraction of the genome and are capable of inducing many types of mutations. L1-seq has since been used to evaluate several types of cancer to establish the level of retrotransposition occurring in colon cancer, lung cancer, breast cancer, and many other cancers [7, 8]. Additional sequencing techniques have confirmed the L1-seq data and demonstrated that L1 elements are active in many cancer types [8–13]. The results have demonstrated that L1s are active in a subset of patients with cancer; in addition, L1 elements are active in all epithelial cancers tested. The L1-seq technique consists of a DNA library prep as well as the validation of the predicted new insertions detected in the samples used. Although few of the insertions may be directly responsible for the development of the disease, it should be possible to utilize known insertions present in a cancer sample for monitoring the cancer's progression to metastasis. To detect metastasis using a new L1 insertion, a PCR would be performed on serum DNA from a patient to determine whether or not the insertion was detectable in the blood and therefore potentially in a floating cancer cell. This technique is useful both for evaluating the overall complement of L1 elements in a genome as well as looking for new insertion events. L1-seq utilizes unique nucleotides, "ACA" 91–93 nucleotides from the 3′ end of the element, to selectively amplify the young and active subset of elements in the human genome [6]. Following the initial five cycles of the PCR, wherein the linear amplification of L1 elements occurs, degenerate primers are added to the mixture to exponentially amplify both polymorphic and potentially somatic insertions present in the genome.

2 Materials

Store all reagents as specified by manufacturers. Diligently follow all waste disposal regulations when disposing of waste materials. All primers need to be diluted to 100 μM upon receipt in diethylpurocarbonate (DEPC) water. Primers will be further diluted as specified later in the protocol.

2.1 DNA Isolation

1. DNeasy Blood and Tissue Kit (Qiagen).
2. 500 mL of absolute ethanol (200 proof).
3. Qubit™ dsDNA BR Assay kit (Life Technologies).

2.2 Library Preparation

1. Promega GoTaq Flexi.
2. 25 mM MgCl2.
3. 10 mM dNTPs.
4. Diethylpurocarbonate (DEPC) water.

5. 100 % DMSO.
6. Pfu polymerase.
7. 1 μg of good quality DNA per sample at a concentration of 100 ng/μL (*see* **Note 1**).
8. LE agarose.
9. 1× TAE; Prepare 50× solution by dissolving 242 g Tris base in 750 mL of deionized water. Carefully add 57.1 mL glacial acetic acid, and 100 mL of 0.5 M EDTA, pH 8.0 and adjust solution to final volume of 1 L. Dilute the 50× solution to 1× in deionized water.
10. Ethidium bromide (10 mg/mL).
11. QIAquick Gel Extraction Kit (Qiagen).
12. MinElute PCR Purification Kit (Qiagen).
13. Isopropanol (200 proof).
14. 500 mL of Ethanol (200 proof).
15. Agilent DNA 1000 kit or high sensitivity DNA kit (choose as needed, *see* **Note 1**).
16. L1-seq primers (order HPLC grade for library preparation). *See* Table 1.

2.3 Next-Generation Sequencing Data Analysis

1. Server with at least 4GB of RAM and L1-seq scripts properly formatted (https://github.com/adamewing/l1seq).
2. Bowtie2 (http://bowtie-bio.sourceforge.net/bowtie2/index.shtml) and all relevant human genome files and indices as per Bowtie2 instructions.

2.4 Data Validation

1. GoTaq Green Master Mix (2×).
2. Diethylpurocarbonate (DEPC) water.
3. DNA at concentration of 12.5 ng/μL.
4. LE agarose.
5. 1× TAE; Prepare 50× solution by dissolving 242 g Tris base in 750 mL of deionized water. Carefully add 57.1 mL glacial acetic acid, and 100 mL of 0.5 M EDTA (pH 8.0) and adjust solution to final volume of 1 L. Dilute the 50× solution to 1× in deionized water.
6. Ethidium bromide (10 mg/mL).
7. QIAquick Gel Extraction Kit (Qiagen).
8. Isopropanol (200 proof).
9. 500 mL of Ethanol (200 proof).
10. L1SP1A2 primer, L1nt112out, L1 "G" primer (*see* Table 1).

Table 1
Primers used for L1-seq

	Sequences 5′ to 3′
Primers with Illumina adapters:	
Adap 1 L 1HsG	CAAGCAGAAGACGGCATAOGAGCTCTTCCGATC TTGCACATGAOCCTAAAACTTAG
Adap2Seq1	AATGATACGGCGACCACCGAGATCTACACTTTOCC TACACGACGACGCTCTTCCGATCT
L1 specific primers:	
L1HsSP1A2	GGGAGATATACCTAATGCTAGATGACAC (specific for L 1Hs subset)
L1 "G" primer	TGCACATGTACOCTAAAACTTAG (specific for L 1Hs subset)
L1nt112out	GATGAACCCGTACCTCAGA
Degenerate primers:	
DEG Seq 1N5TCTGT	ACACTCTTTCCCTACACGACGACGCTCTTCCGA TCTNNNNNTCTGT
DEGSeq1N5CTTCT	ACACTCTTTCCCTACACGACGCTCTTCCGATCTNN NNNCTTCT
DEGSeq1N5TGCCT	ACACTCTTTCCCTACACGACGACGCTCTTCCGAT CTNNNNNTGCCT
DEGSeq1NTCTCA	ACACTCTTTCCCTACACGACGCTCTTCCGATCTN NNNNTCTCA
DEGSeq1N5CAGAG	ACACTCTTTCCCTACACGACGCTCTTCCGATCTNN NNNCAGAG
DEGSeq1N5TTGAA	ACACTCTTTCCCTACACGACGCTCTTCCGATCTNN NNNTTGAA
DEGSeq1N5CTTTG	ACACTCTTTCCCTACACGACGCTCTTCCGATCTNN NNNCTTTG

3 Methods

3.1 Embedding and Cryosectioning Tissue

1. To begin, embed each piece of tissue to be assayed in OCT freezing medium. You can simply put a thin layer of the media onto a pre-chilled (≤20 °C) chuck.
2. Quickly placing the thawed tissue section onto the OCT.
3. Immediately cover the tissue in more OCT until it is barely visible through the medium. The OCT medium will change from clear to white, when the entire block of tissue/OCT is completely frozen, you can begin to cryosection the tissue for DNA extraction. It is best for the freezing to occur as rapidly as possible, to this end, a heat extractor can be used to enhance and shorten the freezing process and adherence to the chuck on which the tissue is being embedded in the OCT freezing medium.

4. Set the cryostat to slice sections of tissue between 10 and 30 μm.
5. During sectioning, carefully remove each roll of tissue and place 10–20 slices into a pre-chilled (≤20 °C) 1.5 mL microtube.
   See **Note 2**.

3.2 Extracting DNA from Sectioned Tissue

1. Remove microtubes with tissue slices from −80° freezer and place them on ice.
2. Add 360 μL of Buffer ATL (DNeasy Blood and Tissue kit). There is no need to further homogenize the tissue. Add 40 μL of proteinase K and mix thoroughly by vortexing (*see* **Note 1**).
3. Incubate at 55 °C overnight until the tissue is completely lysed. Vortex occasionally during incubation to help disperse samples.
4. Vortex for 15 s.
5. Add 400 μL of buffer AL (DNeasy Blood and Tissue Kit) to the sample and mix thoroughly by vortexing.
6. Immediately add 400 μL of ethanol (200 proof) and mix again thoroughly by vortexing (notes).
7. Pipette 750 μL of the mixture (including any precipitate) into the DNeasy Mini spin column placed in a 2 mL collection tube (provided in the kit).
8. Centrifuge at ≥6000*g* for 1 min. Discard flow through.
9. Repeat **steps** 7 and **8** until all of mixture has been run though the same column for each sample.
10. After the final spin with the aforementioned mixture, discard the collection tube and replace it with a new 2 mL collection tube.
11. Add 500 μL of Buffer AW1. (Ensure that ethanol has been added to Buffer AW1 before use.)
12. Centrifuge for 1 min at ≥6000*g*.
13. Discard flow-through and collection tube and place the spin column in a new 2 mL collection tube.
14. Add 500 μL of Buffer AW2. (Ensure ethanol has been added to Buffer AW2 before use.)
15. Centrifuge for 3 min at 20,000*g* to dry the DNeasy membrane and then discard flow-through and collection tube.
16. Place the DNeasy Mini spin column in a clean 1.5 mL microcentrifuge tube and pipet 100 μL of pre-warmed (55 °C) Buffer AE onto the DNeasy membrane.
17. Incubate at room temperature for 10 min.
18. Centrifuge for 1 min at ≥6000*g* to elute.

19. Pipette an additional 50 μL of pre-warmed AE buffer directly onto DNeasy membrane. (There is no need to replace the micro-centrifuge tube. The additional DNA elution can be collected into the same tube up to a volume of no more than 150 μL.)
20. Incubate for 10 min at room temperature.
21. Centrifuge for 1 min at 20,000*g*.

This protocol is adapted from the Qiagen handbook for the DNeasy Blood and Tissue Kit.

3.3 Measuring DNA Concentration (Qubit™)

Use the Qubit™ fluorometer to measure DNA concentration because it is one of the most accurate methods. Follow manufacturer protocols exactly and *see* **Note 1**. (http://www.ebc.uu.se/digitalAssets/176/176882_3qubitquickrefcard.pdf).

3.4 L1-seq Library Preparation (see Note 3)

1. Before beginning the relevant library prep PCRs, it is necessary to determine which samples will be pooled together in the library preparation. As many as ten samples can be pooled together without using barcoding (notes). Equal amounts of each sample must be put into the DNA pool to be used in the library. A total of 24 μL of pooled DNA at a concentration of 100 ng/μL is needed. If there is not enough DNA available from one or more samples, a modified protocol can be used (notes).
2. Round 1 PCR: Linear amplification of L1 flanks followed by a hemi-specific PCR incorporating the Illumina sequencing primer (Fig. 1). To prevent running out of master mix, make enough for nine reactions even though only eight reactions will be assembled. Master mix (per 1 reaction): 10 μL of Promega Go-Taq flexi buffer (5×), 6 μL of 25 mM $MgCl_2$, 2 μL of 20 μM L1SP1A2 primer (the second primer for the reaction, the degenerate primers previously mentioned, will be added following the completion of five cycles of the PCR which consists of the linear amplification step), 0.5 μL of DMSO 100 %, 0.5 μL of FlexiTaq, 1 μL of 10 mM dNTPs, 2 μL of pooled DNA (at 100 ng/μL), 24 μL of DEPC water. The reaction should total 46 μL before the addition of 4 μL of the degenerate primers to be added after linear amplification is finished. When the linear amplification is finished, add 1 of each of the degenerate primers at 5 μM (e.g. DEGSeq1N5TCTGT) to each of the eight reactions. Use the following cycling program:

   L1-Seq PCR 1

   (a) 95 °C for 2 min 30 s.

   (b) 95 °C for 30 s.

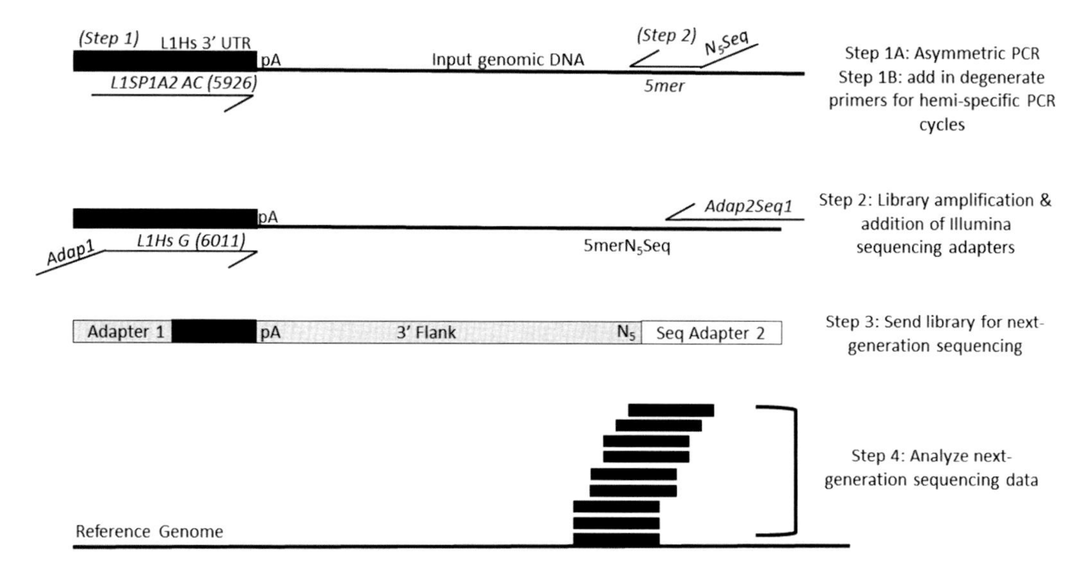

Fig. 1 Scheme of L1-seq. The first PCR consists of five cycles which enrich for sequences containing human-specific L1 sequences via primer extension with the L1Sp1A2 primer using diagnostic nucleotides for the human specific subfamily of L1. After enrichment for human-specific L1 flanks in the first five cycles, each reaction has one degenerate primer added. There are eight degenerate primers, each with a specified 5-mer at the 3′ end preceded by five degenerate bases (NNNNN) and a sequencing primer used for the Illumina platform. Eight different reactions are performed, each with a unique 5mer. The second PCR enriches for human-specific L1 3′ flanks by utilizing another diagnostic nucleotide in the 3′ UTR of L1 and adds the necessary adapter sequences via primer overhangs. The products of this reaction are mixed in equimolar ratios before sequencing on the Illumina 2500 platform. Following sequencing and initial processing, the reads are aligned to the human genome (hg19) and L1 insertions are identified for validation

(c) 58 °C for 1 min.

(d) 72 °C for 2 min.

(e) Go to **step b** (5×).

(f) 60 °C (pause and add 4 μL of degenerate primer into each of the eight reactions, one primer per reaction for each of the eight different degenerate primers).

(g) 95 °C for 30 s.

(h) 55 °C for 30 s.

(i) 72 °C for 1 min and 30 s.

(j) Go to **step g** (14×).

(k) 72 °C for 10 min.

(l) 4 °C hold.

3. Purify all eight reactions on eight separate Qiagen PCR Clean-up columns following the Qiagen protocol, eluting in 50 μL of pre-warmed (55 °C) EB (do a 10 min final incubation before elution to optimize DNA eluted from column).

4. L1-Seq PCR 2: Amplification of library and addition of the Illumina sequencing adapters (Fig. 1). Master mix (for 1 reaction): 12.5 μL of master mix (Promega GoTaq Green 2×), 1.5 μL of 20 mM primer Adap1L1HsG, 1.5 μL of 20 μM Adap2Seq1, 2.5 μL of purified round 1 product (1 degenerate primer per reaction), 7 μL of DEPC ddH_2O. Again, make enough for nine reactions to prevent running out of master mix for the samples to be amplified. The reactions will each have a total of 25 μL. Use the following cycling program:

   L1-Seq PCR 2

   (a) 95 °C for 2 min.

   (b) 95 °C for 30 s.

   (c) 62 °C for 30 s.

   (d) 72 °C for 1 min.

   (e) Go to **step b** (19×).

   (f) 72 °C for 5 min.

   (g) 4 °C Hold.

5. Resolve products on a 1 % TAE gel.
6. Excise the constellation of bands between 200 and 500 nucleotides with a sterile scalpel (using a different scalpel for each reaction) and purify the DNA using the Qiagen Gel Purification protocol.
7. Elute the library in 50 μL of pre-warmed EB buffer (55 °C with a 10 min incubation before elution).
8. Run each DNA sample on the Agilent Bioanalyzer with the DNA 1000 kit to get an accurate measure of concentration and the average size of the DNA amplified. Using the concentration and average size of the molecules, calculate how to add the DNA from all eight reactions in equimolar ratios to one tube. *See* **Note 4**.
9. After mixing the DNAs together, purify the entire mixture with the Qiagen MinElute PCR purification kit eluting in 50 μL of pre-warmed EB (55 °C and 10 min incubation at room temperature before elution).
10. End-polishing must be performed on the library because Taq leaves adenine overhangs which could cause problems when the library is annealed to the Illumina flow cell. To accomplish the end-polishing: mix 6 μL of 10× Pfu buffer, 2.5 μL of Pfu polymerase, 2.5 μL of 10 mM dNTPs, and 49 μL of library. Incubate for 1 h at 72 °C.
11. Purify reaction on a Qiagen MinElute column and elute in 10 μL of pre-warmed (55 °C) EB following a 10 min room-temperature incubation before elution.

12. Measure final DNA concentration with Qubit™ fluorometer to get an accurate concentration for sending samples for next-generation sequencing on the Illumina HiSeq 2500. Opt for single end sequencing and at least 100 bp reads. *See* **Notes 3** and **5**.

3.5 Analyzing the Next-Generation Sequencing Data (see Note 6)

1. Obtain the sequencing reads from the core or center where the samples were sequenced and transfer them into the L1-seq directory on the server in which all the correctly formatted L1-seq scripts reside. These scripts can be acquired from https://github.com/adamewing/l1seq.
2. Obtain the contents of a database with all reference L1 insertions and polymorphic L1 insertions which have been previously published to use for filtering sequencing data. A database of L1 insertions may be obtained at: http://nar.oxfordjournals.org/content/43/D1/D43 [14].
3. Once the documents are downloaded through the terminal, they must be unzipped. To unzip the fastq.gz files type "gunzip –d FILE_NAME.fastq.gz &" (the & symbol allows the unzipping to run in the background so that you can set all the files and can unzip simultaneously by typing this command for each file in turn.)
4. After the files are unzipped, use the script to run bowtie and create indices for your data. You can execute this process with the command "./run_bowtie.py/whatever_fastq/directions/to/bowtie directions/to/hg19.fa &". For any process which takes more than 10 min, it is helpful to use the screen function by typing "screen –rAad" which will allow for the monitoring of all the processes simultaneously running. It also enables the user to monitor the total memory being used for analysis and the length of time each process has been running.
5. Once the run_bowtie script finishes, run the l1seq.py script as follows: "./l1seq.py –bam whatever.ba, > whatever.l1seq.txt &".
6. Once all of the L1-seq.txt files are made, all the files need to be compressed for sorting. To compress the files, type "bgzip whatever_l1seq.txt &".
7. To sort the files, type "tabix –s 1 –b 3 –e 4 watever.l1seq.txt.gz &".
8. All the files must be compared to one another (e.g. normal compared with tumor, etc.) to do this analysis, type "./compare.py group1_L1seq.txt.gz group2_l1seq.txt.gz group3_l1seq.txt.gz > filename_for_comparisons.tsv &".
9. Finally, after the comparison file has been made, primers must be designed for validation of the data. It is best to run the

makeprimers.pl script on the entire comparison file before looking at the data because the script does not take long to run and having the primer sequences ready to order is very useful. To run this final script, type "./makeprimers.pl filename_for_comparisons.tsv > filename_for_comparisons_with_primers.tsv &". Use sftp to transfer the files back to your local computer if desired.

3.6 Validating the Predicted Insertions from L1-Seq with Site-Specific PCR

1. The presence of nonreference insertions is validated with site-specific PCR (Fig. 2). If the samples are not barcoded (*see* **Note 6**), all samples in a pool must be evaluated for the presence or absence of a predicted insertion. The DNA from each input sample from a pool needs to be at 12.5 ng/μL and 2 μL of DNA used per reaction. The primers will be named by the makeprimers.pl script as "filled site" or FS and "empty site" ES refer to Fig. 2 for orientation of primers with regard to the potential insertion. For each validation to be complete, the FS and L1SP1A2 primer reaction needs to be performed on all samples in the pool from which the prediction came. If comparing two states of the same tissue such as tumor and normal

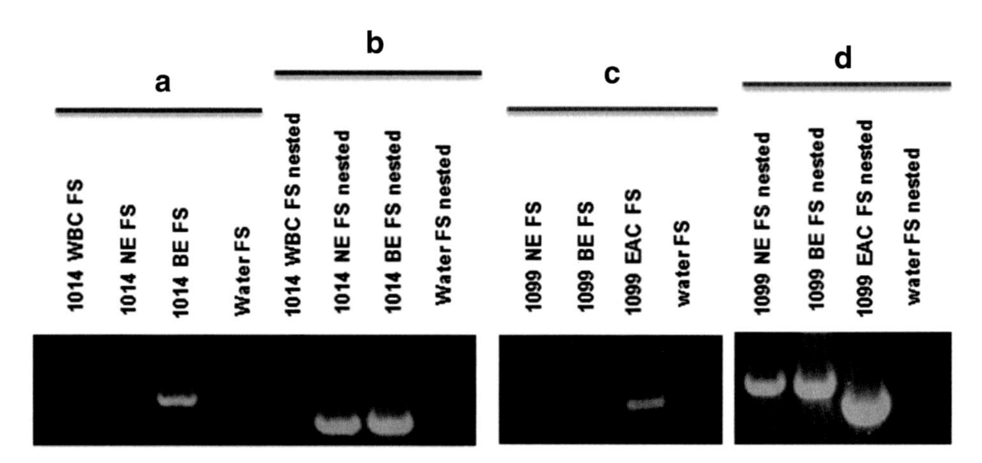

Fig. 2 Validation of results. (**a**) Diagram of the PCR validation scheme for putative insertions, the 3′ end of the LINE-1 insertion is pictured adjacent to a poly A tail. The nested empty site and filled site primers are flanking the empty and filled site primers. In a nested PCR, the nested primers are used in the first of two reactions. One and a half μL of product from the first reaction (with ES and FS primers) are used as template in a second PCR with the nested primers to amplify difficult or rare products. (**b**) Two examples of validations for insertions present in only tumor and absent from normal DNA. On the left, a PCR result depicting both the empty site (ES) and filled site (FS) products for both the normal and tumor DNA samples from patients. Only in the tumor of patient 11 is a filled site band present confirming the insertion is present. In the image on the right side on **b**, a PCR depicting another validation of a somatic insertion present in BE and absent from normal esophageal and white blood cell DNA. There is only a band present in the BE sample for the FS PCR; however, the ES PCR has bands for all three DNA samples as a positive control. (**c**) An insertion sequence with the unique genomic DNA (*blue*), target site duplications (*purple*), LINE-1 sequence (*red*), and the poly-A tail sequence (*orange*) [15]

and the insertion is predicted only in one, the reaction must also be performed on both DNA samples. A control reaction can also be performed with the "empty site" or ES primer and the FS primer. Both the FS and ES primers are genomic and will produce a product of predetermined size in any DNA sample regardless of presence or absence of an insertion.

2. For the FS/L1SP1A2 (filled site PCR) use the following master mix (1×): 12.5 μL of Promega GoTaq green (2×) master mix, 0.8 μL of 20 μM FS primer, 1.6 μL of 20 μM L1SP1A2, 2 μL of genomic DNA (12.5 ng/μL), 8.2 μL of DNA-free H_2O. For the FS/ES primers (empty site PCR) use the following master mix (1×): 12.5 μL of Promega GoTaq green (2×) master mix, 1 μL of 20 μM FS primer, 1 μL of 20 μM ES primer, 2 μL of genomic DNA (12.5 ng/μL), 9.5 μL of DNA-free H_2O. Use the following parameters for the PCR:

3′ L1 Validation PCR

(a) 95 °C for 2 min.

(b) 95 °C for 30 s.

(c) 57 °C for 30 s.

(d) 72 °C for 1 min and 30 s.

(e) Go to **step b** (29×).

(f) 72 °C for 5 min.

(g) 4 °C Hold.

3. Run the PCR products on a 1.5 % TAE gel to resolve the products (*see* **Note 7**). Take images of the gel while it is exposed to UV to visualize the products. Excise fragments which are unique to only one of the samples upon which the PCR was run (Fig. 2b). Isolate the DNA from the band and send for sequencing. If no clear filled site band is uniquely present in one of the samples tested, a nested PCR may be necessary (Fig. 2a). Alternatively, the PCR conditions can be further optimized to attempt to amplify the insertion (*see* **Notes 7** and **8**).

4. Finally, send the DNA for Sanger sequencing to ensure it is the correct product. Sequence the product with both the FS primer as well as the L1SP1A2 primer. When the sequence from the FS is aligned to the genome with BLAT or another alignment algorithm, part of the sequence should align to the genome and a poly T tract should also be visible adjacent to the aligning sequence. For the sequence from the reaction performed with the L1-specific primer, the 3′ end of the L1 should be visible in addition to the poly-A tail (Fig. 2c) (*see* **Notes 7** and **8**).

5. To find the 5′ end of the insertion, several different methods can be utilized. Because many new L1 insertions are truncated

on the 5′ end of the element, it is frequently possible to detect the 5′ end of the element by using the reverse complement of the L1SP1A2 primer (L1 GTG primer) with the ES primer. To do this, make the master mix as follows: mix (1×): 12.5 μL of Promega GoTaq green (2×) master mix, 0.8 μL of 20 μM ES primer, 1.6 μL of 20 μM L1 GTG primer, 2 μL of genomic DNA (12.5 ng/μL), 8.2 μL of DNA-free H_2O. For insertions with a longer 5′ end present, this PCR will likely fail; however, it is possible to tile across the L1 element with various primers at different locations (e.g. L1nt112out) in the element accompanied by the empty site primer to find the 5′ end. For this PCR, use the following master mix (1×): 12.5 μL of Promega GoTaq green (2×) master mix, 0.8 μL of 20 μM ES primer, 1.6 μL of 20 μM L1 internal primer, 2 μL of genomic DNA (12.5 ng/μL), 8.2 μL of DNA-free H_2O.

5′ L1 GTG PCR Parameters

(a) 95 °C for 2 min.
(b) 95 °C for 30 s.
(c) 57.5 °C for 1 min and 30 s.
(d) 72 °C for 3 min.
(e) Go to **step b** (29×).
(f) 72 °C for 5 min.
(g) 4 °C Hold.

5′ L1 Internal Primer (e.g. L1nt112out) PCR Parameters

(a) 95 °C for 2 min.
(b) 95 °C for 30 s.
(c) 57 °C for 30 s.
(d) 72 °C for 45 s.
(e) Go to **step b** (29×).
(f) 72 °C for 5 min.
(g) 4 °C Hold.

4 Notes

1. If very little DNA is available for both library prep and validation PCRs, L1-seq can still be successfully performed. L1-seq has successfully been executed with as little as 25 ng of input per sample for the library prep. For the steps following next-generation sequencing, whole genome amplification can be used (e.g. the Qiagen Repli-G kit) to provide more DNA to use for the validation PCRs. If adjusting the amount of DNA used, be sure to account for volume changes and the concen-

trations of the other reagents to ensure all final concentrations are the same as described in the original technique.

2. If the tissue sample is large enough, more than one tube of tissue slices can be made. Following the sectioning, tissue slices should be stored at −80 °C until it is time to isolate DNA. Embedding tissue in OCT freezing medium is only one way of extracting DNA from frozen tissue.
3. When first performing L1-seq it is prudent to execute a TA cloning step after completing the libraries and mixing them in equimolar ratios, but before completing the end-polishing step. To do this, simply take 1 μL from the mixed libraries and use it in a Topo TA cloning reaction. Follow kit instructions and after growing colonies overnight, select 12 or more from each plate for colony PCR. Following colony PCR, run the product on a gel to be sure the cloning worked effectively, select some or all of the successful clones for Sanger sequencing. When analyzing the Sanger sequencing, look for different L1Ta elements from many different areas of the genome. Essentially, this is a step to check that the library does not consist of amplicons of only a handful of LINE-1 elements in the genome and that elements in the genome are equally represented in the library. This step does not need to be performed for every library prep; however, if a problem occurs with next-generation sequencing, this step could consequently be taken to determine whether or not overrepresentation of a few elements precluded successful sequencing.
4. Occasionally, one of the degenerate primer reactions will not be as robust as the other reactions and when the libraries are run on a gel, the amount of DNA present is variable between reactions. This may not be an issue if there is enough DNA present after the gel purification for the samples to easily be mixed in equal amounts. However, if the concentration of the DNA isolated from the gel purification step is too little to continue without grossly diminishing the amount of total DNA in the combined library, simply repeat the second reaction of L1-seq and combine the isolated DNA from both gel purifications and concentrate the DNA. If the DNAs from the respective degenerate primer reactions are run on the Bioanalyzer and produce very different size distributions of products, it may be necessary to repeat the second L1 PCR again on that DNA sample as well. Ideally, the average product size for each degenerate primer reaction should be within one standard deviation of 350 nucleotides. If the size varies more than one standard deviation from 350, the reaction should be repeated and rerun on a gel. If the size is wrong, it is likely that the excision was initially imprecise.

5. If the DNA being measured at any point in the library prep is at a low concentration and undetectable with the standard Qubit™ broad range kit or the Agilent 1000 DNA chip, there are low concentration versions of these reagents available.
6. Barcoding may also be utilized with this technique; however, results may vary. In 2012, Evrony et al. performed L1-seq using barcoding and were able to validate some new LINE-1 insertions following sequencing analysis. However, other groups have had more difficulty getting the technique to work well and seem to have more success with pooling samples without barcodes. Pooling samples without barcodes does create more work for the validation steps of the technique; however, it seems to have more reproducible results.
7. If validation PCRs are unsuccessful after many attempts, be sure to check the specificity of the primers being used in the amplification. Oftentimes, it is helpful to perform a nested PCR following the first conventional PCR to amplify difficult or low-copy insertions which may have been easily detectable with next-generation sequencing and not with Sanger sequencing. You can nest both the filled site primers as well as the L1Ta-specific primers to increase the specificity of the reaction greatly. Nested PCR along with an increase in cycle numbers and/or altering the melting temperature of the PCR often alleviates validation PCR issues.
8. With regard to choosing predicted insertions for validation, one of two main methods may be employed. A random number generator can be used to select putative somatic insertions for validation which will potentially give a good estimate of the number of true somatic insertions in the data set. Alternatively, putative somatic insertions with unique read counts above 5, map scores of 1, and alignment windows of at least 100 base pairs can be selected for validation. Depending on the validation rate with the primary insertions selected, the level of stringency can be altered until the ideal validation rate is achieved. A validation rate above 60 % is generally acceptable for this technique; however, PCR optimization, good primer design, and good DNA are key to successful validations.

Acknowledgements

This work was funded by a P-50 grant awarded to H.H.K. Jr.

References

1. Voytas DF, Cummings MP, Konieczny A, Ausubel FM, Rodermel SR (1992) *copia*-like retrotransposons are ubiquitous among plants. Proc Natl Acad Sci 89(15):7124–7128
2. Hancks DC, Kazazian HH Jr (2012) Active human retrotransposons: variation and disease. Curr Opin Genet Dev 22(3):191–203
3. Dewannieux M, Esnault C, Heidmann T (2003) LINE-mediated retrotransposition of marked Alu sequences. Nat Genet 35(1): 41–48
4. Ostertag EM, Goodier JL, Zhang Y, Kazazian HH Jr (2003) SVA elements are nonautonomous retrotransposons that cause disease in humans. Am J Hum Genet 73(6):1444–1451
5. Esnault C, Maestre J, Heidmann T (2000) Human LINE retrotransposons generate processed pseudogenes. Nat Genet 24(4):363–367
6. Ewing AD, Kazazian HH Jr (2010) High-throughput sequencing reveals extensive variation in human specific L1 content in individual human genomes. Genome Res 20(9):1262–1270
7. Solyom S, Ewing AD, Rahrmann EP, Doucet T, Nelson HH, Burns MB, Harris RS, Sigmon DF, Casella A, Erlanger B, Wheelan S, Upton KR, Shukla R, Faulkner GJ, Largaespada DA, Kazazian HH Jr (2012) Extensive somatic L1 retrotransposition in colorectal tumors. Genome Res 22(12):2328–2338
8. Lee E, Iskow R, Yang L, Gokcumen O, Haseley P, Luquette LJ, Lohr JG, Harris CC, Ding L, Wilson RK, Wheeler DA, Gibbs RA, Kucherlapati R, Lee C, Kharchenko PV, Park PJ (2012) Landscape of somatic retrotransposition in human cancers. Science 337(6097): 967–971
9. Shukla R, Upton KR, Muñoz-Lopez M, Gerhardt DJ, Fisher ME, Nguyen T, Brennan PM, Baillie JK, Collino A, Ghisletti S, Sinha S, Iannelli F, Radaelli E, Dos Santos A, Rapoud D, Guettier C, Samuel D, Natoli G, Carninci P, Ciccarelli FD, Garcia-Perez JL, Faivre J, Faulkner GJ (2013) Endogenous retrotransposition activates oncogenic pathways in hepatocellular carcinoma. Cell 153(1):101–111
10. Helman E, Lawrence MS, Stewart C, Sougnez C, Getz G, Meyerson M (2014) Somatic retrotransposition in human cancer revealed by whole-genome and exome sequencing. Genome Res 24(7):1053–1063
11. Baillie JK, Barnett MW, Upton KR, Gerhardt DJ, Richmond TA, De Sapio F, Brennan PM, Rizzu P, Smith S, Fell M, Talbot RT, Gustincich S, Freeman TC, Mattick JS, Hume DA, Heutink P, Carninci P, Jeddeloh JA, Faulkner GJ (2011) Somatic retrotransposition alters the genetic landscape of the human brain. Nature 479(7374):534–537
12. Evrony GD, Cai X, Lee E, Hills LB, Elhosary PC, Lehmann HS, Parker JJ, Atabay KD, Gilmore EC, Poduri A, Park PJ, Walsh CA (2012) Single-neuron sequencing analysis of L1 retrotransposition and somatic mutation in the human brain. Cell 151(3):483–496
13. Tubio JM, Li Y, Ju YS, Martincorena I et al (2014) Mobile DNA in cancer. Extensive transduction of nonrepetitive DNA mediated by L1 retrotransposition in cancer genomes. Science 345(6196):1251343
14. Mir AA, Philippe C, Cristofari G (2014) euL1db: the European database of L1HS retrotransposon insertions in humans. Nucleic Acids Res 43(Database issue):D43–D47
15. Doucet-O'Hare T, Rodic N, Sharma R, Darbari I, Abril G, Choi JA, Young Ahn J, Cheng Y, Anders RA, Burns KH, Meltzer SJ, Kazazian HH Jr. (2015) LINE-1 expression and retrotransposition in Barrett's esophagus and esophageal carcinoma. Proc. Natl. Acad Sci. 112(35): 4894–4900

Chapter 6

Combining Amplification Typing of L1 Active Subfamilies (ATLAS) with High-Throughput Sequencing

Raheleh Rahbari and Richard M. Badge

Abstract

With the advent of new generations of high-throughput sequencing technologies, the catalog of human genome variants created by retrotransposon activity is expanding rapidly. However, despite these advances in describing L1 diversity and the fact that L1 must retrotranspose in the germline or prior to germline partitioning to be evolutionarily successful, direct assessment of de novo L1 retrotransposition in the germline or early embryogenesis has not been achieved for endogenous L1 elements. A direct study of de novo L1 retrotransposition into susceptible loci within sperm DNA (Freeman et al., Hum Mutat 32(8):978–988, 2011) suggested that the rate of L1 retrotransposition in the germline is much lower than previously estimated (<1 in 400 individuals versus 1 in 9 individuals (Kazazian, Nat Genet 22(2):130, 1999). Based on these revised estimates of the L1 retrotransposition rate, we modified the ATLAS L1 display technique (Badge et al., Am J Hum Genet 72(4):823–838, 2003) to investigate de novo L1 retrotransposition in human genomes. In this chapter, we describe how we combined a high-coverage ATLAS variant with high-throughput sequencing, achieving 11–25× sequence depth per single amplicon, to study L1 retrotransposition in whole genome amplified (WGA) DNAs.

Key words Non-LTR retrotransposon, Retroelement, LINE-1, L1, Retrotransposition, Polymorphism

1 Introduction

In 2003, Brouha et al. suggested that although there are 90 full-length L1s with intact ORFs in the reference human genome, only six "hot-L1s" are responsible for the majority (86 %) of the total L1 retrotransposition activity [4]. Since then, studies of human-specific L1 retrotransposition using different approaches such as comparative bioinformatics analyses, or transposon display techniques such as ATLAS ([2, 3, 4]; reviewed in [5]) have revealed many more active elements segregating in human populations. However, it has proven very difficult to find de novo L1 retrotransposition events, largely due to the low copy number of active L1s, the low frequency of such events ([1]), and the lack of high-resolution and high-coverage molecular genomic techniques. Recently, high-throughput

Jose L. Garcia-Pérez (ed.), *Transposons and Retrotransposons: Methods and Protocols*, Methods in Molecular Biology, vol. 1400, DOI 10.1007/978-1-4939-3372-3_6, © Springer Science+Business Media New York 2016

sequencing approaches using, as well as array-based systems have begun enable high-throughput analysis of L1s as genomic structural variants. Currently, there are several approaches to study L1 retrotransposition in the genome. One set of approaches are PCR-based transposon display techniques used to characterize polymorphic human L1s in the human genome [3]. Generally, these techniques rely on the selective amplification of groups of retrotransposon sequences based on diagnostic nucleotide polymorphisms specific for each subfamily (e.g., the trinucleotide sequence ACA at positions 5954–6 of the reference element L1.3 (Accession L19088) discriminates the Ta subfamily from older subfamilies). In these methods, selective PCR is applied to gDNA libraries, and amplicons that show presence/absence variation between individuals are isolated for characterization by sequencing.

One such PCR-based display technique used to study active and polymorphic L1s is ATLAS [3]. Using the ATLAS technique, Badge et al. (2003) identified seven novel full-length L1 insertions, of which three were classified as "hot" (highly active) L1s in cell culture retrotransposition assays [3]. The ATLAS technique has several advantages compared to other transposon amplification techniques as it can analyze L1s at many loci genome-wide, and most importantly it can target the polymorphic, and so likely active, subset of L1 elements. Moreover, this technique is versatile, allowing researcher to adapt it to study different aspects of L1 biology; for example the Transduction Specific variant (TS-ATLAS) allows specific lineages of active elements that share a 3′ transduction, to be studied exhaustively [6]. In addition, as library production only involves the ligation of appropriate linkers, covalent epigenetic modifications, such as cytosine methylation are preserved, making it possible to study these modifications at L1 promoters genome-wide: differential digestion with methylation sensitive restriction enzymes during library preparation reveals methylation status by comparison of the pattern of insertions amplified relative to undigested gDNA libraries (Rahbari et al., unpublished). In the current chapter, we present a high-resolution, high-coverage ATLAS variant, which combined with Roche 454 sequencing platform, enables the analysis of near full-length L1 retrotransposon insertions. In this new approach, genomic coverage is improved to ~80 % of the genome, by using the *NIa*III restriction enzyme for library construction. Since higher genome coverage increases the complexity of the amplicon distribution, it becomes unlikely that different insertions can be distinguished by amplicon size alone, necessary to characterize single-molecule L1 retrotransposon insertions. However, we overcome this problem by combining high-coverage ATLAS variant with high-throughput sequencing. This technique not only identifies polymorphic L1s in the genome, but also ables to identify very rare de novo L1 retrotransposition insertions. This technique is able to recover single molecules carrying

L1 sequences using small pools of diluted DNA representing 10–20 sperm genomes (unpublished data). A full description of materials and methods for conducting this technique is detailed below. Finally, we conclude this chapter with some notes on high-throughput data analysis.

2 Materials

2.1 Chemical Reagents and Laboratory Equipment

All chemicals were supplied by local suppliers.

2.2 Restriction Enzymes

New England Biolabs supplied the restriction enzyme, *NIa*III. *Taq* DNA polymerase and T4 DNA ligase can be obtained from other commercial sources.

2.3 Molecular Weight Markers

50 bp, 100 bp, and 1 kb molecular weight markers were supplied by local suppliers.

2.4 Standard Solutions

Southern blot solutions (denaturing and neutralizing), 20× Sodium Chloride Sodium-Citrate (SSC) buffer and 10× Tris-borate/EDTA (TBE) electrophoresis buffer, were made following standard recipes.

Other solutions and buffers are listed below:

1. 1× PCR buffer: 45 mM Tris pH 8.8, 11 mM NH_4SO_4, 4.5 mM $MgCl_2$, 6.7 mM β-mercaptoethanol, 113 μg/ml BSA, and 1 mM dNTPs.

2.5 Oligonucleotides

DNA oligonucleotides were synthesized and HPLC-purified by local suppliers.

3 Methods

All steps of the ATLAS procedure were performed in a Class II laminar flow hood that had been decontaminated by UV exposure for at least 30 min prior to use. All reagents were PCR clean (i.e. opened only in the Class II hood and used only for PCR). For all buffers 18 MΩ water was used, either from a laboratory distillation unit or supplied by Sigma-Aldrich.

Below we describe the methodology for high-coverage ATLAS, starting with genomic DNA preparation for ATLAS library construction, followed by the generation of amplicons for next-generation sequencing. The schematic diagram (Fig. 1) summarizes all the steps of library preparation, amplification, sequencing, and data analysis.

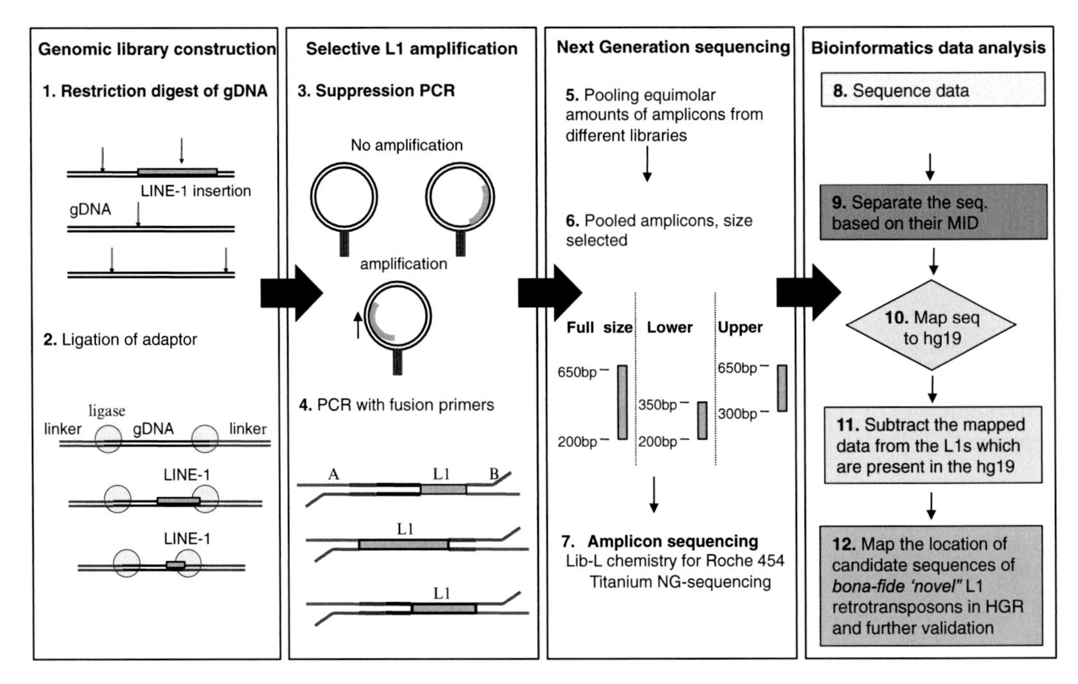

Fig. 1 Summary of the steps involved in modified ATLAS combined with NGS to isolate bona fide de novo LINE-1 insertions from human WGA gDNA

3.1 Genomic DNA Purification

Standard procedures can be applied to purify genomic DNA. If required, and in order to retain molecules evidencing de novo L1 retrotransposition events for further validation by PCR, we recommend performing WGA on the input genomic DNA. The WGA method described by Spits et al. [7] can be used without any additional modifications. Following WGA, samples with a DNA concentration of higher than 3 ng/μl and whose negative controls for WGA are lower than 1 ng/μl should be selected. A Nanodrop or a spectrophotometer can be used for DNA quantification. Next, selected samples are equally divided between three Eppendorf microcentrifuge tubes (preferably DNA LoBind), and an additional round of WGA should be performed on each. By subsequently pooling these three independent WGA reactions allele dropout, resulting from biased amplification in the initial WGA reaction, can be minimized.

3.2 ATLAS Library Construction

200 ng of WGA DNA or genomic DNA is incubated with *NIa*III for 3 h at 37 °C in a final reaction volume of 20 μl (enzyme concentration 20U). Controls should be included in the digestion step including: DNA negative (DNA replaced by H_2O); digestion enzyme negative (enzyme replaced by 50 % glycerol); and a DNA positive. Following digestion, inactivate the enzyme by heating the reaction at 65 °C for 20 min. Note that after this step the digested DNA can be aliquot into smaller volumes and stored at −80 °C until required (Fig. 1, part 1).

3.3 ATLAS Linker Preparation and Ligation

Prepare the linkers by mixing an equal volume of each linker primer RRNBOT2:5′-ACTGGTCTAGAGGGTTAGGTTCCTGCTACAT CTCCAGCCTCATG-3′ and RRNDUP1: 5′-AGGCTGGAGATG TAGCAG-3′ at a concentration of 50 μmol. Denature and anneal the mixed linkers by heating to 65 °C for 10 min and cooling to room temperature at the rate of 1 °C every 15 s (Fig. 1, part 2). Note that in the standard ATLAS protocol [3], 100 ng of the digested DNA is ligated to a 40-fold molar excess of the annealed suppression linker. This amount of linker is calculated by assuming the enzyme completely digests the genome into "X" number of fragments with two ligatable ends, and 3 pg of DNA represents one haploid genome equivalent. "X" varies with respect to the enzyme's cutting frequency (but all calculations are necessarily approximate). For *NIa*III, 2.7 μl of the 25 μmol annealed linker was used for each ligation, in a final volume of 20 μl.

Next ligate 100 ng of genomic DNA with the annealed linker overnight at 15 °C, in a final reaction volume of 20 μl. Linker negative (H_2O), and enzyme negative (50 % glycerol) and reaction positives should be included as controls for the ligation step. The 20 μl ligation reaction final volume should consist of: 100 ng digested DNA, 2.7 μl annealed linker, 1.34 μl (4 Weiss units) T4 ligase, 2 μl 10× ligase buffer, and 8.96 μl H_2O.

Inactivate the ligation by incubating the reaction at 70 °C for 10 min. Remove the excess linkers and short gDNA fragments (<100 bp) using the Qiaquick PCR purification system (Qiagen), according to the manufacturer's instructions. Elute the purified DNA in PCR clean 5 mM Tris-pH 7.5 to a final volume of 30 μl. Note at this stage the ligated DNA can be aliquot into three volumes of 10 μl and stored at −80 °C (Fig. 1, part 2).

3.4 ATLAS Primary PCR

This stage involves a standard suppression PCR. A 15 μl final reaction volume, consisting of 13.5 μl PCR mix and 1.5 μl of constructed library DNA, is assembled under PCR clean conditions (Fig. 1, part 3). The PCR mix comprises 1× PCR buffer, 0.5 μl of 50 μM RVECPA1 (L1-specific linker primer): RVECPA1 5′ ACTGGTCTAGAGGGTTAGG 3′ and RV5SB2 (L1 5′ UTR internal primer) RV5SA2 5′-ATGGAAATGCAGAAATCACCG T-3′, 0.4 units of *Taq* polymerase.

Perform PCR using the following cycle conditions: an initial denaturing step at 96 °C for 30 s, followed by 25 cycles of 96 °C for 30 s, 62 °C for 2 min, and then extension at 72 °C for 10 min.

3.5 ATLAS Secondary PCR

The rationale for fusion primer design for the secondary PCR (Subheading 3.5.1) is explained below. The design is based on the assumption that the PCR products are to be sequenced using the Roche 454 platform, incorporating the 454-specific adapters by PCR. However, the sequencing adaptors can be modified based on the preferred sequencing platform. In the following, we have explained the procedure to conduct ATLAS secondary PCR (Subheading 3.5.2) upstream of 454 sequencing using the GS FLX Titanium chemistry.

3.5.1 Fusion Primers Design

For this technique, the amplicon length needs to be given careful consideration. We selected the Roche 454 technology to achieve long read lengths but other technologies such as Illumina and PacBio could be considered. To extract maximum information, the amplicons need to be fully sequenced to enable their accurate mapping to the reference genome, and the determination of their structure. The estimated sequence read length of the 454 Sequencing System with the GS FLX Titanium chemistry used here is about 450 nucleotides, but some of the amplicons generated by the full-length L1-specific suppression PCR are bigger than 450 bp (up to 750 bp). As a result 450 bp bidirectional reads were used to ensure sufficient overlap in the middle of the larger amplicon sequences, such that long amplicons will be reliably covered.

We allowed up to ~50 nucleotides from each end of the amplicon for the fusion primers. For example, for a read to cover an amplicon entirely, it must traverse its key (4 nucleotides) at the proximal end and the template-specific primers (20 nt) and MID (Multiplex IDentifier), sequences (10 nucleotides each) at both ends.

Each pair of the fusion primers consisted of forward and reverse primers. All the forward fusion primers (5′ to 3′ direction) were constructed of the following segments: Roche-LibL-primer A: 5′-CGTATCGCCTCCCTCGCGCCATCAG-3′, a 10 nucleotide MID, and a linker-specific primer, RVECPA2 5′-CCTGCTACAT CTCCAGCC-3′. All the reverse fusion primers were constructed from the following segments: Roche-LibL-primer B: 5′-CTATGC GCCTTGCCAGCCCGCTCAG-3′, a 10 nucleotide MID and an L1-specific primer RV5SB2 5′-CTTCTGCGTCGCTCACGCT-3′. The schematic diagram in Fig. 2 indicates the primers orientation. The use of 10 bp MIDs is optional. Shorter MIDs can enable unambiguous de-covolution of multiplexed experiments, but 10 bp MIDs enable higher levels of multiplexing, with less chance of read miss-assignment.

3.5.2 ATLAS Secondary PCR

Perform the secondary PCR in a 50 μl final reaction volume consisting of 45 μl PCR mix and 5 μl of the primary PCR reaction (Fig. 1, part 4). The PCR mixture comprises 1× PCR buffer, 0.125 μM fusion primer A (containing a linker-specific primer) and 0.125 μM fusion primer B (containing an L1 internal primer), and 0.4 units of *Taq* polymerase. Use the following thermocycling conditions: an initial denaturing step at 96 °C for 30 s; followed by 25 cycles of 96 °C for 30 s, 75 °C for 2 min; followed by an extension step at 72 °C for 10 min. Figure 3 shows an example of ATLAS secondary PCR products, fractionated by agarose gel electrophoresis.

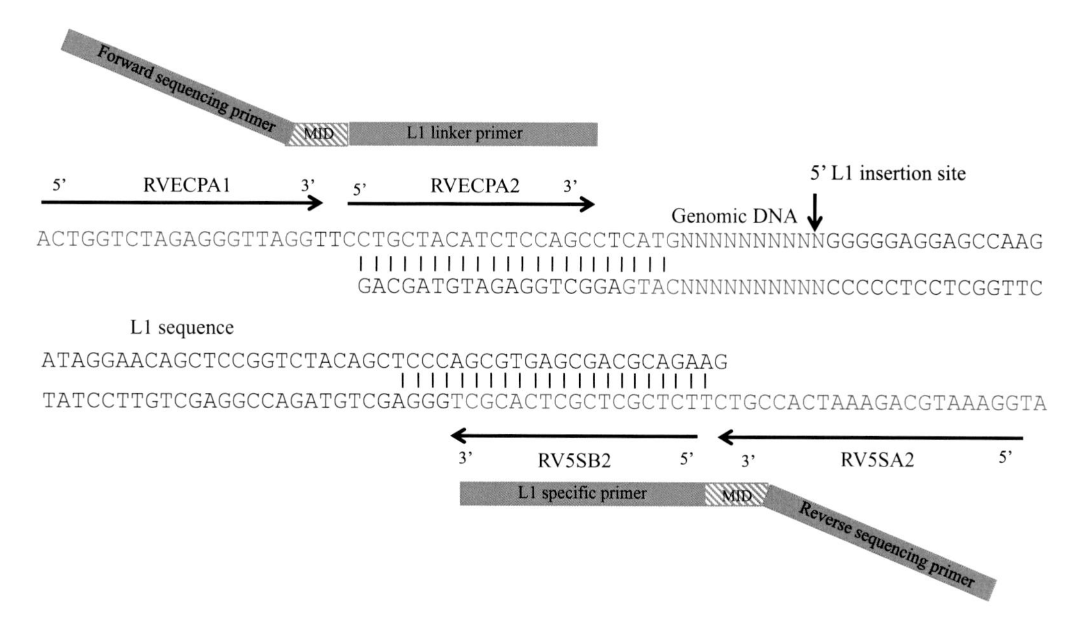

Fig. 2 Schematic diagram of the arrangement of suppression PCR L1-specific, linker-specific, and sequencing fusion primers relative to an example full-length L1 insertion site

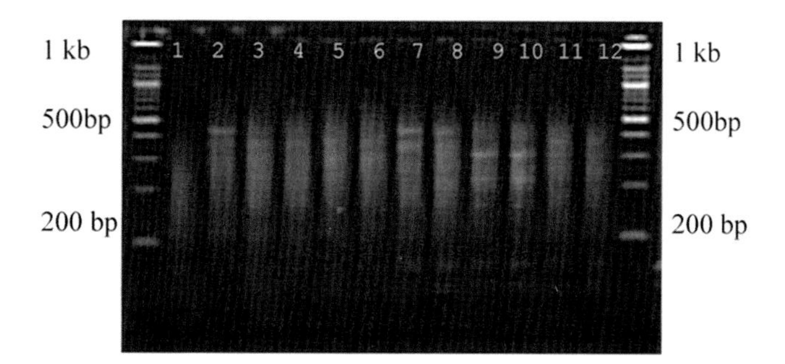

Fig. 3 An example of ATLAS secondary PCR products (RVECPA2 + RV5SB2) (2 μl) from 12 individuals (*Nla*III libraries) separated on a 1.5 % agarose gel (100 bp molecular weight markers, NEB). The secondary products range from 200 to 750 bp in length, including the 454 fusion primers

3.6 Pooling the Barcoded Amplicons

Prior to sequencing, an equimolar concentration of the secondary PCR products from each library should be pooled together. Note that in order to obtain an equal coverage from all the barcoded samples, it is important to quantify all the secondary PCR products accurately prior to the pooling. Methods for accurately quantifying the range of amplicons are picogreen analysis [8] as well as using an Agilent Bioanalyzer (Invitrogen) (Fig. 1, part 5).

3.7 Size Selection of the Pooled Samples

The ATLAS secondary products contain a range of PCR products with variable lengths (200–750 bp). To minimize length biasing during the emulsion phase PCR (emPCR), we recommend dividing pooled libraries into two size-fractionated batches, with different ranges of amplicon length (Fig. 1, part 6). Each batch can be sequenced separately on physically separate regions of a picotiter plate (Roche 454) or different lanes for Hiseq. One batch (pool number two) contains smaller amplicons ranging from 200 to 350 bp and the other batch contains the longer length amplicons ranging from 300 to 750 bp. The lower and upper range products were made by fractionating the pooled secondary PCR products on an agarose gel and extracting the lower range sizes >350 bp and the upper range products <300 bp. A 50 bp size range overlap was allowed between the upper and lower ranges to avoid losing products at the size fraction junction (300–350 bp).

To achieve fractionation, load two equal volumes (100 μl) of the pooled samples on a 2.5 % agarose gel and run at 120 V for 2 h. Transfer the gel onto a Dark Reader visible light transilluminator (Clare Chemical Research) and cut gel blocks to divide the pooled products into two different size ranges: 200–350 bp and 300–750 bp. Extract DNA from the gel blocks using the Qiaquick gel extraction system (Qiagen) according to the manufacturer's instructions. Elute the purified DNA in 30 μl of PCR clean 5 mM Tris-pH 7.5. The purified amplicons can be aliquot for sequencing and remaining aliquots stored at −80 °C. To verify that the pooled samples are size-selected correctly, run 1 μl of each batch on an Agilent Bioanalyser (high sensitivity DNA kit, Invitrogen). An example of size-selected libraries is shown in Fig. 4.

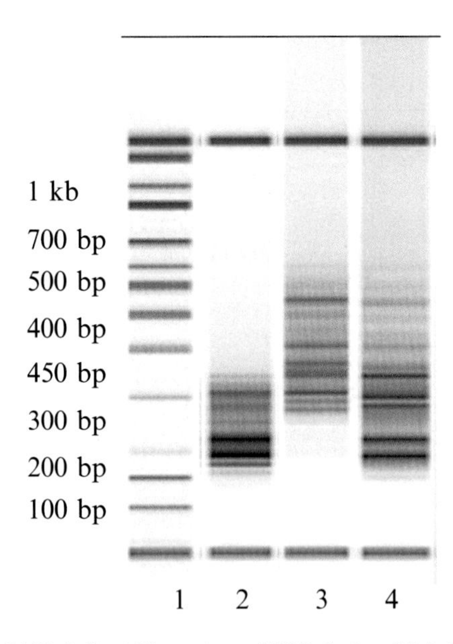

Fig. 4 Use of the 2100 Agilent Bioanalyser 2100 device (High Sensitivity DNA kit, Invitrogen) to check the size selection of the pooled amplicons at different size ranges. Lane 1: Marker, lane 2: lower range product size (200–350 bp), lane 3: upper range product size (300–750 bp), lane 6: No-size selected pooled libraries (control); product size ranges from 200 to 750 bp

3.8 Sequence Coverage Calculation

Prior to sequencing, the required sequence coverage should be calculated to be sure to generate enough reads to confidently characterize single-molecule events. To calculate this, we mapped the L1Hs Ta-specific oligonucleotides used in the primary suppression PCR to the human reference genome resulting in the identification of ~3000 discrete potential priming sites. Data from exhaustive fosmid sequencing studies [9] enable an estimate of the number of novel (i.e. not previously characterized) L1s per screened genome, as between 4 and 6 insertions. Since this is a small fraction of the ~3000 oligo binding sites shared by the majority of human genomes (determined by in silico mapping), failing to account for these in the coverage estimates will only result in a very small overestimation. By contrast, the proportion of polymorphic L1 Ta elements (in any genome) is about 30 % [10], making ~3000 L1 amplicons per average genome a substantial overestimate (as many insertions will be absent from a given genome). By this logic, our simplifying assumptions can only lead to an underestimate of the coverage required. In the current protocol, the *NIa*III restriction enzyme is used to construct the genomic DNA library. Knowing (from in silico digestion Badge and Rouillard, personal communication) that about 80 % of the human genome is within 1 kb of a *NIa*III site, the number of accessible L1 loci for this experiment would be 80 % of ~3000, i.e. ~2400 L1 loci, assuming a random distribution of L1 insertion loci and restriction sites. Based on this knowledge and the given sequencing coverage for the experiment, it is possible to calculate the minimum expected number of reads per each de novo L1 retrotransposition. For example, we used ¾ of a picotiter plate and the number of beads per quarter plate should be around ~160,000 (according to the manufacturer's data). Thus the total number of the reads expected from all three regions is $160{,}000 \times 3 = 480{,}000$ reads. When fourteen libraries are sequenced, the number of expected reads per amplicon would be $480{,}000/3000 \times 14$, or ~11 reads per amplicon. Therefore, it has been estimated that for a single molecule present in one library we should detect about 11 reads, but the coverage would be much higher for constitutive L1 loci present in all libraries. This calculation is very useful in adjusting the data analysis filters (Subheading 4).

3.9 Post-sequencing Data Analysis (see Note 1)

Initially, the bulk reads in a FASTA or FASTQ format should be separated according to their MIDs. Following this, it is recommended to either trim off the sequences introduced by PCR (linkers, fusion primers) from both sides except for the sequence of the L1-specific primer (RV5BS2), which should be retained in the sequence structure.

Following trimming, map the sequences to the human genome reference. While various alignment algorithms can be used, one option is to use the public instance of the Galaxy web service [11]: http://main.g2.bx.psu.edu/ and the included LastZ tool to map the FASTA files [11]. Next, the coordinates of the mapped reads should be compared to the coordinates of the L1 oligo data set. The L1 oligo data set can be generated, by mapping the L1-specific primer (RV5BS2) to the reference genome, also using LastZ.

Reads whose coordinates overlap with the L1 oligo data set correspond to the subset of L1 loci that are already present in the reference genome; to select candidate de novo/novel L1 insertion in the remaining sequences, the coordinates of the candidate novel insertion should be checked against the reported polymorphic L1 insertions in published data (previously characterized, but that are absent from the reference genome [12]). Figure 5 shows an example of an L1 sequence isolated from ATLAS libraries, which is absent from the reference genome. However, it has been previously characterized as a polymorphic insertion. We recommend to check the latest published L1 data sources, as euL1DB [12] and others. Further filters such as minimum and maximum number of reads required from an amplicon can be set, depending on the sequencing experiment design.

3.10 Characterizing Candidate Novel and De Novo L1-Mediated Retrotransposition Events Using Site-Specific PCR

Having identified a subset of reads which are confirmed to be absent from the human genome reference sequence and all the available L1 databases, these reads can be proposed as candidate novel L1 retrotransposons, and subjected to independent validation.

The presence of non-reference insertions should be verified via site-specific PCR. In this procedure the 5′ end and flanking regions of non-reference L1s can be amplified using an L1-specific primer and a primer specific for the 5′ flanking genomic region. Amplification of the "empty" site, using primers specific for the 5′ and 3′ flanking genomic regions from gDNA of an individual unrelated to the sequenced donor can be used to verify the PCR primers function, as well as the ability to amplify the insertion region. Amplification of the 5′ and 3′ ends of the insertion enables characterization of the insertion as a canonical (full-length, carrying Target Site Duplications and terminating in a 3′ poly A-tail) insertion. Long-range PCR of the entire insertion and characterization by direct sequencing from WGA gDNA, even from single cells, is feasible in our hands.

4 Notes

1. Post-sequencing data analysis to identify de novo L1 insertion from the ATLAS data can be done in many different ways, using different software platforms.

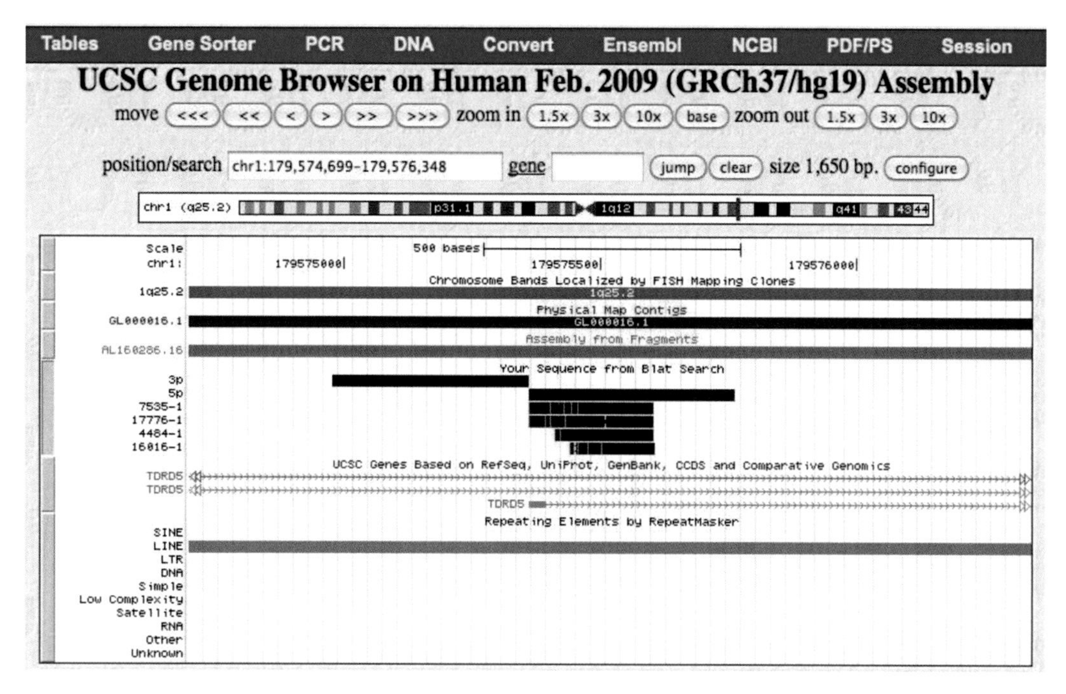

Fig. 5 ATLAS and high-throughput sequencing can capture known polymorphic L1 insertions. The screen shot shows the result of a BLAT search using 454 traces (*numbered black rectangles*) that co-locate with the 5′ flanking DNA (black rectangle labeled "5p") of a known polymorphic L1 element that is absent from hg19. This novel L1 insertion was previously reported

References

1. Freeman P, Macfarlane C, Collier P, Jeffreys AJ, Badge RM (2011) L1 hybridization enrichment: a method for directly accessing de novo L1 insertions in the human germline. Hum Mutat 32(8):978–988. doi:10.1002/humu.21533
2. Kazazian HH Jr (1999) An estimated frequency of endogenous insertional mutations in humans. Nat Genet 22(2):130
3. Badge RM, Alisch RS, Moran JV (2003) ATLAS: a system to selectively identify human-specific L1 insertions. Am J Hum Genet 72(4):823–838
4. Brouha B, Schustak J, Badge RM, Lutz-Prigge S, Farley AH, Moran JV, Kazazian HH Jr (2003) Hot L1s account for the bulk of retrotransposition in the human population. Proc Natl Acad Sci U S A 100(9):5280–5285
5. Beck CR, Garcia-Perez JL, Badge RM, Moran JV (2011) LINE-1 elements in structural variation and disease. Annu Rev Genomics Hum Genet 12:187–215. doi:10.1146/annurev-genom-082509-141802
6. Macfarlane CM, Collier P, Rahbari R, Beck CR, Wagstaff JF, Igoe S, Moran JV, Badge RM (2013) Transduction-specific ATLAS reveals a cohort of highly active L1 retrotransposons in human populations. Hum Mutat 34(7):974–985. doi:10.1002/humu.22327
7. Spits C, Le Caignec C, De Rycke M, Van Haute L, Van Steirteghem A, Liebaers I, Sermon K (2006) Whole-genome multiple displacement amplification from single cells. Nat Protoc 1(4):1965–1970. doi:10.1038/nprot.2006.326
8. Ahn SJ, Costa J, Emanuel JR (1996) PicoGreen quantitation of DNA: effective evaluation of samples pre- or post-PCR. Nucleic Acids Res 24(13):2623–2625
9. Beck CR, Collier P, Macfarlane C, Malig M, Kidd JM, Eichler EE, Badge RM, Moran JV (2010) LINE-1 retrotransposition activity in human genomes. Cell 141(7):1159–1170
10. Boissinot S, Chevret P, Furano AV (2000) L1 (LINE-1) retrotransposon evolution and amplification in recent human history. Mol Biol Evol 17(6):915–928

11. Taylor J, Schenck I, Blankenberg D, Nekrutenko A (2007) Using galaxy to perform large-scale interactive data analyses. Curr Protoc Bioinformatics Chapter 10:Unit 10.15. doi:10.1002/0471250953.bi1005s19
12. Mir AA, Philippe C, Cristofari G (2015) euL1db: the European database of L1HS retrotransposon insertions in humans. Nucleic Acids Res 43(Database issue):D43–D47. doi:10.1093/nar/gku1043

Chapter 7

RNA-Seq Analysis to Measure the Expression of SINE Retroelements

Ángel Carlos Román, Antonio Morales-Hernández, and Pedro M. Fernández-Salguero

Abstract

The intrinsic features of retroelements, like their repetitive nature and disseminated presence in their host genomes, demand the use of advanced methodologies for their bioinformatic and functional study. The short length of SINE (short interspersed elements) retrotransposons makes such analyses even more complex. Next-generation sequencing (NGS) technologies are currently one of the most widely used tools to characterize the whole repertoire of gene expression in a specific tissue. In this chapter, we will review the molecular and computational methods needed to perform NGS analyses on SINE elements. We will also describe new methods of potential interest for researchers studying repetitive elements. We intend to outline the general ideas behind the computational analyses of NGS data obtained from SINE elements, and to stimulate other scientists to expand our current knowledge on SINE biology using RNA-seq and other NGS tools.

Key words SINE, Retrotransposon, RNA-seq, Next-generation sequencing, Bioinformatics

1 Introduction

Roughly 40 % of the human genome is composed of retrotransposons, including SINE (short interspersed elements), LINE (long interspersed elements), and LTR (long terminal repeats) elements [1]. These elements can increase their copy number in the host genome by a transposition mechanism that requires its own transcription to generate an intermediary RNA molecule that will be ultimately integrated into a different genomic location [2]. This amplification process might provoke important changes in the stability and function of the genome due to structural alterations and increased recombination and to the addition of genetic variability. Moreover, recent studies have shown that the expression of the genes in which those elements are located can be affected by several means including the activation of transcriptional enhancers or silencers [3, 4] and/or the generation of small noncoding RNAs (ncRNAs) involved in pre-mRNA processing [5, 6].

Jose L. Garcia-Pérez (ed.), *Transposons and Retrotransposons: Methods and Protocols*, Methods in Molecular Biology, vol. 1400, DOI 10.1007/978-1-4939-3372-3_7,

From an evolutionary perspective, it is generally accepted that the retroelements have contributed to the remodeling of the human transcriptional landscape by adding thousands of novel regulatory elements in the Primate lineage [7, 8].

SINEs are a group of retroelements between 100 and 500 base pairs (bp) in size [9]. An important subgroup within SINEs has been originated by the amplification of the 7SL RNA (signal recognition particle RNA) [10] and includes human Alus [11, 12] and murine B1 and B2 elements [13]. The RNA Pol III complex [14], which recognizes specific DNA sequences known as A- and B-boxes by the transcription factors III-C (TFIIIC) and III-B (TFIIIB) and the catalytic subunit RPC32, normally transcribes these SINEs. Transcription starts upstream the A-box and continues along the SINE sequence until the presence of a stop signal composed by a repetition of at least four thymine residues [15]. As opposed to LINEs, SINEs do not code for any protein, and so their mobilization relies on the LINE machinery. For instance, it is known that SINE insertion sites are directed by LINE endonucleases [16], and that human Alus end in a poly-A sequence recognized by the LINE-1 reverse transcriptase [17].

Most copies of SINEs in the genome have been genetically inactivated by mutations that neutralize their promoter function, and only a small subgroup of those elements maintains its transcriptional capacity. For that, SINE-derived transcripts are generally hardly detected or even undetected in somatic tissues. An exception takes place during the maturation of spermatogonia and oocytes and in early stage embryo development, in which SINE transcription is allowed. Actually, germ cells can be depicted as a battlefield between retrotransposons and the host genome, in which novel DNA insertions, potentially deleterious, might be transmitted to a new generation. Other important players repressing unscheduled SINE transcription and amplification are histone and chromatin modifiers acting through reorganization of the local chromatin [18, 19]. Finally, small RNAs like siRNAs and piRNAs, also actively expressed by mammalian germ cells, are important regulators of the retrotransposon silencing process that occurs during gametogenesis [20]. Interestingly, many of these siRNAs and piRNAs seem to be originated from repetitive sequences. SINE transcription may produce double-stranded RNA molecules (dsRNAs) with a secondary structure resulting from intramolecular folding. These dsRNAs might be detected and processed by endonucleases using a mechanism similar to that employed by DICER to generate siRNAs [21]. Supporting this hypothesis, a decrease in DICER expression leads to an increase in Alu transcript levels in retinal pigmented epithelial cells and DICER can degrade Alu-derived dsRNAs in vitro [22].

In summary, the regulation and control of SINE transcription is likely a key process to preserve the physiology and homeostasis of specific tissues and organs (Fig. 1). This implies that the analysis

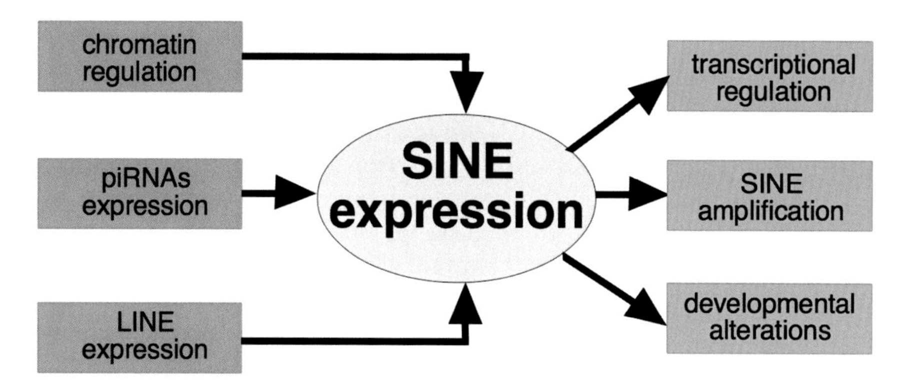

Fig. 1 Schematic representation of the importance of transcription of SINE elements to biological processes

of SINE expression is relevant to understand the mechanisms by which these elements affect cell functioning. SINE transcription can be measured by several available methods.

In vitro techniques, such as in vitro transcription or retrotransposition colony formation assays, are useful for the dissection of general regulatory pathways involving a specific family of SINEs [17]. However, these methods do not give enough mechanistic insight if we are interested in studying the transcription of specific SINE instances along the genome or in detecting novel SINE amplification events during development. Quantitative real-time PCR protocols can be designed to perform these specific analyses; however, NGS tools need to be applied for whole genome analysis of SINEs-derived transcription, where information involving millions of sequence reads need to be processed. Although a great effort has been invested in recent years to develop these techniques, there are still important limitations that make difficult the application of NGS to the study of retrotransposon expression. Particularly relevant are the reduced length and the repetitive nature of SINE elements, two properties that jeopardize the identification and functional validation of single SINE instances regulating cell functions. In this chapter, we propose and discuss several methods to overcome these constraints using an RNA-seq analytic protocol specifically designed for SINE detection.

2 Materials

The main equipment for RNA-seq analysis of SINE expression consists in one of the commercially available pyrosequencers for NGS. Illumina sequencers ranging in sequencing power from the MiSeq to the more potent HiSeq 2500 allow the reading of millions of sequences with maximum read lengths of 2×300 bp. Other technical alternatives like the Applied Biosystems SOLiD system, IonTorrent

PGM, or the Roche-454 sequencers offer similar capabilities with differences in efficiency and/or price.

Additional equipment necessary for the sequencing of SINE-derived small RNA transcripts includes both a DNA electrophoretic system and a NanoDrop spectrophotometer, which are used for size fractionation and for quality assessment of SINE RNA transcripts. In addition, to isolate and to purify small RNAs from tissues and cell cultures, several commercial products can be used such as the QIAGEN miRNeasy kit. A real-time thermocycler can be also used for the validation of the results generated by the RNA-seq analysis.

Finally, very important items needed for this protocol are the hardware and software to analyze the raw data obtained from the sequencer. Our recommendation is to use a dedicated computer with at least 8 GB RAM, 1 TB of hard disk and running a distribution of Linux as Operating System. Software utilized in this protocol comprises Perl, MATLAB, SamTools, BWA, and Blast. Other tools like Python or R can be alternatives to some of the latter.

3 Methods

3.1 Preparation of RNA Extracts from Cell Cultures

1. 5×10^6 cells cultured in 100 mm plates are used as input material. *See* **Note 1** for additional comments about the use of fresh or frozen tissues as input as well as other commentaries regarding RNA extraction.
2. RNA extraction is performed with the QIAGEN miRNeasy kit. This method allows the isolation of total RNA molecules with a size above 18 nucleotides.
3. The quality of purified RNA is assessed by spectrophotometric (NanoDrop, 260/280 ratio of ~2) and electrophoretic (RNA Integrity Number, RIN > 8, using Agilent RIN software [23]) methods.

3.2 Sequencing SINE RNAs

1. The following steps are normally done in a next-generation sequencing service. *See* **Note 2** for additional comments on the process.
2. RNA is converted into cDNA using random hexamers, and the second strand is synthesized using Illumina TruSeq protocols.
3. Proprietary Illumina adaptors (120 nucleotides in length) are ligated to the RNA sequences. This process prepares the RNA fragments to be sequenced in an Illumina machine. It also allows the multiplexing capability of the RNA-seq to share a single Illumina run for several distinct samples.
4. The samples are separated electrophoretically in agarose gels and the fragments of interest are recovered and purified using

commercially available products like the QIAGEN Gel Extraction kit, taking into account that the fragments are 120 nucleotide longer due to the addition of the Illumina adaptors. As an example, fragments from 180 to 400 nucleotides can be excised and purified for Alu-derived RNA transcripts.

5. After the libraries have been prepared, RNA-seq can be performed using different approaches like the 1×75 runs in a Next-Generation Sequencer Illumina MiSeq.

3.3 Pre-processing the Raw Sequence Data

1. The raw data from a NGS service usually consist into separated FASTQ files for each sample used in the experiment. These sequence files lack adaptor sequences and usually contain several million reads. For that, files are big in size and the standard procedure to download them to our personal computer or server is through the File Transferring Protocol (FTP) from the sequencing service. *See* **Note 3** for additional comments about pre-processing of data.

2. In some cases, FASTQ files must be converted into FASTA format. FASTA is an standard in bioinformatic studies, and it is necessary for subsequent analyses using Blast or other additional tools. There are several methods to convert FASTQ in FASTA; one command in a Perl/Linux environment is:

```
cat /path/to/file.fastq | perl -e '$i=0;while(<>){if(/^\@/ && $i==0){s/^\@/\>/;print;}elsif($i==1){print;$i=-3}$i++;}'> path/to/new/
file.fasta
```

3. For certain analyses, FASTQ reads will have to be placed into the positions of a particular genome. Some tools that will be of help for the process are the BWA and SamTools. To do that, we first index a reference genome or a transcriptome:

```
bwa index -c /path/to/genome.fasta
```

We then align our reads to the genome:

```
bwa aln -c /path/to/genome.fasta path/to/file.fastq >
   /path/to/align.sai
```

Transform the binary into a more readable SAM file:

```
bwa samse -c path/to/genome.fasta path/to/file.fastq
   /path/to/align.sai /path/to/file.fastq > /path/to/
   file.sam
```

This SAM file will be sorted and indexed:

```
samtools -sU /path/to/file.sam -o /path/to/file.bam
samtools -sort /path/to/file.bam
samtools -index /path/to/file.sorted.bam
```

To generate a final file that will be file.sorted.bam.bai, small enough to be copied and transferred using regular methods such as e-mail.

3.4 Analysis of the Processed Data

1. Now that the files of our sequenced samples have been processed, we can use them to infer novel data from such results (Fig. 2). We will briefly explain two different protocols intended to obtain the expression profiles of specific SINE families, but there are other potentially new methods that can benefit from variations of these protocols. In **Note 4**, we underline some of the updates and comment tips for the computational analysis of SINE RNA-seq.
2. Quantification of SINE RNAs families using aligned reads and RSEM or eXpress. The amount of SINE RNA can be measured with a program like RSEM [24]. In this case, a crucial step is the selection of the reference transcriptome FASTA file (*see* **step 4** in Subheading 3.3). A general transcriptome FASTA dataset obtained from a database like Ensembl can be used, or you can prepare a custom-made file. This is an important issue because if a FASTA file is created with, for example, a general representation of SINE families, a quantification of the expression of these elements can be performed. In addition, the aligning process in the BWA align command can be adjusted to permit sequence mismatches (with *-M* and *-n* parameters). Finally, the parameter (*-o* 0) should be used in the same command to avoid gaps in the alignment, as RSEM do not allow

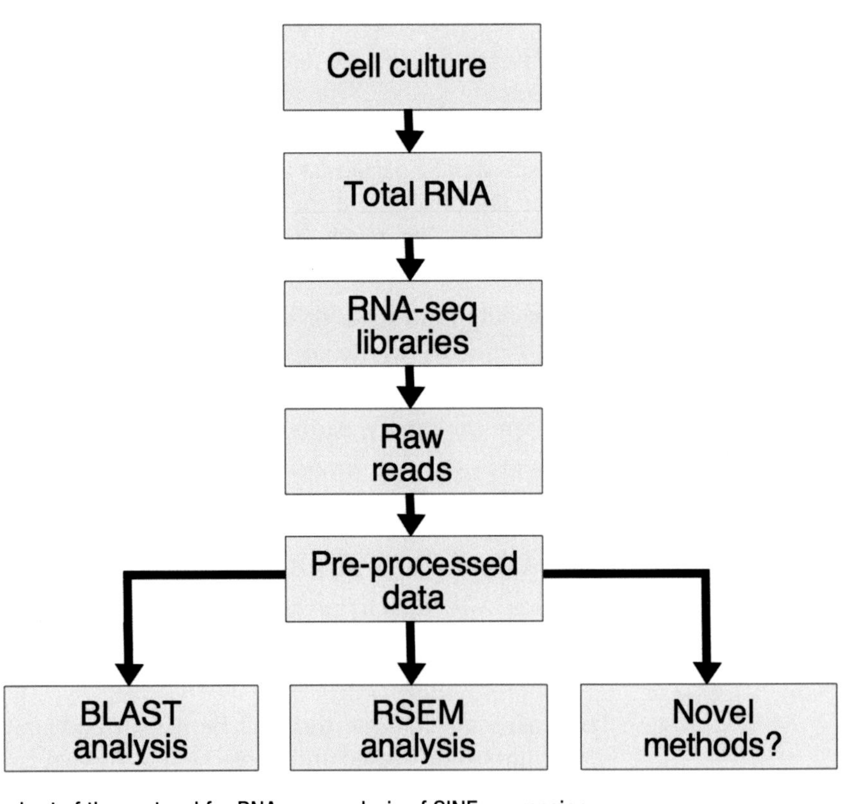

Fig. 2 Flowchart of the protocol for RNA-seq analysis of SINE expression

for gaps in the sequences. In this way, two different approaches can be used: (1) few SINE families in the reference transcriptome with flexibility in read alignment, or (2) more detailed SINE subfamilies in the reference transcriptome with additional post-processing to summarize the values in the families. An example command is:

```
rsem-calculate-expression -bam --no-bam-output /
path/to/file.sorted.bam.
bai /path/to/transcriptome.fa name_of_sample
```

Important resulting files from the analysis are "name_of_sample.genes.results" and "name_of_sample.isoforms.results". These are tab-separated files easy to use for post-processing and for their analysis in programs like R or MATLAB. These programs have several parameters to indicate the expression level of each gene/isoform, among them, FPKM (Fragments per Kilobase of gene/isoform per Million Reads). If the expression of SINE families between different samples are to be compared, ad hoc standard post tests for genome-wide studies can be used if the number of elements in the transcriptome is high enough. In addition, the usage of permutation approaches for the generation of a p-value quantifying the probability of a significant difference in SINE expression is also recommended. In brief, that method consists in permuting the result values in such a way that each value is assigned to a sample. As a result, data for differences in expression are again obtained. If this process is repeated *N* times (e.g. 10,000 times), the number of instances in which the difference in expression in permutations is equal or higher than the experimental results can be determined, and *P* will be calculated dividing this number by *N*. There are many other approaches to quantify the expression of RNA-seq data such as the eXpress software [25]. This is similar to RSEM, and thus produces a FPKM expression measure for genes and isoforms. A common advantage of these software tools is that they can be run under not only Linux but also Windows and Mac operating systems.

3. Quantification of SINE expression using raw FASTA reads. Another possibility for the analysis of SINE RNA expression consists in the direct use of the raw reads in FASTA format to infer new data of SINE expression levels. In this case, a simple shell script can be prepared to run BLAST under Linux. BLAST [26] is a classic bioinformatic tool which detects database sequences similar to the query. Its main advantages with respect to other more recent programs are its simplicity and flexibility. Although initially designed for evolutionary purposes, BLAST can be customized far enough to be adapted for other needs. For example, some useful commands could be the following:

```
makeblastdb -in /path/to/file.fasta -dbtype nucl
```

```
makeblastdb -in /path/to/query.fasta -dbtype nucl
blastn -db ./path/to/file.fasta -query /path/to/query.
  fasta -outfmt "7 qacc qstart qend evalue" -out /path/
  to/results.txt
```

The first two commands index both the sequence and the query databases. That is necessary for BLAST, and it is done only once. The second command searches the database of our raw reads with a list of query sequences (also in FASTA format). The results obtained will be a tabulated list of BLAST outputs (one for each query sequence) concatenated, with the Query accession, the Query Start, the Query End, and the E-value for each positively identified target. Again, as in Subheading 2 above, prediction can be modulated using gap and mismatch penalties, in order to maintain sensitivity without decreasing specificity. These tabulated files are fitted for statistical post-processing with programs like R and MATLAB. In addition to assessing expression differences between samples or SINEs, interesting data about the relative position of transcripts within a SINE can be obtained with this analysis. For instance, we can study if a SINE is expressed producing similar levels of different transcripts along the element, or if there are transcriptional peaks originated from specific regions of the element. Those possible outcomes might point to the existence of different isoforms of SINE transcripts, like in the case of human Alus that can produce small cytoplasmic and full Alu-derived sequences [27]. As in the prior protocol, standard or custom statistical tests are required to evaluate the significance of the results.

4 Notes

1. Fresh or frozen tissues can be also processed for RNA extraction; commercial RNA purification kits fit specific requirements for RNA isolation. We recommend the use of homogenizers such as motor-driven grinders for soft tissues and an IKA Ultra-Turrax apparatus for harder or more difficult tissues. Life Technologies' RNA later or similar alternatives for RNA stabilization are highly recommended for the analysis of frozen tissues.
2. The steps indicated in Subheading 3.2 are normally performed in a next-generation sequencing facility. Nevertheless, it is very useful to know and to compare the technical specifications of several sequencing platforms to understand which one is the best for our specific experiments. The analysis of small RNAs from SINE elements is currently not very common in most of these facilities. For those experiments, it is particularly important to establish a constant flow of information between technical assistants and scientists in order to select the best steps to

be followed regarding library generation, adapters, and sequencing protocol.

3. There are additional computational tools to pre-process the data. We select the set formed by Perl/BWA/Samtools because of their wide implementation. New users of the Linux environment will normally experience some difficulties and, in that context, previous information from other users deposited in Internet could be of great help. We therefore strongly recommend the use of a general Linux distribution, like Fedora or Ubuntu, for similar reasons. Finally, we also advise to learn the basics of shell scripting to be able to save, comment, and finally reuse the commands that were previously run. These will also warrantee the standardization of the protocols (similar to molecular ones), a rapid error detection capability and the potential to modify methods, altogether saving much time.
4. The quantification of SINE expression needs to be carefully assessed for the presence of errors. The nature of SINE elements (small, repetitive) makes them prone to errors. We suggest a few controls for the proposed analyses to avoid the misdetection of SINEs. The more precise is the detection, the higher is the probability of errors. In this context, to quantify the expression of a single SINE instance in the genome, control sequences with little modifications should be analyzed in parallel. For example, another instance of the same subfamily or the general sequence of its subfamily could be used. Then, the results obtained can be compared with the previous ones got with your original query to assess if there were due to a subfamily effect. Even when we suggest two different methods for RNA-seq analysis of SINE expression, there are other approaches that can be used for similar targeted studies. Modified versions of these protocols can be adapted to the analysis of, for example, the detection of SINE amplification in specific cellular conditions. Aligning our reads to a reference genome could allow us to find novel neighboring regions to SINE retrotransposons.

References

1. Deininger PL, Batzer MA (2002) Mammalian retroelements. Genome Res 12:1455–1465
2. Bennett EA, Keller H, Mills RE et al (2008) Active Alu retrotransposons in the human genome. Genome Res 18:1875–1883
3. Roman AC, Benitez DA, Carvajal-Gonzalez JM et al (2008) Genome-wide B1 retrotransposon binds the transcription factors dioxin receptor and Slug and regulates gene expression in vivo. Proc Natl Acad Sci U S A 105: 1632–1637
4. Wang T, Zeng J, Lowe CB et al (2007) Species-specific endogenous retroviruses shape the transcriptional network of the human tumor suppressor protein p53. Proc Natl Acad Sci U S A 104:18613–18618
5. Borchert GM, Holton NW, Williams JD et al (2011) Comprehensive analysis of microRNA

genomic loci identifies pervasive repetitive-element origins. Mob Genet Elements 1:8–17
6. Gu TJ, Yi X, Zhao XW et al (2009) Alu-directed transcriptional regulation of some novel miRNAs. BMC Genomics 10:563
7. Cowley M, Oakey RJ (2013) Transposable elements re-wire and fine-tune the transcriptome. PLoS Genet 9, e1003234
8. Jacques PE, Jeyakani J, Bourque G (2013) The majority of primate-specific regulatory sequences are derived from transposable elements. PLoS Genet 9, e1003504
9. Singer MF (1982) SINEs and LINEs: highly repeated short and long interspersed sequences in mammalian genomes. Cell 28:433–434
10. Weiner AM (1980) An abundant cytoplasmic 7S RNA is complementary to the dominant interspersed middle repetitive DNA sequence family in the human genome. Cell 22:209–218
11. Deininger PL, Jolly DJ, Rubin CM et al (1981) Base sequence studies of 300 nucleotide renatured repeated human DNA clones. J Mol Biol 151:17–33
12. Rubin CM, Houck CM, Deininger PL et al (1980) Partial nucleotide sequence of the 300-nucleotide interspersed repeated human DNA sequences. Nature 284:372–374
13. Kramerov DA, Grigoryan AA, Ryskov AP et al (1979) Long double-stranded sequences (dsRNA-B) of nuclear pre-mRNA consist of a few highly abundant classes of sequences: evidence from DNA cloning experiments. Nucleic Acids Res 6:697–713
14. Okada N (1991) SINEs. Curr Opin Genet Dev 1:498–504
15. Geiduschek EP, Kassavetis GA (2001) The RNA polymerase III transcription apparatus. J Mol Biol 310:1–26
16. Jurka J (1997) Sequence patterns indicate an enzymatic involvement in integration of mammalian retroposons. Proc Natl Acad Sci U S A 94:1872–1877
17. Dewannieux M, Esnault C, Heidmann T (2003) LINE-mediated retrotransposition of marked Alu sequences. Nat Genet 35:41–48
18. Adeniyi-Jones S, Zasloff M (1985) Transcription, processing and nuclear transport of a B1 Alu RNA species complementary to an intron of the murine alpha-fetoprotein gene. Nature 317:81–84
19. Ichiyanagi K, Li Y, Watanabe T et al (2011) Locus- and domain-dependent control of DNA methylation at mouse B1 retrotransposons during male germ cell development. Genome Res 21:2058–2066
20. Watanabe T, Totoki Y, Toyoda A et al (2008) Endogenous siRNAs from naturally formed dsRNAs regulate transcripts in mouse oocytes. Nature 453:539–543
21. Kim W, Benhamed M, Servet C et al (2009) Histone acetyltransferase GCN5 interferes with the miRNA pathway in Arabidopsis. Cell Res 19:899–909
22. Kaneko H, Dridi S, Tarallo V et al (2011) DICER1 deficit induces Alu RNA toxicity in age-related macular degeneration. Nature 471:325–330
23. Schroeder A, Mueller O, Stocker S et al (2006) The RIN: an RNA integrity number for assigning integrity values to RNA measurements. BMC Mol Biol 7:3
24. Li B, Dewey CN (2011) RSEM: accurate transcript quantification from RNA-seq data with or without a reference genome. BMC Bioinformatics 12:323
25. Roberts A, Pachter L (2013) Streaming fragment assignment for real-time analysis of sequencing experiments. Nat Methods 10:71–73
26. Altschul SF, Gish W, Miller W et al (1990) Basic local alignment search tool. J Mol Biol 215:403–410
27. Maraia RJ, Driscoll CT, Bilyeu T et al (1993) Multiple dispersed loci produce small cytoplasmic Alu RNA. Mol Cell Biol 13:4233–4241

Chapter 8

Qualitative and Quantitative Assays of Transposition and Homologous Recombination of the Retrotransposon Tf1 in *Schizosaccharomyces pombe*

Maya Sangesland, Angela Atwood-Moore, Sudhir K. Rai, and Henry L. Levin

Abstract

Transposition and homologous recombination assays are valuable genetic tools to measure the production and integration of cDNA from the long terminal repeat (LTR) retrotransposon Tf1 in the fission yeast (*Schizosaccharomyces pombe*). Here we describe two genetic assays, one that measures the transposition activity of Tf1 by monitoring the mobility of a drug resistance marked Tf1 element expressed from a multicopy plasmid and another assay that measures homologous recombination between Tf1 cDNA and the expression plasmid. While the transposition assay measures insertion of full-length Tf1 cDNA mediated by the transposon integrase, the homologous recombination assay measures levels of cDNA present in the nucleus and is independent of integrase activity. Combined, these assays can be used to systematically screen large collections of strains to identify mutations that specifically inhibit the integration step in the retroelement life cycle. Such mutations can be identified because they reduce transposition activity but nevertheless have wild-type frequencies of homologous recombination. Qualitative assays of yeast patches on agar plates detect large defects in integration and recombination, while the quantitative approach provides a precise method of determining integration and recombination frequencies.

Key words Tf1, *Schizosaccharomyces pombe*, Fission yeast, Transposition assay, Homologous recombination assay, Quantitative assay, LTR-retrotransposon

1 Introduction

The long terminal repeat (LTR) retrotransposon Tf1 isolated from *Schizosaccharomyces pombe* encodes four proteins: Gag, protease (PR), reverse transcriptase (RT), and integrase (IN), and coding sequences are flanked by two LTRs (*see* Fig. 1a). Tf1 mobilizes in a manner similar to retroviruses, through particle formation and reverse transcription followed by integration of the cDNA into the genome of the host cell. As integration into new genomic locations can have deleterious effects on the host by damaging genetic content, determining the targeting mechanism of these elements is of

Jose L. Garcia-Pérez (ed.), *Transposons and Retrotransposons: Methods and Protocols*, Methods in Molecular Biology, vol. 1400, DOI 10.1007/978-1-4939-3372-3_8,

significant importance [1, 2]. Previous studies have shown that Tf1 preferentially targets promoters of RNA pol II-transcribed genes, and notably insertions are observed at high levels within promoters of stress response genes. However, it remains largely unclear which host factors play a role in the observed integration preference [3–8]. Transposition and homologous recombination assays therefore provide a useful genetic tool to examine these questions and to analyze the role of host factors in mediating transposon integration. These genetic assays function by monitoring integration events into new genomic locations and by measuring homologous recombination of Tf1 cDNA with plasmid sequences [9]. The transposition and recombination activities of Tf1 are measured by expressing a plasmid-encoded copy of Tf1 under the control of a heterologous promoter (*nmt1*) which is activated in the absence of thiamine. The Tf1 sequence in the expression plasmid contains a drug resistance gene in the opposite orientation with respect to the transposon. This drug resistance gene is interrupted by an artificial intron that due to its orientation is spliced out of the Tf1 transcript but not from the transcript of the resistance gene (*see* Fig. 1).

Cells with integration and recombination events are selected through the expression of the drug resistance marker [9–12]. The qualitative assays of yeast patched on solid media provide a general indication of the transposition and recombination activities in the cell, whereas the quantitative assays provide a highly reproducible measure of cDNA integration and recombination frequencies. The genetic assays outlined here provide a particularly useful genetic tool to study transposon biology.

2 Materials

Here we present details for two genetic assays for Tf1 containing either the neomycin (Neo) or nourseothricin (Nat) drug selection markers. *S. pombe* can be grown on either complex or synthetic minimal media depending on the needs of the experiment. Complex yeast extract with supplements (YES) medium is made from yeast extract and includes all amino acids, including uracil and adenine supplements. Minimal medium is used to select for auxotrophic markers, and we recommend the use of the pombe glutamate medium (PMG) as it provides high transposition activity and is compatible for selection of the Neo marker by geneticin (G418) resistance. G418 selection is also effective in YES media, and selection of Nat resistance is equally effective in both complex and minimal media. All media and stock solutions should be filter-sterilized (liquid) or autoclaved (liquid and agar solutions) prior to use for only 15 min in order to avoid glucose caramelization [13]. Solid medium is made by the addition of an equal volume of 4 % (w/v) Difco Bacto Agar to 2× concentrated liquid media. Sterile liquid media can be kept for months at room temperature; plates

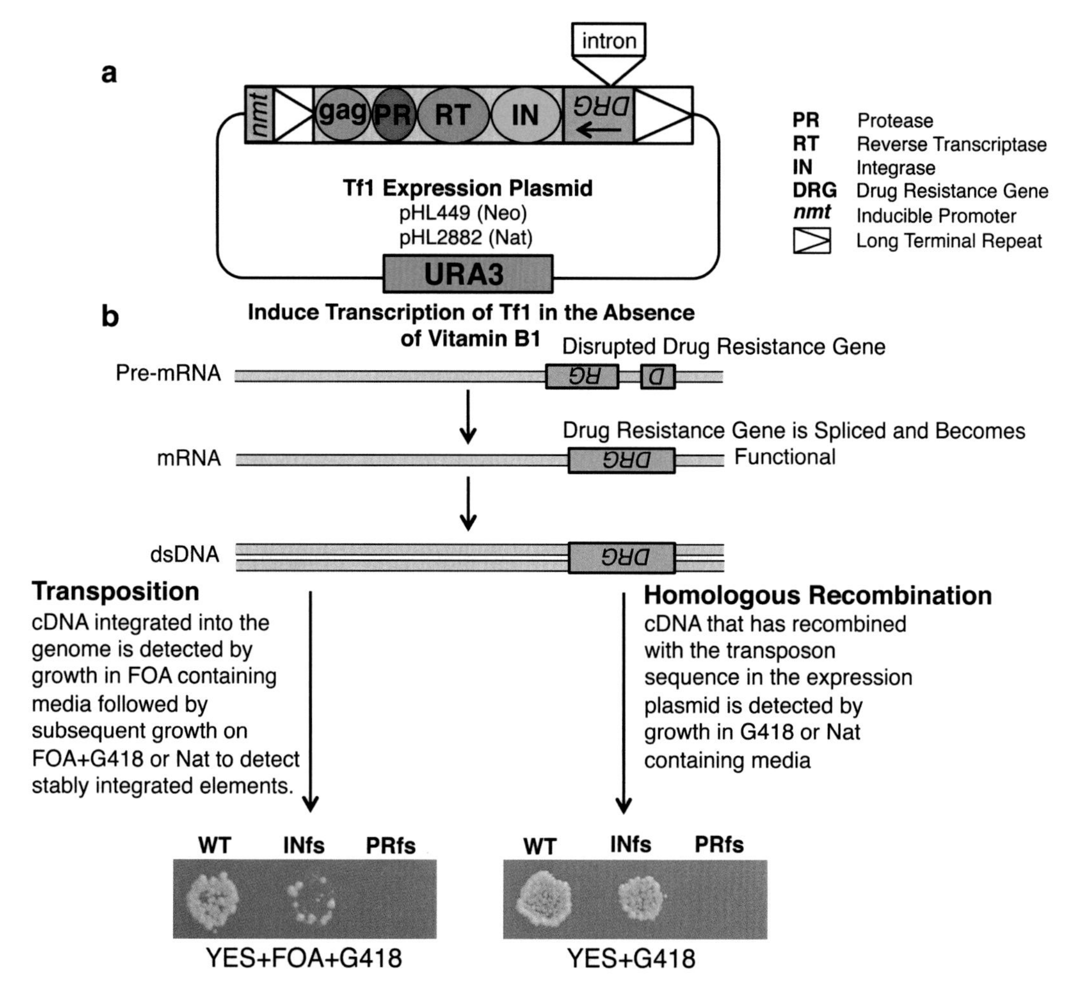

Fig. 1 Genetic assays for transposition and homologous recombination. (**a**) Tf1, encoding the proteins Gag, PR, RT, and IN, is expressed on a plasmid that is transformed into *S. pombe* cells. Tf1 expression is under the control of an inducible promoter, *nmt1*. The expression vector also contains the selectable marker *URA3*. (**b**) In the absence of vitamin B1, the Tf1 transcript is expressed and forms a functional drug resistance gene following splicing of the artificial intron. Transposition and recombination activity can be detected as patches on the appropriate selection media. Wild-type (WT) Tf1 generates confluent growth; IN frameshift (INfs) displays significantly reduced growth in the transposition assay (*left panel*) and approximately twofold less grown in the recombination assay (*right panel*). The PR frameshift (PRfs) lacks growth on selection media for both assays

can be stored when wrapped, to protect against drying for up to 4 months at 4 °C.

2.1 Preparation of Fission Yeast Media

1. YES media (per liter): 5 g yeast extract, 30 g glucose, and 2 g complete dropout powder (*see* **Note 1**).
2. PMG media (per liter): 3 g potassium hydrogen phthalate [14.7 mM], 2.2 g Na_2PO_4 [15.5 mM], 3.75 g l-glutamic acid, monosodium salt, 20 g glucose (2 % w/v), 20 mL of 50× salt

stock, 1 mL of 1000× vitamin stock, 0.1 mL of 10,000× mineral stock, 2 g of dropout powder lacking uracil and leucine (*see* **Note 1**).

3. YES media containing 5-fluoroorotic acid or 5-FOA (YES + 5-FOA): To YES medium add (to a final concentration) 1 mg/mL 5-FOA (US Biologicals) (*see* **Note 2**).
4. YES media containing NAT (YES + NAT): To YES medium, add (to a final concentration) 100 μg/mL nourseothricin (Jena Bioscience).
5. YES media containing G418 (YES + G418): To YES medium, add (to a final concentration) 500 μg/mL G418 (*see* **Note 3**).
6. YES media containing 5-FOA and nourseothricin (YES + 5-FOA + NAT): To YES medium add (to a final concentration) 1 mg/mL of 5-FOA and 100 μg/mL nourseothricin (*see* **Note 2**).
7. YES media containing 5-FOA and G418 (YES + 5-FOA + G418): To YES medium add (to a final concentration) 1 mg/mL of 5-FOA and 500 μg/mL G418 (*see* **Notes 2** and **3**).
8. PMG containing supplements, 5-FOA and vitamin B1 (PMG + U + L + 5-FOA + B1): To PMG medium add (to a final concentration) 250 μg/mL l-leucine, 50 μg/mL uracil, 10 μM vitamin B1, and 1 mg/mL 5-FOA (*see* **Note 4**).
9. PMG containing supplements leucine and vitamin B1 (PMG-U + L + B1): To PMG medium, add (to a final concentration) 250 μg/mL l-leucine, and 10 μM vitamin B1 (*see* **Note 5**).
10. PMG containing the supplement leucine (PMG-U + L-B1): To PMG medium, add (to a final concentration) 250 μg/mL l-leucine (*see* **Note 6**).
11. PMG media containing the supplements leucine, uracil, and vitamin B1 (PMG + U + L + B1): To PMG medium, add (to a final concentration) 250 μg/mL l-leucine, 250 μg/mL uracil, and 10 μM vitamin B1.

2.2 Preparation of Stock Solutions/ Powder

1. Complete dropout powder: 5 g adenine SO_4 and 2 g each of the following amino acids: alanine, arginine HCl, aspartic acid, asparagine H_2O, cysteine $HCl{\cdot}H_2O$, glutamic acid, glutamine, glycine, histidine $HCl{\cdot}H_2O$, isoleucine, leucine, lysine HCl, methionine, phenylalanine, proline, serine, threonine, tryptophan, tyrosine, uracil, and valine. Mix thoroughly (*see* **Note 7**). For addition to PMG media, exclude uracil and leucine components as these will be supplemented or expressed by exogenous plasmid markers (*see* **Note 1**).
2. 50× salt stock (per liter): 2 g Na_2SO_4 [14.1 mM], 50 g KCL [0.67 M], 0.735 g $CaCl_2{\cdot}2H_2O$ [4.99 mM], and 52.5 g

$MgCl_2{\cdot}6H_2O$ [0.26 M]. Dissolve in deionized water and autoclave.

3. 1000× vitamin stock (per liter): 1 g pantothenic acid [4.20 mM], 10 g nicotinic acid [81.2 mM], 10 g inositol [55.5 mM], and 10 mg biotin [40.8 μM]. Dissolve in deionized water and filter sterilize.
4. 10,000× mineral stock (per liter): 5 g boric acid [80.9 mM], 4 g $MnSO_4$ [23.7 mM], 4 g $ZnSO_4{\cdot}7H_2O$ [13.9 mM], 2 g $FeCl_2{\cdot}6H_2O$ [7.40 mM], 0.4 g molybdic acid [2.47 mM], 1 g KI [6.02 mM], 0.4 g $CuSO_4{\cdot}5H_2O$ [1.60 mM], and 10 g citric acid [47.6 mM]. Dissolve in deionized water and filter sterilize.

2.3 Plasmids and Strains

1. The following plasmids are transformed into *S. pombe* to assay the NeoAI marked version of Tf1. Wild-type Tf1: pHL449-1, Tf1 with IN frameshift: pHL476-3, and Tf1 with PR frameshift: pHL490-80.
2. The following plasmids are transformed into *S. pombe* to assay the NatAI marked version of Tf1. Wild-type Tf1: pHL2882 Tf1 with IN frameshift: pHL2884, and Tf1 with PR frameshift: pHL2883.
3. Strains of *S. pombe* we commonly use are YHL912 (*h–*, *ura4-294*, *leu1-32*) and YHL1101 (*h+*, *ura4-D18*, *leu1-32*, *ade6-M210*). Strains used with NatAI-expressing Tf1 were derived from the Bioneer collection (*h+*, *ura4-D18*, *leu1-32*, *ade6-M210* or *ade6-M216*). All of these strains were derived from Leupold's original isolates of *S. pombe* and therefore lack endogenous copies of Tf1.

3 Methods

Genetic assays for transposition and recombination can be conducted as a qualitative assay with yeast patched on solid media (*see* Figs. 1b and 2a) and as a quantitative assay in liquid culture (*see* Fig. 2b, c). All *S. pombe* strains should be grown at 32 °C.

3.1 Qualitative Transposition Assay (Fig. 2a)

1. Freshly revive experimental strains (four independent transformants of each Tf1 plasmid are tested with Tf1 controls containing frameshift mutations) from frozen perm stocks for 3 days on PMG-U + L + B1 solid media (*see* **Note 8**).
2. Establish a master plate containing all experimental strains and controls by removing a small portion of the cell mass with a sterile toothpick from the thick part of the streak and spotting a pinhead-sized amount onto solid PMG-U + L + B1 media (*see* **Note 9**). Incubate for 3 days.

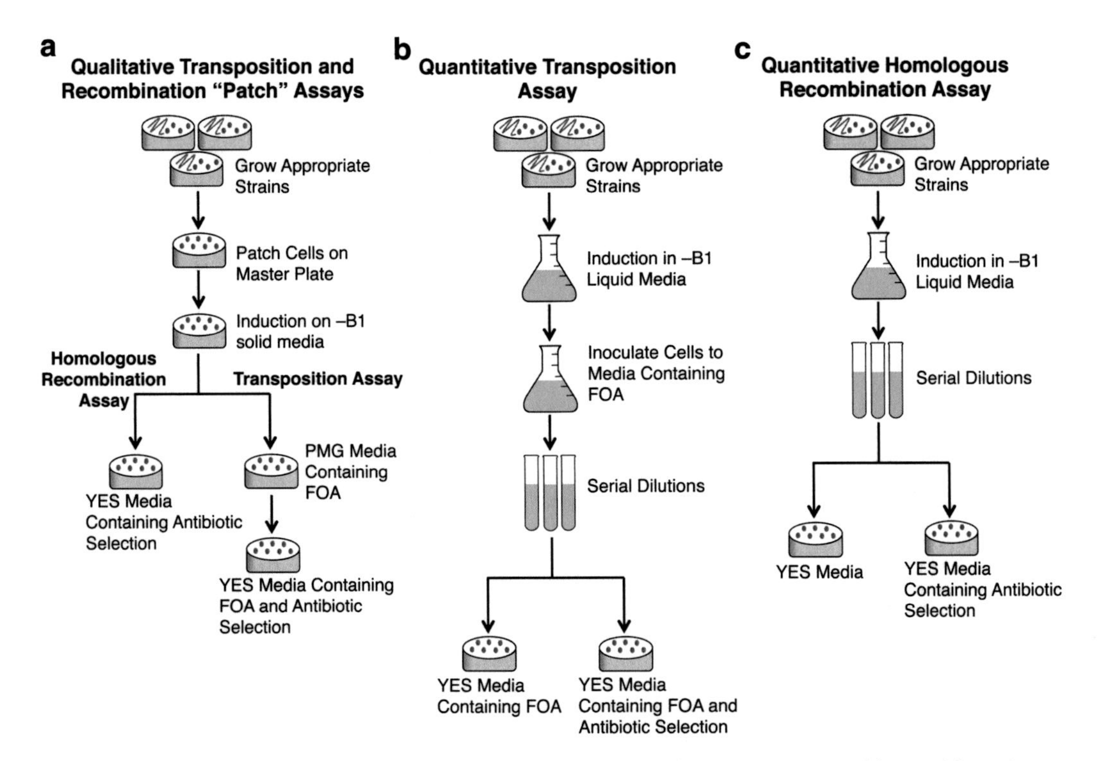

Fig. 2 Flow diagrams for the qualitative and quantitative assays. Qualitative transposition and homologous recombination assays can be conducted with patches of yeast on solid media (**a**) or as a more accurate quantitative assay in liquid media (**b**, **c**)

3. Induce activity of the retrotransposon by replica printing the master plate to solid media lacking vitamin B1 (PMG-U + L-B1) using sterile velvets (*see* **Note 10**). The master plate can also be propagated at this point by making a second print onto a second non-inducing + B1-containing plate. Incubate the cells on PMG-U + L-B1 for 4 days to ensure sufficient induction of the *nmt1* promoter.
4. Replica print cells to PMG + 5-FOA + U + L + B1 solid media using sterile velvets and incubate for 3 days (*see* **Notes 11** and **12**).
5. Replica print cells to YES + 5-FOA + desired antibiotic-containing media (G418 or Nat) using sterile velvets (*see* **Note 13**). Allow growth for 24–48 h (*see* **Note 14** and Fig. 1b).

3.2 Qualitative Homologous Recombination Assay (Fig. 2a)

1. Freshly revive desired experimental strains (four independent transformants of each Tf1 plasmid are tested with Tf1 controls containing frameshift mutations) from frozen perm stocks by streaking a match-head's worth of cells onto PMG-U + L + B1 solid media and incubating for 3 days (*see* **Note 8**).
2. Establish a master plate containing all desired experimental and control strains by removing a small portion of the cell mass

with a sterile toothpick from the thick part of the streak and spotting a pinhead-sized amount onto solid PMG-U + L + B1 media (*see* **Note 9**). Incubate for 3 days.

3. Induce transposition activity of the retrotransposon by replica printing the master plate to solid media lacking vitamin B1 (PMG-U + L-B1) using sterile velvets (*see* **Note 10**). The master plate can also be propagated at this point by making a second print onto a second non-inducing + B1-containing plate. Incubate cells on PMG-U + L-B1 for 4 days to ensure sufficient induction of the *nmt1* promoter.
4. Print cell mass to YES + antibiotic (G418 or Nat)-containing media (*see* **Note 15**). Incubate for 24 h (*see* **Note 16** and Fig. 1b).

3.3 Quantitative Transposition Assay (Fig. 2b)

1. Freshly revive desired experimental strains (each strain is tested in triplicate including controls containing Tf1 frameshift mutations) from frozen perm stocks by streaking a match-head's worth of cells onto PMG-U + L + B1 solid media and incubating for 3 days (*see* **Note 8**). In addition to testing strains in triplicate, multiple transformants of each Tf1 plasmid must be assayed. It is therefore best to combine these requirements by testing three independent transformants of the Tf1 plasmid in each strain of interest.
2. Scrape a match-head amount of cell mass from the thick part of the streak and resuspend cells into 1 mL of PMG-U + L-B1 media (*see* **Note 9**). Vortex cells for 10 s or until cells are visibly and evenly resuspended in solution.
3. Harvest cells by centrifugation at (3400 RCF) for 1 min and pour off the supernatant. Blot off excess supernatant on a clean paper towel.
4. Resuspend cells again in 1 mL PMG-U + L-B1. Vortex, centrifuge, and remove supernatant. Repeat with four washes to ensure adequate removal of the vitamin B1 for efficient promoter induction (*see* **Note 17**).
5. Set up 5 mL induction cultures in PMG-U + L-B1 media using 15 mL disposable glass culture tubes at a starting optical density (OD 600 nm) of 0.05. Grow cultures for 4 days with turning on a roller drum (for optimal aeration of cells).
6. Measure the OD (600 nm) of the induction cultures and use them to inoculate 5 mL cultures in PMG + 5-FOA + U + L + B1 media using 15 mL disposable class culture tubes with a starting OD 0.1. Grow for 36 h on a roller drum.
7. Measure the OD (600 nm) of cells after 36 h of growth and dilute samples to OD 1 (2×10^7 cells/mL) in PMG + U + L + B1 media.

8. Prepare tenfold serial dilutions for each sample starting at 2×10^7 cells/mL (OD 1) and ending with 2×10^4 cells/mL in PMG + U + L + B1 media (*see* **Note 18**).
9. Plate 100 μL of dilutions containing 10^7, 10^6, and 10^5 cells/mL onto YES + 5-FOA + antibiotic-containing media (*see* **Note 13**) using glass beads (~3 mm). Plate 100 μL of dilutions containing 10^6, 10^5, and 10^4 cells/mL on YES + 5-FOA-containing media (*see* **Note 19**).
10. Allow cells to grow for exactly 3 days (72 h). Count the number of colonies formed on YES + 5-FOA and YES + 5-FOA + antibiotic-containing plates (*see* **Note 20**).
11. Calculate the frequency of Tf1 transposition by comparing the ratio of cells grown on 5-FOA media to cells grown on 5-FOA + antibiotic-containing media (*see* **Notes 21** and **24**).
12. The results of wild-type Tf1 and the frameshift mutations are shown graphically in Fig. 3a and the quantities are listed in Table 1.

3.4 Quantitative Homologous Recombination Assay (Fig. 2c)

1. Freshly revive desired experimental strains (each strain is tested in triplicate including Tf1 controls containing frameshift mutations) from frozen perm stocks by streaking a match-head's worth of cells onto PMG-U + L + B1 solid media and growing for 3 days (*see* **Note 8**). In addition to testing strains in triplicate,

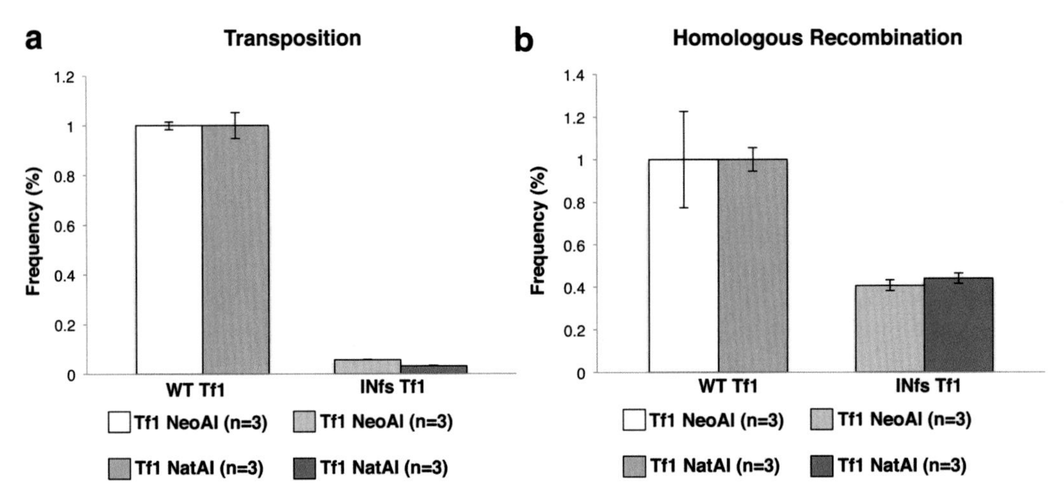

Fig. 3 Quantitative transposition and homologous recombination assays under G418 and nourseothricin selection. Transposition levels (**a**) and homologous recombination levels (**b**) in cells containing the wild-type Tf1 plasmids with either G418 or Nat selection are compared to levels in cells containing an INfs Tf1 expression vector. Transposition levels are reduced roughly 20-fold in the INfs controls (**a**) and homologous recombination levels are reduced roughly twofold in INfs controls when compared to WT (**b**). All samples are normalized to WT-Tf1 transposition or recombination levels. Each strain was assayed in triplicate ($n = 3$) and the results averaged

Table 1
Raw and normalized transposition frequencies from strains tested in triplicate

	Strain	Transposition frequency (%)	Standard deviation (%)	Normalized transposition frequency (%)	Standard deviation (%)
Tf1 NeoAI	WT Tf1	1.30	0.021	1.00	0.016
	INfs Tf1	0.074	0.001	0.057	0.001
	PRfs Tf1	0.00	0.00	0.00	0.00
Tf1 NatAI	WT Tf1	0.532	0.028	1.00	0.053
	INfs Tf1	0.017	0.001	0.033	0.002
	PRfs Tf1	0.00	0.00	0.00	0.00

multiple transformants of each Tf1 plasmid must be assayed. It is therefore best to combine these requirements by testing three independent transformants of the Tf1 plasmid in each strain of interest.

2. Scrape a small amount match-head's worth of cell mass from the thick part of the streak and resuspend cells into 1 mL of PMG-U + L-B1 (*see* **Note 9**). Vortex cells for at least 10 s or until cells are evenly resuspended in solution.
3. Harvest cells with centrifugation at (3400 RCF) for 1 min and pour off the supernatant.
4. Resuspend cells again in 1 mL PMG-U + L-B1. Vortex, centrifuge, and remove supernatant. Repeat with four washes to ensure adequate removal of vitamin B1 for sufficient induction of the *nmt1* promoter (*see* **Note 17**).
5. Set up 5 mL induction cultures of PMG-U + L-B1 media in disposable 15 mL glass culture tubes at a starting optical density (OD 600 nm) of 0.05. Incubate for 4 days with rolling for adequate aeration on a roller drum.
6. Measure OD (600 nm) of cells and dilute samples to OD 1 (2×10^7 cells/mL) in PMG-U + L + B1 medium.
7. Set up tenfold serial dilutions for each sample ranging from 2×10^7 cells/mL (OD 1) to 2×10^4 cells/mL in PMG-U + L + B1 media (*see* **Note 18**).
8. Plate 100 μL of dilutions containing 10^7, 10^6, and 10^5 cells/mL onto YES + antibiotic-containing media (*see* **Note 15**) using glass beads to spread cells (~3 mm). Plate 100 μL of dilutions containing 10^6, 10^5, and 10^4 cells/mL onto YES media (*see* **Note 22**).

Table 2
Raw and normalized homologous recombination frequencies from strains tested in triplicate

	Strain	Recombination frequency (%)	Standard deviation (%)	Normalized recombination frequency (%)	Standard deviation (%)
Tf1 NeoAI	WT Tf1	0.589	0.133	1.00	0.227
	INfs Tf1	0.239	0.015	0.407	0.026
	PRfs Tf1	0.00	0.00	0.00	0.00
Tf1 NatAI	WT Tf1	1.96	0.109	1.00	0.056
	INfs Tf1	0.862	0.049	0.439	0.025
	PRfs Tf1	0.00	0.00	0.00	0.00

9. Incubate cells for exactly 3 days (72 h). Count the number of colonies formed on YES and YES + antibiotic-containing media (*see* **Note 20**).
10. The frequency of homologous recombination is the ratio of colonies grown on YES + antibiotic-containing media to colonies grown on YES-containing media (*see* **Notes 23** and **25**).
11. The results of wild-type Tf1 and the frameshift mutations are shown graphically in Fig. 3b and the quantities are listed in Table 2.

4 Notes

1. Dropout mixture contains equal quantities of all amino acids except for adenine, whose levels are increased to 2.5 times the amount of the other components. The complete dropout mixture should be used in YES media. For the experiments described here, the PMG dropout mixture excludes leucine and uracil as these will be supplemented based on selection needs, or with the amino acid selection markers on the Tf1-expressing plasmids. We find that using this dropout mixture yields far better transposition activity than the traditional five amino acid supplements typically recommended for *S. pombe*.
2. 5-FOA is used to select against *URA3* activity which is expressed from our Tf1 plasmids. Because 5-FOA has low solubility in water, it helps to incubate and shake media at 37 °C for 30–45 min following 5-FOA addition. However, 5-FOA should be added to media only after sterilization of liquid media, since it is unstable at high temperatures.

3. The purity of commercially available G418 ranges between 60 and 90 %. Therefore, to determine accurate concentrations and weighing for stock solutions, it is necessary to correct for purity. Our 500 μg/mL concentration is the corrected value.
4. When using 5-FOA plates containing uracil, it is necessary to reduce the amount of uracil from 250 to 50 μg/mL to ensure maximal selection efficiency against *URA3*. The conversion of 5-FOA to a toxic 5-fluorouridine monophosphate compound by the *URA3* enzyme is only effective in media lacking excess amounts of uracil, as uracil will dilute toxicity.
5. Uracil is excluded from media to select for cells containing the *URA3* selection marker in the Tf1 expression vector (*see* Fig. 1a).
6. Vitamin B1 (thiamine) is excluded from media to induce Tf1 expression via the *nmt1* promoter. All measures to reduce even small amounts of contaminating B1 should be taken, such as using disposable plasticware.
7. Effective methods include mixing via a coffee grinder or in a stone roller with steel balls.
8. Controls include strains containing wild-type Tf1 plasmids, as well as versions of Tf1 containing a frameshift stop codon in the beginning of the IN-coding region (Tf1 INfs) and in the PR-coding region (Tf1 PRfs). Tf1 INfs mutants will not produce IN protein and will lack transposition activity. PRfs mutants produce no RT or IN proteins and, as a result, are unable to perform reverse transcription or integration resulting in a lack of cDNA production, homologous recombination activity, and transposition activity (*see* Fig. 1b).
9. Using cells from the thick part of the streak from a frozen stock is preferable to single colony selection as this will reduce issues associated with suppressor mutations from colony populations. All experimental cells should be revived from frozen stocks that have previously been single-colony purified and frozen in 15 % glycerol prior to use.
10. It is helpful after printing plates to sterile velvet, to print a sterile plate lid to the cell mass before printing the velvet to new media. This technique will reduce excess carryover of cell mass and creates more homogeneously phenotypic patches.
11. Between printings, unwanted cell mass or contamination can occur. These can easily be removed using a sterile scalpel prior to printing to new media.
12. This intermediate PMG + 5-FOA + U + L + B1 step selects against cells with the expression plasmid as demonstrated by the reduction of the INfs control in experiments. When using nourseothricin as a selection marker, two sequential rounds of

5-FOA selection are required, one round with 3 days of incubation time, followed by a round of 2 days of incubation. G418 is more effective under 5-FOA selection and will need to undergo only a single round of 5-FOA selection for the 3 days (as written in the protocol).

13. Antibiotics used for selection are either G418 (Neo) or nourseothricin (Nat). These drugs in combination with 5-FOA will allow for selection of cells that have acquired new genomic integrations and have also lost the Tf1 expression plasmid.
14. The optimal incubation time is determined by observing cell growth; normally this occurs at the point when there is the greatest difference between the wild-type Tf1 and INfs controls. At this point, INfs control patches will exhibit approximately 1/20th the growth/density of WT patches. PRfs patches should exhibit no growth (*see* Fig. 1b). If incubated for too long, background levels of homologous recombination can occur between Tf1 cDNA and genomic copies of Tf2. This causes the INfs to have similar levels of growth to the wild-type Tf1.
15. Antibiotic selection is either G418 (Neo) or nourseothricin (Nat). This step will select for cells where Tf1 cDNA has successfully recombined with the expression plasmid, thereby replacing the non-spliceable drug resistance gene with a version that has undergone transcription, splicing, and reverse transcription, creating an active and spliced copy of the drug selection gene on the plasmid.
16. Confluent growth should be evident after 24 h. A slight but notable (twofold) reduction in growth on the IN frameshift control patches should be evident when compared to WT patches. Single-amino acid mutations in IN that block integration do not exhibit this twofold reduction in homologous recombination indicating that the recombination process is stimulated twofold by IN protein. The PRfs control patches should exhibit no growth (*see* Fig. 1b).
17. This step removes residual B1 which will inhibit *nmt1* induction of Tf1, resulting in significantly reduced transposition and recombination activity [9].
18. To ensure the most accurate serial dilutions, we find it ideal to perform dilutions in 1 mL volumes. All samples should be vortexed well (20 s) between each serial dilution. This step is critical to ensure reproducibility between replicates.
19. This media will allow for selection of cells that have successfully lost the *URA3* containing Tf1-expression plasmid.

20. Micro-colonies are abundant on YES + 5-FOA + antibiotic-containing media. To ensure reproducibility in transposition and recombination frequencies, it is necessary to establish a limit based on colony size. We find it easiest to count all large- to midsize colonies that are well formed and round. Including very small colonies in the count generally does not negatively impact the analysis or decrease experimental reproducibility as long as size limits are consistent through all steps of the experiment.
21. Calculate the transposition frequency with the following equation:

$$\text{Transposition frequency}\,(\%) = \frac{(\#\text{ of colonies on YES} + 5 - \text{FOA} + \text{antibiotic media}) \times 100}{(\#\text{ of colonies on YES} + 5 - \text{FOA} \times \text{dilution differential})}$$

22. Plating cells to YES will determine concentration of viable cells in the culture.
23. Calculate the homologous recombination frequency with the following equation:

$$\text{Homologous recombination frequency}\,(\%) = \frac{(\#\text{ of colonies on YES} + \text{antibiotic media}) \times 100}{(\#\text{ of colonies on YES} \times \text{dilution differential})}$$

24. The frequency of transposition should be roughly 20-fold reduced in INfs samples when compared to WT controls (*see* Fig. 3a and Table 1).
25. The frequency of homologous recombination should be roughly twofold reduced in INfs when compared to WT controls (*see* Fig. 3b and Table 2).

References

1. Levin HL, Moran JV (2011) Dynamic interactions between transposable elements and their hosts. Nat Rev Genet 12(9):615–627. doi:10.1038/nrg3030, nrg3030 [pii]
2. Beck CR, Garcia-Perez JL, Badge RM, Moran JV (2011) LINE-1 elements in structural variation and disease. Annu Rev Genomics Hum Genet 12:187–215. doi:10.1146/annurev-genom-082509-141802
3. Singleton TL, Levin HL (2002) A long terminal repeat retrotransposon of fission yeast has strong preferences for specific sites of insertion. Eukaryot Cell 1:44–55
4. Bowen NJ, Jordan I, Epstein J, Wood V, Levin HL (2003) Retrotransposons and their recognition of pol II promoters: a comprehensive survey of the transposable elements derived from the complete genome sequence of Schizosaccharomyces pombe. Genome Res 13(Sept):1984–1997
5. Leem YE, Ripmaster TL, Kelly FD, Ebina H, Heincelman ME, Zhang K, Grewal SIS, Hoffman CS, Levin HL (2008) Retrotransposon Tf1 is targeted to pol II promoters by transcription activators. Mol Cell 30(1):98–107
6. Guo Y, Levin HL (2010) High-throughput sequencing of retrotransposon integration provides a saturated profile of target activity in Schizosaccharomyces pombe. Genome Res 20(2):239–248

7. Feng G, Leem YE, Levin HL (2013) Transposon integration enhances expression of stress response genes. Nucleic Acids Res 41(2):775–789. doi:10.1093/nar/gks1185
8. Chatterjee AG, Esnault C, Guo Y, Hung S, McQueen PG, Levin HL (2014) Serial number tagging reveals a prominent sequence preference of retrotransposon integration. Nucleic Acids Res 42(13):8449–8460. doi:10.1093/nar/gku534
9. Atwood A, Choi J, Levin HL (1998) The application of a homologous recombination assay revealed amino acid residues in an LTR-retrotransposon that were critical for integration. J Virol 72(2):1324–1333
10. Levin HL (1995) A novel mechanism of self-primed reverse transcription defines a new family of retroelements. Mol Cell Biol 15:3310–3317
11. Levin HL, Boeke JD (1992) Demonstration of retrotransposition of the Tf1 element in fission yeast. EMBO J 11:1145–1153
12. Boeke JD, Garfinkel DJ, Styles CA, Fink GR (1985) Ty elements transpose through an RNA intermediate. Cell 40(491):491–500
13. Forsburg SL, Rhind N (2006) Basic methods for fission yeast. Yeast 23(3):173–183

Chapter 9

LINE Retrotransposition Assays in *Saccharomyces cerevisiae*

Axel V. Horn and Jeffrey S. Han

Abstract

Long interspersed nuclear element (LINE) retrotransposons make up significant parts of mammalian genomes. They alter host genomes by direct mutagenesis through integration of new transposon copies, by mobilizing non-autonomous transposons, by changes in host gene activity due to newly integrated transposons and by recombination events between different transposon copies. As a consequence, LINEs can contribute to genetic disease. Simple model systems can be useful for the study of basic molecular and cellular biology of LINE retrotransposons. Here, we describe methods for the analysis of LINE retrotransposition in the well-established model organism *Saccharomyces cerevisiae*. The ability to follow retrotransposition in budding yeast opens up the possibility of performing systematic screens for evolutionarily conserved interactions between LINE retrotransposons and their host cells.

Key words L1 retrotransposon, Retrotransposition assay, Zorro3, LINE retrotransposition, Yeast retrotransposition

1 Introduction

Saccharomyces cerevisiae is a key model organism for molecular and genetic research. The conservation of important cellular processes between yeast and more complex eukaryotes has made yeast an attractive system for unraveling basic understanding about how cells and chromosomes replicate and divide. Some of the advantages of using yeast as a model organism are its fast division, dispersed cells, compact genome, well-defined and manipulable sexual cycle, ability to stably maintain and shuffle plasmids, and ease of modifying yeast chromosomes via homologous recombination. These traits have made yeast a premier model system for genetics and the proving ground for most new genomic technologies. Additionally, the ease of yeast chromosome manipulation has led to the construction of various useful libraries for the scientific community, such as the yeast knockout collection [1] or the yeast GFP-fusion collection [2].

Jose L. Garcia-Pérez (ed.), *Transposons and Retrotransposons: Methods and Protocols*, Methods in Molecular Biology, vol. 1400, DOI 10.1007/978-1-4939-3372-3_9, © Springer Science+Business Media New York 2016

In order to utilize the technical advantages of *S. cerevisiae* for the study of LINE retrotransposition, we designed a system to mobilize LINE retrotransposons in budding yeast. Since the *S. cerevisiae* genome does not encode any known LINEs [3], we reengineered a known active LINE from *Candida albicans* called Zorro3 (Z3) [4]. Zorro3 is a member of the L1 clade [5, 6], and structurally similar to the currently active human and mouse L1 retrotransposons. Zorro3 retrotransposition in budding yeast is dependent on known critical amino acid residues in ORF1 and ORF2, and Zorro3 retrotransposition events cloned from the genome resemble mammalian retrotransposition events. Thus, it is likely that the mechanism of Zorro3 replication is similar to mammalian L1s, and essential host factors involved in L1 retrotransposition are conserved in budding yeast. This provides a simple tool to investigate the mechanisms by which retrotransposons and host cells interact. Additionally, the Zorro3 assay in *S. cerevisiae* allows experiments in a genetic background without endogenous retrotransposons of the same or closely related retrotransposon families.

Here we present a protocol to monitor Zorro3 retrotransposition in *S. cerevisiae*. This assay is easily adapted to different culture conditions. We also describe how to analyze and isolate retrotransposition events.

2 Materials

2.1 Yeast Strains

The described assay has been performed in the *S. cerevisiae* strains BY4741 ([7]; ATCC number 201388; genotype *MATa*, *his3Δ1*, *leu2Δ0*, *met15Δ0*, *ura3Δ0*) and GRF167 ([4, 8]; ATCC number 90849; genotype *MATα*, *his3Δ200*, *ura3-167*, *GAL+*) or strains derived from either of both.

2.2 Plasmids

1. pSCZorro3mHIS3AI [4] contains a full-length Z3 sequence adapted to *S. cerevisiae* codon usage (scZ3). scZ3 is under the control of the galactose-inducible GAL1 promoter. Inserted into the 3′ untranslated region of scZ3 is an mHIS3AI cassette. This auxotrophic marker allows the selection for retrotransposition events on SC-His plates.
2. pSCZorro3mHIS3AIRTmut was derived from pSCZorro3mHIS3AI by introducing the DD674,675AA mutation in the reverse transcriptase domain of ORF2. This inactive Z3 mutant was used as negative control in retrotransposition reporter assays.

2.3 Media

1. YP broth: For 1 L of broth mix 10 g yeast extract, 20 g peptone, and 20 g appropriate sugar and autoclave for 15 min at 120 °C. Use dextrose for YPD broth, or galactose for YPGal.
2. YP plates: Mix 1 L of YPD or YPGal broth and add 20 g bacto agar. Autoclave for 15 min at 120 °C. Let the mix cool down and pour into 10 cm petri dishes.

3. Synthetic complete (SC) media broth: For 1 L mix 6.7 g yeast nitrogen base (w/o amino acids), 1.4 g appropriate dropout mix, and 20 g appropriate carbon source (dextrose or galactose) and autoclave for 15 min at 120 °C.
4. SC plates: Prepare 500 mL 2× SC broth and 500 mL 4 % bacto agar. Autoclave for 15 min at 120 °C. Mix, cool down, and pour into 10 cm petri dishes (*see* **Note 1**).
5. Dropout mix: The complete mix contains 1 g adenine, 1 g uracil, 1 g l-tryptophan, 1 g l-histidine, 1 g l-arginine, 1 g l-methionine, 3 g l-tyrosine, 4 g l-leucine, 4 g l-isoleucine, 2.5 g l-phenylalanine, 3 g l-lysine, 5 g l-glutamic acid, 5 g l-aspartic acid, 7.5 g l-valine, 10 g l-threonine, and 20 g l-serine. Use 1.4 g mix per liter of media. To make the appropriate dropout mix, omit amino acids as needed from the mix (e.g., for SC-ura omit the uracil),
6. LB broth: For 1 L of broth mix 5 g yeast extract, 10 g NaCl, and 10 g bacto-tryptone. Autoclave for 15 min at 120 °C.
7. LB agar plates: Prepare 1 L LB broth and add 20 g bacto agar. Autoclave for 15 min at 120 °C. Let the media cool down, add the appropriate antibiotic, and pour into 10 cm petri dishes.

2.4 Chemical Stock Solutions

1. 50 % PEG: Add 250 g polyethylene glycol (PEG) 3350 to 250 mL H_2O. Mix and heat until the PEG is completely dissolved. Add H_2O to a final volume of 500 mL. Autoclave for 15 min at 120 °C. Aliquot and store at room temperature.
2. 1 M lithium acetate: Add 51 g lithium acetate (Acros Organics) to 450 mL H_2O. Mix until everything is dissolved. Add H_2O to a final volume of 500 mL. Autoclave for 15 min at 120 °C. Aliquot and store at room temperature.
3. Herring sperm DNA: Add 250 mg herring sperm DNA to 25 mL TE buffer (10 mM Tris pH 8.0, 1 mM EDTA) for a final concentration of 10 mg/mL. Mix until the DNA is completely dissolved. Aliquot and store at −20 °C.
4. Acid-washed beads (Sigma).
5. Lysis buffer: 100 mM NaCl, 10 mM Tris–HCl pH 7.5, 1 mM EDTA pH 8.0, 2 % Triton X-100, 1 % SDS.

Other routinely used chemicals and reagents used are obtained from local suppliers.

3 Methods

3.1 Yeast Cultivation

Yeast should be streaked from frozen stock cultures on YPD plates and incubated at 30 °C for 2–3 days before transformation. If necessary, re-streak them on another YPD plate to obtain single colonies. In general, all assays should be performed with freshly grown yeast. All strains and transformants can be stored on the workbench for several days or at 4 °C for a few weeks.

3.2 Lithium Acetate Yeast Transformation

1. Inoculate 4 mL YPD with freshly grown yeast. Incubate overnight at 30 °C.
2. The next morning, measure OD_{600}. Inoculate 4 mL YPD with OD_{600} = 0.1/mL. Prepare one 4 mL culture for every transformation. Incubate at 30 °C until OD_{600} is between 0.4 and 0.5 (*see* **Note 2**).
3. Centrifuge in a tabletop centrifuge for 5 min at 1000 × *g*. Discard media. Resuspend pellet in 1 mL H_2O and transfer into 1.5 mL tube.
4. Centrifuge for 5 min at 1000 × *g*. Discard H_2O. Resuspend pellet in 1 mL H_2O.
5. Centrifuge for 5 min at 1000 × *g*. Discard H_2O. Resuspend pellet in 1 mL 0.1 M lithium acetate.
6. Centrifuge for 5 min at 1000 × *g*. Discard lithium acetate. Add transformation mix and mix well: 240 μL 50 % PEG, 36 μL 1 M lithium acetate, 41 μL H_2O, 5 μL boiled herring sperm DNA (10 μg/mL), 2 μL plasmid DNA (0.1–0.2 μg). Incubate for 30 min at 30 °C on a rotator.
7. Add 36 μL DMSO to the transformation. Mix well. Incubate for 10 min at 42 °C.
8. Centrifuge for 2 min at 4000 × *g*. Discard supernatant. Resuspend pellet in 100 μL H_2O.
9. Plate 10 μL and 90 μL on selective plates (SC-ura for pZorro3mHIS3AI). Incubate at 30 °C.
10. Three to four days later pick several transformants from every transformation and patch them on a new SC-ura plate. Incubate at 30 °C. Use these for retrotransposition assays.

3.3 Zorro3 Retrotransposition Assay

1. Inoculate 4 mL SC-ura + galactose with freshly grown yeast transformants. Incubate at 23 °C for 3 days on a roller drum (*see* **Notes 3** and **4**).
2. From the induced cultures, prepare at least a 10^{-5} dilution in H_2O and plate 100 μL of this dilution on a YPD plate (*see* **Note 5**). The dilution should be made in serial steps. Incubate at 30 °C for 3 days. The colonies on this plate indicate the colony-forming units in the culture.
3. Take a 1 mL aliquot of the induced culture and spin for 5 min at 1000 × *g*. Discard the media and resuspend the pellet in 150 μL H_2O. Plate everything on an SC-his plate. Incubate at 30 °C for 5 days. Colonies on this plate represent retrotransposition events.

4. Document the plates and count the colonies (*see* **Note 6**).
5. Calculate the Z3 retrotransposition frequencies (*see* **Note 7**). The formula is (# colonies on SC-his plate)/(# colonies on YPD plate × 10 × YPD dilution).

3.4 Isolation of De Novo Zorro3 Retrotransposition Events

1. Take a colony or colonies of freshly transformed with pSCzorro3mHIs3AI and patch the cells uniformly over the entire surface of a 100 mm YPGal plate. A smooth flat sterile toothpick is best for spreading the cells. Induce Z3 transcription by incubating for at least 3 days at 23 °C (*see* **Note 8**).
2. Replica plate to SC-his and incubate for 5 days at 30 °C.
3. Pick colonies from the plate to isolate genomic DNA. Inoculate 4 mL SC-his cultures with the colonies and incubate for 2 days at 30 °C. Harvest the yeast by centrifuging the cultures for 5 min at 1000 × *g*. Remove the media. Add 0.3 g of acid-washed beads. Resuspend the pellet in 0.2 mL lysis buffer and 0.2 mL of phenol/chloroform by vortexing for 4 min. Add 0.2 mL TE buffer and vortex briefly. Spin for 5 min at 18,000 × *g* and keep the top aqueous layer. Add two volumes of 100 % ethanol, mix well, and spin for 2 min at 18,000 × *g*. Remove the supernatant and wash the pellet with 0.75 mL 70 % ethanol. Spin again for 2 min at 18,000 × *g*. Remove the supernatant and air-dry the pellet for 5–10 min. Resuspend each pellet in 50 μL TE buffer. Quantify the genomic DNA by your method of choice (*see* **Note 9**).
4. Genomic DNA digestion: Mix 0.25 μg genomic DNA, 10 U *Eco*RI, and 2 μL 10× *EcoRI* buffer in a total volume of 20 μL. Incubate overnight at 37 °C.
5. Adaptor annealing: Mix 10 μL primer JH1 (0.5 mM), 10 μL primer JH122 (0.5 mM), 1 μL 5 M NaCl, and 30 μL TE buffer. Incubate for 2 min at 95 °C in a heat block. Remove the heat block and allow to gradually cool to room temperature. For primer sequences (*see* **Note 10**).
6. Adaptor ligation: Add to the digested genomic DNA 3 μL 10× ligation buffer, 6 μL adaptor, and 1 μL (400 U) T4 DNA ligase. Ligate overnight at 4 °C.
7. PCR amplification: A single reaction mix contains 2 μL ligation mix, 5 μL 10× ExTaq buffer, 4 μL dNTPs (2.5 mM each), 2.5 μL primer JH4 (5 mM), 2.5 μL JH102 (5 mM), and 0.2 μL (1 U) ExTaq DNA polymerase (Takara). Add H_2O to a final volume of 50 μL. Cycling parameters: denaturation for 2 min at 94 °C, 40 cycles of 15 s at 94 °C, 15 s at 55 °C, 10 min at 72 °C, final elongation for 20 min at 72 °C (*see* **Note 11**).
8. 3′ junction cloning: Run the PCR reaction on an agarose gel and cut out the band(s). Extract the DNA with a gel extraction kit according to the manufacturer's protocol. Elute DNA in

30 μL H_2O. This product can be cloned and sequenced using any commercial or homemade TA vector (*see* **Note 12**).

9. Once the 3′ genomic flank of the Zorro3 events are determined, the entire insertion can be PCR isolated by with primers designed to be complementary to the 3′ flank, and roughly 500 bp upstream of the 3′ flank, based on the sequenced yeast genome (*see* **Note 13**).

4 Notes

1. Due to the low pH of SC media, SC and bacto agar are autoclaved separately to prevent hydrolysis of agar and occasional batches of "soft" plates.
2. It will take 4–6 h to grow wild-type BY4741/GRF167 from OD_{600} = 0.1 to 0.4.
3. Induction of Zorro3 transcription at temperatures higher than 23 °C results in lower retrotransposition activity.
4. The retrotransposition assay should be done in at least a quadruplicate for every construct or condition. The low frequency of retrotransposition can lead to large variability.
5. The end goal is to have roughly more than 100 but less than 500 colonies in the YPD plate, such that colonies can be reliably counted.
6. An easy documentation and counting method is to first scan the plates. Make sure to provide a dark background that gives a good contrast to the colonies. Secondly, use an image-processing program to count the colonies. For example ImageJ is available for free and provides a handy multi-point selection tool. Always visually verify that the program is correctly marking the colonies.
7. The average Z3 retrotransposition rate in the GRF167 strain is around 2×10^{-6} events/plated cells [4]. Variations are possible due to the assay conditions and the strains used.
8. If the Z3 transposition frequency is much lower than in wild-type yeasts (2×10^{-6} events/plated cells) then the induction time should be extended.
9. This protocol does not remove RNA, so UV absorbance will not give an accurate reading for genomic DNA. RNA can be removed with RNaseA followed by phenol/chloroform extraction.
10. Adaptor primer sequences: JH1 5′-GTAATACGACTCACTAT AGGGCTCCGCTTAAGGGAC-3′, JH122 5′Phos-AATTG TCCCTTAAGCGGAG-3′NH2. Different enzymes can be used to digest genomic DNA if the 5′ overhang of JH122 is altered accordingly. It is unlikely that any single enzyme/adaptor combination can be used to isolate all events.

11. PCR primer sequences: JH4 5′-GTAATACGACTCACTATAGGGC-3′, JH102 5′-AGAGCTCCCGGCATCCTCTCGA-3′. JH4 hybridizes to the adaptor sequence. JH102 hybridizes to the 3′ end of scZorro3mHIS3AI. An aliquot of the PCR reaction should be loaded and separated on an agarose gel. If there are no distinct bands then a second PCR can be performed with either the same primers or nested primers and 2 μL of the first PCR as template.
12. Sometimes the polyA tail will be too long to sequence through. In this case, sequence adjacent to the polyA tail can be obtained by sequencing with anchored polyA primers $(A)_{19}T$, $(A)_{19}C$, and $(A)_{19}G$.
13. Note that the majority of retrotransposition events produced in the Zorro3 assay result in episomal circles [9]. These circles tend to be unstable under nonselective conditions. To enrich for chromosomally integrated events, the SC-ura + galactose plates can be replica plated to YPD plates and incubated for 48 h at 30 °C prior to replica plating to SC-his plates. Chromosomally integrated events will tend to form the larger colonies on the SC-his plates.

Acknowledgements

This work was supported by NIH grant GM090192 to J.S.H.

References

1. Winzeler EA, Shoemaker DD, Astromoff A, Liang H, Anderson K, Andre B, Bangham R, Benito R, Boeke JD, Bussey H, Chu AM, Connelly C, Davis K, Dietrich F, Dow SW, Bakkoury El M, Foury F, Friend SH, Gentalen E, Giaever G, Hegemann JH, Jones T, Laub M, Liao H, Liebundguth N, Lockhart DJ, Lucau-Danila A, Lussier M, M'Rabet N, Menard P, Mittmann M, Pai C, Rebischung C, Revuelta JL, Riles L, Roberts CJ, Ross-MacDonald P, Scherens B, Snyder M, Sookhai-Mahadeo S, Storms RK, Véronneau S, Voet M, Volckaert G, Ward TR, Wysocki R, Yen GS, Yu K, Zimmermann K, Philippsen P, Johnston M, Davis RW (1999) Functional characterization of the *S. cerevisiae* genome by gene deletion and parallel analysis. Science 285:901–906
2. Huh W-K, Falvo JV, Gerke LC, Carroll AS, Howson RW, Weissman JS, O'Shea EK (2003) Global analysis of protein localization in budding yeast. Nature 425:686–691
3. Goffeau A, Barrell BG, Bussey H, Davis RW, Dujon B, Feldmann H, Galibert F, Hoheisel JD, Jacq C, Johnston M, Louis EJ, Mewes HW, Murakami Y, Philippsen P, Tettelin H, Oliver SG (1996) Life with 6000 genes. Science 274:546–563
4. Dong C, Poulter RT, Han JS (2009) LINE-like retrotransposition in *Saccharomyces cerevisiae*. Genetics 181:301–311
5. Goodwin TJ, Ormandy JE, Poulter RT (2001) L1-like non-LTR retrotransposons in the yeast Candida albicans. Curr Genet 39:83–91
6. Goodwin TJ, Busby JN, Poulter RT (2007) A yeast model for target-primed (non-LTR) retrotransposition. BMC Genomics 8:263
7. Brachmann CB, Davies A, Cost GJ, Caputo E, Li J, Hieter P, Boeke JD (1998) Designer deletion strains derived from *Saccharomyces cerevisiae* S288C: a useful set of strains and plasmids for PCR-mediated gene disruption and other applications. Yeast 14(2):115–132
8. Boeke JD, Garfinkel DJ, Styles CA, Fink GR (1985) Ty elements transpose through an RNA intermediate. Cell 40:491–500
9. Han JS, Shao S (2012) Circular retrotransposition products generated by a LINE retrotransposon. Nucleic Acids Res 40:10866–10877

Chapter 10

LINE-1 Cultured Cell Retrotransposition Assay

Huira C. Kopera, Peter A. Larson, John B. Moldovan, Sandra R. Richardson, Ying Liu, and John V. Moran

Abstract

The Long INterspersed Element-1 (LINE-1 or L1) retrotransposition assay has facilitated the discovery and characterization of active (i.e., retrotransposition-competent) LINE-1 sequences from mammalian genomes. In this assay, an engineered LINE-1 containing a retrotransposition reporter cassette is transiently transfected into a cultured cell line. Expression of the reporter cassette, which occurs only after a successful round of retrotransposition, allows the detection and quantification of the LINE-1 retrotransposition efficiency. This assay has yielded insight into the mechanism of LINE-1 retrotransposition. It also has provided a greater understanding of how the cell regulates LINE-1 retrotransposition and how LINE-1 retrotransposition impacts the structure of mammalian genomes. Below, we provide a brief introduction to LINE-1 biology and then detail how the LINE-1 retrotransposition assay is performed in cultured mammalian cells.

Key words LINE-1, Alu, Retrotransposition assay, *trans*-complementation assay, Mammalian cultured cells

1 Introduction

Sequences derived from Long INterspersed Element-1 (LINE-1 or L1) comprise approximately 17 % of human genomic DNA [1]. Although greater than 99.9 % of human L1s have been rendered inactive by mutational processes, a small cohort of active L1s can move to new genomic locations by a "copy-and-paste" process known as retrotransposition (reviewed in Ref. [2]). Here, we refer to these L1s as retrotransposition-competent or RC-L1s.

A full-length RC-L1 is ~6 kb in length. Human RC-L1s contain a 5′ untranslated region (UTR), two nonoverlapping open reading frames (ORFs) that are separated by a short 63 bp inter-ORF sequence, and a 3′ UTR that ends in an adenosine-rich tract [3, 4]. The human L1 5′ UTR contains an internal RNA polymerase II promoter that directs the initiation of transcription at or near the first base of the element [5–7]; it also has an ill-defined antisense

Jose L. Garcia-Pérez (ed.), *Transposons and Retrotransposons: Methods and Protocols*, Methods in Molecular Biology, vol. 1400, DOI 10.1007/978-1-4939-3372-3_10, © Springer Science+Business Media New York 2016

promoter [8]. L1 ORF1 encodes an ~40 kDa RNA binding protein (ORF1p) that has nucleic acid chaperone activity [9–17]. ORF2 encodes an ~150 kDa protein (ORF2p) that has endonuclease (EN) and reverse transcriptase (RT) activities that are critical for retrotransposition [18–25]. ORF2p also contains a cysteine-rich domain (C) that is required for retrotransposition, although its function requires elucidation [24, 26]. The L1 3′ UTR contains a guanosine-rich polypurine tract, a weak RNA polymerase II polyadenylation signal, and ends in an A-rich tract [27–30]. Experiments in cultured cells have demonstrated that the L1 3′ UTR is not strictly required for retrotransposition [18, 24].

L1 elements retrotranspose by target site-primed reverse transcription (TPRT) [20, 31, 32]. As a result of TPRT, genomic L1s typically have the following structural hallmarks: (1) they are truncated at their 5′ ends; (2) they are flanked by target-site duplications (TSDs) that vary both in their length and sequence; (3) they end in an A-rich tract; and (4) they generally insert into an L1 ORF2p endonuclease consensus cleavage site (5′-TTTT/A-3′ and variations of that sequence, where "/" denotes the endonucleolytic nick) [1, 33–40].

The proteins encoded by L1s (ORF1p and/or ORF2p) occasionally act *in trans* to retrotranspose Short INterspersed Element (SINE) RNAs (e.g., the RNAs transcribed from Alu and SINE-VNTR-Alu (SVA) elements), noncoding RNAs (e.g., uracil-rich small nuclear RNAs (snRNAs) and small nucleolar RNAs (snoRNAs)), and RNA polymerase II transcribed messenger RNAs (mRNAs) [41–51]. In aggregate, these sequences comprise at least ~11 % of human genomic DNA ([1], reviewed in Ref. [2]). Thus, LINE-1 mediated retrotransposition events account for approximately one-third, or one billion base pairs of human genomic DNA. Ongoing LINE-1 retrotransposition continues to affect both interindividual and intraindividual human genetic diversity and has caused ~100 sporadic cases of genetic disease ([52], reviewed in Refs. [2, 53]).

The development of the cultured cell retrotransposition assay represented a seminal moment in LINE-1 biology because it allowed the experimental study of LINE-1 retrotransposition in "real time" [24]. The rationale of the assay is built upon experimental strategies to examine retrotransposon mobility in yeast and mammalian cells [54–57]. Briefly, a retrotransposition indicator cassette is introduced into the L1 3′ UTR in the opposite orientation of the L1 transcript [24]. The reporter cassette consists of a reporter gene (e.g., neomycin phosphotransferase in the *mneoI* reporter cassette) equipped with a heterologous promoter and polyadenylation signal (Fig. 1a). The reporter gene is interrupted by an intron that is in the same orientation with respect to L1 transcription. Thus, the reporter gene can only be expressed upon transcription of the L1, splicing of the intron in the reporter cassette, and the reverse transcription of the L1 RNA to introduce a

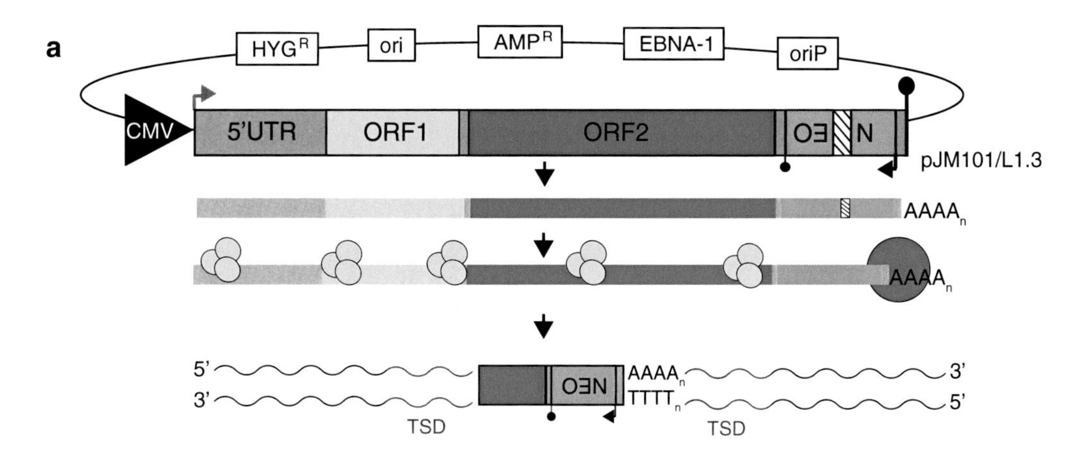

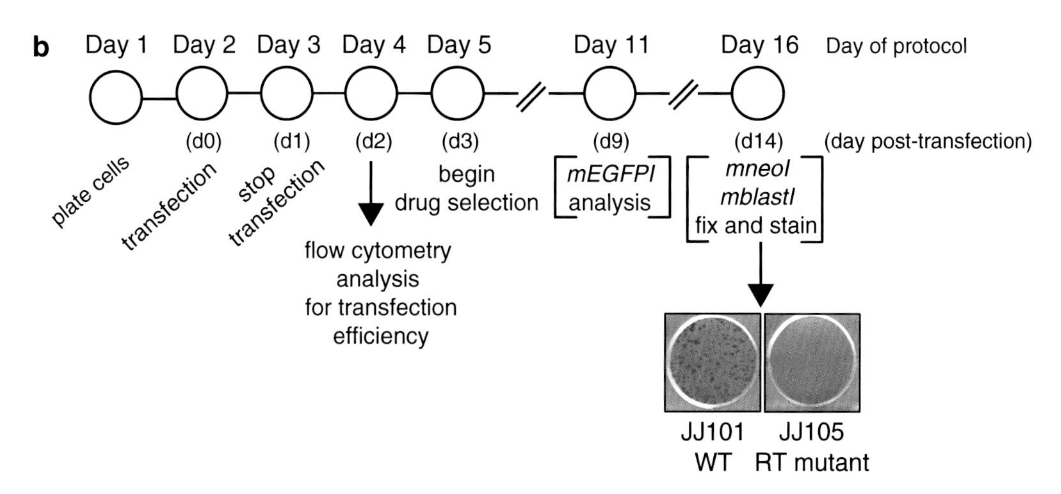

Fig. 1 LINE-1 retrotransposition assay. (**a**) A full-length retrotransposition-competent LINE-1 contains the *mneoI* reporter cassette (*orange box*) in the 3′ UTR in the opposite orientation with respect to LINE-1 (pJM101/L1.3) transcription. The reporter gene, neomycin phosphotransferase (*backwards* NEO) is interrupted by an intron (*hatched box*), which is in the same orientation with respect to LINE-1 transcription. The reporter cassette has its own promoter (*upside down black arrow*) and polyadenylation signal (*upside down small black lollipop*). The pCEP4 plasmid backbone encodes for the EBNA-1 (EBNA-1) viral protein and contains an origin of viral replication (oriP) and hygromycin B-resistance gene (HYG^R) for plasmid replication and hygromycin-selection, respectively, in mammalian cultured cells. The backbone also has a bacterial origin of replication (ori) and an ampicillin-resistance gene (AMP^R) for replication and ampicillin-selection, respectively, in *E. coli*. The LINE-1 is transcribed from the CMV promoter (*black triangle* labeled "CMV") or its promoter in the 5′ UTR (*gray arrow*) and transcription is terminated at an SV40 polyadenylation signal (*large black lollipop*). Once transcribed, the intron is spliced out and L1 ORF1p (*yellow circles*) and ORF2p (*blue circle*) are translated from the L1 mRNA (*gray*, *yellow*, *blue*, and *orange line* followed by "AAAA$_n$"). Only upon reverse transcription and integration into a genomic locus (5′ truncated *blue* and *orange box*), which is flanked by target-site duplications (TSD, *red wavy lines*), can the NEO gene be expressed to confer drug-resistance. (**b**) A timeline of the assay is depicted and described in Subheading 3. Days of the protocol are noted above and the corresponding days post-transfection (d0–14) are noted below. The end result of the assay is depicted under d14. The pJJ101/L1.3 construct is a wild type LINE-1 with a blasticidin deaminase (*mblastI*) reporter cassette and its retrotransposition results in many drug-resistant colonies. The pJJ105/L1.3 construct contains a LINE-1 reverse transcriptase mutant, cannot complete retrotransposition, and does not result in any drug-resistant colonies

new L1 copy and the reporter cassette into the genome [24]. Once integrated into the genome, the cells harboring new retrotransposition events can be identified by either selecting or screening for reporter gene expression. The number of colonies or cells expressing the reporter genes allows the quantification of the LINE-1 retrotransposition efficiency.

Since the initial publication of the LINE-1 retrotransposition assay [24], several adaptations have made the assay more efficient [58] and applicable to study an array of biological questions. For example, the *mneoI* cassette and a derivative of the cassette, neo-*Tet*, which allows the detection of a retrotransposed RNA polymerase III transcribed RNA, have been used to study the retrotransposition of Alu and SVA SINE RNAs, U6 snRNA, and cellular mRNAs [33, 43, 45–47, 49, 51]. Modified versions of the *mneoI* cassette have been used to recover LINE-1 retrotransposition events from genomic DNA, enabling detailed analyses of how LINE-1 retrotransposition events impact the genome [33, 34, 38]. Indeed, these studies revealed that de novo retrotransposition events from engineered L1 constructs resemble endogenous L1 insertions in their structure. Moreover, they revealed that L1 is not simply an insertional mutagen, but that LINE-1 retrotransposition events also can generate intrachromosomal deletions, intrachromosomal duplications, and perhaps interchromosomal translocations [33, 34, 38].

Other variations on the retrotransposition reporter cassette include incorporation of an enhanced green fluorescent protein (EGFP) reporter gene (*mEGFPI*), which has been used to assay LINE-1 retrotransposition both in vitro and in vivo [59–66]. The subsequent development of the blasticidin S-resistance reporter cassette (*mblastI*) was used to examine DNA repair deficient Chinese hamster ovary (CHO) cell lines for retrotransposition [36, 67–69]. Blasticidin S kills cells more efficiency than G418 and may be used for cell lines that have a high tolerance for G418. More recently, luciferase-based retrotransposition indicator cassettes, which may be amenable for high-throughput screening, have been developed to assay for retrotransposition [70].

In sum, the LINE-1 cultured cell retrotransposition assay has been used successfully in multiple human and other mammalian cell lines (summarized in Ref. [71]). The following protocol has been adapted from previously published assays [24, 36, 43, 58, 66] and is optimized for HeLa cells. Importantly, the necessary experimental controls required to correctly interpret retrotransposition assay results (e.g., transfection efficiency, cell viability, and off-target effects) are highlighted in Subheading 4 of this protocol.

2 Materials

2.1 Cell Culture Media and Transfection Reagents

1. Two clonal HeLa cell lines support retrotransposition. HeLa-JVM cells are used to assay LINE-1 retrotransposition [24]. HeLa-HA are used to assay either LINE-1 or Alu retrotransposition [72].
2. Cultured cell growth medium for HeLa cells.
   (a) To assay LINE-1 retrotransposition (*in cis*), HeLa-JVM cells are grown in DMEM (4.5 g/L d-glucose) containing 10 % FBS, and 1× Pen–Strep, glutamine (100 U/mL penicillin, 100 μg/mL streptomycin, and 292 μg/mL glutamine). This is called HeLa-JVM DMEM growth medium in the protocol below.
   (b) To assay Alu retrotransposition (*in trans*), HeLa-HA cells are grown in MEM (with Earle's salts) containing 10 % FBS, 1× MEM nonessential amino acids, and 1× Pen–Strep, glutamine (100 U/mL penicillin, 100 μg/mL streptomycin, and 292 μg/mL glutamine). This is called HeLa-HA MEM growth medium in the protocol below.
3. 1× Phosphate-buffer saline (PBS), pH 7.4, sterilized.
4. 0.05 % Trypsin–EDTA.
5. A cell counter (e.g., Countess® Automated Cell Counter, Life Technologies) or a hemocytometer.
6. Tissue culture dishes or flasks (6-well plates, T-75 flasks, or 10 cm dishes).
7. FuGENE® 6 transfection reagent (Promega).
8. Opti-MEM® I (Life Technologies).
9. Geneticin (G418, 50 mg/mL stock), blasticidin S-HCl (10 mg/mL stock), or puromycin (10 mg/mL stock).
10. A flow cytometer (e.g., the Accuri C6 Flow Cytometer).

2.2 Plasmids

1. To assay LINE-1 retrotransposition, *in cis* (Fig. 1):
   (a) The pCEP4 mammalian expression episomal plasmid (Life Technologies), which generally is the backbone of the LINE-1 expression plasmids.
   (b) A LINE-1 expression plasmid that is tagged with a retrotransposition reporter cassette (e.g., pJM101/L1.3 (*mneoI*), pJJ101/L1.3 (*mblastI*), or pLRE3-mEGFPI (*mEGFPI*) [67, 73–76]).
   (c) A reporter plasmid (e.g., hrGFP (Agilent)) to monitor transfection efficiency.
   (d) A plasmid expressing a cDNA of interest. This plasmid should not contain the same selectable marker as the L1 reporter cassette.

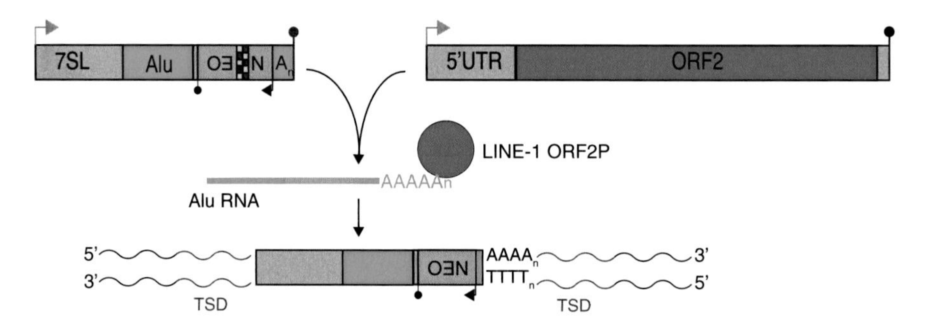

Fig. 2 Alu retrotransposition (*trans*-complementation) assay. Retrotransposition of Alu requires the LINE-1 ORF2 protein [43]. For the Alu *trans*-complementation assay, the reporter plasmid contains the Alu sequence (*light blue box*) and is tagged with a modified NEO retrotransposition indicator cassette (neo*Tet*) interrupted by a self-splicing group I intron (*checkered box*) [81]. The neo*Tet* cassette is followed by a variably sized poly A tract (A_n) and the RNA pol III terminator sequence (*red lollipop*). The Alu plasmid must be co-transfected with a plasmid (pCEP5′UTR-ORF2pΔneo) expressing LINE-1 ORF2p (*blue circle*) to detect retrotransposition in HeLa cells. The resulting G418-resistant colonies generally contain de novo full-length Alu retrotransposition events flanked by target-site duplications (TSD, *red wavy lines*)

2. To assay Alu retrotransposition, *in trans* (Fig. 2):
   (a) A "reporter" retrotransposition plasmid that contains an Alu element and a modified *mneoI* cassette (e.g., pAlu-neo*Tet*) that allows the detection of a retrotransposed RNA polymerase III transcribed RNA [43].
   (b) A "driver" LINE-1 expression plasmid (i.e., either a full-length LINE-1 or a plasmid that expresses LINE-1 ORF2p (e.g., pJM101/L1.3Δneo or pCEP5′UTR-ORF2pΔneo [51, 77])) that lacks the retrotransposition reporter cassette.
   (c) A reporter plasmid (e.g., hrGFP) to monitor transfection efficiency.
3. To test if cellular proteins affect LINE-1 retrotransposition (Fig. 3; [78–80]):
   (a) A LINE-1 expression plasmid that is tagged with a retrotransposition reporter cassette (e.g., *mneoI*, *mblastI*, or *mEGFPI*).
   (b) A non-LINE-1 control plasmid (e.g., pU6iNEO [79]) that expresses an intact (i.e., intronless) version of the same selectable or screenable marker present in the LINE-1 retrotransposition reporter cassette.
   (c) An expression vector containing the cDNA that is being tested for its effect on retrotransposition (e.g., pK_A3A that expresses APOBEC3A [78, 79]).
   (d) A negative control cDNA expression vector that expresses a protein that does not significantly affect retrotransposition (e.g., pK_β-arrestin [78, 79]).

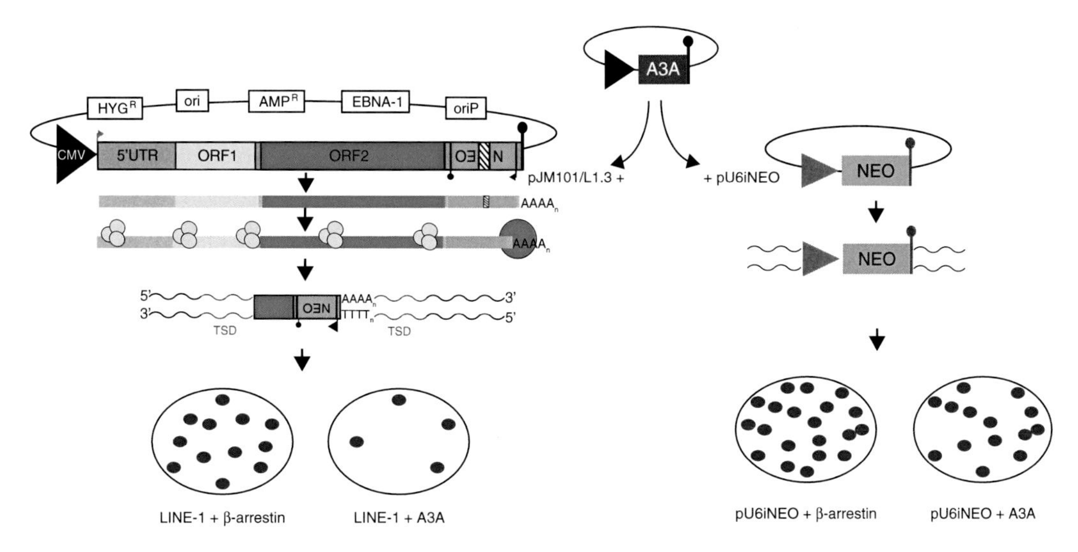

Fig. 3 Co-transfection of LINE-1 with cDNAs. The LINE-1 retrotransposition assay is carried out as described in Fig. 1 except that a plasmid expressing pK_A3A (*purple box*) is co-transfected with a LINE-1 expression construct (pJM101/L1.3, *see* Fig. 1a for details). In parallel, pK_A3A is co-transfected with a reporter control, pU6iNEO, which expresses the neomycin phosphotransferase gene without the requirement for retrotransposition (i.e., the neomycin phosphotransferase gene lacks an intron). Both sets of co-transfected cells are grown under drug-selection and the resulting drug-resistant colonies (*dark purple circles*) are counted. The parallel pU6iNEO experiment is essential to determine to what extent the effects of A3A overexpression are LINE-1 specific. Note the greater decrease in G418-resistant colonies in "LINE-1 + A3A" than "pU6iNEO + A3A" [79]

2.3 Fixing and Staining

1. Fix: 2 % formaldehyde, 0.2 % glutaraldehyde, in 1× PBS, pH 7.4.
2. Stain (can use any one of the three):
   (a) 0.1 % bromophenol blue, weight to volume (w/vol) in H_2O.
   (b) 0.1 % crystal violet blue, w/vol in H_2O.
   (c) 0.4 % Giemsa, w/vol in H_2O.

3 Methods

Standard practices for handling cultured cells should be applied, such as use of laminar flow biosafety hoods and sterile techniques for all reagents. The transfection efficiency and the concentration of drugs used for genetic selection should first be optimized for each cell type used in this assay. The rationale and a timeline of the assay are presented in Fig. 1. Cultured cell lines are routinely verified/authenticated by short tandem repeat (STR) analyses and checked for mycoplasma to ensure their integrity and lack of contamination, respectively.

3.1 LINE-1 Retrotransposition Assay in HeLa-JVM Cells

1. Day 1—Plate cells: Seed 2×10^4 HeLa-JVM cells in each well of a 6-well tissue culture plate in HeLa-JVM DMEM growth media. Cells are grown in a humidified incubator at 37 °C with 7 % CO_2 (*see* **Note 1**).
2. Day 2—Transfect cells: Cells typically are transfected 14–16 h post-plating, day zero (d0) (Fig. 1b), using the FuGENE® 6 transfection reagent following the manufacturer's instructions. Every retrotransposition assay should include the following transfection conditions: (1) a vector-only (pCEP4) or mock transfection; (2) a wild type LINE-1 retrotransposition plasmid (e.g., pJM101/L1.3, pJJ101/L1.3, or pLRE3-*mE*GFPI), which serves as a positive control; and (3) a mutant LINE-1 plasmid (e.g., pJM105/L1.3 has a mutation in the RT domain of L1 [51]), which serves as a negative control. To assay for retrotransposition, prepare a transfection mix in a 1.5 mL microcentrifuge tube containing 1 μg pCEP4 or a LINE-1 expression plasmid (Fig. 1a) and 3–4 μL of FuGENE® 6 in 100 μL of Opti-MEM® I. Incubate the solution at room temperature for 20 min. Add the transfection mix to the growth medium of one well of cells in a 6-well tissue culture plate. To determine the transfection efficiency, prepare an additional transfection mix(es) in a 1.5 mL microcentrifuge tube containing 1 μg pCEP4 or a LINE-1 expression plasmid, 0.5 μg of an hrGFP expression plasmid, and 4–6 μL of FuGENE® 6 in 100 μL of Opti-MEM® I. Incubate the solution for 20 min at room temperature. Then, add the transfection mix to the growth medium of one well of cells in a 6-well tissue culture plate; these transfections will be used to calculate transfection efficiencies. Transfect three wells in each plate for each transfection condition to yield three technical replicates. It is important to calculate transfection efficiency and adjust retrotransposition assay results for each L1 plasmid individually, as subtle differences in L1 plasmid size and transfection mix composition can affect transfection efficiency. These differences could impact the final result of the retrotransposition assay (*see* **Note 2**).
3. Day 3—Stop the transfection: Approximately 16–24 h post-transfection, 1 day post-transfection (d1) (Fig. 1b), aspirate the medium from the cells and add fresh HeLa-JVM DMEM growth medium to the cells.
4. Day 4—Determine the transfection efficiency: 2 days post-transfection (d2) (Fig. 1b), trypsinize the hrGFP-transfected cells and collect the cells from each well in separate microcentrifuge tubes (*see* **Note 3**). Spin the cells at $2000 \times g$ at 4 °C for 5 min and aspirate the medium. Wash the cells in 1× PBS, spin at $2000 \times g$ at 4 °C for 5 min, then aspirate the PBS. Resuspend the cell pellet in 250–500 μL 1× PBS. Determine the number (i.e., percentage) of hrGFP-expressing cells, gating for live

cells, on a flow cytometer (Accuri Flow Cytometer or similar; *see* **Note 4**). The number of live cells that express hrGFP serves as an indication of the percentage of cells successfully transfected with plasmids (i.e., transfection efficiency).

5. Days 5–16—Select cells for retrotransposition events: Begin drug selection 3 days post-transfection (d3) and continue until 14 days post-transfection (d14) (Fig. 1b).
   (a) For *mneoI*-based assays, add 400 μg/mL G418 to the HeLa-JVM DMEM growth medium. Change the G418-containing medium every day until 14 days post-transfection (d14).
   (b) For *mblastI*-based assays, add 10 μg/mL blasticidin S-HCl to HeLa-JVM DMEM growth medium and change the blasticidin S-HCl-containing medium once at 8 days post-transfection (d8) (*see* **Note 5**).
   (c) For *mEGFPI*-based assays, add 5 μg/mL puromycin to HeLa-JVM DMEM growth medium. Change the puromycin-containing medium every other day until 9 days post-transfection (d9) to select for cells that have the L1 expression plasmid.
6. Quantification of the LINE-1 retrotransposition assay.
   (a) For *mneoI* and *mblastI* assays, 14 days post-transfection (d14) rinse the cells with 1× PBS and fix the cells (using Fix solution, Subheading 2.3, **item 1**) for 30 min to 1 h at room temperature or longer at 4 °C. Rinse the cells in water and stain (using one of the three Stain solutions, Subheading 2.3, **item 2**) at room temperature for 1 h. Rinse the cells with water and let dry (*see* **Note 5**). Count the stained foci in each well.
   (b) For *mEGFPI* assays, 9 days post-transfection (d9), trypsinize the cells and collect the cells from each well in separate microcentrifuge tubes. Collect the cells by centrifugation at 2000 × *g* at 4 °C for 5 min. Aspirate the medium. Rinse the cells with 1× PBS and spin the cells again at 2000 × *g* at 4 °C for 5 min. Aspirate the PBS and resuspend the cell pellet in 250–500 μL 1× PBS. Analyze the number of EGFP-expressing cells, gating for live cells, on a flow cytometer. The number of live cells that express EGFP serves as an indication of the number of cells that have successfully undergone a round of de novo retrotransposition.
7. To calculate the retrotransposition efficiency, drug-resistant colonies or EGFP-expressing cells are counted and adjusted for transfection efficiency (*see* **Note 6**).
   (a) For G418- or blasticidin S-resistant colonies, calculate the mean colony counts (from **step 6a**) for the three wells of

the same transfection condition (three technical replicates). To calculate the adjusted retrotransposition mean, divide the mean colony counts and the standard deviation by the transfection efficiency (calculated in **step 4**) (*see* **Note 6**). To express the adjusted retrotransposition values as a percentage of the wild type control (i.e., the retrotransposition efficiency), divide the adjusted retrotransposition mean of an experimental sample (e.g., the number of retrotransposition events generated from a mutant L1 expression construct) by the adjusted retrotransposition mean of wild type L1 (e.g., pJM101/L1.3) and then multiply by 100 (*see* **Note 7**).

(b) For EGFP-expressing cells, divide the mean percentage of EGFP-expressing cells (calculated in **step 6b**) and the standard deviation from three wells per transfection condition (three technical replicates) by the transfection efficiency (calculated in **step 4**) to calculate the adjusted retrotransposition mean (*see* **Note 6**). The adjusted retrotransposition mean for *mEGFPI* retrotransposition assays represents the percentage of puromycin-resistant cells that express EGFP. Again, at least three biological replicates, each containing three technical replicates, should be done.

3.2 Alu Retrotransposition Assay in HeLa-HA Cells (in trans)

Retrotransposition *in trans* occurs at a lower frequency than *in cis* [43, 51]. Therefore, the retrotransposition assay detailed above is scaled-up. Please note that transfection conditions need to be optimized when using larger tissue culture plates or flasks (*see* Fig. 2 for rationale).

1. Day 1—Plate cells: Seed 5×10^5 HeLa-HA cells in 10 cm tissue culture dish or T-75 flask in HeLa-HA MEM growth medium.
2. Day 2—Transfect cells: Cells typically are transfected 14–16 h post-plating, day zero (d0) (Fig. 1b), using the FuGENE® 6 transfection reagent following the manufacturer's instructions. To assay for Alu retrotransposition, prepare a transfection mix containing 4 μg of a reporter plasmid (e.g., pAlu-neo*Tet*), 4 μg driver plasmid (e.g., pCEP5′UTR-ORF2pΔneo; Fig. 2), and 24–32 μL of FuGENE® 6 in 500 μL of Opti-MEM® I. Incubate the solution at room temperature for 20 min. Add the transfection mix to the growth medium of the cells in the tissue culture dish or flask. To determine transfection efficiencies, prepare a transfection mix containing 0.5 μg each of the reporter plasmid, the driver plasmid, and hrGFP plasmid and 4–6 μL of FuGENE® 6 in 100 μL of Opti-MEM® I. Incubate the transfection mix at room temperature for 20 min. Transfect cells with the hrGFP-containing transfection mix in 6-well plates (as stated in Subheading 3.1, **step 2**).

3. Day 3—Stop the transfection: Approximately 24 h post-transfection, 1 day post-transfection (d1) (Fig. 1b), aspirate the medium from the cells and add fresh HeLa-HA MEM growth medium to the cells.
4. Day 4—Determine the transfection efficiency: same as Subheading 3.1, **step 4**.
5. Days 5–16—Select the cells for retrotransposition events: same as Subheading 3.1, **step 5a**.
6. Quantitate the LINE-1 retrotransposition assay: same as Subheading 3.1, **step 6a**.
7. Calculate the retrotransposition efficiency: same as Subheading 3.1, **step 7a**.

3.3 LINE-1 Retrotransposition Assay Co-transfected with cDNAs

The experimental design is similar to the assay above except cells are transfected with an L1 and a cDNA expression construct. When co-transfecting an L1 plasmid with a cDNA expression construct, it is essential to monitor side effects of cDNA overexpression. In order to monitor potential cDNA off-target effects, HeLa cells should also be co-transfected with the cDNA expression plasmid and a control reporter plasmid that expresses an intact copy of the same selectable marker (i.e., no intron disrupting the reporter) as the LINE-1 retrotransposition plasmid. The LINE-1 retrotransposition assay and control reporter assay must be done in parallel (i.e., at the same time) [79, 80]. For example, the effect of APOBEC3A (A3A) on LINE-1 retrotransposition (*mneoI* reporter cassette) and on a control reporter plasmid (pU6iNEO), as described in Richardson et al., is described below [79]. Retrotransposition efficiency is corrected to the reporter control assay (*see* Fig. 3 and **Note 8**).

1. Day 1—Plate cells: Seed 2×10^4 HeLa-JVM cells in each well of a 6-well tissue culture plate in HeLa-JVM DMEM growth medium (as stated above in Subheading 3.1, **step 1**) (*see* **Note 1**).
2. Day 2—Transfect cells: Cells typically are transfected 14–16 h post-plating, day zero (d0) (Fig. 1b), using the FuGENE® 6 transfection reagent following the manufacturer's instructions. Experimental conditions for L1 transfection include: (1) pCEP4 plus pK_β-arrestin; (2) pJM101/L1.3 plus pK_β-arrestin, (which has no significant effect on LINE-1 retrotransposition and is a control for the co-transfected cDNA) [78]; (3) pJM101/L1.3 plus pK_A3A; (4) pJM105/L1.3 plus pK_β-arrestin or pK_A3A; and (5) mock transfected cells or cells transfected only with pCEP4. In parallel, experimental conditions for the reporter control plasmid include: (1) pU6iNEO plus pK_β-arrestin and (2) pU6iNEO plus pK_A3A. Prepare a transfection mix in a 1.5 mL microcentrifuge tube containing either an L1 expression or control vector

(0.5 μg of the pCEP4 vector, 0.5 μg of an L1 expression plasmid, or 0.5 μg pU6iNEO) and a cDNA expression vector (0.5 μg pK_β-arrestin or 0.5 μg of an pK_A3A expression plasmid) with 3–4 μL of FuGENE® 6 in 100 μL of Opti-MEM® I. Incubate the solution at room temperature for 20 min. Add the transfection mix to the growth medium of one well of cells in a 6-well tissue culture plate. To calculate the transfection efficiency, prepare a transfection mix in a 1.5 mL microcentrifuge tube containing the above reagents and 0.5 μg of an hrGFP expression plasmid (as described above in Subheading 3.1, **step 2**).

3. Day 3—Stop the transfection: Approximately 24 h post-transfection, 1 day post-transfection (d1) (Fig. 1b), aspirate the medium from the cells and add fresh HeLa-JVM DMEM growth medium to the cells.
4. Day 4—Determine the transfection efficiency: same as Subheading 3.1, **step 4**.
5. Days 5–16—Select the cells for retrotransposition events: same as Subheading 3.1, **step 5a**.
6. Determine the LINE-1 retrotransposition efficiency and the effect of cDNA expression on the ability of the reporter construct (pU6iNEO) to form drug resistant foci: same as Subheading 3.1, **step 6a**.
7. To calculate the corrected retrotransposition mean, the adjusted retrotransposition mean of L1 plus pK_β-arrestin is divided by the adjusted retrotransposition mean of pU6iNEO plus pK_β-arrestin (*see* **Note 8**). Similarly, the adjusted retrotransposition mean of L1 plus pK_A3A is divided by the adjusted retrotransposition mean of pU6iNEO plus pK_A3A. This corrected calculation accounts for any cytotoxic and/or off-target effects pK_A3A may have on the cell.
8. To calculate the corrected retrotransposition efficiency, divide the corrected retrotransposition mean of L1 plus pK_A3A expression vector by the corrected retrotransposition mean of L1 plus pK_β-arrestin and multiply by 100 (*see* **Note 9**). The retrotransposition efficiency reported reflects the specific inhibition of pK_A3A on LINE-1 retrotransposition (*see* **Note 10**).

4 Notes

1. Each cell line must be optimized for cell plating density, plasmid concentrations, transfection reagents, and the drug concentration needed for selection. The experiment also can be scaled up into flasks:
   (a) For 75 cm² flasks: transfect 2×10^6 cells with 8 μg DNA, 32 μL FuGENE® 6, and 500 μL of Opti-MEM® I.

(b) For 175 cm^2 flasks: transfect 6×10^6 cells with 10–25 μg DNA, 40–100 μL FuGENE® 6, and 1 mL of Opti-MEM® I.

2. Assays are done with three technical replicates in order to calculate a standard deviation for each experimental variable. Each assay is repeated at least three independent times, yielding three biological replicates.
3. Typically, co-transfect 1 μg L1 plasmid and 0.5 μg hrGFP plasmid for each well of a 6-well plate to calculate transfection efficiency. A mock-transfected cell sample should always be included to properly gate true GFP-expressing and non-expressing cells. Using *mEGFPI* does not interfere with detection of hrGFP expression because GFP expression from *mEGFPI* resulting from retrotransposition is detectable between 7 and 9 days post-transfection (d7–d9) [66]. GFP expression from hrGFP is detectable 1–3 days post transfection (d1–d3). Importantly, GFP expression from the retrotransposed *mEGFPI* reporter cassette is analyzed in cells transfected without hrGFP, as these transfections are done in parallel. Alternatively, a plasmid expressing mCherry can be used to determine transfection efficiency.
4. Retrotransposition efficiency can only be calculated from cells that are transfected with an L1 construct. It is important to account for slight variations in transfection efficiencies when calculating retrotransposition efficiencies. Use of Lipofectamine® (Life Technologies) also has been reported for L1 studies, following manufacturer's directions [24]. Transfection efficiencies will vary among cell lines.
5. Blasticidin S selection was previously reported as starting 5 days post-transfection (d5) in Chinese hamster ovary cells [36]. Selection with blasticidin S should be optimized when using different cell lines. HeLa-JVM cells die sooner under blasticidin S-selection than G418-selection. Blasticidin S-resistant colonies can be fixed and stained 10–12 days post-transfection (d10–d12).
6. Adjusted retrotransposition mean = (average number of drug-resistant colonies)/(fraction of hrGFP-positive cells).
7. Retrotransposition efficiency = 100 × (adjusted retrotransposition mean)/(adjusted retrotransposition mean of the wild type L1). For example, if the adjusted retrotransposition mean of wild type L1 is 88, then the retrotransposition efficiency is 100 × (88/88) = 100 %. Similarly, if the adjusted retrotransposition efficiency of an EN mutant L1 from the same experiment is 1, then the retrotransposition efficiency of the EN mutant L1 is 100 × (1/88) = 1.1 %.

8. Corrected retrotransposition mean = (adjusted L1 plus expression plasmid retrotransposition mean)/(mean number of colonies from the reporter control plus the same expression plasmid). For example, the corrected retrotransposition mean of L1 + pK_β-arrestin is calculated by dividing the adjusted retrotransposition mean of L1 + pK_β-arrestin by the mean number of colonies from pU6iNEO+ pK_β-arrestin. Similarly, the corrected retrotransposition mean of L1 + pK_A3A is calculated by dividing the adjusted retrotransposition mean of L1 + pK_A3A by the mean number of colonies from pU6iNEO+ pK_A3A.
9. Corrected retrotransposition efficiency = 100 × (corrected retrotransposition mean for L1 plus cDNA expression plasmid)/(corrected retrotransposition mean of the wild type L1 plus an empty vector or pK_β-arrestin expression plasmid control).
10. Co-transfections involving cDNAs expressing host factors must be assayed for off-target effects (e.g., cell toxicity) before interpreting the effects on retrotransposition. For example, overexpression of A3A decreases the number of G418-resistant colonies in cells co-transfected with a control G418-resistance plasmid (pU6iNEO), but dramatically decreases the number of G418-resistant colonies in cells co-transfected with an L1/*mneoI* plasmid [79] (Fig. 3). A possible interpretation of these data is that the decrease in LINE-1 retrotransposition is due to the cytotoxicity of A3A overexpression. However, the cytotoxic effects of A3A overexpression are not solely responsible for the reduction in LINE-1 retrotransposition (LINE-1/*mneoI* plus A3A) because the retrotransposition efficiency is corrected to the control plasmid (pU6iNEO plus A3A) to account for the A3A cytotoxic contribution. This correction of the data reveals a clearer representation of the specific effect of A3A on LINE-1 retrotransposition by accounting for the non-specific, off-target effects of A3A overexpression. Testing the same conditions on a non-L1 plasmid containing the same selectable or screenable marker (i.e., drug-resistance or GFP, respectively) would account for any indirect, off-target effects from data interpretation.

Acknowledgements

The authors would like to thank Nancy Leff for helpful comments during the preparation of this manuscript. This work was supported in part by NIH grant GM060518 to J.V.M. The authors were supported in part by fellowships from the American Cancer Society #PF-07-059-01GMC (H.C.K.), NIGMS #5-T32-GM07544 (P.A.L. and S.R.R.), and NIGMS #T32-GM007315 (J.B.M.). J.V.M. is an investigator of the Howard Hughes Medical Institute.

Conflict of Interest

J.V.M. is an inventor on the patent: "Kazazian, H.H., Boeke, J.D., Moran, J.V., and Dombrowski, B.A. Compositions and methods of use of mammalian retrotransposons. Application No. 60/006,831; Patent No. 6,150,160; Issued November 21, 2000." J.V.M. has not made any money from this patent and voluntarily discloses this information.

References

1. Lander ES, Linton LM, Birren B et al (2001) Initial sequencing and analysis of the human genome. Nature 409:860–921
2. Beck CR, Garcia-Perez JL, Badge RM, Moran JV (2011) LINE-1 elements in structural variation and disease. Annu Rev Genomics Hum Genet 12:187–215
3. Dombroski BA, Mathias SL, Nanthakumar E, Scott AF, Kazazian HH Jr (1991) Isolation of an active human transposable element. Science 254:1805–1808
4. Scott AF, Schmeckpeper BJ, Abdelrazik M, Comey CT, O'Hara B, Rossiter JP, Cooley T, Heath P, Smith KD, Margolet L (1987) Origin of the human L1 elements: proposed progenitor genes deduced from a consensus DNA sequence. Genomics 1:113–125
5. Athanikar JN, Badge RM, Moran JV (2004) A YY1-binding site is required for accurate human LINE-1 transcription initiation. Nucleic Acids Res 32:3846–3855
6. Becker KG, Swergold GD, Ozato K, Thayer RE (1993) Binding of the ubiquitous nuclear transcription factor YY1 to a cis regulatory sequence in the human LINE-1 transposable element. Hum Mol Genet 2:1697–1702
7. Swergold GD (1990) Identification, characterization, and cell specificity of a human LINE-1 promoter. Mol Cell Biol 10:6718–6729
8. Speek M (2001) Antisense promoter of human L1 retrotransposon drives transcription of adjacent cellular genes. Mol Cell Biol 21:1973–1985
9. Callahan KE, Hickman AB, Jones CE, Ghirlando R, Furano AV (2012) Polymerization and nucleic acid-binding properties of human L1 ORF1 protein. Nucleic Acids Res 40:813–827
10. Hohjoh H, Singer MF (1996) Cytoplasmic ribonucleoprotein complexes containing human LINE-1 protein and RNA. EMBO J 15:630–639
11. Hohjoh H, Singer MF (1997) Ribonuclease and high salt sensitivity of the ribonucleoprotein complex formed by the human LINE-1 retrotransposon. J Mol Biol 271:7–12
12. Holmes SE, Singer MF, Swergold GD (1992) Studies on p40, the leucine zipper motif-containing protein encoded by the first open reading frame of an active human LINE-1 transposable element. J Biol Chem 267:19765–19768
13. Khazina E, Weichenrieder O (2009) Non-LTR retrotransposons encode noncanonical RRM domains in their first open reading frame. Proc Natl Acad Sci U S A 106:731–736
14. Kolosha VO, Martin SL (1997) In vitro properties of the first ORF protein from mouse LINE-1 support its role in ribonucleoprotein particle formation during retrotransposition. Proc Natl Acad Sci U S A 94:10155–10160
15. Kolosha VO, Martin SL (2003) High-affinity, non-sequence-specific RNA binding by the open reading frame 1 (ORF1) protein from long interspersed nuclear element 1 (LINE-1). J Biol Chem 278:8112–8117
16. Martin SL (1991) Ribonucleoprotein particles with LINE-1 RNA in mouse embryonal carcinoma cells. Mol Cell Biol 11:4804–4807
17. Martin SL, Bushman FD (2001) Nucleic acid chaperone activity of the ORF1 protein from the mouse LINE-1 retrotransposon. Mol Cell Biol 21:467–475
18. Doucet AJ, Hulme AE, Sahinovic E, Kulpa DA, Moldovan JB, Kopera HC, Athanikar JN, Hasnaoui M, Bucheton A, Moran JV, Gilbert N (2010) Characterization of LINE-1 ribonucleoprotein particles. PLoS Genet 6:e1001150
19. Ergun S, Buschmann C, Heukeshoven J, Dammann K, Schnieders F, Lauke H, Chalajour F, Kilic N, Stratling WH, Schumann GG (2004) Cell type-specific expression of LINE-1 open reading frames 1 and 2 in fetal and adult human tissues. J Biol Chem 279:27753–27763
20. Feng Q, Moran JV, Kazazian HH Jr, Boeke JD (1996) Human L1 retrotransposon encodes a conserved endonuclease required for retrotransposition. Cell 87:905–916
21. Goodier JL, Ostertag EM, Engleka KA, Seleme MC, Kazazian HH Jr (2004) A poten-

tial role for the nucleolus in L1 retrotransposition. Hum Mol Genet 13:1041–1048
22. Hattori M, Kuhara S, Takenaka O, Sakaki Y (1986) L1 family of repetitive DNA sequences in primates may be derived from a sequence encoding a reverse transcriptase-related protein. Nature 321:625–628
23. Mathias SL, Scott AF, Kazazian HH Jr, Boeke JD, Gabriel A (1991) Reverse transcriptase encoded by a human transposable element. Science 254:1808–1810
24. Moran JV, Holmes SE, Naas TP, DeBerardinis RJ, Boeke JD, Kazazian HH Jr (1996) High frequency retrotransposition in cultured mammalian cells. Cell 87:917–927
25. Taylor MS, Lacava J, Mita P, Molloy KR, Huang CR, Li D, Adney EM, Jiang H, Burns KH, Chait BT, Rout MP, Boeke JD, Dai L (2013) Affinity proteomics reveals human host factors implicated in discrete stages of LINE-1 retrotransposition. Cell 155:1034–1048
26. Fanning T, Singer M (1987) The LINE-1 DNA sequences in four mammalian orders predict proteins that conserve homologies to retrovirus proteins. Nucleic Acids Res 15:2251–2260
27. Moran JV (1999) Human L1 retrotransposition: insights and peculiarities learned from a cultured cell retrotransposition assay. Genetica 107:39–51
28. Moran JV, DeBerardinis RJ, Kazazian HH Jr (1999) Exon shuffling by L1 retrotransposition. Science 283:1530–1534
29. Perepelitsa-Belancio V, Deininger P (2003) RNA truncation by premature polyadenylation attenuates human mobile element activity. Nat Genet 35:363–366
30. Usdin K, Furano AV (1989) The structure of the guanine-rich polypurine:polypyrimidine sequence at the right end of the rat L1 (LINE) element. J Biol Chem 264:15681–15687
31. Luan DD, Korman MH, Jakubczak JL, Eickbush TH (1993) Reverse transcription of R2Bm RNA is primed by a nick at the chromosomal target site: a mechanism for non-LTR retrotransposition. Cell 72:595–605
32. Cost GJ, Feng Q, Jacquier A, Boeke JD (2002) Human L1 element target-primed reverse transcription in vitro. EMBO J 21:5899–5910
33. Gilbert N, Lutz S, Morrish TA, Moran JV (2005) Multiple fates of L1 retrotransposition intermediates in cultured human cells. Mol Cell Biol 25:7780–7795
34. Gilbert N, Lutz-Prigge S, Moran JV (2002) Genomic deletions created upon LINE-1 retrotransposition. Cell 110:315–325
35. Jurka J (1997) Sequence patterns indicate an enzymatic involvement in integration of mammalian retroposons. Proc Natl Acad Sci U S A 94:1872–1877
36. Morrish TA, Gilbert N, Myers JS, Vincent BJ, Stamato TD, Taccioli GE, Batzer MA, Moran JV (2002) DNA repair mediated by endonuclease-independent LINE-1 retrotransposition. Nat Genet 31:159–165
37. Grimaldi G, Skowronski J, Singer MF (1984) Defining the beginning and end of KpnI family segments. EMBO J 3:1753–1759
38. Symer DE, Connelly C, Szak ST, Caputo EM, Cost GJ, Parmigiani G, Boeke JD (2002) Human l1 retrotransposition is associated with genetic instability in vivo. Cell 110:327–338
39. Myers JS, Vincent BJ, Udall H, Watkins WS, Morrish TA, Kilroy GE, Swergold GD, Henke J, Henke L, Moran JV, Jorde LB, Batzer MA (2002) A comprehensive analysis of recently integrated human Ta L1 elements. Am J Hum Genet 71:312–326
40. Beck CR, Collier P, Macfarlane C, Malig M, Kidd JM, Eichler EE, Badge RM, Moran JV (2010) LINE-1 retrotransposition activity in human genomes. Cell 141:1159–1170
41. Buzdin A, Gogvadze E, Kovalskaya E, Volchkov P, Ustyugova S, Illarionova A, Fushan A, Vinogradova T, Sverdlov E (2003) The human genome contains many types of chimeric retrogenes generated through in vivo RNA recombination. Nucleic Acids Res 31:4385–4390
42. Buzdin A, Ustyugova S, Gogvadze E, Vinogradova T, Lebedev Y, Sverdlov E (2002) A new family of chimeric retrotranscripts formed by a full copy of U6 small nuclear RNA fused to the 3′ terminus of l1. Genomics 80:402–406
43. Dewannieux M, Esnault C, Heidmann T (2003) LINE-mediated retrotransposition of marked Alu sequences. Nat Genet 35:41–48
44. Dewannieux M, Heidmann T (2005) L1-mediated retrotransposition of murine B1 and B2 SINEs recapitulated in cultured cells. J Mol Biol 349:241–247
45. Esnault C, Maestre J, Heidmann T (2000) Human LINE retrotransposons generate processed pseudogenes. Nat Genet 24:363–367
46. Garcia-Perez JL, Doucet AJ, Bucheton A, Moran JV, Gilbert N (2007) Distinct mechanisms for trans-mediated mobilization of cellular RNAs by the LINE-1 reverse transcriptase. Genome Res 17:602–611
47. Hancks DC, Goodier JL, Mandal PK, Cheung LE, Kazazian HH Jr (2011) Retrotransposition

of marked SVA elements by human L1s in cultured cells. Hum Mol Genet 20:3386–3400
48. Hancks DC, Mandal PK, Cheung LE, Kazazian HH Jr (2012) The minimal active human SVA retrotransposon requires only the 5′-hexamer and Alu-like domains. Mol Cell Biol 32:4718–4726
49. Raiz J, Damert A, Chira S, Held U, Klawitter S, Hamdorf M, Lower J, Stratling WH, Lower R, Schumann GG (2012) The non-autonomous retrotransposon SVA is trans-mobilized by the human LINE-1 protein machinery. Nucleic Acids Res 40:1666–1683
50. Weber MJ (2006) Mammalian small nucleolar RNAs are mobile genetic elements. PLoS Genet 2:e205
51. Wei W, Gilbert N, Ooi SL, Lawler JF, Ostertag EM, Kazazian HH, Boeke JD, Moran JV (2001) Human L1 retrotransposition: cis preference versus trans complementation. Mol Cell Biol 21:1429–1439
52. Kazazian HH Jr, Wong C, Youssoufian H, Scott AF, Phillips DG, Antonarakis SE (1988) Haemophilia A resulting from de novo insertion of L1 sequences represents a novel mechanism for mutation in man. Nature 332:164–166
53. Hancks DC, Kazazian HH Jr (2012) Active human retrotransposons: variation and disease. Curr Opin Genet Dev 22:191–203
54. Boeke JD, Garfinkel DJ, Styles CA, Fink GR (1985) Ty elements transpose through an RNA intermediate. Cell 40:491–500
55. Curcio MJ, Garfinkel DJ (1991) Single-step selection for Ty1 element retrotransposition. Proc Natl Acad Sci U S A 88:936–940
56. Heidmann T, Heidmann O, Nicolas JF (1988) An indicator gene to demonstrate intracellular transposition of defective retroviruses. Proc Natl Acad Sci U S A 85:2219–2223
57. Freeman JD, Goodchild NL, Mager DL (1994) A modified indicator gene for selection of retrotransposition events in mammalian cells. Biotechniques 17:46, 48-49, 52
58. Wei W, Morrish TA, Alisch RS, Moran JV (2000) A transient assay reveals that cultured human cells can accommodate multiple LINE-1 retrotransposition events. Anal Biochem 284:435–438
59. Coufal NG, Garcia-Perez JL, Peng GE, Marchetto MC, Muotri AR, Mu Y, Carson CT, Macia A, Moran JV, Gage FH (2011) Ataxia telangiectasia mutated (ATM) modulates long interspersed element-1 (L1) retrotransposition in human neural stem cells. Proc Natl Acad Sci U S A 108:20382–20387
60. Coufal NG, Garcia-Perez JL, Peng GE, Yeo GW, Mu Y, Lovci MT, Morell M, O'Shea KS, Moran JV, Gage FH (2009) L1 retrotransposition in human neural progenitor cells. Nature 460:1127–1131
61. Garcia-Perez JL, Morell M, Scheys JO, Kulpa DA, Morell S, Carter CC, Hammer GD, Collins KL, O'Shea KS, Menendez P, Moran JV (2010) Epigenetic silencing of engineered L1 retrotransposition events in human embryonic carcinoma cells. Nature 466:769–773
62. Kubo S, Seleme MC, Soifer HS, Perez JL, Moran JV, Kazazian HH Jr, Kasahara N (2006) L1 retrotransposition in nondividing and primary human somatic cells. Proc Natl Acad Sci U S A 103:8036–8041
63. Muotri AR, Chu VT, Marchetto MC, Deng W, Moran JV, Gage FH (2005) Somatic mosaicism in neuronal precursor cells mediated by L1 retrotransposition. Nature 435:903–910
64. Muotri AR, Marchetto MC, Coufal NG, Oefner R, Yeo G, Nakashima K, Gage FH (2010) L1 retrotransposition in neurons is modulated by MeCP2. Nature 468:443–446
65. Ostertag EM, DeBerardinis RJ, Goodier JL, Zhang Y, Yang N, Gerton GL, Kazazian HH Jr (2002) A mouse model of human L1 retrotransposition. Nat Genet 32:655–660
66. Ostertag EM, Prak ET, DeBerardinis RJ, Moran JV, Kazazian HH Jr (2000) Determination of L1 retrotransposition kinetics in cultured cells. Nucleic Acids Res 28:1418–1423
67. Kopera HC, Moldovan JB, Morrish TA, Garcia-Perez JL, Moran JV (2011) Similarities between long interspersed element-1 (LINE-1) reverse transcriptase and telomerase. Proc Natl Acad Sci U S A 108:20345–20350
68. Morrish TA, Garcia-Perez JL, Stamato TD, Taccioli GE, Sekiguchi J, Moran JV (2007) Endonuclease-independent LINE-1 retrotransposition at mammalian telomeres. Nature 446:208–212
69. Goodier JL, Zhang L, Vetter MR, Kazazian HH Jr (2007) LINE-1 ORF1 protein localizes in stress granules with other RNA-binding proteins, including components of RNA interference RNA-induced silencing complex. Mol Cell Biol 27:6469–6483
70. Xie Y, Rosser JM, Thompson TL, Boeke JD, An W (2011) Characterization of L1 retrotransposition with high-throughput dual-luciferase assays. Nucleic Acids Res 39:e16
71. Rangwala SH, Kazazian HH Jr (2009) The L1 retrotransposition assay: a retrospective and toolkit. Methods 49:219–226

72. Hulme AE, Bogerd HP, Cullen BR, Moran JV (2007) Selective inhibition of Alu retrotransposition by APOBEC3G. Gene 390:199–205
73. Brouha B, Schustak J, Badge RM, Lutz-Prigge S, Farley AH, Moran JV, Kazazian HH Jr (2003) Hot L1s account for the bulk of retrotransposition in the human population. Proc Natl Acad Sci U S A 100:5280–5285
74. Sassaman DM, Dombroski BA, Moran JV, Kimberland ML, Naas TP, DeBerardinis RJ, Gabriel A, Swergold GD, Kazazian HH Jr (1997) Many human L1 elements are capable of retrotransposition. Nat Genet 16:37–43
75. Brouha B, Meischl C, Ostertag E, de Boer M, Zhang Y, Neijens H, Roos D, Kazazian HH Jr (2002) Evidence consistent with human L1 retrotransposition in maternal meiosis I. Am J Hum Genet 71:327–336
76. Garcia-Perez JL, Marchetto MC, Muotri AR, Coufal NG, Gage FH, O'Shea KS, Moran JV (2007) LINE-1 retrotransposition in human embryonic stem cells. Hum Mol Genet 16:1569–1577
77. Alisch RS, Garcia-Perez JL, Muotri AR, Gage FH, Moran JV (2006) Unconventional translation of mammalian LINE-1 retrotransposons. Genes Dev 20:210–224
78. Bogerd HP, Wiegand HL, Hulme AE, Garcia-Perez JL, O'Shea KS, Moran JV, Cullen BR (2006) Cellular inhibitors of long interspersed element 1 and Alu retrotransposition. Proc Natl Acad Sci U S A 103:8780–8785
79. Richardson SR, Narvaiza I, Planegger RA, Weitzman MD, Moran JV (2014) APOBEC3A deaminates transiently exposed single-strand DNA during LINE-1 retrotransposition. Elife 3:e02008
80. Moldovan JB, Moran JV (2015) The zinc-finger antiviral protein ZAP inhibits LINE and Alu retrotransposition. PLoS Genet 11(5): e1005121
81. Esnault C, Casella JF, Heidmann T (2002) A Tetrahymena thermophila ribozyme-based indicator gene to detect transposition of marked retroelements in mammalian cells. Nucleic Acids Res 30:e49

Chapter 11

L1 Retrotransposition in Neural Progenitor Cells

Alysson R. Muotri

Abstract

Long interspersed nucleotide element 1 (LINE-1 or L1) is a family of non-LTR retrotransposons that can replicate and reintegrate into the host genome. L1s have considerably influenced mammalian genome evolution by retrotransposing during germ cell development or early embryogenesis, leading to massive genome expansion. For many years, L1 retrotransposons were viewed as a selfish DNA parasite that had no contribution in somatic cells. Historically, L1s were thought to only retrotranspose during gametogenesis and in neoplastic processes, but recent studies have shown that L1s are extremely active in the mouse, rat, and human neuronal progenitor cells (NPCs). These de novo L1 insertions can impact neuronal transcriptional expression, creating unique transcriptomes of individual neurons, possibly contributing to the uniqueness of the individual cognition and mental disorders in humans.

Key words LINE-1, L1, Retrotransposition, Neural stem cells, Neural progenitor cells, Somatic mosaicism, Brain

1 Introduction

Neural stem cells (NSC) reside in discrete neurogenic regions of the adult brain. NSC can remain multipotent and continue to replicate in the neurogenic niche. Upon stimulus, NSC can differentiate into glial progenitors, which will mature into astrocytes or oligodendrocytes, or into neuronal progenitor cells (NPCs). When a committed to the neuronal lineage, a specific gene expression profile is activated. Interestingly, L1 retrotransposon is one of the transcripts upregulated upon neuronal commitment [1]. And since the L1 element is an autonomous retrotransposons [2], does it actually retrotranspose in NPCs?

Previous studies have indicated that L1 retrotransposition can occur in germ cells or in early embryogenesis, before the germ line becomes a distinct lineage [3, 4], whereas a cultured cell retrotransposition assay has revealed that human and mouse L1 elements can retrotranspose in a variety of transformed or immortalized cultured cell lines [5–7]. Now we know L1 is capable of high levels retrotransposition

Jose L. Garcia-Pérez (ed.), *Transposons and Retrotransposons: Methods and Protocols*, Methods in Molecular Biology, vol. 1400, DOI 10.1007/978-1-4939-3372-3_11,

in NPCs, generating a neuronal genetic mosaicism in differentiated networks [8].

Initial studies on this field had take advantage of an engineered active L1 retrotransposition cassette, a marker that only expresses if the element inserts back into the genome [5, 9]. For example, if the marker in the L1 cassette is eGFP, then a cell expressing eGFP indicates the engineered L1 successfully retrotransposed. Utilizing the L1-eGFP cassette, the ability of NPCs to support L1 retrotransposition was first reported in rat hippocampal NPCs in vitro [1]. Since then, the L1-eGFP cassette has proven L1 can also retrotranspose in human NPCs in vitro and mouse NPCs in vivo [1, 10–13]. Determining the integration sites of de novo L1 sequences in NPCs would yield insights into the potential effects of neuronal retrotransposition. Early sequencing efforts in neurons discovered L1 could integrate into introns of active genes [1, 10]. L1 integration events in rat NPCs have been shown to alter expression of nearby genes by promoter enhancement and epigenetic silencing [1].

Here, we show how to optimize the protocol to test L1 retrotransposition in neural progenitor cells using a tagged L1-eGFP element in a plasmid. Subsequent investigations may include cellular phenotypic assays and genomic analyses for de novo L1 insertions.

2 Materials (*See* Note 1)

1. A fresh maxi-prep of the L1-eGFP (Fig. 1) plasmid in a concentration of 1 μg/μL diluted in water. We generally use a Endo-Free plasmid Maxi kit to purify this DNA. *See* **Note 2** for a detailed description of this construct.
2. A retrotransposition-defective L1 construct (JM111-eGFP) that contains two missense mutations in ORF1 [5, 14, 15] can be used as a negative control (*see* **Note 3**).
3. A positive using the same plasmid backbone (pCEP4, Invitrogen) containing the eGFP driven by the CMV promoter.
4. A fresh culture of NPCs prepared and cultured as described for rodent adult NPCs [16] or iPSC-derived NPCs [11, 17]. *See* **Note 4**.
5. Dulbecco's modified Eagle medium (DMEM/F12).
6. N2 supplement (Invitrogen).
7. L-glutamine in solution.
8. Basic fibroblast growth factor (FGF-2). Prepare it following manufacturer's instructions.
9. Poly-L-ornithine and laminin coated tissue culture grade dishes.
10. A transfection system. Because of the large size of the L1 indicator plasmids and the sensitivity of the NPC, we have successfully used the Nucleofector technology (Lonza). Make sure

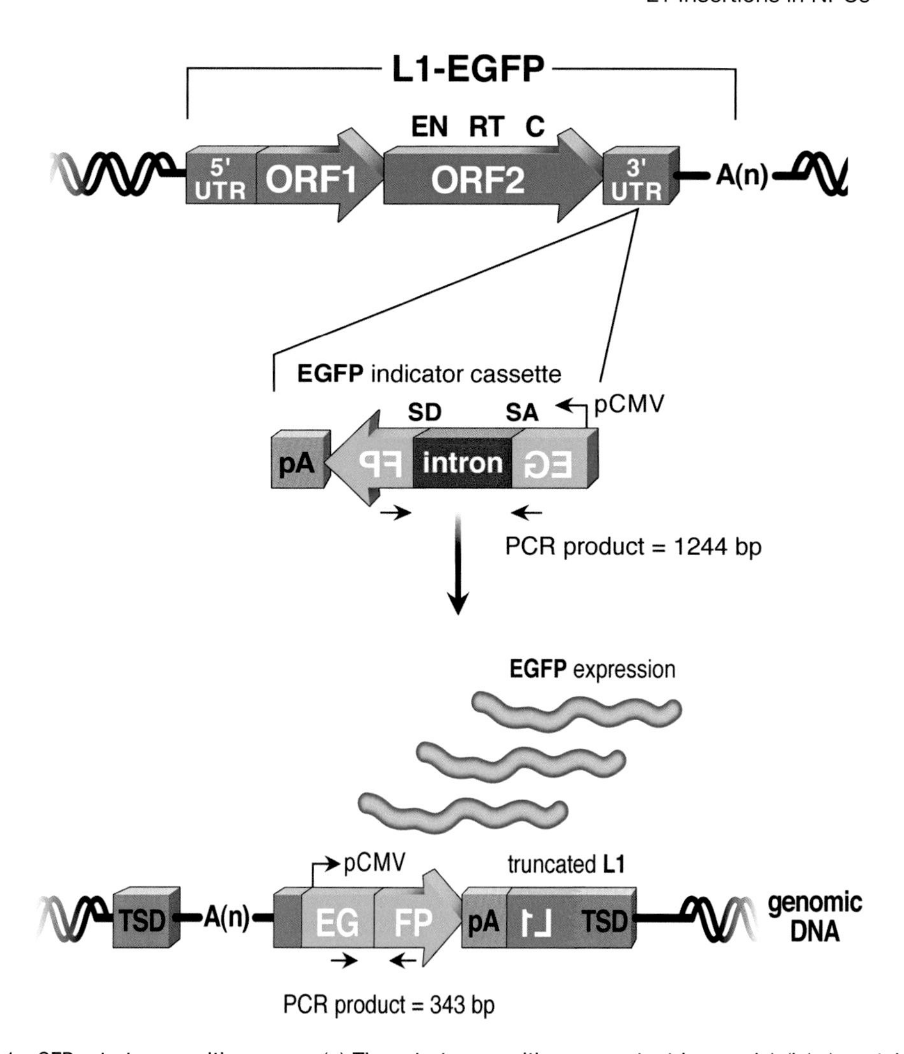

Fig. 1 L1-eGFP retrotransposition assay. (**a**) The retrotransposition-competent human L1 ($L1_{RP}$) contains a 5′ untranslated region (UTR) that harbors an internal promoter, two open reading frames (ORF1 and ORF2; not drawn to scale), and a 3′ UTR that ends in a poly (A) tail. ORF2 contains an endonuclease (EN) and reverse transcriptase (RT) domain as well as a cysteine-rich 3′ end (C). The eGFP retrotransposition indicator cassette consists of a *backward copy* of the *eGFP* gene whose expression is controlled by the human cytomegalovirus major immediate early promoter (pCMV) and the herpes simplex virus thymidine kinase polyadenylation sequence (pA). This arrangement ensures that eGFP expression will only become activated upon L1 retrotransposition. The *black arrows* indicate PCR primers flanking the intron present in the *eGFP* gene. The 343-bp PCR product, diagnostic for the loss of the intron, is indicative of a retrotransposition event. Sequencing of the 343-bp PCR product should validate the precise splicing of the intron

you have the right kit for your cells (rodent or human NPC). *See* **Note 5**.

11. Puromycin.
    (a) Warm (37 °C) the required volume of N2-medium, add fresh FGF-2 just before you start the experiment (*see* **Note 4**).

(b) Working solution of puromycin or other specific antibiotic to select the L1 reporter in the target cells.

3 Methods (*See* Note 6)

1. Prepare plasmid aliquots (1–5 μL) in no more than 10 μL volume in sterile 1 mL microtubes.
2. Prepare 100 μL of Nucleofector solution in sterile 1 mL microtubes by mixing the cell type specific Nucleofector solution with the supplement reagent provided by the kit. The final solution can be store at 4 °C for a month.
3. Harvest the target cell types using appropriated methods, and gently centrifuge (1000 × *g*) to final volume of 100 μL. *See* **Note 7**.
4. Mix the cells with the Nucleofector solution to a final volume of 200 μL.
5. Add the right amount of plasmid DNA (4 μg) to the mixture. Gently pipette up and down to homogenize the mixture.
6. Immediately, transfer the solution to the electroporation cuvette, avoiding bubbles. Select the cell-type specific program and proceed with the electroporation. *See* **Note 8**.
7. Remove the cuvette from the machine; slowly add 500 μL of pre-warmed media with FGF-2 to the cuvette.
8. Remove the total volume from the cuvette and immediately plate the electroporated cells into pre-warm culture dishes with media. Make sure they are well distributed along the tissue culture dish. *Optional*: you can add extra add 500 μL of pre-warmed media with FGF-2 to the cuvette to recovery remaining cells.
9. Wait 3–8 h for cells to attach to the bottom of the plate. Time will depend on cell type and viability. Next, change the entire media from the culture dishes. *See* **Note 9**.
10. Cells harboring the L1 expression constructs can be selected by the addition of puromycin (1 μg/mL) 48 h after electroporation to the culture medium. Puromycin-resistant cells can be screened for EGFP expression by flow cytometry.
11. After 3–7 days, eGFP expression can be detected in NPCs (*see* **Note 10**). After 7 days, transfected puromycin-resistant cells can be analyzed with a Becton Dickenson FACStar Plus containing a blue argon laser (488 nm) and fluorescein filter sets (530/30 bandpass). A total of 10,000 events were analyzed based on forward scatter versus side scatter profiles. Dead cells were excluded by propidium iodide gating. Live cells were analyzed for fluorescence intensity.

12. PCR can confirm the presence of the retrotransposed (i.e., spliced) eGFP gene in NPC-positive cells; sequencing of the PCR products will confirm the precise splicing of the intron (Fig. 1). The eGFP PCR primers were previously described [15]. *See* **Note 11**.

4 Notes

1. Prepare all cell media and washing solutions using ultrapure water (prepared by purifying deionized water to attain a sensitivity of 18 MΩ cm at 25 °C) and analytical grade reagents. Prepare and store all reagents at room temperature (unless indicated otherwise). We do not add sodium azide to the reagents. Diligently follow all waste disposal regulations when disposing waste materials.
2. The indicator cassette consists of the enhanced green fluorescence protein gene (eGFP) in the opposite orientation of the L1 transcript, a heterologous promoter (pCMV), and a polyadenylation signal (pA) in the 3′ UTR region of the element (Fig. 1). The eGFP gene is interrupted by an intron (IVS2 of the γ-globin gene) in the same transcriptional orientation as the L1 transcript. This arrangement ensures that eGFP-positive cells will arise only when a transcript initiated from the promoter driving L1 expression is spliced, reverse-transcribed, and integrated into chromosomal DNA, thereby allowing expression of the retrotransposed eGFP gene from the pCMV promoter [5, 15].
3. Make sure you sequence or use restriction enzymes to verify the identity of your plasmids maxi-prep. The L1-eGFP or other L1 indicator plasmid is available upon request from pioneers in the L1 retrotransposon field, Drs. Haig Kazazian, Joef Boeke, and John Moran. Here, we used the eGFP to exemplify the retrotransposition assay in neural progenitor cells, but other genes as L1 indicators should work just fine.
4. Briefly, NPCs were maintained on poly-l-ornithine and laminin coated flasks in N2-medium. Make sure you wash your coated plates at least three times with PBS before use. Coating left over materials may affect cell survival and proliferation.

   N2 media is comprised of Dulbecco's modified Eagle medium (DMEM/F12) (1:1), N2 supplement (Invitrogen), 2 mM l-glutamine and 20 ng/mLFGF-2. Change the N2/FGF-2 media every 2–3 days. Passage the cells when they reach about 90 % confluency.
5. Other reagents such as Lipofectamine (Invitrogen), Fugene 6 (Promega/Roche) or calcium-based transfection can also work and needs to be optimized for the target cell type. Previously optimize your transfection efficiency using a reporter system

(such as the pCEP-eGFP) in your target cells. We observed some variability with different cell preparations, species, and passage number.

6. Carry out all procedures at room temperature unless otherwise specified. All tissue culture experiments should be performed inside safety cabinets, following sterile working practices.
7. We avoid using enzymatic methods but if you do, make sure to inhibit the enzyme or wash the cell pellet before proceed. Also, the number of cells to be electroporated needs to be optimized. We find that 50,000 rat NPCs to be ideal in our conditions. Too much cells may inhibit the reaction and too little may decrease survival.
8. Generally, we perform one reaction a time, and have observed toxicity in cells that stay in the cuvette for more than 10 min.
9. The Nucleofector solution is quite toxic. Thus, the earlier you can replace the culture medium, the better.
10. If you experience cell death in the first hours post-transfection, you may try to use 24–48 h conditioned media (50:50) from healthy-growing cell cultures, instead pure fresh media. Some cells benefit from released factors and survive better in this condition.
11. Puromycin-resistant eGFP-negative NPC clones may also harbor a PCR product that corresponded in size to the retrotransposed eGFP gene. This observation may represent a truncation of the 5′ end or the retrotransposed eGFP gene may undergo epigenetic silencing either during or soon after L1 retrotransposition.

Acknowledgements

The work was supported by grants from the National Institutes of Health (NIH) R01 MH094753-01, and the NIH Director's New Innovator Award Program, 1-DP2-OD006495-01.

References

1. Muotri AR, Chu VT, Marchetto MC, Deng W, Moran JV, Gage FH (2005) Somatic mosaicism in neuronal precursor cells mediated by L1 retrotransposition. Nature 435:903–910
2. Muotri AR, Marchetto MC, Coufal NG, Gage FH (2007) The necessary junk: new functions for transposable elements. Hum Mol Genet 16(Spec No. 2):R159–R167. doi:10.1093/hmg/ddm196, 16/R2/R159 [pii]
3. Ostertag EM, DeBerardinis RJ, Goodier JL, Zhang Y, Yang N, Gerton GL, Kazazian HH Jr (2002) A mouse model of human L1 retrotransposition. Nat Genet 32:655–660
4. Prak ET, Dodson AW, Farkash EA, Kazazian HH Jr (2003) Tracking an embryonic L1 retrotransposition event. Proc Natl Acad Sci U S A 100:1832–1837
5. Moran JV, Holmes SE, Naas TP, DeBerardinis RJ, Boeke JD, Kazazian HH Jr (1996) High

frequency retrotransposition in cultured mammalian cells. Cell 87:917–927

6. Morrish TA, Gilbert N, Myers JS, Vincent BJ, Stamato TD, Taccioli GE, Batzer MA, Moran JV (2002) DNA repair mediated by endonuclease-independent LINE-1 retrotransposition. Nat Genet 31:159–165
7. Han JS, Szak ST, Boeke JD (2004) Transcriptional disruption by the L1 retrotransposon and implications for mammalian transcriptomes. Nature 429:268–274
8. Muotri AR, Gage FH (2006) Generation of neuronal variability and complexity. Nature 441:1087–1093. doi:10.1038/nature04959, nature04959 [pii]
9. Freeman JD, Goodchild NL, Mager DL (1994) A modified indicator gene for selection of retrotransposition events in mammalian cells. Biotechniques 17:46, 48-49, 52
10. Coufal NG, Garcia-Perez JL, Peng GE, Yeo GW, Mu Y, Lovci MT, Morell M, O'Shea KS, Moran JV, Gage FH (2009) L1 retrotransposition in human neural progenitor cells. Nature 460:1127–1131. doi:10.1038/nature08248, nature08248 [pii]
11. Muotri AR, Marchetto MC, Coufal NG, Oefner R, Yeo G, Nakashima K, Gage FH (2010) L1 retrotransposition in neurons is modulated by MeCP2. Nature 468:443–446. doi:10.1038/nature09544, nature09544 [pii]
12. Muotri AR, Zhao C, Marchetto MC, Gage FH (2009) Environmental influence on L1 retrotransposons in the adult hippocampus. Hippocampus 19:1002–1007. doi:10.1002/hipo.20564
13. Garcia-Perez JL, Marchetto MC, Muotri AR, Coufal NG, Gage FH, O'Shea KS, Moran JV (2007) LINE-1 retrotransposition in human embryonic stem cells. Hum Mol Genet 16:1569–1577
14. Brouha B, Meischl C, Ostertag E, de Boer M, Zhang Y, Neijens H, Roos D, Kazazian HH Jr (2002) Evidence consistent with human L1 retrotransposition in maternal meiosis I. Am J Hum Genet 71:327–336
15. Ostertag EM, Prak ET, DeBerardinis RJ, Moran JV, Kazazian HH Jr (2000) Determination of L1 retrotransposition kinetics in cultured cells. Nucleic Acids Res 28:1418–1423
16. Gage FH, Ray J, Fisher LJ (1995) Isolation, characterization, and use of stem cells from the CNS. Annu Rev Neurosci 18:159–192
17. Marchetto MC, Carromeu C, Acab A, Yu D, Yeo GW, Mu Y, Chen G, Gage FH, Muotri AR (2010) A model for neural development and treatment of Rett syndrome using human induced pluripotent stem cells. Cell 143:527–539. doi:10.1016/j.cell.2010.10.016

Chapter 12

Characterization of Engineered L1 Retrotransposition Events: The Recovery Method

David Cano, Santiago Morell, Andres J. Pulgarin, Suyapa Amador, and Jose L. Garcia-Pérez

Abstract

Long Interspersed Element class 1 retrotransposons (LINE-1 or L1) are abundant Transposable Elements in mammalian genomes and their mobility continues to impact the human genome. The development of engineered retrotransposition assays has been instrumental to understand how these elements are regulated and to identify domains involved in the process of retrotransposition. Additionally, the modification of a retrotransposition indicator cassette has allowed developing straightforward approaches to characterize the site of new L1 insertions in cultured cells. In this chapter, we describe a method termed "L1-recovery" that has been used to characterize the site of insertion on engineered L1 retrotransposition events in cultured mammalian cells. Notably, the recovery assay is based on a genetic strategy and avoids the use of PCR and thus reduces to a minimum the appearance of false positives/artifacts.

Key words LINE-1, Retrotransposon, Recovery, Engineered, Insertion, Deletion, Target site duplication

1 Introduction

Most mammalian genomes are characterized for the high prevalence of repeated DNA sequences. Among repeated DNA sequences, Transposable Elements (TEs) are repeated DNA sequences that can move within genomes (reviewed in Refs. [1–3]). TEs are very diverse in their structure and abundance depending on the genome that is examined [3]. In humans, between 45 and 70 % of the human genome is made of TEs [4, 5], and active TEs continue to impact our genome [1–3]. Long Interspersed Element class 1 retrotransposons (LINE-1 or L1) are very prevalent sequences in the human genome and up to 21 % of our genome is made of LINE derived sequences [5]. Although most L1 copies are inactive fossils accumulated during human genome evolution, an average human genome contains between 80 and 100 active

Jose L. Garcia-Pérez (ed.), *Transposons and Retrotransposons: Methods and Protocols*, Methods in Molecular Biology, vol. 1400, DOI 10.1007/978-1-4939-3372-3_12,

L1s (termed Retrotransposition Competent L1s or RC-L1s) [6, 7]. RC-L1s are retrotransposons that move using a copy and paste mechanism termed Target Primed Reverse Transcription (TPRT) [1–3, 8]. Active RC-L1s are 6-kb in length elements containing a 900-bp long 5′ Untranslated Region (UTR) that contains conserved sense and antisense promoter activities [9–11], two non-overlapping Open Reading Frames (ORF1 and ORF2) and end in a short 3′ UTR region with a weak polyadenylation sequence [12–15]. L1-ORF1p codes for a 40 kDa protein with RNA binding and nucleic acid chaperone activity [16–19]. L1-ORF2p codes for a 150 kDa protein with demonstrated Endonuclease (EN) and Reverse Transcriptase (RT) activities [20–22]. Both ORFs are strictly required for the mobilization of RC-L1, as demonstrated using an engineered L1 retrotransposition assay in cultured cells [14]. Retrotransposition starts with the generation of a full-length polyadenylated L1 mRNA from an active L1 located elsewhere in the genome. The L1 mRNA is exported to the cytoplasm where L1-ORF1p translation takes place by a cap-dependent mechanism [23]. Notably, the L1 mRNA is unusual because is bi-cistronic and studies in cultured cells have revealed that L1-ORF2p is translated by an unconventional termination/reinitiation mechanism [24]. L1-ORF2p translation is very inefficient when compared to L1-ORF1p translation, and although translation of the L1 mRNA might produce hundreds of L1-ORF1p molecules, as little as one L1-ORF2p molecule might be translated from the same L1 mRNA [1–3, 24]. Remarkably, both proteins seem to bind back strongly to the same L1 mRNA used as a template for translation, a concept termed *cis*-preference [25, 26]. The L1 mRNA and both encoded proteins form a high weight molecular complex termed L1 ribonucleoprotein particle (L1-RNPs) that is supposed to be a retrotransposition intermediate [27]. L1-RNPs can be visualized in cultured cells using epitope tags, and several studies have revealed that L1-RNPs are often located in dense cytoplasmic foci [27–29]. During retrotransposition, L1-RNPs enter the nucleus where, using the intermediate L1 mRNA as a template, a new insertion is generated in a different genomic location, by TPRT. During TPRT, the EN activity of L1-ORF2p is thought to mediate a single strand break on DNA, releasing a free 3′ OH group that is then used by the RT activity of L1-ORF2p as a primer to generate the first cDNA copy of the L1 mRNA attached to the genome of the cell [8]. Second strand cDNA synthesis is thought to occur using a similar mechanism, resulting in the insertion of a new L1 copy in a different genomic place. Because of the L1-EN cleaving mechanism, most de novo L1 insertions are flanked by 2–20 bp Target Site Duplication (TSD) sequences. However, L1 insertions can also generate small deletions or no duplication of sequence (i.e., blunt insertion), depending on how L1-EN cleaves both strands at the insertion site [30]. Notably, most de novo L1 insertions are 5′

truncated, by an ill-defined mechanism [31]; additionally, some L1 insertions generated inverted/deleted structures by a mechanism termed Twin Priming [32]. These characteristics associated with de novo L1 retrotransposition events are bona fide hallmarks often used to characterize a new L1 insertion in a given genome. Because of their repeated nature and their high prevalence in genomes, identifying new L1 insertions in genomes is a very complicated task equivalent to identify a needle in a haystack.

However, the mechanism of L1 retrotransposition by TPRT allowed the developing of a genetic based L1 retrotransposition assay that is based on the activation of a reporter gene (REP) only after a round of L1-retrotransposition in transfected cultured cells (Fig. 1) [14, 33].

Briefly, in 1996 Moran and colleagues exploit the existence of an intermediate L1 mRNA during L1 retrotransposition to design a reporter gene using an engineered intron with a configuration that allowed expressing a functional reporter product only after a round of bona fide retrotransposition ([14], a method in this book and reviewed in Refs. [34, 35]). Briefly, the modified reporter cassette consists of the ORF that codes for a given reporter gene equipped with an exogenous promoter and polyadenylation sequences cloned in an antisense configuration in the 3′ UTR of an RC-L1. Additionally, the reporter ORF is interrupted by an intron cloned in the same transcriptional orientation as the RC-L1 (Fig. 1). With this configuration, only mRNAs generated from the L1 promoter can remove the intron by *cis*-splicing and the resulting chimeric L1 mRNA go trough a round of L1 retrotransposition, resulting in the insertion of a reporter gene lacking the intron and thus allowing efficient expression of the reporter.

The retrotransposition assay has been instrumental to increase our knowledge of L1 biology. The original assay used the sequence of the neomycin phosphotransferase gene as a reporter (*mneoI* cassette [14, 33]) and cells harboring a new retrotransposition event could be selected in culture using the mammalian antibiotic neomycin or G418 (Fig. 1). Since then, a number of reporter ORFs have been used to developed different retrotransposition markers (also covered in this Book and reviewed extensively elsewhere). Notably, using the *mneoI* retrotransposition indicator cassette, new L1 insertions in the genome of cultured cells are tag with a unique spliced *mneoI* sequence, which further allowed to characterize L1 insertions using conventional library construction/screening or inverse PCR methods [14, 26, 36]. Indeed, the characterization of de novo engineered L1 insertions using this assay in cultured cells has allowed to demonstrated how L1 can delete genomic DNA upon insertion, can mediate major alterations at the insertion site, can retrotranspose in mammalian neuronal cells, how L1 can insert by a new mechanism of insertion independent of its endonuclease, etc. [13, 30, 37–42].

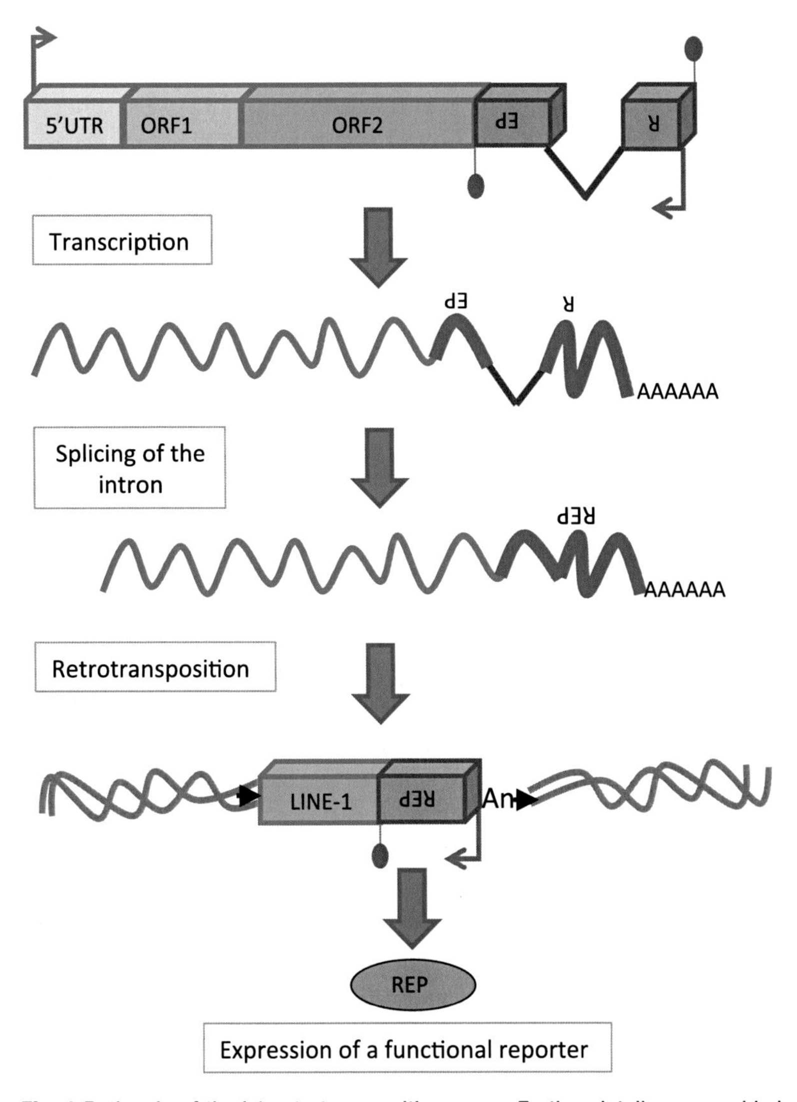

Fig. 1 Rationale of the L1-retrotransposition assay. Further details are provided in the text

In 2002, Gilbert and colleagues developed a modified *mneoI-based* retrotransposition indicator cassette that allows the recovery of engineered L1 insertions as autonomously replicating plasmids in bacteria [39] (Fig. 2).

Briefly, as the codifying sequence NEO confers G418 resistance in mammalian cells and kanamycin resistance in bacteria, Gilbert and colleagues modified the *mneoI* retrotransposition indicator cassette and included a prokaryotic promoter (EM7) and a Shine–Dalgarno sequence upstream the initiating AUG codon of NEO. Additionally, Gilbert and colleagues added a bacterial origin of replication (ColE1) downstream of the modified *mneoI* cassette and upstream of the L1 polyadenylation sequence. With this

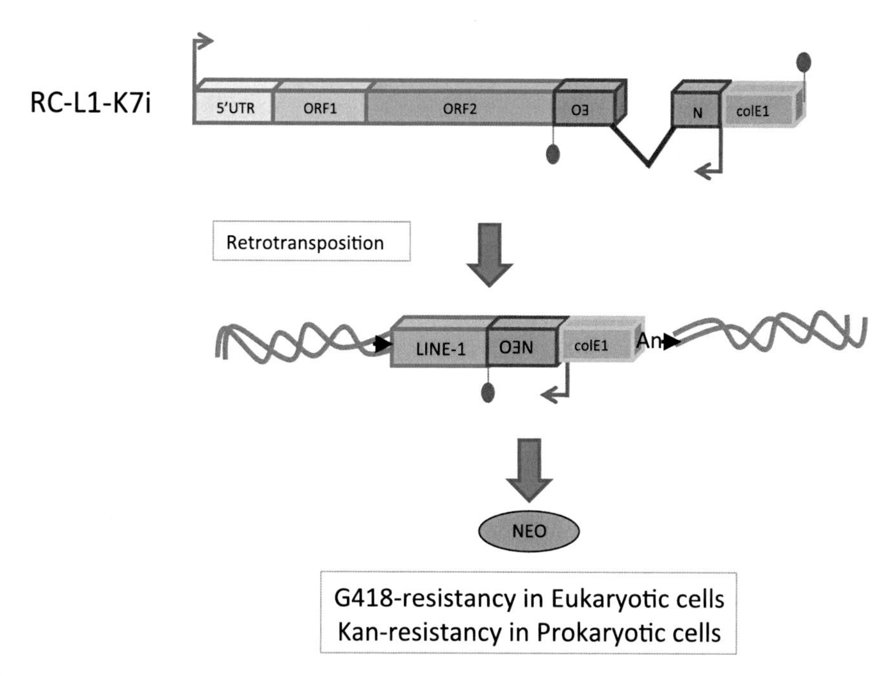

Fig. 2 Structure of the engineered L1-recovery plasmid. Shown is a scheme of a RC-L1 containing a recovery retrotransposition indicator cassette (termed K7i)

configuration, a new retrotransposition event will deliver into the genome of cells a spliced modified *mneoI* and a colE1 sequence, which is the basis of the generation of a prokaryotic plasmid (Fig. 2).

Thus, cells are first transfected with a plasmid containing an RC-L1 tagged with the recovery indicator cassette and cells containing a new retrotransposition evens can be selected with G418. Next, foci are expanded (either as single colonies or pools) and cells allowed to grow to obtain enough cells to isolate at least 20 μg of genomic DNA (gDNA). The gDNA of a colony is next digested with a restriction enzyme that cleaves frequently in the genome of the host cells (for human cells *Ssp*I, *Hind*III, *Bgl*II, etc.) and that ideally does not cut within the sequence of the transfected plasmid (or if it does, an enzyme that cleaves as closer as possible to nucleotide 1 of the transfected L1). The digested DNA is next ligated in very diluted conditions to favor intramolecular ligation events, resulting in the generation of a prokaryotic plasmid structure (Fig. 3).

This mix of ligated DNAs is next transformed to ultracompetent *E. coli* cells, and after recovery of transformed cells, the engineered L1 insertion plasmid is selected using kanamycin in bacteria (Fig. 3). These colonies will have replicated the engineered retrotransposed product plus flanking genomic DNA as a plasmid that can be extracted and purified using conventional miniprep purification of DNA. Finally, these plasmids can be sequenced

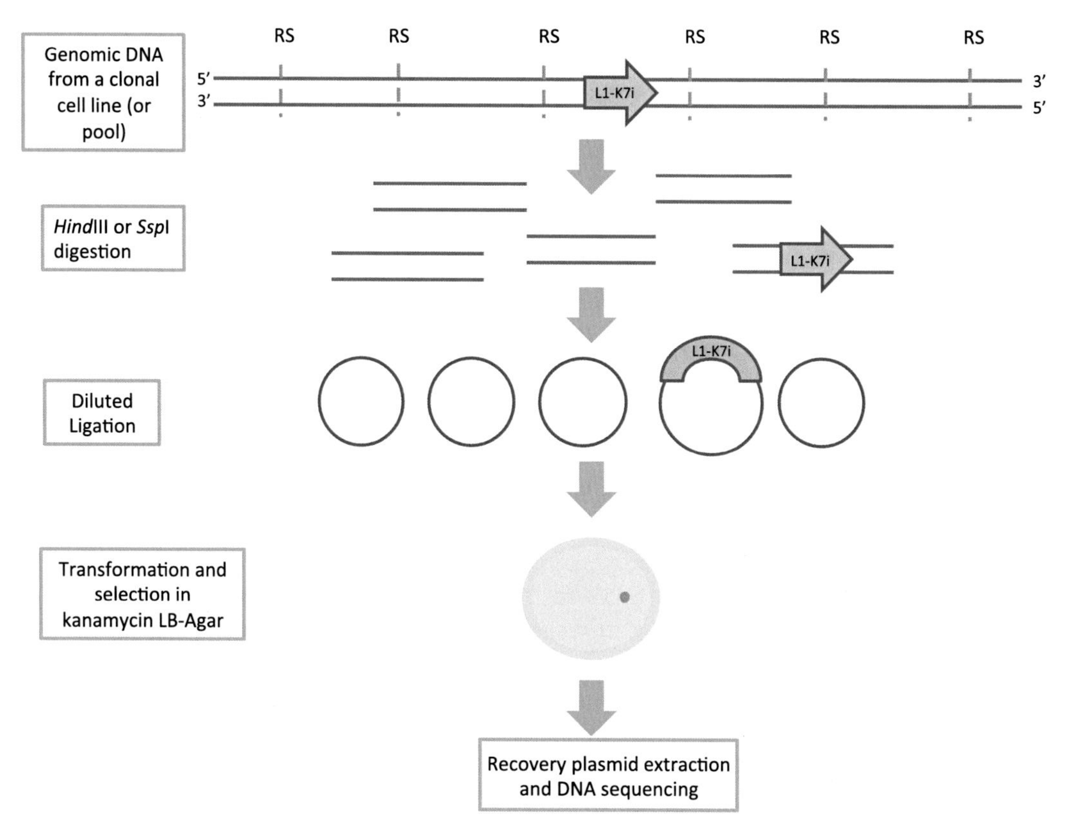

Fig. 3 Rationale of the recovery assay. Details are included in the text. R.S., restriction site

using Sanger methodology and the characteristics associated with new L1 retrotransposition events inferred at the nucleotide level (Fig. 3 and *see* [39]). In this chapter, we cover the recovery process, which is essentially the assay described by Gilbert and colleagues in 2002 with minor modifications that increase the efficiency of the recovery process. Notably, alternative methods based on similar methodologies have been also described [43], although they are not covered in this chapter.

2 Materials

2.1 Retrotrans position Assay and Establishment of Cell Lines

1. Cell culture medium for the cell line to be used (*see* **Note 1**).
2. 10 cm cell culture dishes.
3. Appropriate transfection reagent (Fugene 6 (Promega), Xtreme (Roche), Lipofectamine (Invitrogen), etc.; *see* **Note 1**).
4. 50 mg/ml geneticin (G418).
5. Trypsin 0.05 or 0.25 % (*see* **Note 1**).

6. 70 % ethanol.
7. 0.2 × 0.2 cm 3MM sterilized chromatography paper (Whatman).
8. Sterile small forceps.
9. 24-well tissue culture plates.
10. T75 tissue culture flasks.
11. Cryopreservation tubes and cryopreservation media (*see* **Note 1**).
12. 1× PBS.
13. An engineered L1 construct containing the recovery retrotransposition indicator cassette (as published by the Moran lab) [39–41]. The plasmid might be purified from bacteria using a Midi or Maxiprep purification kit (from local providers). The plasmid used for cell transfection should be in a supercoiled form (80 % or higher).

2.2 Genomic DNA Extraction

1. Blood and Cell Culture DNA midi kit (Qiagen) or a similar kit that provided high quality genomic DNA.
2. Cell lysis buffer: 10 mM Tris–HCl, pH 8.2, 10 mM EDTA, 200 mM NaCl, 0.5 % (w/v) SDS. To prepare 100 ml of lysis buffer: 0.12114 g Tris–HCl; 0.29225 g EDTA; 1.17 g NaCl; 0.5 g SDS (powder) Adjust pH with HCl.
3. 100 % ethanol.
4. 20 mg/ml proteinase K.
5. 25:24:1 phenol–chloroform–isoamyl alcohol.
6. TE buffer, pH 8.
7. Sterilized 1.7 ml microfuge tubes.
8. 70 % ethanol.

2.3 Recue of Retrotransposition Events

1. *Hind* III and *Ssp*I enzymes (New England Biolabs).
2. 10× Buffer 2 and *Ssp*I 10× buffer (New England Biolabs).
3. 100× BSA (New England Biolabs).
4. T4 DNA ligase (New England Biolabs).
5. 10× T4 DNA ligase buffer (New England Biolabs) (*see* **Note 2**).
6. Ultra 0.5 ml Centrifugal Filters Ultracel—100 K (Amicon).
7. Ultracompetent XL1-gold or XL1 Blue bacteria (*see* **Note 3**).
8. 15 ml Falcon tubes.
9. Liquid LB culture medium.
10. 140 mm bacterial plates.
11. 25 mg/ml kanamycin.

12. Wizard Plus SV Miniprep DNA purification system (Promega) or a similar kit from local suppliers.
13. *Eco*RI (New England Biolabs).
14. 10× *Eco*RI buffer (New England Biolabs).
15. 5 ml sterilized bacteria culture tubes.

2.4 Sequencing

1. Primers to sequence recovery plasmids (*see* Table 1).

Table 1
Sequence of primers commonly used to characterize L1 insertions using the recovery method

NEO210as	gaccgcttcctcgtgctttacg
Reco3	ctaaagtatatatgagtaacc
ORF2Jas	ggtgcgctgcacccactaatg
ORF2Ias	cacattttcttaatccagtc
ORF2Has	ctaactggtgtgagatgatatc
ORF2Gas	gaattgatttttgtataaggtg
ORF2Fas	cattttcacgatattgattcttcc
ORF2Eas	gtcccatcaatacctaatttattg
ORF2Das	gaattcggctgtgaatccatctgg
ORF2Cas	ttaattgtgatgttagggtgtc
ORF2Bas	gatttggggtggagagttctg
ORF2Aas	gccttctttgtctcttttgat
ORF1Cas	ctttccatgtttagcgcttcc
ORF1Bas	ggtcttttcacatagtcccat
ORF1Aas	cgttcctttggaggaggagaggc
5′ UTRas	gcaggcaggcctccttgagctg

3 Methods

General note: standard good practices for handling cells should be used, including the use of a certified laminar flow biosafety hood and sterile materials and techniques for any reagent used. Note that mammalian cells are very sensitive to DNA transfection when infected by *Mycoplasma* spp. We routinely test for *Mycoplasma* spp. once a month. We also verified the nature of cell lines using short tandem repeat (STR) analyses to ensure cross-contamination with other cell lines.

3.1 Retrotransposition Assay

1. Trypsinize, count, and seed an appropriate number of cultured cells in a sterile 10 cm cell culture plate (*see* **Notes 1** and **4**).
2. Transfect cells with the L1 recovery plasmid (*see* **Notes 1** and **5**) 12–14 h after seeding.
3. After 12 h, change media in the 10 cm plates.
4. 72 h post-transfection, replace medium and add medium supplemented with G418 (*see* **Note 6**).
5. Allow G418 to select G418-resistant foci; the process take a minimum of 10 days but can change between cell lines (*see* **Notes 1** and **6**). Change the media every other day. *See* also **Note 7**.

3.2 Establishment of G418-Resistant Cell Lines (See Note 8)

1. Take 10–20 small sterilized 0.2 × 0.2 cm 3MM paper squares and soak them in Trypsin using a sterile 10 cm tissue culture plate (*see* **Note 1**).
2. Sterilize forceps by immersion in 70 % ethanol. Air-dry forceps.
3. Aided with the sterilized forceps, put a trypsin soaked 0.2 × 0.2 cm 3MM square paper on top of a colony. Incubate at RT for 1–3 min.
4. Using the forceps, put the piece of 3MM paper in a well of a 24-well plate containing 1 ml of complete media. Shake the plate right/left and up/down 2–3 times and incubate cells overnight in an incubator (*see* **Note 9**).
5. Repeat **steps 2–4** as many times as required.
6. Next day, remove papers using a vacuum sucker device and add 1 ml of complete cell culture medium to each well.
7. Expand G-418 foci following standard procedures until obtaining a T-75 full flask of cells (95 % confluent).
8. From the T-75 flask, trypsinize cells and freeze at least one vial of cells using an appropriate method for the cell line of choice (*see* **Note 1**). Collect the remaining cells using a Falcon 15 ml tube by centrifugation at 1500 × *g* for 5 min at 4 °C. Remove supernatant and store pellets at −80 °C.

3.3 Genomic DNA Isolation for the Rescue of Engineered LINE-1 Retrotransposition Events (See Note 10)

1. Extract genomic DNA from cell lines (*see* **Note 11**). The starting amount of cells required to obtain enough genomic DNA depends on the method used to isolate DNA (*see* **Note 12**).
2. If using a commercial DNA extraction kit, follow the manufacturer's instructions. Jump to **step 4**.
3. If using the homemade protocol, follow these steps:
   (a) Thaw frozen pellets on ice.
   (b) Resuspend the cell pellet in 0.5 ml lysis buffer.
   (c) Add 5 μl of 20 mg/ml proteinase k (*see* **Note 13**).

(d) Incubate for 3 h at 56 °C. If necessary, the digestion can be done overnight at 56 °C (*see* **Note 13**).

(e) Transfer the lysates into clean 1.7 ml tubes or 2 ml tubes depending on the volume of your lysate.

(f) Add 1 volume of phenol–chloroform–isoamyl alcohol and shake tubes vigorously.

(g) Centrifuge for 5 min at max speed at room temperature (RT) using a microfuge.

(h) Keep the aqueous phase (top part) and transfer it to a new 1.7 ml tube (*see* **Note 14**).

(i) Add 2 volumes of 100 % ethanol and place tubes at −20 °C for at least 3–4 h or overnight.

(j) Centrifuge tubes at max speed at 4 °C for 15 min. Discard the supernatant.

(k) Wash the pellet with 500 μl of 70 % ethanol and centrifuge at max speed for 10 min at 4 °C. Discard supernatant.

(l) Air-dry pellets (*see* **Note 15**) but do not over-dry.

(m) Add 200 μl of TE buffer, pH 8.0 and dissolve the pellet. If the pellet is resistant to solution, heat the tube at 56 °C for 1 h with strong agitation. If necessary, incubate it longer.

4. Quantify the concentration of your genomic DNA sample. We recommend using an spectrophotometer as measurements are more accurate that when using a NanoDrop or similar devices.

3.4 The Recovery Protocol (See Note 10)

Below, we described the recovery process divided in days.

Day 1

1. Genomic DNA digestion. Digest 8 μg of genomic DNA with either *Hind*III or *Ssp*I (*see* **Note 16**) at 37 °C overnight as follows (*see* **Note 17**):
   - 5 μl of *Hind*III or *Ssp*I.
   - 10 μl of buffer number 2 or 10 μl of *Ssp*I buffer.
   - 1 μl of BSA 100×.
   - 8 μg of genomic DNA.
   - Complete with DNA-free ddH_2O up to 100 μl (*see* **Notes 10** and **18**).

Day 2

1. In the morning, add 1 μl of either *Hind*III or *Ssp*I and incubate at 37 °C for 2 h. Heat at 65 °C to inactive *Hind*III or *Ssp*I enzymes for 25 min.
2. Set up ligation. Add the following reagents to the 100 μl digested DNA sample:

- 8 μl of T4 DNA ligase.
- 40 μl of 10× T4 DNA ligase buffer.
- 351 μl of ddH_2O (*see* **Notes 10** and **18**).

3. Incubate overnight at 16 °C for *Hind*III digested DNAs and at 25 °C for *Ssp*I digested DNAs (*see* **Note 19**).

Day 3

1. In the morning, add 1 μl of T4 DNA ligase and incubate tubes for at least 4 h at RT.
2. Concentrate DNAs as follows:
   - Transfer each ligation into a clean Amicon.
   - Centrifuge at 5000×*g* for 5 min at RT. Discard the flow-through.
   - Add 450 μl of DNA-free ddH_2O (*see* **Notes 10** and **18**).
   - Centrifuge at 5000×*g* for 5 min at RT. Note that a small volume of liquid will remain in the Amicon. Do not discard it.
   - Place the Amicon upside down and put it inside a new 1.7 ml tube (*see* **Note 20**).
   - Short spin. The volume of the concentrated DNA should be around 50–60 μl. If larger, recentrifuge at 5000×*g* for 5 min at RT.

3 Bacteria transformation.
   - Slowly thaw on ice a 500 μl aliquot of XL1 Blue ultracompetent cells (*see* **Note 21**).
   - Label 15 ml tubes and cool them on ice. Also, prewarm an aliquot of sterile LB medium at 37 °C (*see* **Note 22**).
   - Add 1/3 of the concentrated DNA solution (15–20 μl) to the thawed bacteria. Transfer the mix slowly to a precooled 15 ml tube.
   - Incubate on ice for 25–30 min.
   - Heat shock at 42 °C for 38 s. CRITICAL: Do not exceed this time.
   - Incubate on ice for 2 min.
   - Add 1 ml of prewarmed LB (37 °C). Incubate the mix overnight at RT on a shaker but do not exceed 600 rpm (*see* **Note 23** and *see* Fig. 4).

Day 4

1. Pellet bacteria by centrifugation at 500×*g* for 8 min at RT.
2. Remove supernatant from tubes but keep around 200–300 μl.
3. Resuspend the bacterial suspension carefully with a pipette.

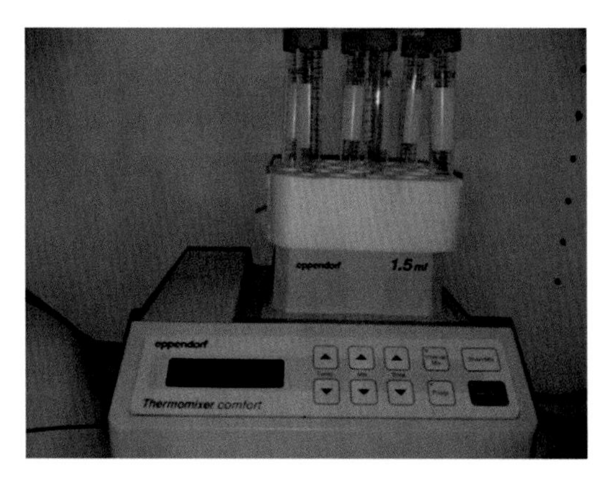

Fig. 4 A system to shake transformed bacteria. See text for further details

4. Plate the suspension of bacteria in 140 mm LB-Agar plates containing 25 μg/ml kanamycin. Incubate at 37 °C overnight (*see* **Note 24**).

Day 5

1. After 24 h, colonies should be already visible. In the afternoon, pick individual colonies using an sterilized tip and inoculate a tube containing 2–4 ml of LB with 25 μg/ml kanamycin. Plates with colonies can be also stored at 4 °C for a week.
2. Incubate tubes overnight at 37 °C on a shaker.

Day 6

1. Extract plasmid DNA from bacteria cultures using a miniprep kit. Follow manufacturer's instructions.
2. Digest a 5 μl aliquot of each recovered plasmid with *Eco*RI using a standard protocol (*see* **Note 25**).
3. Resolve digestions on 1–1.5 % agarose gels and identify particular restriction patterns (Fig. 5 and **Note 25**).

3.5 Sequencing Recovered Plasmids

1. Select the plasmids with a unique restriction pattern and sequence them with oligonucleotides Neo 210AS and Reco3 (*see* Table 1). *See* **Note 26**.

4 Notes

1. The recovery assay can be used in any cell line that support engineered L1 retrotransposition. The media, the transfection conditions, freezing conditions and other general methods must be optimized for each cell line.
2. To avoid degradation of ATP in cycles of thawing/freezing, aliquot T4DNA ligase 10× buffer in 100 μl aliquots.

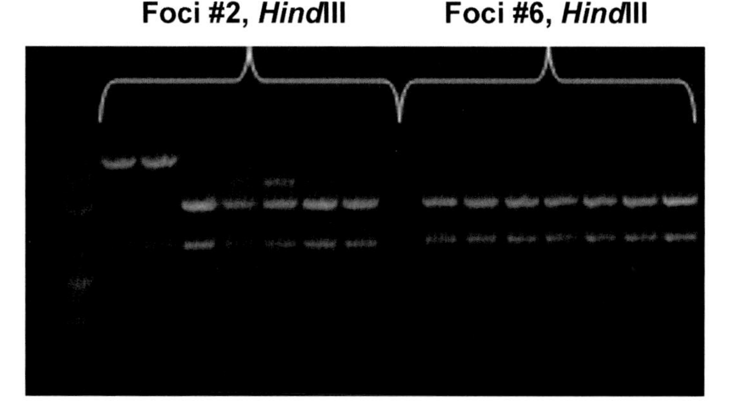

Fig. 5 Representative results from the recovery assay. Shown are results of plasmids isolated from two G418-resistant foci and digested with *Eco*RI. In cell line #2, two patterns can be observed which upon sequencing revealed that this foci contained at least two engineered L1 insertions

3. XL1-gold or XL1-blue ultracompetent bacteria can be prepared in house or purchased from local vendors. It is critical that bacteria are ultracompetent.
4. In our hands, we have been able to recover insertions from any cellular type than can support engineered L1 retrotransposition. The number of cells that need to be plated for each cell line might be determined experimentally. The goal of the assay is to have enough colonies dispersed in the 10 cm plate after G418 selection. For HeLa cells, we routinely seed $4–8 \times 10^4$ cells/10 cm plate.
5. Similarly, the transfection method or reagent that can be used depends on the cellular type used. We routinely use transfection reagents such as Fugene 6 or Xtreme 9, as both reagents allows transfecting cells plated at a low cellular density without elevated toxicity. For each transfection reagent, follow manufacturer's instructions.
6. The concentration of G418 used for selection should be empirically determined for each cell line. For HeLa cells, we routinely use 400 μg/ml.
7. It is informative to use an internal control in the retrotransposition assay, by using alleles of the engineered L1 plasmids containing missense mutations in domains involved in retrotransposition (EN and/or RT missense mutants).
8. To avoid cross-contamination in the establishment of G418 resistant cell lines, G418-foci should be distinct in plates, separated by at least 0.5 cm from nearby foci.
9. The well size depends on the ability of cells to grow at a low cellular density. Some cell lines can grow very well starting from few cells in a large culture dish while others are slower/

incapable of generating foci. Thus, the well size will vary among cell lines and it must be empirically determined.

10. The recovery process is very sensitive to cross-contamination with other plasmids that use kanamycin as selection. Unfortunately, these plasmids are very common in most labs and special attention should be paid to avoid cross-contamination. Use pipettes and a microfuge devoted only for the recovery process; alternatively, clean all used equipment with diluted bleach and 70 % ethanol.
11. DNA extraction can be performed either using a specific kit (i.e., Qiagen) or using a homemade protocol. In our hands, both protocols are equally efficient but, of course, the costs associated with the homemade protocol are lower.
12. If using a kit, just follow the manufacturer's instructions about required number of cells. If using the homemade protocol collect $\approx 5 \times 10^6$ (i.e., half of a T75 flask of cells at a 95 % confluence).
13. After 3 h, check the transparency of the lysate. If the lysate is dense/cloudy and it is difficult to pipette up and down, leave it overnight but re-add 0.5 ml of lysis buffer and 5 μl of 20 mg/ml proteinase k.
14. Two phases form: the top aqueous phase (contains the genomic DNA) and the lower phase (organic phase, undigested cellular components will be there). Be careful when collecting the top phase pipetting since it is very easy to take part of the organic phase. Pipette very carefully and do not try to recover the whole upper phase. 2/3 or 3/4 at the most might be sufficient to extract enough but purified genomic DNA.
15. In order to quickly dry pellets, if possible, leave tubes open inside a fuming hood to efficiently remove the remaining ethanol.
16. In our lab, we routinely conduct independent *Hind*III and SspI digestions to increase the efficiency of the recovery process.
17. We recommend starting in the afternoon (16.00 PM).
18. Be especially careful to avoid cross-contamination with kanamycin resistant-containing plasmids. We recommend acquire commercial ddH_2O purified water and prepare 5 ml aliquots.
19. *Hind*III produces overhang ends while *Ssp*I generates blunt ends.
20. We routinely use the collection tubes that are provided with the Amicon tubes.
21. It is important to use XL1-Blue cells, as they can replicate methylated genomic DNAs. Other strains of *E. coli* can be used, but make sure they are ultracompetent and that they can replicate methylated mammalian DNAs.

22. Avoid any cross-contamination with kanamycin resistant-containing plasmids. We recommend acquire commercial LB and prepare 5 ml aliquots.
23. In our lab, we use a very simple system: a polyspam rack attached to a Thermomixer with tape. A conventional shaker can also be used, but make sure temperature is not higher than 25 °C.
24. We recommend to seed transformations either late in the morning or soon in the afternoon, since recovery colonies need about 24 h to be visible by eye.
25. An *Eco*RI digestion will allow you to distinguish among different insertions. In fact, sequence those that do not have the same restriction pattern upon *Eco*RI digestion as they might be different insertions that occurred in the same colony.
26. Sequencing with Reco3 will reach the poly A tail of the L1 insertion. If the polyA tail it is not too long, the inferred DNA sequence might be sufficient to identify the 3′ region of the insertion. If Neo 210AS does not allow you to reach the genomic region at the 5′ end of the insertion, keep sequencing upstream with the rest of the primers that anneal on the LINE-1 sequence (Table 1).

Acknowledgments

David Cano, Santiago Morell, and Andres J. Pulgarin have contributed equally to this study and are listed alphabetically. We acknowledge current members of the J.L.G.-P. lab (Genyo) for valuable input during the course of the project. We also acknowledge Dr. John Moran (University of Michigan, USA) for helpful discussions. D.C. is supported by a FPU fellowship from the Government of Spain (MINECO, Ref AP2010-0135). J.L.G.P's lab is supported by CICE-FEDER-P09-CTS-4980, CICE-FEDER-P12-CTS-2256, Plan Nacional de I+D+i 2008-2011 and 2013-2016 (FIS-FEDER-PI11/01489 and FIS-FEDER-PI14/02152), PCIN-2014-115-ERA-NET NEURON II, the European Research Council (ERC-Consolidator ERC-STG-2012-233764) and by an International Early Career Scientist grant from the Howard Hughes Medical Institute (IECS-55007420).

References

1. Beck CR, Garcia-Perez JL, Badge RM, Moran JV (2011) LINE-1 elements in structural variation and disease. Annu Rev Genomics Hum Genet12:187–215.doi:10.1146/annurev-genom-082509-141802
2. Levin HL, Moran JV (2011) Dynamic interactions between transposable elements and their hosts. Nat Rev Genet 12(9):615–627. doi: 10.1038/nrg3030
3. Macia A, Blanco-Jimenez E, Garcia-Perez JL (2015) Retrotransposons in pluripotent cells: impact and new roles in cellular plasticity. Biochim Biophys Acta 1849(4):417–426. doi:10.1016/j.bbagrm.2014.07.007

4. de Koning AP, Gu W, Castoe TA, Batzer MA, Pollock DD (2011) Repetitive elements may comprise over two-thirds of the human genome. PLoS Genet 7(12):e1002384. doi:10.1371/journal.pgen.1002384
5. Lander ES, Linton LM, Birren B, Nusbaum C, Zody MC, Baldwin J, Devon K, Dewar K, Doyle M, FitzHugh W, Funke R, Gage D, Harris K, Heaford A, Howland J, Kann L, Lehoczky J, LeVine R, McEwan P, McKernan K, Meldrim J, Mesirov JP, Miranda C, Morris W, Naylor J, Raymond C, Rosetti M, Santos R, Sheridan A, Sougnez C, Stange-Thomann N, Stojanovic N, Subramanian A, Wyman D, Rogers J, Sulston J, Ainscough R, Beck S, Bentley D, Burton J, Clee C, Carter N, Coulson A, Deadman R, Deloukas P, Dunham A, Dunham I, Durbin R, French L, Grafham D, Gregory S, Hubbard T, Humphray S, Hunt A, Jones M, Lloyd C, McMurray A, Matthews L, Mercer S, Milne S, Mullikin JC, Mungall A, Plumb R, Ross M, Shownkeen R, Sims S, Waterston RH, Wilson RK, Hillier LW, McPherson JD, Marra MA, Mardis ER, Fulton LA, Chinwalla AT, Pepin KH, Gish WR, Chissoe SL, Wendl MC, Delehaunty KD, Miner TL, Delehaunty A, Kramer JB, Cook LL, Fulton RS, Johnson DL, Minx PJ, Clifton SW, Hawkins T, Branscomb E, Predki P, Richardson P, Wenning S, Slezak T, Doggett N, Cheng JF, Olsen A, Lucas S, Elkin C, Uberbacher E, Frazier M, Gibbs RA, Muzny DM, Scherer SE, Bouck JB, Sodergren EJ, Worley KC, Rives CM, Gorrell JH, Metzker ML, Naylor SL, Kucherlapati RS, Nelson DL, Weinstock GM, Sakaki Y, Fujiyama A, Hattori M, Yada T, Toyoda A, Itoh T, Kawagoe C, Watanabe H, Totoki Y, Taylor T, Weissenbach J, Heilig R, Saurin W, Artiguenave F, Brottier P, Bruls T, Pelletier E, Robert C, Wincker P, Smith DR, Doucette-Stamm L, Rubenfield M, Weinstock K, Lee HM, Dubois J, Rosenthal A, Platzer M, Nyakatura G, Taudien S, Rump A, Yang H, Yu J, Wang J, Huang G, Gu J, Hood L, Rowen L, Madan A, Qin S, Davis RW, Federspiel NA, Abola AP, Proctor MJ, Myers RM, Schmutz J, Dickson M, Grimwood J, Cox DR, Olson MV, Kaul R, Raymond C, Shimizu N, Kawasaki K, Minoshima S, Evans GA, Athanasiou M, Schultz R, Roe BA, Chen F, Pan H, Ramser J, Lehrach H, Reinhardt R, McCombie WR, de la Bastide M, Dedhia N, Blocker H, Hornischer K, Nordsiek G, Agarwala R, Aravind L, Bailey JA, Bateman A, Batzoglou S, Birney E, Bork P, Brown DG, Burge CB, Cerutti L, Chen HC, Church D, Clamp M, Copley RR, Doerks T, Eddy SR, Eichler EE, Furey TS, Galagan J, Gilbert JG, Harmon C, Hayashizaki Y, Haussler D, Hermjakob H, Hokamp K, Jang W, Johnson LS, Jones TA, Kasif S, Kaspryzk A, Kennedy S, Kent WJ, Kitts P, Koonin EV, Korf I, Kulp D, Lancet D, Lowe TM, McLysaght A, Mikkelsen T, Moran JV, Mulder N, Pollara VJ, Ponting CP, Schuler G, Schultz J, Slater G, Smit AF, Stupka E, Szustakowski J, Thierry-Mieg D, Thierry-Mieg J, Wagner L, Wallis J, Wheeler R, Williams A, Wolf YI, Wolfe KH, Yang SP, Yeh RF, Collins F, Guyer MS, Peterson J, Felsenfeld A, Wetterstrand KA, Patrinos A, Morgan MJ, de Jong P, Catanese JJ, Osoegawa K, Shizuya H, Choi S, Chen YJ (2001) Initial sequencing and analysis of the human genome. Nature 409(6822):860–921
6. Beck CR, Collier P, Macfarlane C, Malig M, Kidd JM, Eichler EE, Badge RM, Moran JV (2010) LINE-1 retrotransposition activity in human genomes. Cell 141(7):1159–1170
7. Brouha B, Schustak J, Badge RM, Lutz-Prigge S, Farley AH, Moran JV, Kazazian HH Jr (2003) Hot L1s account for the bulk of retrotransposition in the human population. Proc Natl Acad Sci U S A 100(9):5280–5285
8. Luan DD, Korman MH, Jakubczak JL, Eickbush TH (1993) Reverse transcription of R2Bm RNA is primed by a nick at the chromosomal target site: a mechanism for non-LTR retrotransposition. Cell 72(4):595–605
9. Macia A, Munoz-Lopez M, Cortes JL, Hastings RK, Morell S, Lucena-Aguilar G, Marchal JA, Badge RM, Garcia-Perez JL (2011) Epigenetic control of retrotransposon expression in human embryonic stem cells. Mol Cell Biol 31(2):300–316
10. Speek M (2001) Antisense promoter of human L1 retrotransposon drives transcription of adjacent cellular genes. Mol Cell Biol 21(6):1973–1985
11. Swergold GD (1990) Identification, characterization, and cell specificity of a human LINE-1 promoter. Mol Cell Biol 10(12):6718–6729
12. Boeke JD (1997) LINEs and Alus--the polyA connection. Nat Genet 16(1):6–7
13. Moran JV, DeBerardinis RJ, Kazazian HH Jr (1999) Exon shuffling by L1 retrotransposition. Science 283(5407):1530–1534
14. Moran JV, Holmes SE, Naas TP, DeBerardinis RJ, Boeke JD, Kazazian HH Jr (1996) High frequency retrotransposition in cultured mammalian cells. Cell 87(5):917–927
15. Scott AF, Schmeckpeper BJ, Abdelrazik M, Comey CT, O'Hara B, Rossiter JP, Cooley T, Heath P, Smith KD, Margolet L (1987) Origin of the human L1 elements: proposed progenitor genes deduced from a consensus DNA sequence. Genomics 1(2):113–125

16. Hohjoh H, Singer MF (1996) Cytoplasmic ribonucleoprotein complexes containing human LINE-1 protein and RNA. EMBO J 15(3):630–639
17. Hohjoh H, Singer MF (1997) Sequence-specific single-strand RNA binding protein encoded by the human LINE-1 retrotransposon. EMBO J 16(19):6034–6043
18. Khazina E, Weichenrieder O (2009) Non-LTR retrotransposons encode noncanonical RRM domains in their first open reading frame. Proc Natl Acad Sci U S A 106(3): 731–736
19. Martin SL, Bushman FD (2001) Nucleic acid chaperone activity of the ORF1 protein from the mouse LINE-1 retrotransposon. Mol Cell Biol 21(2):467–475
20. Feng Q, Moran JV, Kazazian HH Jr, Boeke JD (1996) Human L1 retrotransposon encodes a conserved endonuclease required for retrotransposition. Cell 87(5):905–916
21. Martin F, Maranon C, Olivares M, Alonso C, Lopez MC (1995) Characterization of a non-long terminal repeat retrotransposon cDNA (L1Tc) from Trypanosoma cruzi: homology of the first ORF with the ape family of DNA repair enzymes. J Mol Biol 247(1):49–59
22. Mathias SL, Scott AF, Kazazian HH Jr, Boeke JD, Gabriel A (1991) Reverse transcriptase encoded by a human transposable element. Science 254(5039):1808–1810
23. Dmitriev SE, Andreev DE, Terenin IM, Olovnikov IA, Prassolov VS, Merrick WC, Shatsky IN (2007) Efficient translation initiation directed by the 900-nucleotide-long and GC-rich 5′ untranslated region of the human retrotransposon LINE-1 mRNA is strictly cap dependent rather than internal ribosome entry site mediated. Mol Cell Biol 27(13): 4685–4697
24. Alisch RS, Garcia-Perez JL, Muotri AR, Gage FH, Moran JV (2006) Unconventional translation of mammalian LINE-1 retrotransposons. Genes Dev 20(2):210–224
25. Esnault C, Maestre J, Heidmann T (2000) Human LINE retrotransposons generate processed pseudogenes. Nat Genet 24(4): 363–367
26. Wei W, Gilbert N, Ooi SL, Lawler JF, Ostertag EM, Kazazian HH, Boeke JD, Moran JV (2001) Human L1 retrotransposition: cis preference versus trans complementation. Mol Cell Biol 21(4):1429–1439
27. Doucet AJ, Hulme AE, Sahinovic E, Kulpa DA, Moldovan JB, Kopera HC, Athanikar JN, Hasnaoui M, Bucheton A, Moran JV, Gilbert N (2010) Characterization of LINE-1 ribonucleoprotein particles. PLoS Genet 6(10):e1001150
28. Goodier JL, Zhang L, Vetter MR, Kazazian HH Jr (2007) LINE-1 ORF1 protein localizes in stress granules with other RNA-binding proteins, including components of RNA interference RNA-induced silencing complex. Mol Cell Biol 27(18):6469–6483
29. Taylor MS, Lacava J, Mita P, Molloy KR, Huang CR, Li D, Adney EM, Jiang H, Burns KH, Chait BT, Rout MP, Boeke JD, Dai L (2013) Affinity proteomics reveals human host factors implicated in discrete stages of LINE-1 retrotransposition. Cell 155(5):1034–1048. doi:10.1016/j.cell.2013.10.021
30. Gilbert N, Lutz S, Morrish TA, Moran JV (2005) Multiple fates of l1 retrotransposition intermediates in cultured human cells. Mol Cell Biol 25(17):7780–7795
31. Coufal NG, Garcia-Perez JL, Peng GE, Marchetto MC, Muotri AR, Mu Y, Carson CT, Macia A, Moran JV, Gage FH (2011) Ataxia telangiectasia mutated (ATM) modulates long interspersed element-1 (L1) retrotransposition in human neural stem cells. Proc Natl Acad Sci U S A 108(51):20382–20387. doi:10.1073/pnas.1100273108
32. Ostertag EM, Kazazian HH Jr (2001) Twin priming: a proposed mechanism for the creation of inversions in L1 retrotransposition. Genome Res 11(12):2059–2065
33. Freeman JD, Goodchild NL, Mager DL (1994) A modified indicator gene for selection of retrotransposition events in mammalian cells. Biotechniques 17(1):46, 48-49, 52
34. Moran JV (1999) Human L1 retrotransposition: insights and peculiarities learned from a cultured cell retrotransposition assay. Genetica 107(1-3):39–51
35. Rangwala SH, Kazazian HH (2009) The L1 retrotransposition assay: a retrospective and toolkit. Methods 49(3):219–226
36. Wei W, Morrish TA, Alisch RS, Moran JV (2000) A transient assay reveals that cultured human cells can accommodate multiple LINE-1 retrotransposition events. Anal Biochem 284(2):435–438
37. Cost GJ, Feng Q, Jacquier A, Boeke JD (2002) Human L1 element target-primed reverse transcription in vitro. EMBO J 21(21): 5899–5910
38. Coufal NG, Garcia-Perez JL, Peng GE, Yeo GW, Mu Y, Lovci MT, Morell M, O'Shea KS, Moran JV, Gage FH (2009) L1 retrotransposition in human neural progenitor cells. Nature 460(7259):1127–1131

39. Gilbert N, Lutz-Prigge S, Moran JV (2002) Genomic deletions created upon LINE-1 retrotransposition. Cell 110(3):315–325
40. Morrish TA, Garcia-Perez JL, Stamato TD, Taccioli GE, Sekiguchi J, Moran JV (2007) Endonuclease-independent LINE-1 retrotransposition at mammalian telomeres. Nature 446(7132):208–212
41. Morrish TA, Gilbert N, Myers JS, Vincent BJ, Stamato TD, Taccioli GE, Batzer MA, Moran JV (2002) DNA repair mediated by endonuclease-independent LINE-1 retrotransposition. Nat Genet 31(2):159–165
42. Muotri AR, Chu VT, Marchetto MC, Deng W, Moran JV, Gage FH (2005) Somatic mosaicism in neuronal precursor cells mediated by L1 retrotransposition. Nature 435(7044): 903–910
43. Symer DE, Connelly C, Szak ST, Caputo EM, Cost GJ, Parmigiani G, Boeke JD (2002) Human l1 retrotransposition is associated with genetic instability in vivo. Cell 110(3):327–338

Chapter 13

SINE Retrotransposition: Evaluation of Alu Activity and Recovery of De Novo Inserts

Catherine Ade and Astrid M. Roy-Engel

Abstract

Mobile element activity is of great interest due to its impact on genomes. However, the types of mobile elements that inhabit any given genome are remarkably varied. Among the different varieties of mobile elements, the Short Interspersed Elements (SINEs) populate many genomes, including many mammalian species. Although SINEs are parasites of Long Interspersed Elements (LINEs), SINEs have been highly successful in both the primate and rodent genomes. When comparing copy numbers in mammals, SINEs have been vastly more successful than other nonautonomous elements, such as the retropseudogenes and SVA. Interestingly, in the human genome the copy number of Alu (a primate SINE) outnumbers LINE-1 (L1) copies 2 to 1. Estimates suggest that the retrotransposition rate for Alu is tenfold higher than LINE-1 with about 1 insert in every twenty births. Furthermore, Alu-induced mutagenesis is responsible for the majority of the documented instances of human retroelement insertion-induced disease. However, little is known on what contributes to these observed differences between SINEs and LINEs. The development of an assay to monitor SINE retrotransposition in culture has become an important tool for the elucidation of some of these differences. In this chapter, we present details of the SINE retrotransposition assay and the recovery of de novo inserts. We also focus on the nuances that are unique to the SINE assay.

Key words Alu, ORF2 protein, RNA polymerase III, Retrotransposition, SINE

1 Introduction

The ongoing activity of L1 (a *L*ong *In*terspersed *E*lement, or LINE-1) and Alu (a *S*hort *In*terspersed *E*lement, or SINE) currently contributes to genetic diversity and disease through retrotransposition. Due to the significant impact of these retroelements on the human genome, there is great interest in understanding their amplification mechanism and regulation. One of the methods that greatly advanced the field of human retroelement biology was the development of an engineered L1 containing a specially designed reporter cassette [1] that allowed for the evaluation of L1 activity in a tissue culture system [2] (details are shown in Fig. 1). The strategy behind the design of this cassette is the addition of an inverted reporter gene that is disrupted by an intron in the opposite orientation to the 3′

Jose L. Garcia-Pérez (ed.), *Transposons and Retrotransposons: Methods and Protocols*, Methods in Molecular Biology, vol. 1400, DOI 10.1007/978-1-4939-3372-3_13, © Springer Science+Business Media New York 2016

end of the L1. Because the intron is in the "wrong" orientation relative to the reporter gene, expression will not yield a functional reporter gene product. The intron will be spliced from transcripts generated by the L1 promoter. However, because the reporter gene is in the opposite orientation relative to L1, translation of these transcripts will not yield a functional reporter gene product. When the spliced L1RNA undergoes retrotransposition, the new insert will now contain a functional reporter gene that can be evaluated in culture. The first cassette monitored L1 activity through the expression of neomycin resistance [2], which was followed by the generation of other selection cassettes that express blasticidin resistance [3], green fluorescence [4], and firefly luciferase [5].

Unfortunately, the L1 reporter cassette could not be directly applied to Alu, as SINEs and LINEs differ in their requirements for expression and construct development. The majority of the differences stem from the fact that LINEs are transcribed by RNA polymerase II (Pol II), while SINEs are transcribed by RNA polymerase III (Pol III). Figure 1 highlights the differences between the engineered SINE and LINE constructs used to monitor retrotransposition.

There are several limitations that need to be considered when designing a reporter cassette for Alu:

- Limitation 1: Pol III transcripts do not undergo the same processing as Pol II-derived transcripts; therefore Alu transcripts containing the L1 reporter cassette would not be spliced. In 2002, the Heidmann group developed a reporter cassette with a self-splicing intron from *Tetrahymena thermophila* [6] that allowed the monitoring of Alu retrotransposition in culture [7], as well as other SINEs [8, 9]. The design and location of the intron within the marker gene is critical because the catalytic efficiency of the self-splicing intron depends on the flanking sequences [6].
- Limitation 2: T-rich sequences (usually of 4Ts or more) serve as Pol III transcription terminators [10]. Thus, SINE reporter cassettes need to be devoid of internal Pol III terminator signals that would generate of truncated transcripts, rendering the approach useless.
- Limitation 3: Pol III transcripts are usually very short, with the majority being less than 300 base pairs (bp). Although it has not been formally evaluated, it is possible that adding a large reporter cassette to a SINE sequence may reduce transcription efficiency. Furthermore, the introduction of a reporter cassette creates a significantly larger transcript than naturally occurring SINE RNA, which adds an artificial variable in the study of SINE biology. To this date, only one reporter cassette (neo*TET*) is available for monitoring SINE activity. Unfortunately, this precludes studying SINE retrotransposition in any cell lines that already have neomycin resistance (e.g., XPA-complemented cell line (Coriell GM15876)).

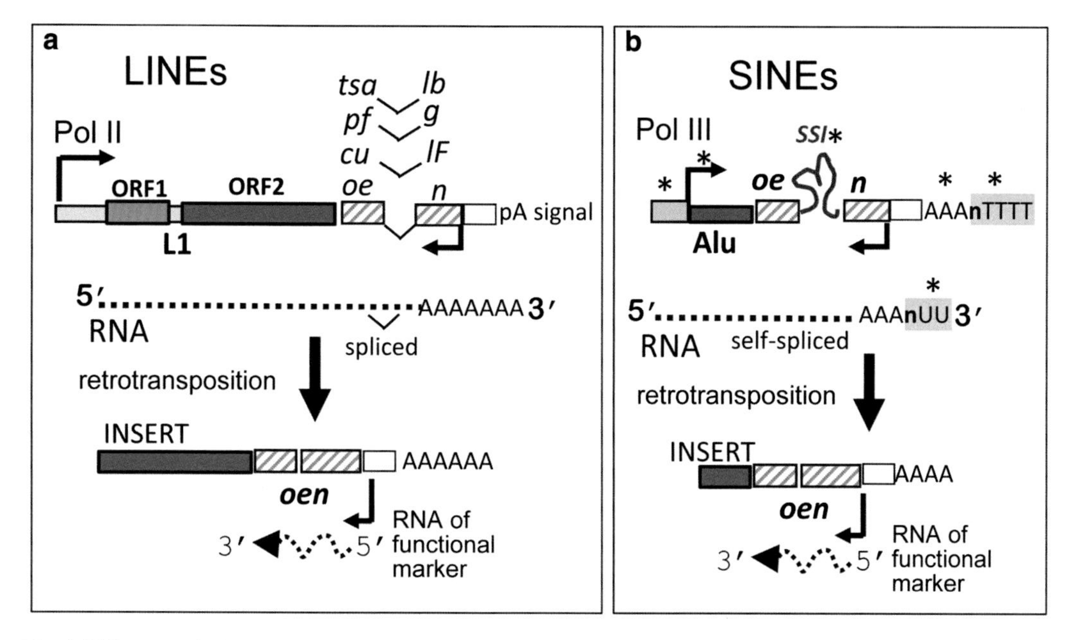

Fig. 1 Differences between reporter cassettes used to monitor retrotransposition of LINEs and SINEs. A schematic of the basic construct of a tagged L1 (**a**) and of a tagged Alu (**b**) and the fundamental steps of the how the retrotransposition assay works is shown. *Asterisks* indicate the components that differ in SINE constructs. In both LINE and SINE constructs, the reporter cassette (*hatched boxes*) is located at the 3′ end in the opposite orientation relative to the retroelement (*green*). The reporter cassette is disrupted by a "regular" intron (L1) or the *Tetrahymena* self-splicing intron (SSI; shown in *red*). However, the intron is in the same orientation as the retroelement, so that only transcripts generated by the promoter driving the retroelement undergo splicing. LINEs are transcribed by RNA polymerase II (Pol II), while SINEs are transcribe by RNA polymerase III (Pol III). SINEs require upstream enhancer sequences (*yellow*) to drive efficient transcription of the internal Pol III promoter. In addition, Pol III transcripts do not undergo polyadenylation like Pol II-derived transcripts. Thus, the A-tail in SINEs has to be encoded in the construct. Pol III transcription terminates at T-rich sequences containing four or more Ts (*light blue*). Sequences located between the A-tail and the terminator (shown as "n" *light blue* area) will be present in the transcript but will not be present in the new insert. *Note*: Although not shown, a source of L1 ORF2 is required for Alu retrotransposition and usually supplemented *in trans* in the assay. Only spliced RNA will have the potential to generate an insert with a functional marker gene. Expression of the marker gene serves as an indicator that retrotransposition of the tagged transcript occurred. There are multiple marker cassettes for tagging L1 elements: neomycin resistance (neo), blasticidin resistance (blast), green fluorescence (gfp), and firefly luciferase (Fluc), but only one (neo) for SINEs

Another tool that provided valuable information about the genomic impact of L1 retrotransposition was the creation of an L1 construct that allowed for the easy recovery of de novo L1 inserts in a culture system [11–14]. Using the L1 construct design as a guide, our lab adapted the available SINE reporter cassette to rescue de novo SINE inserts in a similar manner as described for L1 [15]. In this approach, tagged SINE constructs contain specific sequences that allow de novo SINE insertions to function as a plasmid expressing the kanamycin resistance gene. The resulting plasmid product is created by circularizing digested genomic DNA using restriction enzymes. Subsequent transformation of the circularized DNA into an *E. coli* strain allows for recovery of the DNA for analysis and sequencing [15]. The two components introduced to the SINE reporter cassette consist of the EM7 bacterial promoter that drives the expression of the neomycin/kanamycin gene, and a modified minimal γ origin of replication (305 bp) from the R6K plasmid [16, 17] (*see* Fig. 2a). These components provide the properties needed for the successful recovery of the de novo insert. We further adapted the R6Kγori sequence by eliminating all runs of 4 or more Ts or more, which act as strong Pol III terminators contributing to the generation of truncated tagged SINE transcripts [15]. In this chapter, we describe the protocol for recovery of de novo tagged Alu inserts using this approach in detailed.

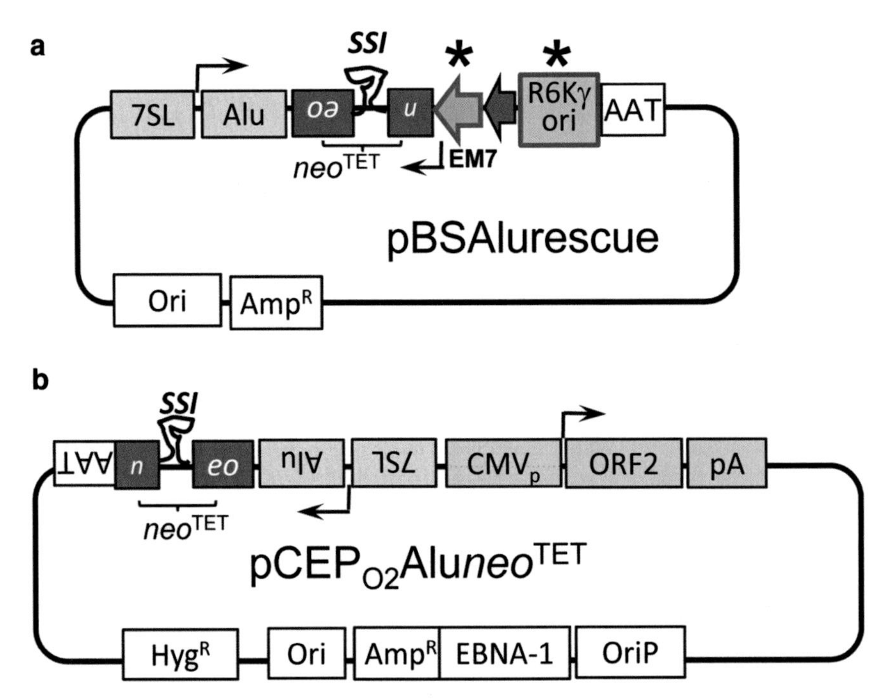

Fig. 2 Schematic of existing SINE constructs. A schematic of the components present in a construct of a tagged Alu used to rescue de novo inserts in culture (**a**) and an episomal construct designed to co-express a tagged Alu and the L1 ORF2p (**b**) is shown. The Alu rescue construct consists of a pBluescript vector containing the basic tagged Alu with the addition of the EM7 bacterial promoter (*blue arrow*) to drive the neomycin gene (expressed as kanamycin resistance in bacteria) and the R6kγ origin of replication (*blue box*). These two components (indicated by *) provide the properties to the DNA containing a retrotransposed de novo insert so that it functions as an independent plasmid. The episomal construct (pCEP-Hyg) carries a cytomegalovirus (CMVp) promoter driving the expression of the L1 ORF2 protein in the opposite orientation relative to the tagged Alu cassette. Amp^R = ampicillin resistance gene; Hyg^R = hygromycin B resistance; pA = polyadenylation signal; AAT = A-tail and Pol III terminator; Ori = pUC origin of replication; OriP = Epstein-Barr virus latent origin of replication; EBNA-1 = Epstein-Barr virus nuclear antigen-1 gene. *Note*: Alu is shown as the example, different sequences will replace the 7SL-Alu region (*light gray boxes*) in other SINE constructs. *Note*: constructs are not drawn to scale

Although SINE retrotransposition shares similar characteristics with LINE retrotransposition, there are several differences that influence how these elements are studied in culture. First, because SINEs are nonautonomous, they require the supplementation of L1 factor(s) *in trans*. However, in contrast to L1 that requires both ORF1 and ORF2 proteins for retrotransposition, Alu is only ORF2 dependent. Supplementation can be achieved by either co-transfecting an additional plasmid expressing the required L1 factor(s), or by using a single plasmid that co-expresses both the SINE and L1 components at the same time (Fig. 2b).

In general, the amount of ORF2 protein expressed directly correlates with SINE retrotransposition efficiency. Co-transfecting plasmids expressing required L1 machinery adds a level of complexity to the system. Protocols that rely on co-expressing either L1 or the ORF2 protein need to be optimized to find the appropriate ratio between SINE and LINE components. For example, during optimization it is important to determine the ideal conditions that promote efficient SINE retrotransposition. However, an excess of L1 or ORF2 may induce cell death due to the toxic effects of the endonuclease and reverse transcriptase activities of the ORF2 protein [18]. These ratios could vary greatly depending on the cell line used. A second difference between Alu and L1 retrotransposition centers on cellular environment requirements. Although the reason is unknown, some cells that efficiently support L1 retrotransposition are unable to support Alu. For example, published data on two different populations of HeLa cells demonstrated that only one supports Alu activity while both support L1 retrotransposition [19]. Retrotransposition efficiency can also vary between experiments [20], which is likely due to the polyclonal nature of most cell lines. Previous data demonstrate that individual clones derived from a human cell line can vary significantly in their capacity to support retrotransposition [20]. These differences are thought to arise from continuously passaging mixed populations of cells, selecting for and accumulating cells with particular genetic and epigenetic changes. Therefore, a reference control is often required to standardize between experimental variation. Finally, many differences between LINEs and SINEs are due to their different transcriptional requirements. As previously described, SINEs are transcribed by RNA polymerase III. Therefore, experimental conditions affecting Pol III transcription will alter retrotransposition results. For example, co-transfection of plasmids driven by the U6 promoter (e.g., shRNA used to reduce expression of target genes) will likely compete with SINE expression [21], effectively reducing the amount of tagged RNA generated and retrotransposition events. Thus, additional controls are needed when performing SINE retrotransposition experiments using these types of approaches. Overall, these observations reveal that important careful experimental design is essential for the study of SINE biology.

2 Materials

2.1 Tissue Culture

1. Appropriate cell line(s) that support retrotransposition (*see* **Note 1**). We mostly use HeLa due to their ability to support SINE retrotransposition very efficiently. However, we have observed SINE retrotransposition in several human and rodent cell lines.
2. Complete and serum free media appropriate for cell lines (*see* **Note 2**). To make complete MEM: add 50 mL FBS to a 500 mL bottle of Minimum Essential Medium, (+) Earle's salts, (+)l-glutamine (MEM). Supplement with 5 mL of nonessential amino acids (NEAA) and 5 mL sodium pyruvate (NaPyr). Keep refrigerated ~4 °C. To make complete DMEM: add 50 mL FBS to 500 mL Dulbecco's Modified Eagle Medium, (+) 4.5 g/L d-glucose, (+)l-glutamine, (+) 110 mg/mL sodium pyruvate (DMEM). Keep refrigerated ~4 °C.
3. Complete medium supplemented with the appropriate antibiotic(s) for selection. Optional: the addition of antibiotics to prevent contamination in experiments requiring long tissue culture incubations.
4. Tissue culture flasks and plates (*see* **Note 3**).
5. Trypsin–EDTA 0.05 %.
6. Sterile phosphate buffer saline (1×) pH 7.4 (PBS).
7. Cell counter (we use a hemocytometer).
8. Trypan blue stain (0.4 % w/v).

2.2 Transient Transfection

1. Plasmids: tagged SINE construct, e.g., pBSAluYa5*neo*^{TET} [9] and a driver plasmid expressing a full L1 or just the L1 ORF2 protein, e.g., pBudORF2$_{CH}$ [15] (*see* **Note 4**).
2. Transfection Reagents. We routinely use Lipofectamine and Plus reagents from Invitrogen. Other transfection reagents are also known to work well [22].
3. Complete and serum free media.
4. Selection medium: complete medium supplemented with geneticin, also known as G418 (*see* **Note 2**).
5. Crystal violet staining solution (0.2 % (w/v) crystal violet in 5 % (v/v) acetic acid and 2.5 % (v/v) isopropanol) (*see* **Note 5**).

2.3 Episomal Transfection

1. Plasmid: tagged SINE episomal construct, e.g., pCEPAlu*neo*^{TET}, and a driver plasmid expressing a full L1 or just the L1 ORF2 protein, e.g., pBudORF2$_{CH}$ (*see* **Note 4**). Alternatively, use a plasmid that co-expresses the ORF2 protein with the tagged SINE, e.g., pCEP$_{O2}$Alu*neo*^{TET}, Fig. 2b.
2. Transfection reagents: Lipofectamine and Plus reagents.

3. Complete and serum free media.
4. Selection media: complete medium supplemented with hygromycin B and complete medium supplemented with geneticin (*see* **Note 2**).
5. 0.05 % trypsin–EDTA.
6. Sterile phosphate buffer saline (1×) pH 7.4 (PBS).
7. Cell counter (hemocytometer).
8. Trypan blue stain (0.4 % w/v).
9. Crystal violet staining solution (*see* **Note 5**).

2.4 Alu Rescue

1. Dedicated reagents and equipment to be used only with L1 and Alu rescues. Contamination with other plasmids routinely used in the laboratory can become a significant problem during any of the steps of this procedure. (A comprehensive list of dedicated materials can be found in **Note 6**).
2. Plasmids: tagged SINE rescue construct, e.g., pBS-Ya5rescue-A70D-SH [15] or tagged SINE rescue episomal construct, e.g., pCEP-Ya5rescue-AT [15] and a driver plasmid expressing a full L1 or just the L1 ORF2 protein, e.g., pBudORF2$_{CH}$ (*see* **Note 4**).
3. DNA-Easy Blood and Tissue kit (Qiagen)or a similar genomic DNA extraction kit.
4. *pir*-116 Electrocompetent *E. coli* (obtained from local providers).
5. Electroporation Cuvettes (0.4 cm Gene Pulser/MicroPulser Bio-Rad or similar).
6. LB media (200–300 μL/transformation).
7. Electroporation apparatus for pulsing of electrocompetent *E. coli*. We use the MicroPulser Electroporator (Bio-Rad) using the default setting for bacteria.
8. Round-bottom polystyrene 5 mL tubes (for growth of electroporated bacteria).
9. Falcon 15 mL conical centrifuge tubes (to collect cells from pooled colonies for DNA extraction).
10. 1.5 mL Eppendorf microfuge tubes (for plasmid and genomic DNA extraction).
11. Centrifuges: one for 1.5 mL Eppendorf microfuge tubes and another for 15 mL conical Falcon tubes.
12. 0.05 % trypsin-EDTA.
13. Micron filter system (Amicon Ultra 0.5 mL Centrifugal Filters Ultracel-50K).
14. Sterile 1× PBS pH.7.4.

15. Bacterial culture tubes (for growth of bacterial colonies to extract plasmid DNA).
16. LB media and agar plates supplemented with 50 μg/mL kanamycin (*see* **Note 7**).
17. A 37 °C incubator for bacterial growth.
18. Plasmid DNA isolation reagents or kit.
19. Enzymes and buffers for digesting the kan^R rescue plasmids or genomic DNA (see Note 6). We use either *Sal*I and *Sfi*I, or *Aat*II and *Avr*II (example shown in Fig. 5).
20. Heat block or incubator set to 37 °C.
21. Standard low mr agarose to make a 1 % gel.
22. Buffer for agarose gel electrophoresis. We use 0.5 % Tris–Borate–EDTA (TBE) with ethidium bromide (*see* **Note 8**).
23. DNA marker (e.g., 1 kb ladder).
24. A gel imager with ethidium bromide fluorescence detection capability.
25. Sequencing primers:
    (a) For the 5′ genomic flank upstream of the Alu insert: RAluneoj primer: 5′-TTCTTCTGAGGGGATTTGAGAC GT-3′.
    (b) For the 3′ A-tail: FAtail230 primer: 5′-CTTATAAAACT TAAAACCTTAGAGGC-3′.
    (c) For the 3′ genomic flank downstream of the Alu insert: primer to be designed after 5′ genomic sequence is obtained (*see* **Note 9**).

3 Methods

3.1 Seeding Cells

1. Cells should be kept at 37 °C throughout all retrotransposition experiments performed.
2. When cells become between 80 and 90 % confluent, wash cells with 4–6 mL of sterile 1× PBS per T75.
3. Add 2 mL of 0.05 % trypsin to each T75 to remove adherent cells. Allow to sit at room temperature for at least 5 min, or until cells have dislodged.
4. Deactivate trypsin using at least 3 mL of media. Triturate the cells to break up clusters of cells.
5. Count cells as directed by the manufacturer's protocol for your cell counting device. Our lab uses a hemocytometer. We add 400 μL of trypan blue to 100 μL of the trypsinized cells, pipetting gently to mix. Add 10 μL of the cell suspension to the hemocytometer to count.

Table 1
Seeding densities (based on experience using HeLa)

Tissue culture flask/dish size	Seeding density for retrotransposition	Transfected Alu DNA	Transfected L1 or ORF2 DNA[a]
		Alu/Alurescue[a]	Alu/Alurescue[b]
6 well (per well)	$0.05–0.1 \times 10^6$	0.5 μg /not done	0.16 μg/not done
T25 flask	$0.1–0.25 \times 10^6$	1 μg/2–3 μg[b]	0.3–1 μg/0.6–1 μg[b]
T75 flask	$0.5–1.0 \times 10^6$	3 μg/6 μg[b]	1 μg/2 μg[b]

[a]We find that ratios of 2:1 or 3:1 Alu:L1 or ORF2 works best in our hands
[b]The plasmid used for Alu rescues is less efficient so higher amount of DNA is used for transfection

6. The amount of cells seeded depends on the size of the tissue culture container. *See* Table 1 for our recommendations for HeLa. Different cell lines may need to be individually evaluated for optimal conditions.
7. Add complete medium to the flasks or plates and incubate overnight at 37 °C.
8. The cells will be ready to transfect the following day approximately 16–18 h post-seeding.

The Alu retrotransposition assay can be performed using two different approaches (Fig. 3). The first approach is a basic transient transfection followed by selection with geneticin to detect the retrotransposition events that occurred. When using this approach, variations in transfection efficiency will directly affect the results. The second approach differs by using an episomal plasmid (Fig. 2b) that contains a resistance marker (hygromycin) that will allow for selecting successfully transfected cells and eliminating the untransfected cells (Fig. 3b). After a week of selection, the cells are reseeded at different cell densities and then grown under geneticin selection to detect the retrotransposition events. This approach is unaffected by variations in transfection efficiency and allows to evaluate retrotransposition rate by using the number of seeded hygromycin resistant (hyg^R) cells as the denominator.

3.2 Transient Transfection

1. A simple schematic of the episomal transfection approach is shown in Fig. 3a.
2. Transfections are performed 18–24 h after seeding.
3. Follow the manufacturer's recommended protocol for setting up and performing transfections. Our lab uses the Lipofectamine and Plus system from Invitrogen (*see* **Note 10**). For this approach, serum free media is recommended. Thus, the cell medium needs to be removed and replaced with serum free medium. In addition, serum free medium should be used when

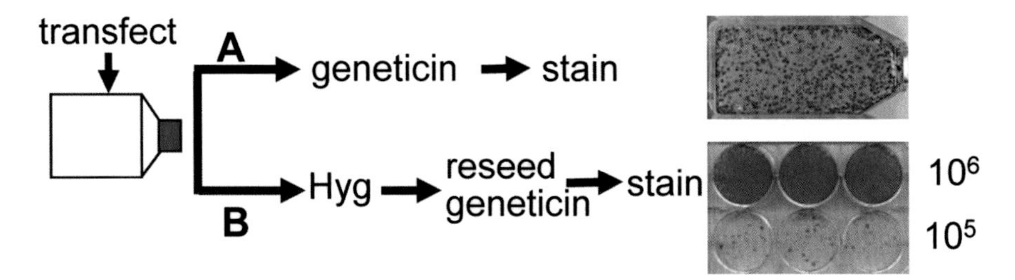

Fig. 3 Schematic of the two approaches used to determine retrotransposition rates. For simplicity we refer to approach *A* as the "transient transfection" method and *B* as the "episomal transfection" method. Cells are transfected and the following day grown under either A geneticin selection to detect tagged retrotransposed inserts or B hygromycin B (Hyg) for a week to select for cells containing the transfected episomal plasmid followed by reseeding at different cell densities and growth under geneticin to detect the tagged retrotransposed inserts. After 2 weeks of selection colonies can be stained to count or expanded to extract DNA for recovery of inserts

setting up the DNA–Lipofectamine/Plus mixtures. Details of our transfection parameters are shown in Table 2.

4. Add the DNA–Lipofectamine/Plus mixture to the cells. Incubate between 3 and 5 h. We routinely incubate transfections for 3 h (*see* **Note 11**).
5. Remove media with transfection solution. Add appropriate media (complete DMEM or complete MEM). Incubate overnight to allow cells to recover.
6. The following day change the cells to selection media. Incubate cells with appropriate selection media for 14 days. Change media as necessary. We usually feed twice a week (*see* **Note 12**).
7. Stain flasks by adding a sufficient amount of crystal violet to each flask to cover all the colonies. Rock at room temperature for a minimum of 15 min.
8. Wash flasks thoroughly with tap water until the water runs clear. Invert to dry.
9. Scan flasks or plates and count the colonies.

3.3 Episomal Transfections

1. We use the pCEP4 plasmid (Invitrogen) as the vector backbone for episomal transfections. This plasmid contains the Epstein-Barr Virus replication origin and nuclear antigen that allows for extrachromosomal replication (the schematic of one of our pCEP based constructs is shown in Fig. 2b). Our pCEP plasmids carry the hygromycin B resistance gene used for stable selection of transfected cells. Thus, only cells that have been successfully transfected with plasmid will grow. A simple schematic of the episomal transfection approach is shown in Fig. 3b.
2. Cells are seeded as the protocol states in Subheading 3.1.
3. The following day, put cells under hygromycin B selection to select for transfected cells. Continue selection until cell death

Table 2
DNA–Lipofectamine/Plus mixtures (*see* Note 13)

Total amount transfected DNA/flask or dish (μg)	Lipofectamine (μL)	Plus reagent (μL)
<1	2	1
1	3	2
2–4	6–9	4–8
5–13	9–12	8–12

is not observed anymore. This process takes 1 week for HeLa but can vary depending on cell type (*see* **Note 14**).

4. Reseed the hyg^R cells at different concentrations to ensure that the final colony number obtained per well is within a linear range for counting. Wells (6-well plate) containing more than 300 colonies are considered out of the linear range and unreliable for quantitative analyses (see an example of results in Fig. 3b). We routinely seed 10^4, 10^5, and 10^6 cells per well in 6-well plates. Seeding more than 10^6 cells is not recommended in a 6-well plate Instead, a tissue culture flask with larger surface area should be used.
5. The next day, change the medium of the hyg^R cells to geneticin selection medium to select for cells with retrotransposition events.
6. Incubate cells with appropriate selection media for 14 days (*see* **Note 12**) and stain colonies with crystal violet.
7. Scan plates and count the colonies.

3.4 Transfections for Recovery of Alu Inserts: "Alu Rescue"

Transfections for the recovery of Alu inserts follow the same protocols as indicated above in Subheadings 3.2 and 3.3 with some minor changes due to the lower retrotransposition efficiency of the Alu tagged with the rescue cassette (*see* Table 1).

1. Follow the protocols above but use more plasmid DNA per transfection (*see* Table 1).
2. Grow cells under geneticin selection to obtain colonies.
3. When distinct, large colonies are visible to the naked eye, proceed to expand the cells for DNA extraction.
4. Ideally, we usually trypsinize and combine 50 or more colonies per pool (*see* **Note 15**). Based on the number of colonies it may be necessary to change to a smaller flask to accommodate pools with low cell numbers (*see* **Note 16**).

5. Continue to grow cells under geneticin selection until the amount of cells needed for genomic DNA isolation is obtained. This will vary based on your experimental needs, as well as, the requirements of the genomic DNA isolation system you are using (*see* **Note 17**).

   From this point on it is critical that all materials and buffers used are dedicated (*see* **Note 6**).

6. Trypsinize cells and count. We use the DNAeasy Blood & Tissue kit that recommends the use of 5×10^6 cells per sample. Add 1× PBS to deactivate the trypsin and transfer the cells to a 15 mL conical Falcon tube. Centrifuge for 5 min at 4 °C at ~2000 × *g*. Discard the liquid and invert the tube for a few minutes to remove excess liquid from the cell pellet. At this stage the cell pellet may be frozen at −20 °C (short term storage) or −80 °C (long term storage) for processing at a later time.
7. Follow the manufacturer's protocol for genomic DNA isolation. We elute the DNA from the column by adding 200 μL of water twice to increase gDNA yield. The 400 μL of pooled elute is sufficient for two genomic digestions.
8. Incubate 200 μL of extracted gDNA (~200 μg) with 200 U of the selected restriction enzyme and appropriate buffer for at least 5 h at 37 °C (*see* **Note 6** for enzyme selection).
9. Heat-inactivate the restriction enzyme by incubating the sample at 65 °C for 20 min.
10. Add 700 μL water, 100 μL 10× T4 DNA ligase buffer, and 1200 U T4DNA ligase to each rescue digestion and incubate overnight (~16 h) at 16 °C.
11. The next day concentrate the 1 mL of ligated genomic DNA using a micron filter (Amicon 50K) by pipetting 500 μL of the ligation reaction at a time into the micron filter. Centrifuge for 10 min at 8000 × *g* (*see* **Note 18**). Empty the collection tube after each spin.
12. Wash the concentrated DNA twice with 500 μL sterile deionized water to remove salts. Centrifuge for 10 min at 8000 × *g*. Empty the collection tube after each spin.
13. Continue centrifugation to reduce the final volume to about 10 to 20 μL (Fig. 4). This usually requires a longer centrifugation time after the final water wash (*see* **Note 19**).
14. While the DNA is being concentrated, chill the electroporation cuvettes on ice until ready to use.
15. Invert the micron filter into a sterile collecting tube (provided by manufacturer) and spin at ~16,000 × *g* for 30 s to collect the concentrated DNA.

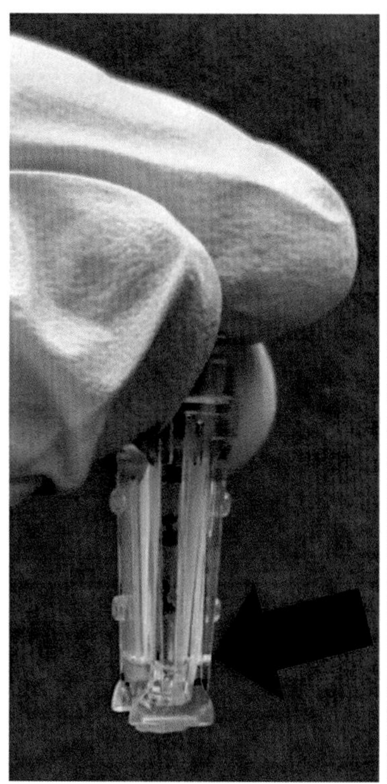
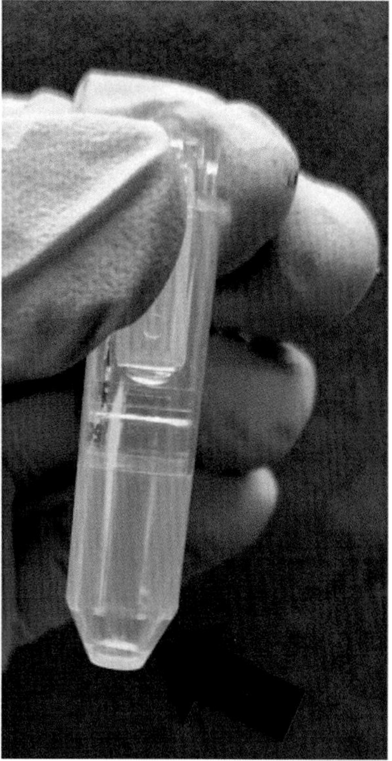

Fig. 4 Example of the ideal final volume in the micron filter after the final centrifugation. The picture on the *left* shows the filter with the final volume; while the picture on the *right* shows the sample after it has been spun out of the filter into a sterile tube. *Arrows* highlight the small volume

16. Allow the *pir*-116 electrocompetent *E. coli* to thaw on ice about 10 min before starting the electroporation (*see* **Note 20**). The electrocompetent *E. coli* should be stored at −80 °C.
17. Add between 30 and 50 μL of the electrocompetent *E. coli* cells to the well of the cuvette located between the two metal plates (*see* **Note 21**). Do NOT pipette cells up and down.
18. Add the 10–20 μL of concentrated DNA to the *E. coli* cells in the cuvette. Flick gently to mix. Keep on ice.
19. Pulse the *E. coli* cells by using the "bacteria" setting of the electroporation apparatus (we use MicroPulser from Bio-Rad).
20. Pipette 200–300 μL of sterile LB media (or any other rich medium available, e.g., SOC) into the cuvette using a stripette. Mix by pipetting up and down several times. Transfer the entire mixture into a culture tube appropriate for bacterial growth (a round-bottom polystyrene 5 mL tubes).
21. Grow the bacteria at 37 °C in a shaking incubator for 1 h.
22. Spread the full volume of bacteria onto a kanamycin LB agar plate. Incubate at 37 °C for at least 18 h. This *pir*-116 strain of

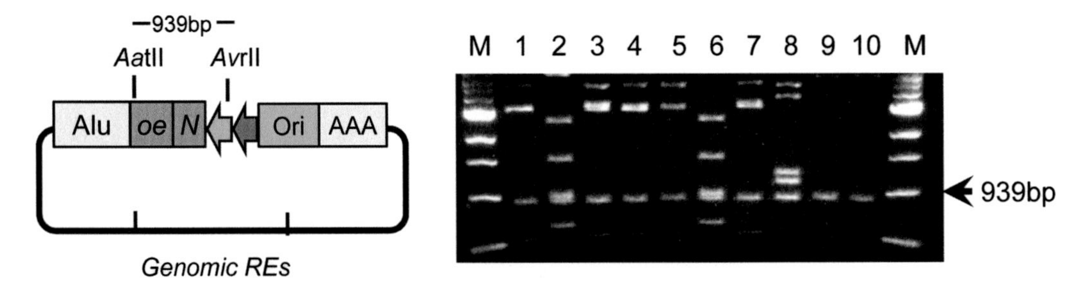

Fig. 5 Evaluation of Alu rescue clones by restriction digest. A schematic representation of the expected recovered DNA with the tagged-Alu insert with the circularized flanking sequence is shown. Restriction sites (*Aat*II and *Avr*II) were selected so that they flank the location of the splice junction to yield a 939 bp fragment. Clones that do not show the 939 bp fragment are likely recovered contaminants of other kan^R plasmids. Alu inserts located in different genomic regions will contain different flanking sequences and will generate unique digest patterns depending on presence/absence of the selected restriction sites (REs). In the example shown, clones 2 and 6 are likely duplicates of the same Alu insert, while clones 1, 2, 8, and 9 are likely recovered Alu inserts located at different genomic sites. M = marker. Similar analysis can be performed using *Sfi*I and *Sal*I digests

E. coli cells might require a longer incubation time for colonies to grow to a size detectable by the naked eye.

23. Pick individual bacterial colonies for growth. We usually grow bacteria in 2 mL of kanamycin LB broth for plasmid extraction (miniprep). Shake overnight at 37 °C.
24. Transfer about two-thirds of the 2 mL culture into a 1.5 mL Eppendorf microfuge tube and centrifuge to pellet the bacterial cells (~16,000 × *g* for 2 min). Discard supernatant. The leftover culture is stored at 4 °C as backup in case more is needed later.
25. Isolate the plasmid DNA following the manufacturer's protocol.
26. Evaluation of the kan^R rescue clones (i.e., plasmids) is performed by a restriction enzyme digestion using a combination of either *Aat*II and *Avr*II or *Sfi*I and *Sal*I. This digestion will monitor potential contamination and provide a visual representation of different recovered retrotransposition events used to guide selection of which plasmids to sequence (Fig. 5 shows an example).
27. Each selected plasmid is sequenced using three primers: RAluneoj primer, FAtail230 primer, and a primer uniquely designed for the 3′ genomic flank of the Alu insert (*see* **Note 9**).
28. Once sequences are received, the genomic position of each rescued Alu insertion is determined by BLAT (http://genome.ucsc.edu). In some occasions, an insert lands in a repetitive sequence, which may be very difficult to precisely map.

4 Notes

1. Cell lines: we have been able to perform SINE retrotransposition in both human, rodent cells and in one chicken embryo fibroblast cell line [23]. Our experience is limited to cell lines that form monolayers, but LINE retrotransposition has been described in chicken DT40 suspension cells using soft agarose medium [24]. The human cell lines we primarily use are HeLa, HCT-116, and HEK293. The rodent cell lines we use are BHK, CHO-K1, and CHOUV20. Cell selection is critical, as some cells cannot support retrotransposition [19]. Cells can also exhibit clonal variation with regard to how well they support retrotransposition [20].
2. Supplements for tissue culture media can be obtained from local vendors: fetal bovine serum (FBS), MEM nonessential amino acids (NEAA) 100×, and sodium pyruvate 100 mM (NaPyr). For antibiotic selection, concentrations vary depending on the cell line. These are the concentrations routinely used in HeLa cells: Geneticin stock of 50 mg/mL active geneticin (4.4 mL stock per 500 mL of medium); Hygromycin B stock in 1× PBS 50 mg/mL (700 μL stock per 500 mL of medium). For prevention against mycoplasma, bacteria, and fungi (optional), the following reagents can be added to the culture media: Normocin (Invivogen), Fungizone, and Penicillin/Streptomycin (Pen/Strep). This is recommended for experiments that take multiple weeks.
3. Tissue culture flasks and plates: we use standard polystyrene vented cap flasks and clear polystyrene flat bottom plates. However, different cell lines may have different requirements such as the need for a surface treatment that enhances cell attachment. Use the materials that are appropriate for the cell line.
4. L1 factors are required to drive Alu retrotransposition events. Therefore, Alu must be co-transfected with a plasmid expressing either the full length L1 or the L1 ORF2 protein as the driver [7]. Several of the plasmids are available in Addgene: http://www.addgene.org/browse/pi/1826/
5. Crystal violet staining solution (0.2 % crystal violet in 5 % acetic acid and 2.5 % isopropanol) is prepared as follows: dissolve and mix 1 g crystal violet, 15 mL isopropanol, 25 mL glacial acetic acid, and purified water to a final volume of 500 mL.
6. Dedicated reagents and materials. *Very Important*: all materials including the buffers and enzymes should be used *exclusively* for rescuing Alu inserts and should *only* come into contact with the dedicated pipettes, pipettors, and tips. This method is highly susceptible to failure due to contamination with other common plasmids present in most laboratories. We recom-

mend finding a lab bench or other dedicated space that is free of exposure from other plasmids, specifically plasmids containing kanamycin resistance. This is a list of dedicated materials: set of pipettors and tips (pipettors can be decontaminated if required); genomic DNA extraction kit with dedicated buffers; 1-mL stripettes; restriction enzymes and buffers (we routinely use *Eco*RI and *Hin*dIII, although any restriction enzyme that is absent within the retrotransposed sequence and can be heat-inactivated will work. We have also successfully used *Spe*I, *BsrG*I, *Nhe*I, or *Nde*I); T4 DNA ligase and ligase buffer; sterile water.

7. We use LB broth, Miller powder, and LB agar, Miller powder. To prepare 1 L of LB broth, dissolve 20 g of powder in 800 mL of water; once dissolved adjust volume to 1 L and autoclave. Add kanamycin (final concentration of 50 μg/mL) to the LB broth once it is completely cooled. Keep refrigerated ~4 °C. To prepare 1 L of LB agar, dissolve 35 g of powder in 800 mL of water; once dissolved adjust volume to 1 L and autoclave. Add kanamycin (final concentration of 50 μg/mL) to the LB agar after it has cooled but not solidified (~55–60 °C) to prevent inactivation of the kanamycin. Pour the plates and let them solidify. For long-term storage, keep refrigerated ~4 °C.
8. 0.5× Tris–Borate–EDTA (TBE) (45 mM Tris–Borate, 1 mM EDTA). To make a 5× Stock mix and dissolve: 54 g of Tris base, 27.5 g of boric acid, 20 mL of 0.5 M EDTA, pH 8.0, and purified water to a final volume of 1 L. Dilute 1/10 and add ethidium bromide to a final concentration of ~0.2–0.5 μg/mL.
9. Design of the primer to sequence the 3′ genomic flank of the Alu rescue insert: the primer should anneal ~150–200 bases from the predicted insertion site based on the data obtained from the sequence from the 5′ genomic flank. The primer should be evaluated to make sure it does not anneal to a repetitive sequence. In addition, the location of the sequence corresponding to the restriction site used to digest the genomic DNA (e.g., *Eco*RI) should be noted to avoid designing a primer to a sequence that may not be present in the circularized DNA. For example, if there is an *Eco*RI site 180 bp downstream from the predicted insertion site a primer that anneals 210 bp downstream will not work as the circularized DNA will not include anything downstream of where the enzyme cut (180 bp).
10. Depending on the total number of transfections and experimental repeats, transfection reagents can be mixed in 24-, 12-, or 6-well plates instead of individual tubes. The DNA transfection parameters given in Table 1 are our recommendations for what concentrations work in our cell lines. However, each laboratory may need to optimize the total amount of DNA and/or transfection reagents for the best results. We have

found that a 1:3 ratio of driver to Alu plasmid DNA, respectively, works well for any Alu retrotransposition assay regardless of total amount of DNA transfected.

11. Individual cell lines can be sensitive to the transfection mixture, resulting in premature cell death. Optimization of the incubation time may need to be tailored to certain cell lines. Our observations indicate that an incubation period of 3 h is sufficient and well tolerated by all the cell lines we have evaluated.
12. The time of selection has been chosen arbitrarily, using colony size as the parameter to determine when to stain cells. HeLa cells form colonies that allow for easy visualization and counting after 2 weeks of selection with geneticin. However, cells that grow faster, such as BHK, will require less time for growth under selection. On the other hand, slow growing cells may require longer periods of time. If media becomes turbid due to large numbers of floating dead cells, change the media more frequently.
13. The indicated Lipofectamine–Plus ratios are the most commonly used in our laboratory. However, transfection efficiencies will vary between cell lines and further optimization of transfection conditions may be needed.
14. The concentration of hygromycin B required for effective selection varies between cell lines. Slow growing cells may require longer time of growth under selection to ensure the death of the untransfected cells.
15. The number of recovered inserts per pool is directly proportional to the number of colonies in the pool. We routinely recover between a fourth and a third of the inserts from any given pool. The recovery efficiency can be improved by processing the same pool multiple times with different restriction enzymes when digesting the genomic DNA.
16. Cells may not grow well at very low densities. It might be necessary to reduce the surface area to grow the cells. For example, if only one rescue colony is recovered, it might be necessary to move the colony into a well of a 6-well or 12-well plate.
17. We routinely expand pooled cells in 150 cm^2 dishes to obtain a large amount of material that can be used for multiple DNA extractions. However, if less DNA is needed a T75 would suffice.
18. The time and speed for spinning can vary, as long the parameters selected comply within those suggested by the Amicon 50K manufacturer's protocol.
19. It is imperative not to dry the membrane out. Please refer to Fig. 4 for an example of an appropriate volume of concentrated genomic DNA.

20. Commonly used electrocompetent *E. coli* strains will not support replication of the R6Kγori. A specific bacterial strain is required: *E. coli pir*-116 [*F*- *mcrA* Δ(*mrr-hsdRMS-mcrBC*) *φ80dlacZΔM15 ΔlacX74 recA1 endA1 araD139* Δ(*ara*, *leu*)*7697 galU galK λ- rpsL* (*StrR*) *nupG pir-116*(*DHFR*)].
21. A regular 1000 μL pipette tip cannot be used for this step because it is too wide to fit into the groove of the cuvette. Longer, thinner pipette tips should be used (see an example http://www.usascientific.com/200ul-extra-long-filtertip.aspx).
22. *Rescue Digestions*: mix reagents as follows:

    5 μL rescued plasmid

    2 μL 10× appropriate buffer (we used the one supplied by manufacturer)

    0.2 μL *Sal*I + 0.2 μL *Sfi*I or

    0.2 μL *Aat*II + 0.2 μL *Avr*II

    12.6 μL H_2O (final volume of 20 μL)

    Incubate at 37 °C for 1 h and run on a 1 % agarose gel.

    The restriction sites (*Aat*II and *Avr*II or *Sal*I and *Sfi*I) were selected so that they flank the location of the splice junction to yield a 939 bp or 1413 bp fragment, respectively. Clones that do not show the expected fragment are likely recovered contaminants of other kan^R plasmids. Alu inserts located at different genomic regions will have different flanking sequences generating unique digest patterns depending on presence/absence of the selected restriction sites (REs). This will allow for the elimination of duplicates from the same Alu insert and avoiding sequencing the same insert multiple times.

Acknowledgement

The protocols detailed here were developed from work supported by grants from the National Institutes of Health (NIH) P20GM103518/P20RR020152 and R01GM079709A to AMR-E.

References

1. Freeman JD, Goodchild NL, Mager DL (1994) A modified indicator gene for selection of retrotransposition events in mammalian cells. Biotechniques 17:46, 48-49, 52
2. Moran JV, Holmes SE, Naas TP et al (1996) High frequency retrotransposition in cultured mammalian cells. Cell 87:917–927
3. Goodier JL, Zhang L, Vetter MR, Kazazian HH Jr (2007) LINE-1 ORF1 protein localizes in stress granules with other RNA-binding proteins, including components of RNA interference RNA-induced silencing complex. Mol Cell Biol 27:6469–6483
4. Ostertag EM, Prak ET, DeBerardinis RJ et al (2000) Determination of L1 retrotransposi-

tion kinetics in cultured cells. Nucleic Acids Res 28:1418–1423

5. Xie Y, Rosser JM, Thompson TL et al (2011) Characterization of L1 retrotransposition with high-throughput dual-luciferase assays. Nucleic Acids Res 39:e16. doi:10.1093/nar/gkq1076, gkq1076 [pii]
6. Esnault C, Casella JF, Heidmann T (2002) A Tetrahymena thermophila ribozyme-based indicator gene to detect transposition of marked retroelements in mammalian cells. Nucleic Acids Res 30:e49
7. Dewannieux M, Esnault C, Heidmann T (2003) LINE-mediated retrotransposition of marked Alu sequences. Nat Genet 35:41–48
8. Dewannieux M, Heidmann T (2005) L1-mediated retrotransposition of murine B1 and B2 SINEs recapitulated in cultured cells. J Mol Biol 349:241–247
9. Kroutter EN, Belancio VP, Wagstaff BJ et al (2009) The RNA polymerase dictates ORF1 requirement and timing of LINE and SINE retrotransposition. PLoS Genet 5:e1000458
10. Orioli A, Pascali C, Quartararo J et al (2011) Widespread occurrence of non-canonical transcription termination by human RNA polymerase III. Nucleic Acids Res 39:5499–5512. doi:10.1093/nar/gkr074, gkr074 [pii]
11. Symer DE, Connelly C, Szak ST et al (2002) Human l1 retrotransposition is associated with genetic instability in vivo. Cell 110:327–338
12. Gilbert N, Lutz-Prigge S, Moran JV (2002) Genomic deletions created upon LINE-1 retrotransposition. Cell 110:315–325
13. Gilbert N, Lutz S, Morrish TA et al (2005) Multiple fates of L1 retrotransposition intermediates in cultured human cells. Mol Cell Biol 25:7780–7795
14. Ostertag EM, Kazazian HH Jr (2001) Twin priming: a proposed mechanism for the creation of inversions in l1 retrotransposition. Genome Res 11:2059–2065
15. Wagstaff BJ, Hedges DJ, Derbes RS et al (2012) Rescuing Alu: recovery of new inserts shows LINE-1 preserves Alu activity through A-tail expansion. PLoS Genet 8:e1002842. doi:10.1371/journal.pgen.1002842, PGENETICS-D-11-02609 [pii]
16. Stalker DM, Kolter R, Helinski DR (1982) Plasmid R6K DNA replication: I. Complete nucleotide sequence of an autonomously replicating segment. J Mol Biol 161:33–43
17. Shafferman A, Helinski DR (1983) Structural properties of the beta origin of replication of plasmid R6K. J Biol Chem 258:4083–4090
18. Wallace NA, Belancio VP, Deininger PL (2008) L1 mobile element expression causes multiple types of toxicity. Gene 419:75–81
19. Hulme AE, Bogerd HP, Cullen BR et al (2007) Selective inhibition of Alu retrotransposition by APOBEC3G. Gene 390:199–205
20. Streva VA, Faber ZJ, Deininger PL (2013) LINE-1 and Alu retrotransposition exhibit clonal variation. Mob DNA 4:16. doi:10.1186/1759-8753-4-16, 1759-8753-4-16 [pii]
21. Roy AM, West NC, Rao A, Adhikari P et al (2000) Upstream flanking sequences and transcription of SINEs. J Mol Biol 302:17–25
22. Bennett EA, Keller H, Mills RE et al (2008) Active Alu retrotransposons in the human genome. Genome Res 18:1875–1883
23. Wallace N, Wagstaff BJ, Deininger PL et al (2008) LINE-1 ORF1 protein enhances Alu SINE retrotransposition. Gene 419:1–6
24. Honda H, Ichiyanagi K, Suzuki J et al (2007) A new system for analyzing LINE retrotransposition in the chicken DT40 cell line widely used for reverse genetics. Gene 395:116–124. doi:10.1016/j.gene.2007.02.017, S0378-1119(07)00103-5 [pii]

Chapter 14

The Engineered SVA *Trans*-mobilization Assay

Anja Bock and Gerald G. Schumann

Abstract

Mammalian genomes harbor autonomous retrotransposons coding for the proteins required for their own mobilization, and nonautonomous retrotransposons, such as the human SVA element, which are transcribed but do not have any coding capacity. Mobilization of nonautonomous retrotransposons depends on the recruitment of the protein machinery encoded by autonomous retrotransposons. Here, we summarize the experimental details of SVA *trans*-mobilization assays which address multiple questions regarding the biology of both nonautonomous SVA elements and autonomous LINE-1 (L1) retrotransposons. The assay evaluates if and to what extent a noncoding SVA element is mobilized in *trans* by the L1-encoded protein machinery, the structural organization of the resulting marked de novo insertions, if they mimic endogenous SVA insertions and what the roles of individual domains of the nonautonomous retrotransposon for SVA mobilization are. Furthermore, the highly sensitive *trans*-mobilization assay can be used to verify the presence of otherwise barely detectable endogenously expressed functional L1 proteins via their marked SVA *trans*-mobilizing activity.

Key words SVA, LINE-1, Nonautonomous retrotransposon, *Trans*-mobilization, Retrotransposition reporter assay

1 Introduction

Human non-LTR retrotransposons comprise long interspersed nuclear elements (LINEs), short interspersed nuclear elements (SINEs) and SINE-VNTR-*Alu* (SVA) elements which remain active and are currently mobilized in the genome (Fig. 1). The human genome harbors approximately 80–100 retrotransposition-competent LINE-1 (L1) retrotransposons which are termed autonomous because they encode the protein machinery required for their own mobilization (Fig. 1). In contrast, nonautonomous non-LTR retrotransposons, like the *Alu* and SVA elements, are noncoding and it was demonstrated by utilizing *trans*-mobilization assays in tissue culture that these elements co-opt the L1-encoded proteins for their own mobilization [1–4] (Fig. 1b). By applying an alternative retrotransposition reporter assay in cell culture, Esnault and coworkers demonstrated that human L1 proteins can also mobilize cellular

Jose L. Garcia-Pérez (ed.), *Transposons and Retrotransposons: Methods and Protocols*, Methods in Molecular Biology, vol. 1400, DOI 10.1007/978-1-4939-3372-3_14,

a

Autonomous

- **LINE1**

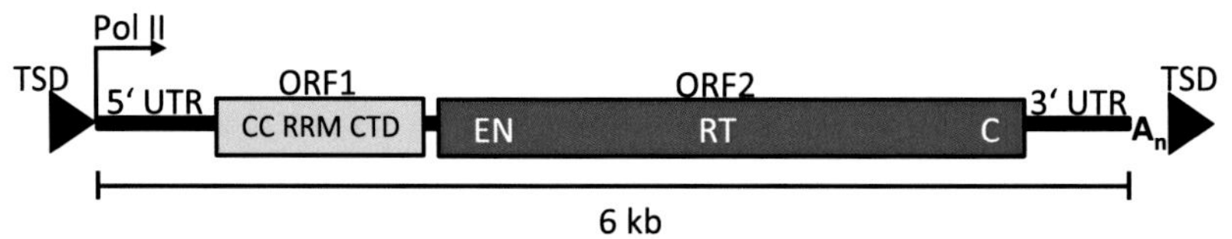

Non Autonomous

- ***Alu***

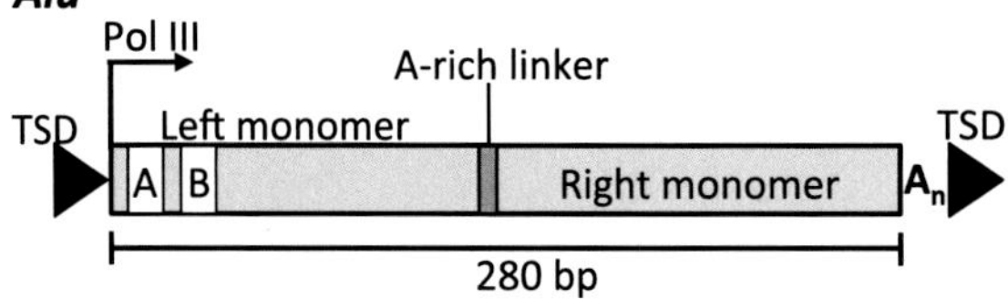

- **SVA**

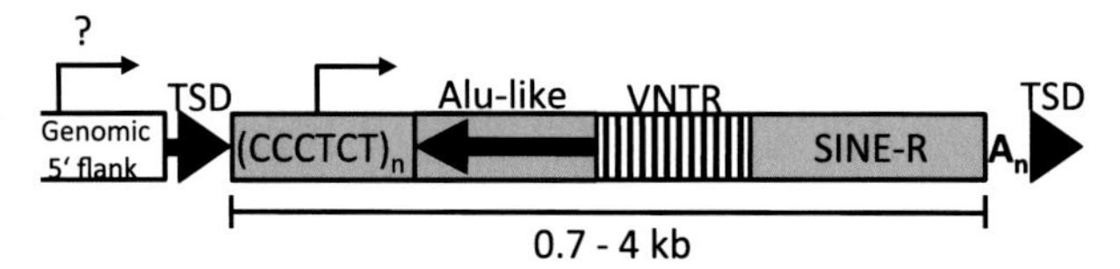

- **Processed pseudogene**

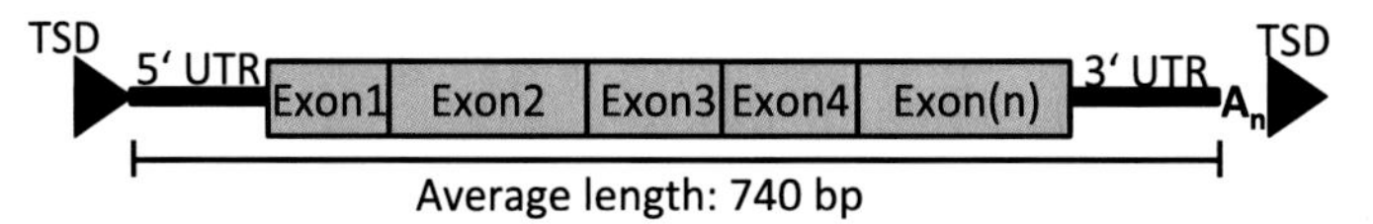

b

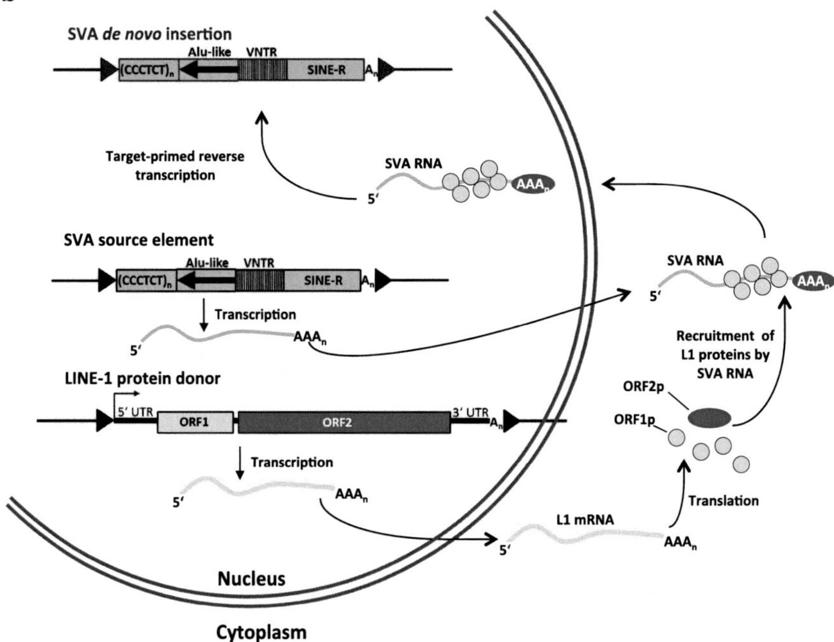

mRNA in *trans* [5] resulting in the formation of new copies of RNA polymerase II (PolII) genes that are termed "processed pseudogenes" and disclose the following features (Fig. 1): loss of intron and promoter, acquisition of a poly(A) 3′ end and presence of target site duplications (TSDs) of varying length [6, 7]. Both L1-encoded proteins are necessary for processed pseudogene formation.

SVA elements are hominid-specific composite nonautonomous retrotransposons ranging from ~700 to ~4000 bp [8, 9], and represent the youngest family of hominid non-LTR retrotransposons [10] encompassing approximately 2700 elements per human genome [10–14]. SVA elements display the hallmarks of L1-mediated retrotransposition [10, 15, 16] via target-primed reverse transcription (TPRT) [17–19] including (1) 5′ truncations, (2) 5′ inversions, (3) insertion at DNA sites resembling the L1 EN consensus site (5′-TTT/A-3′), (4) insertions flanked by target site duplications of various lengths (4–20 bp), (5) insertions ending in a 3′ poly(A) tail, and (6) insertions often containing 3′ transductions as a consequence of transcriptional readthrough. Full-length SVA elements (Fig. 1a) vary greatly in size because of repeat variation [8, 20] and the presence or absence of transductions [9, 20, 21], but they primarily consist of four domains, in order from the 5′ end: (1) a CT-rich repeat, with CCCTCT being the most common motif (hexamer repeat), (2) a sequence sharing homology to two antisense *Alu*-like fragments (*Alu*-like), (3) a *v*ariable *n*umber of GC-rich *t*andem *r*epeats (VNTR), with a unit length of

Fig. 1 Autonomous and nonautonomous non-LTR retrotransposons are mobilized by L1 proteins. (**a**) Organization of autonomous nonautonomous human non-LTR retrotransposons demonstrated to be mobilized by the human L1 protein machinery. Members of the L1 family of retrotransposons represent the only autonomous transposable elements that are currently functional in humans. Full-length and intact L1 elements are 6 kb in length and encompass a 5′ untranslated region (5′ UTR) which includes an internal bidirectional DNA polymerase II (Pol II) promoter, open reading frame 1 (ORF1) and ORF2, a 3′ UTR and a poly(A) tail (A*n*). ORF1 and ORF2 encode an RNA-binding protein (L1 ORF1p; cc, coiled coil domain; RRM, RNA recognition motif; CTD, carboxy-terminal domain) and a polyprotein (L1 ORF2p) including an endonuclease (EN) domain and a reverse transcriptase (RT) domain, respectively. Genomic insertions mediated by the L1 protein machinery result in target site duplications (TSDs) which flank new retrotransposition events. *Alu* and SVA elements which express noncoding RNAs, and Pol II-transcripts encoded by host genes, recruit the L1 protein machinery in order to retrotranspose. *Alu* elements are SINEs. An *Alu* element is a dimer of two nonidentical sequences ancestrally derived from the 7SL RNA gene separated by a middle A-rich linker region [49]. The left monomer of an *Alu* contains the internal RNA polymerase III (Pol III) promoter A and B boxes [50]. SVA elements are composite retrotransposons containing CCCTCT hexameric repeats, an *Alu*-like region, a VNTR region and a SINE-R region followed by a poly(A) sequence. Processed pseudogenes are derived from spliced mRNAs that were reverse transcribed and inserted in the genome by L1 protein activity. (**b**) Schematic of *trans*-mobilization of SVA elements by the L1-encoded protein machinery. Functional L1 retrotransposons are transcribed in the nucleus and the resulting L1 mRNA is translated into L1 ORF1p and L1 ORF2p. The majority of L1 proteins assemble in cytoplasmic L1 ribonucleoprotein (RNP) complexes containing a full-length L1 RNA molecule, L1 ORF1p and L1ORF2p. A small subset of ORF1 and ORF2 proteins can be hijacked by SVA RNAs resulting in SVA RNP complexes which initiate target-primed reverse transcription of the SVA RNA and finally to an SVA de novo insertion

either 36–42 or 49–51 bp [10], and (4) an ~490-bp sequence derived from the envelope gene and 3' long terminal repeat (LTR) of an extinct HERV-K10 (SINE-R) [11] that contains a canonical poly(A) signal (AATAAA).

Although SVAs lack a defined transcriptional unit [8, 9, 20, 21], firefly luciferase assays indicate that the SVA 5′ end (CT hexamer and Alu-like domain) has some promoter activity [22]. It was shown that SVAs can act as a classical transcriptional regulatory domain in the context of a reporter gene construct both in vitro in a human neuroblastoma cell line and in vivo in a chick embryo model [23]. Many different SVA structural variant classes exist in the human genome [20], with some elements belonging to multiple classes: Approximately 63 % of SVAs in the human genome reference sequence are full length [20] (Fig. 1a). Some SVAs contain (1) 5′ transductions [9, 20, 24], (2) 3′ transductions [10, 21], (3) new 5′ ends acquired via pre-mRNA splicing [9, 20, 24], or (4) 3′ truncations [8, 20] as a consequence of premature transcriptional termination at noncanonical poly(A) sites in the SINE-R. SVAs are polymorphic in humans [8, 20, 25], and hundreds of insertions are unique to each hominid lineage [26–29].

1.1 The SVA Trans-complementation Assay

Retrotransposition cell culture assays can be divided into *cis*- and *trans*-complementation [5, 30, 31, 33] assays which are alternatively termed *cis*- and *trans*-mobilization assays. The *cis* assay is based on the transfection of cells with a plasmid containing an autonomous retrotransposon (Fig. 2a), i.e., L1, marked with a retrotransposition indicator cassette.

Fig. 2 (continued) Retrotransposition frequencies resulting from cotransfection of the passenger pCEPneo and the driver pJM101/L1$_{RP}$Δneo indicate the background frequency of processed pseudogene formation in this experimental setup. To investigate if the *trans*-mobilization frequency of the SVA reporter element exceeds processed pseudogene formation frequency, SVA reporter plasmid and L1 driver pJM101/L1$_{RP}$Δneo are cotransfected in HeLa cells. By comparing retrotransposition frequencies obtained after separate cotransfection of the SVA reporter plasmid with the L1 driver plasmids pJM101/L1$_{RP}$Δneo and pJM101/L1$_{RP}$ΔneoΔORF1, it can be clarified, if in addition to L1 ORF2p, L1 ORF1p is also essential for SVA mobilization in *trans*. (**b**) *Trans*-mobilization frequencies refer to the *cis* retrotransposition frequency of the L1 retrotransposition reporter element pJM101/L1$_{RP}$ [48] which is set as 100 %. pA, SV40 polyadenylation signal; hygro, hygromycin resistance gene; *Method 1*: Hygromycin selection for the presence of passenger construct and L1 driver plasmid with subsequent G418^R selection. SVA reporter plasmids or pCEPneo are each cotransfected with either of the two L1 driver plasmids into HeLa cells which were subsequently selected for hygromycin resistance for 14 days. Resulting Hyg^R cell populations were assayed for retrotransposition events by selecting for 11–12 days for G418^R colonies. *Method 2*: Transient cotransfection of passenger and driver plasmid with subsequent G418^R selection. (**c**) Giemsa-stained retrotransposition reporter assays in T75-flasks 15-cm dishes. After completion of the G418-selection, G418-resistant HeLa-HA colonies are stained with Giemsa solution, and stained colonies can be counted. The number of colonies is a measure for the retrotransposition activity of the *mneoI*-tagged reporter element that was expressed in HeLa-HA cells. The SVA *trans*-mobilization assay can be performed in T75-flasks or in 15-cm plates. Control: HeLa-HA cells cotransfected with pJM101/L1$_{RP}$Δneo and pCEP4 are used to control for a successful G418 selection procedure; SVA-*trans*: HeLa-HA cells cotransfected with pJM101/L1$_{RP}$Δneo and SVA reporter plasmid; L1-*cis*(1:10): HeLa-HA cells cotransfected with pJM101/L1$_{RP}$ and pCEP4 were seeded at a1:10 dilution; Images of stained assays are obtained from [2]

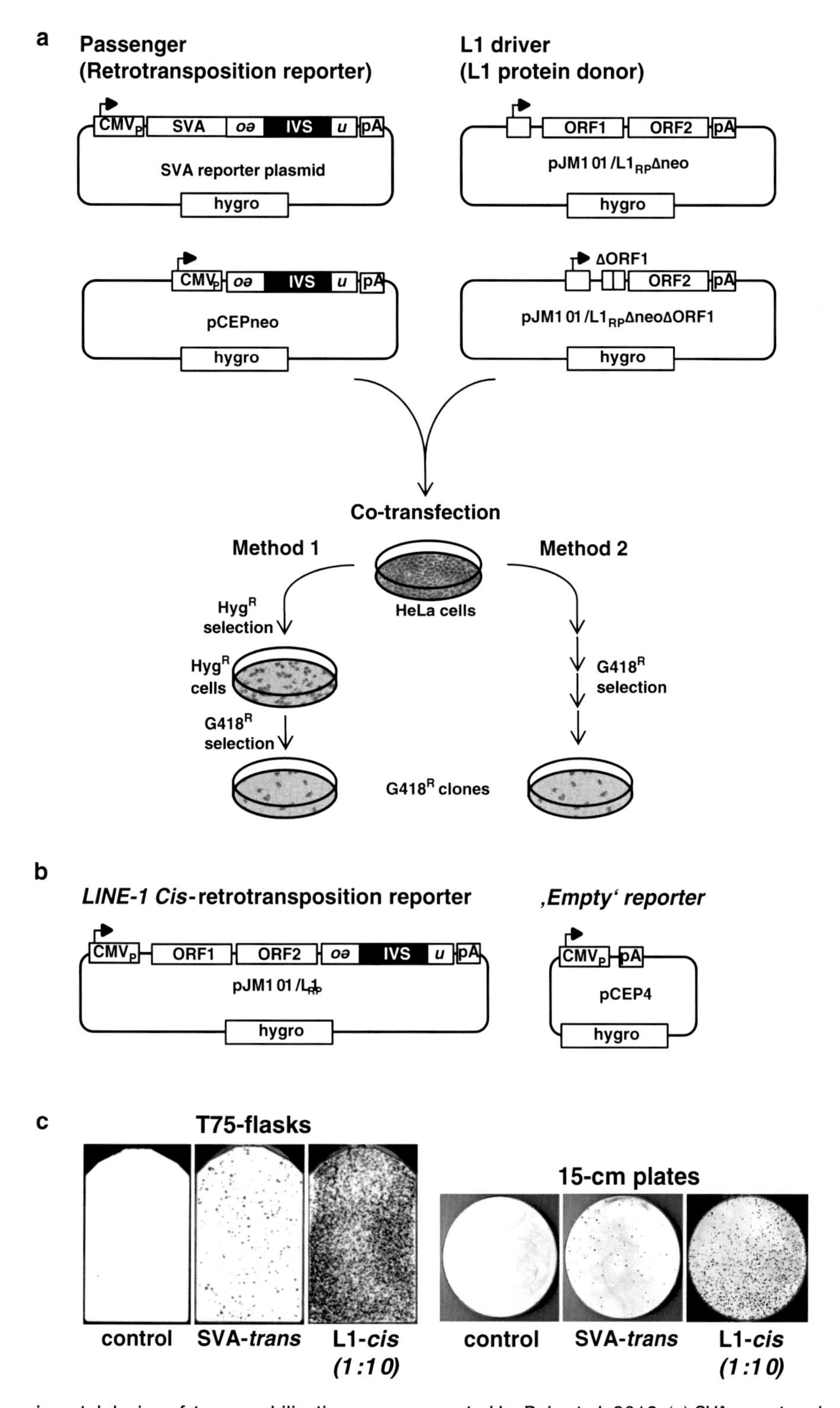

Fig. 2 Experimental design of *trans*-mobilization assays reported by Raiz et al. 2012. (**a**) SVA reporter elements and pCEPneo (Passengers) are marked with the *mneo*I retrotransposition indicator cassette [33] and cloned into the episomal expression vector pCEP4. The *mneo*I expression cassette consists of a neomycin resistance gene (*neo*^{R}) that is reverse oriented relative to SVA, with an SV40 promoter and thymidine kinase polyA signal. The *neo*^{R} gene is interrupted by an intron (IVS) in the sense orientation. Thus only after splicing of an SVA transcript followed by reverse transcription and integration into the genome will G418 resistance (G418^R) be conferred upon the transfected cell. Each passenger plasmid (SVA reporter plasmid or pCEPneo) is cotransfected with each of the L1 driver plasmids (pJM101/L1$_{RP}$$\Delta$neo or pJM101/L1$_{RP}$$\Delta$neo$\Delta$ORF1) separately.

SVA *trans*-mobilization assays are cell culture assays that are based on the coexpression of a driver plasmid coding for the entire L1 protein machinery, or its two components (L1 ORF1p or L1 ORF2p) separately, and an engineered SVA element located on a passenger construct which is marked with reporter genes (*mneo*I [32, 33] or mEGFP [34]), also referred to as retrotransposition indicator cassettes (Fig. 2a). If the RNA expressed from the

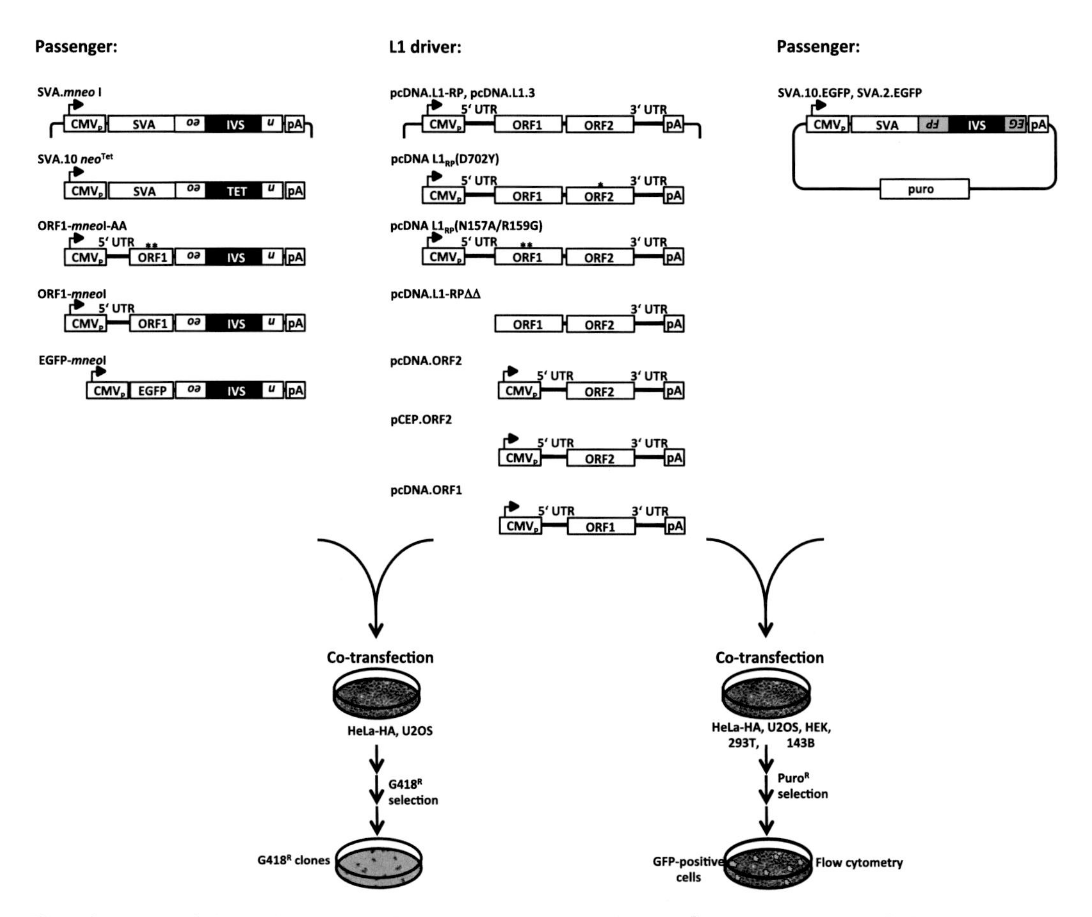

Fig. 3 Overview of alternative *trans*-mobilization assays including $Puro^R$ selection and the EGFP retrotransposition indicator cassette. The Kazazian laboratory developed alternative SVA *trans*-mobilization assays [3, 4] which differ from the assays described above in the utilization of additional passenger and driver plasmids. To determine the retrotransposition frequency of the SVA.*mneo*I reporter construct (*left* "Passenger" panel), HeLa-HA and/or U2OS cells were transiently cotransfected with the driver plasmid pcDNA.L1-RP [48] or pcDNA.L1.3 (*middle* "L1 driver" panel). To ensure that the resulting $G418^R$ foci represented L1-mediated retrotransposition and not plasmid–plasmid recombination events, the SVA reporter was cotransfected in a separate experiment with pcDNA.L1-RPΔΔ (*middle panel*) which differs from pcDNA.L1-RP (*middle panel*) exclusively by the absence of both CMV promoter (CMV_P) and L1 5′ UTR. A driver devoid of any promoter cannot generate $G418^R$ colonies unless recombination occurs between the plasmid lacking the promoter and the construct carrying the promoter sequence [31] or endogenous RT activity is employed. To compare the SVA retrotransposition frequency with that of an *Alu* element encoded by Alu.*neo*^{Tet} [51], the *mneo*I reporter in SVA.10 *mneo*I was replaced with the *neo*^{Tet} retrotransposition indicator cassette [52] resulting in the construct SVA.10 *neo*^{Tet} (*left panel*). *neo*^{Tet} differs from *mneo*I in that the *neo*^{R} gene is interrupted by a self-splicing group1 intron instead of a nuclear mRNA intron cassette [52]. The number of $G418^R$ foci for each passenger is

engineered SVA reporter element is recognized by the L1 protein machinery as substrate, cotransfection with plasmids expressing the L1 proteins will result in retrotransposition of the SVA reporter element (*trans*-complementation) which again will result in the expression of the respective indicator cassette. Briefly, the initial orientation of the reporter gene, which is antisense relative to the transcriptional direction of the nonautonomous SVA retrotransposon, remains inactive due to interruption by an intron that has the same direction as the retrotransposon (Fig. 2a, b). Reporter expression is activated following one round of SVA reporter element transcription, splicing, and cDNA integration via TPRT, into genomic DNA (Fig. 1b). This leads to either neomycin-resistant foci (Neo^r) (Figs. 2 and 3) or EGFP-positive cells (Fig. 3).

SVA *trans*-mobilization assays were developed in both the Kazazian laboratory and the Schumann laboratory to study the biology of nonautonomous SVA retrotransposons. They can be applied to address the objectives listed in the next paragraph. The two approaches established in the Schumann laboratory are outlined schematically in Fig. 2 and described in detail in this work. Principles and constructs, the SVA *trans*-mobilization assays established in the Kazazian laboratory are based on, are summarized in Fig. 3. Preferred cell lines in which SVA *trans*-mobilization assays

Fig. 3 (continued) normalized to the number of $G418^R$ foci resulting from processed pseudogene formation by *trans*-mobilization of the EGFP-*mneoI* construct (*left panel*). *Trans*-mobilization of ORF1*mneoI*-AA (*left panel*), a construct expressing L1 ORF1 coding sequence with two point mutations and marked with *mneoI*, can also be used to determine the processed pseudogene formation frequency. In this experimental setup, retrotransposition frequency of the marked SVA was reported to exceed processed pseudogene formation rate by up to 54-fold. To determine role and relevance of L1 ORF2p in *trans*-mobilization, SVA reporter plasmids were cotransfected with the L1 driver pcDNA.$L1_{RP}$(D702Y) (*middle panel*) carrying a point mutation in the RT domain (D702Y) abolishing L1 RT activity. To investigate the potential requirement of the RNA-binding L1 ORF1p for SVA mobilization, the SVA.*mneoI* reporter was cotransfected with pcDNA.ORF2 (*middle panel*), a construct containing 5′ UTR and ORF2 coding sequence, or with pCEP.ORF2 (*middle panel*) which is a self-replicating ORF2 expressing plasmid. To further investigate the role of ORF1p in SVA mobilization, experiments with the driver pcDNA.$L1_{RP}$(N157A/R159G) (*middle panel*) containing the double mutation in the RNA recognition motif (RRM) domain of ORF1p (Fig. 1a), were carried out. This mutant is relevant, because the mutations have been demonstrated to abolish engineered L1 retrotransposition in *cis*, to affect L1 RNP formation and disrupt both cytoplasmic ORF1p foci formation and the accruement of cytoplasmic L1 foci carrying ORF2p. As a positive control for *trans*-mobilization, pcDNA.$L1_{RP}$(N157A/R159G) was transfected with ORF1 *mneoI* [31] because this construct does not require a functional ORF1p in *trans* [46] for its mobilization. Since the use of the EGFP (enhanced green fluorescent protein) retrotransposition indicator cassette [34] has been shown to reduce the amount of time necessary to perform retrotransposition reporter assays and has been useful in monitoring phenomena intractable to the neo^R assay [46], SVA-EGFP reporter constructs (SVA.10.EGFP, SVA.2.EGFP) were generated, in which the *mneoI* cassette was replaced by the EGFP retrotransposition indicator cassette (*right* "Passenger" panel). Puro, puromycin resistance gene

were performed so far, are the cervical carcinoma cell line HeLa-HA and the human osteosarcoma cell line U2OS [2, 3]. To date, the majority of *cis*- and *trans* mobilization assays have been executed in HeLa cells [1–3, 30, 31, 33, 35, 36], which exhibit low or even undetectable endogenous L1 expression levels. However, recently U2OS cells were used for SVA cell culture retrotransposition assays successfully because these cells also support engineered retrotransposition and display the highest levels of retrotransposition [3, 4, 37] among all cell lines tested to date. U2OS cells were also used for SVA *trans*-mobilization assays to define SVA domains that are important for L1-mediated retrotransposition [4]. Additional human cell lines which also demonstrated to support *trans*-mobilization of nonautonomous non-LTR retrotransposons are HEK293T (human embryonic kidney cell line 293T; ATCC®CRL-11268™) and the human bone osteosarcoma cell line 143B (ATCC®CRL-8303™) [3] (Fig. 3).

1.2 Objectives of the Trans-mobilization Assay

To date, *trans*-mobilization assays have been utilized to study the mobilization of human nonautonomous retrotransposons *Alu* and SVA, and gibbon LAVA elements [38], and the retrotransposition of random PolII transcripts by the *trans*-activity of the L1-encoded protein machinery. Objectives to be addressed by *trans*-mobilization assays are:

1. To evaluate if a noncoding repetitive element and potential retrotransposon is mobilized by the L1-encoded protein machinery in *trans*, and if the encoded RNA is a preferred substrate of the L1 or LINE-like protein machinery before random PolII-mRNAs.
2. To evaluate if engineered de novo insertions of the potential nonautonomous non-LTR retrotransposon display hallmarks of L1-mediated retrotransposition, mimic endogenous genomic and recent disease-causing insertions.
3. Determine the role of the individual domains, the nonautonomous transposable element is composed of.
4. Test for the expression of functional endogenous L1-encoded ORF2p or the entire L1 protein machinery. The *trans*-mobilization assay provides an extremely sensitive means to test for the presence of endogenous L1 ORF2p activity [30] mobilizing nonautonomous reporter elements *in trans*. It was reported that endogenous ORF2 may be expressed in the human osteosarcoma cell line U2OS that mobilizes engineered SVA elements in the absence of any functional driver plasmid [4]. In order to answer the question if endogenous L1 activity is significantly upregulated in certain tissues (e.g., brain regions) [39, 40] or in diseases like cancer [41, 42], *trans*-

mobilization assays could be performed in the appropriate cell line or primary cells in the absence of any driver plasmid. To date, the primary way to detect endogenous L1 activity has been targeted high-throughput sequencing for new insertions [39, 41, 43, 44]. Since the ORF2 protein is difficult to detect [45] due to its comparatively low endogenous expression levels, and ORF2p can be expressed from 5′ truncated L1 copies that do not encode ORF1p, a genetic *trans*-complementation assay could be applied to monitor if functional endogenous ORF2p is expressed: The assay can monitor ORF2p expression of endogenous wild-type or mutant L1 elements to *trans*-mobilize a reporter mRNA consisting of a transcribed SVA reporter element which is marked with a spliced retrotransposition reporter gene (e.g., *mneo*I or EGFP; Figs. 2a, b and 3) in the presence of transiently expressed ORF1p. The constructs ORF1*mneo*I and pCEPneo (Figs. 2a and 3), which have been useful in studying mechanisms of L1 retrotransposition [30, 31, 46] and processed pseudogene formation frequencies [2], respectively, and passenger expression plasmids encoding marked *Alu* or SVA elements, are useful in detecting endogenous ORF2p activity in tissues of interest, provided that the respective corresponding cell lines are available or can be derived.

2 Materials

2.1 Cell Culture Material

1. The clonal HeLa cell line HeLa-HA [47] is used to assay both L1 retrotransposition in *cis* and SVA retrotransposition in *trans* (Fig. 2a, b).
2. Cultured cell growth medium for HeLa-HA cells (DMEM growth medium): To assay L1 retrotransposition in *cis* (Fig. 1a) or SVA mobilization in *trans*, HeLa-HA cells are grown in DMEM High Glucose supplemented with 10 % FCS, 100 μg/ml streptomycin, 100 U/ml penicillin, and 2 mM l-glutamine.
3. 1× Phosphate Buffered Saline (1× PBS): 137 mM NaCl, 3 mM KCl, 16.5 mM Na_2HPO_4, 1.5 mM KH_2PO_4.
4. Hygromycin: 50 mg/ml dissolved in 1× PBS.
5. G418/Geneticin: 500 mg/ml.
6. FuGENE HD Transfection Reagent (Promega).
7. Opti-MEM I Reduced Serum Medium, GlutaMAX supplement.
8. 0.25 % trypsin–EDTA (0.25 % trypsin, 1 % EDTA in 1× PBS).
9. Hemocytometer.

10. Giemsa's azur eosin methylene blue solution (Giemsa solution).
11. Cell culture plates: 150 × 20 mm; 24-well plates; 6-well plates.
12. Cell culture flasks, 75 cm^2.
13. Colony counter for adherent mammalian cell colony forming assays.

2.2 Plasmids

1. To assay LINE-1 retrotransposition in *cis* (Fig. 2b):
   (a) The mammalian episomal expression plasmid pCEP4 (Life Technologies) is the backbone of all used expression plasmids and serves as negative control construct.
   (b) A LINE-1 expression plasmid that is tagged with the *mneo*I retrotransposition reporter cassette (e.g., pJM101/$L1_{RP}$) [48].
   (c) A reporter plasmid (e.g., pEGFP-N1 (Clontech)) to monitor transfection efficiency.
2. To assay SVA retrotransposition in *trans* (Fig. 2b):
   (a) A LINE-1 protein donor ("driver") plasmid expressing either a functional full-length LINE-1 (e.g., pJM101/$L1_{RP}$Δneo) or exclusively L1 ORF2p (e.g., pJM101/$L1_{RP}$ΔneoΔORF1) [2] and lacks the retrotransposition reporter cassette (e.g., *mneo*I).
   (b) A retrotransposition "reporter" plasmid expressing a full-length SVA element or its fragments encoding SVA components, and an *mneo*I cassette (e.g., pAD3/SVA_E, pSC4/SVA_{F1}, pAD4/SVA_E, pSC3/SVA_{F1}) [2].
   (c) A control retrotransposition reporter plasmid which differs from the SVA reporter plasmids exclusively in the absence of any SVA retrotransposon sequence (e.g., pCEPneo) [2]. This plasmid is used to determine processed pseudogene formation frequency which represents the *trans*-mobilization background observed for any polyadenylated RNA-Polymerase II-transcript in this specific assay.
   (d) A reporter plasmid (e.g., pEGFP-N1) to monitor transfection efficiency.

2.3 Reagents for the Analyses of L1 Protein Levels

1. To control for expression of the driver plasmid-encoded L1 proteins, a 12 % SDS-polyacrylamide gel has to be generated:
   (a) To generate the resolving gel, add 10.6 ml H_2O, 4 ml 30 % acrylamide mix, 5 ml 1.5 M Tris Base (pH 8.8), 0.2 ml 10 % SDS, 0.2 ml 1 % (w/v) ammonium persulfate, and 16 μl TEMED to a final volume of 20 ml.
   (b) To generate the stacking gel, add 4.1 ml H_2O, 1 ml 30 % acrylamide mix, 0.75 ml 1 M Tris Base (pH 6.8), 60 μl

10 % SDS, 60 μl 10 % (w/v) ammonium persulfate, and 6 μl TEMED to a final volume of 6 ml.

2. Immunoblotting Components:
   (a) Lysis buffer: 20 mM Tris–HCl, pH 7.5, 150 mM NaCl, 1 % (w/v) sodium deoxycholate, 1 % (v/v) Triton-X100, 10 mM EDTA, 0.1 % (w/v) SDS, 1× and protease inhibitors.
   (b) 3× SDS Sample Buffer (from a local provider).
   (c) Nitrocellulose transfer membrane (from a local provider).
   (d) 1× PBS-T buffer: 137 mM NaCl, 3 mM KCl, 16.5 mM Na_2HPO_4, 1.5 mM KH_2PO_4, 0.05 % (v/v) Tween 20.
   (e) 5 % (w/v) solution of nonfat milk powder in 1× PBS-T.
3. Primary antibodies
   (a) Polyclonal rabbit-anti-L1ORF1p antibody #984 [2]; 1:2000 dilution in 1× PBS-T containing 5 % milk powder.
   (b) Anti-β-actin antibody (at the manufacturer indicated dilution) in 1× PBS-T containing 5 % milk powder.
4. Secondary antibodies
   (a) The secondary antibody used for anti-L1ORF1p is HRP-conjugated (anti-rabbit IgG antibody) and used at the manufacturer indicated dilution in 1× PBS-T containing 5 % milk powder.
   (b) The secondary antibody specific for the anti-β-actin antibody is also HRP-conjugated and used at the manufacturer indicated dilution in 1× PBS-T containing 5 % milk powder.
   (c) Pierce ECL 2 Western Blotting Substrate (Fisher Scientific).

3 Methods

The use of laminar flow biological safety cabinets, humid CO_2 incubators and sterile techniques that guarantee controlled conditions of temperature, humidity, acidity and sterility of cultured cells are essential requirements for the successful application of the outlined methods. Growth conditions, transfection efficiency, and appropriate concentrations of the antibiotics (Geneticin/G418, Hygromycin) to be used for genetic selection have to be optimized for the used cell line before the assay is started.

Before the actual assay is initiated, cultivate the cell line HeLa-HA [47] in DMEM growth medium in a humidified incubator at 37 °C with 5 % CO_2. Grow cells until sufficient numbers of cells are available to proceed with Subheading 3.1 or 3.2 (*see* **Note 1**).

3.1 SVA Retrotransposition Reporter Assay with Initial Hygromycin Selection for the Presence of SVA Retrotransposition Reporter Plasmid ("Passenger") and L1 Protein Donor Plasmid ("Driver")

1. Day 1: Plate 2×10^6 cells per T75-flask in 20 ml DMEM growth medium.
2. Day 2: Cotransfect cells with 3 μg of an SVA retrotransposition reporter plasmid or pCEPneo (control for processed pseudogene formation frequency) and 3 μg of the L1 *driver* plasmid pJM101/L1$_{RP}$Δneo, pJM101/L1$_{RP}$ΔneoΔORF1 or pCEP4 (negative control for *trans*-mobilization by transiently overexpressed L1 protein machinery). 3 μg of the L1 retrotransposition reporter plasmid pJM101/L1$_{RP}$ [48] and 3 μg of pCEP4 are cotransfected to determine *cis*-retrotransposition frequency of the L1 reporter element (*see* **Note 2**).
   (a) For each cotransfection, 3 μg SVA reporter plasmid (or pCEPneo) and 3 μg driver plasmid are mixed in 800 μl Opti-MEM medium. Next, 18 μl FuGENE HD reagent (FuGENE® HD:DNA ratio of 3:1 [volume:weight]) are added and the suspension is incubated at RT for 15 min. Subsequently, 2×10^6 HeLa-HA cells are cotransfected according to the instructions of the manufacturer of the FuGENE-HD reagent.
   (b) Each cotransfection is performed at least twice in quadruplicate: in each case, three out of four cotransfection experiments are used to quantify retrotransposition frequencies resulting from coexpression of the SVA passenger and the L1 driver plasmid or processed pseudogene formation frequencies resulting from coexpression of pCEPneo and the L1 driver plasmid. Cells obtained as a result of the fourth cotransfection experiment will be used to analyze expression of both driver and passenger plasmid after 14 days of hygromycin selection.
3. Day 3: Subject cotransfected cells to hygromycin selection (200 μg/ml) for the following 14 days, starting 24 h posttransfection. To this end, replace DMEM growth medium on transfected cells plated the day before by fresh DMEM growth medium containing 200 μg/ml hygromycin (*see* **Note 3**).
4. Day 17: Trypsinize cells of each flask after 14 days of hygromycin selection, reseed them immediately on a new flask in 20 ml DMEM growth medium containing 400 μg/ml G418, and start G418-selection (*see* **Note 4**).
   (a) Harvest cells of one of the four T75-flasks with HeLa cells that resulted from each *combination of cotransfected driver and passenger construct* after transient cotransfection and isolate cell lysates and total RNA.
   (b) Select for L1-mediated retrotransposition events causing G418^R HeLa colonies by growing cells in DMEM growth medium containing 400 μg/ml G418 for the following 11–12 days, and change medium every 2–3 days (*see* **Note 5**).

(c) G418^R selection is completed when all Hyg^R cells that resulted from cotransfection of pCEP4 and pJM101/$L1_{RP}\Delta$neo (which do not encode any *mneo*I gene) and were subsequently cultivated in the presence of G418, are eradicated.

5. Day 28 or day 29: G418^R selection is accomplished. Fixation and staining of G418^R colonies:

   (a) Remove growth medium from flasks and wash G418^R colonies with 15 ml 1× PBS. After removal of 1× PBS, add 5 ml Giemsa solution per flask and incubate colonies in Giemsa solution for 10–15 min at room temperature (*see* **Note 6**).

   (b) Subsequently, rinse flasks/plates with water to remove Giemsa solution that was not absorbed by HeLa colonies, and count stained colonies. If the density of stained colonies is too high to be counted without any device, use an inverted microscope or apply a colony counter for adherent mammalian cell colony forming assays (*see* **Note 7**).

   (c) The numbers of G418^R colonies per plate or flask can then be used to quantify retrotransposition events [2, 33] of the different SVA reporter cassettes per 10^6 transfected cells (*see* **Note 8**).

6. Assess L1 ORF1 protein (ORF1p) (Fig. 1a) expression levels in cotransfected Hyg^R-selected HeLa cell cultures at the end of the Hyg selection process (*see* **Note 9**).

   (a) All cells of one flask of each cotransfection were harvested at day 17; half of these cells is lysed in lysis buffer (the other half is used for RNA isolation), and lysates are cleared by centrifugation at 1000 × *g*. 20 μg of each protein lysate are boiled in SDS buffer, loaded on a 12 % polyacrylamide gel, subjected to SDS-PAGE, and blotted onto a nitrocellulose membrane.

   (b) After protein transfer, the membrane is blocked for 5 h at room temperature in a 5 % solution of nonfat milk powder in 1× PBS-T, and incubated overnight with anti-L1 ORF1p antibody #984 [2]. The membrane was washed six times for 10 min in 1× PBS-T and incubated with an HRP-conjugated, secondary anti-rabbit IgG antibody for 1 h at RT. Subsequently, the membrane was washed six times for 10 min in 1× PBS-T.

   Immunocomplexes were visualized using Pierce ECL2 Western blotting substrate.

   (c) After stripping of the membrane, this is blocked for 5 h at room temperature in 1× PBS-T containing 5% milk powder before it is incubated with an anti-β-actin antibody.

The membrane was washed six times for 10 min in 1× PBS-T and incubated with the secondary anti-mouse HRP-linked species-specific antibody. After washing the membrane, immunocomplexes are visualized as described before.

7. Apply qRT-PCR to quantify the relative amounts of mRNA coding for the spliced *mneo*I cassette expressed from the passenger plasmids after 14 days of hygromycin selection [2] (*see* **Note 10**).

3.2 SVA Retrotransposition Reporter Assay after Transient Cotransfection of SVA Passenger Plasmid and L1 Driver Plasmid

1. Day 1: Seed 2.8×10^6 cells per 15-cm cell culture dish in 30 ml DMEM growth medium (*see* **Note 11**).
2. Day 2: Cotransfect cells with 8 μg of an SVA retrotransposition reporter plasmid ("*passenger*") or pCEPneo (control for processed pseudogene formation frequency) and 8 μg of the L1 protein donor plasmid ("*driver*") pJM101/L1$_{RP}$Δneo, pJM101/L1$_{RP}$ΔneoΔORF1 or pCEP4 (negative control for *trans*-mobilization by transiently overexpressed L1 protein machinery). 8 μg of the L1 retrotransposition reporter plasmid pJM101/L1$_{RP}$ [48] and 8 μg of pCEP4 are cotransfected to determine *cis*-retrotransposition frequency of the L1 reporter element (*see* **Note 2**).
   (a) For each cotransfection, 8 μg reporter plasmid (or pCEPneo) and 8 μg driver plasmid are mixed in 800 μl Opti-MEM medium. Next, 48 μl FuGENE HD reagent are added and incubated at RT for 15 min. Subsequently, 2.8×10^6 HeLa-HA cells are cotransfected according to the instructions of the manufacturer.
   (b) Perform each cotransfection at least twice in quadruplicate (*see* **Note 12**).
   (c) One of the eight cotransfections per individual driver/passenger combinations has to be used to determine individual transfection efficiencies. Determine transfection efficiency by cotransfecting 4 μg of pEGFP-N1 (Clontech), 4 μg of the SVA driver and 8 μg of the L1 donor plasmid into 2.8×10^6 HeLa-HA cells using 48 μl FuGENE HD transfection reagent. Quantify the percentage of EGFP expressing cells 24 h post-transfection by flow cytometry. Use the percentage of green fluorescent cells to determine the transfection efficiency of each sample [31, 34].
3. Day 5: Subject cotransfected cells to G418 selection for the following 11–12 days, starting 3 days post-transfection.
   (a) To this end, trypsinize cells of three out of the four plates (resulting from cotransfection experiments per passenger/driver combination) 3 days after cotransfection, and

reseed, and cultivate them in DMEM growth medium containing 400 μg/ml G418 (*see* **Note 4**).

(b) Harvest cells of one of the four 15-cm dishes with HeLa cells that resulted from each combination of cotransfected driver and passenger construct after transient cotransfection and isolate cell lysates and total RNA (*see* **Note 13**).

(c) Select for L1-mediated retrotransposition events causing $G418^R$ HeLa colonies by growing cells in DMEM growth medium containing 400 μg/ml G418 for the following 11–12 days, and change medium every 2–3 days.

(d) $G418^R$ selection is completed when all HeLa-HA cells that were cotransfected with pCEP4 and pJM101/$L1_{RP}$Δneo (which do not encode any *mneo*I gene) and were subsequently cultivated in the presence of G418, are eradicated.

4. Day 16 or 17: $G418^R$ selection is accomplished. There are two alternative treatments of the plates with $G418^R$-colonies, depending on the questions asked by the experimenter:

(a) Alternative 1: If *trans*-mobilization frequencies of the passenger elements should be determined, colonies have to be fixed, stained and counted as described in Subheading 3.1, **step 5**.

(b) Calculation of *trans*-mobilization frequencies (*see* **Note 14**).

(c) Alternative 2: If single $G418^R$ colonies should be expanded to isolate individual genomic *mneo*I-tagged SVA de novo insertions from these colonies, the following procedure has to be performed: aspirate DMEM growth medium from the plates; wash cells on the plate with 30 ml 1× PBS; briefly submerge Whatman filter paper squares (Size: 3 mm × 3 mm) in 0.25 % Trypsin-EDTA, apply a single paper square on each $G418^R$ colony to be isolated, and leave it on the colony for 2 min; transfer the Whatman paper square with attached cells in a single well of a 24-well plate containing 1 ml DMEM growth medium (without G418). Rinse cells in the well from the filter paper and discard the paper. Replace DMEM growth medium in each well containing cells every 2–3 days until the well is confluent. Trypsinize cells of a confluent well using 150 μl 0.25 % Trypsin-EDTA and transfer cells to a single well of a 6-well plate containing 3 ml DMEM growth medium. Expand the clone in the 6-well plate until a sufficient number of cells per clone is generated.

5. Assess L1 ORF1p (Fig. 1a) expression levels in cotransfected HeLa cell cultures at the day of initiation of G418 selection (*see* Subheading 3.1, **step 6**).

(a) Use total RNA isolated at the day G418^R selection for *trans*-mobilization events was initiated, to quantify relative amounts of mRNA coding for the spliced *mneo*I cassette expressed at the day of G418 selection initiation, by qRT-PCR [2] (*see* **Note 10**).

4 Notes

1. For optimization of the L1-mediated *trans*-mobilization assay, HeLa-HA cells should be cultivated for at least 2 passages (≥7 days) after thawing of the cryopreserved cells, but not more than 11 passages (≤38 days) before they are seeded for transfection with retrotransposition reporter plasmids. During cultivation, HeLa-HA cells should be split two to three times per week at a dilution of 1:5. Do not split at dilutions exceeding the ratio of 1:5.
2. In order to be able to compare data on retrotransposition frequencies and expression resulting from transfections of different reporter and expression plasmids, all transfections that are part of the same study in which resulting data have to be compared with each other, have to be performed in parallel at the same day, using the same reagents and the same experimenter. pJM101/L1$_{RP}$ and pCEP4 are cotransfected to quantify L1 *cis* retrotransposition frequency which will be set as 100 % relative retrotransposition frequency. Relative *trans*-mobilization frequencies of nonautonomous reporter elements will refer to the 100 % L1 *cis* activity.
3. Generate DMEM growth medium containing 200 μg/ml hygromycin (Hyg) one day before you start Hyg selection. After 14 days of Hyg selection, plates with resulting populations of Hyg-resistant (Hyg^R) Hela cells will be confluent. Change Hyg-containing DMEM growth medium every 2–3 days during these 14 days.
4. To ensure that countable results are obtained for L1 retrotransposition in *cis*, trypsinized cells cotransfected with pJM101/L1$_{RP}$ and pCEP4 have to be reseeded at 1:10, 1:100, and 1:200 dilutions for G418 selection.
5. The length of the G418 selection process can vary among different sets of retrotransposition assays between 11 and 12 days, but is clearly completed as soon as all Hyg^R cells that resulted from cotransfection of pCEP4 and pJM101/L1$_{RP}$Δneo, are eradicated.
6. HeLa colonies can be incubated with Giemsa solution even longer than 15 min. However, make sure that after decantation of the staining solution, the plate is rinsed thoroughly to remove Giemsa solution from the flask in order to be able to

generate presentable images of the flasks with colonies. To this end, scan the flasks on a flatbed color scanner to generate images of each flask carrying G418^R colonies.

7. If you use an inverted microscope to count stained HeLa colonies, place a copier film with a grid printed on it between flask or plate carrying stained colonies, and the sample stage of the microscope. Locate and count colonies with the help of the grid.
8. To calculate *trans*-mobilization frequencies, G418-resistant colonies are counted. Calculate the mean colony counts for the six flasks of the same transfection condition (six technical replicates). To express the adjusted retrotransposition values as a percentage of the L1 *cis* retrotransposition frequency, divide the retrotransposition mean of the passenger sample (e.g., the number of retrotransposition events generated from the passenger expression construct) by the retrotransposition mean of the L1 reporter element (e.g., pJM101/L1$_{RP}$ cotransfected with pCEP4) and then multiply by 100 to calculate the *trans*-mobilization frequency of the passenger.
9. L1 protein expression from the driver plasmid should be analyzed during the G418 selection procedure. This analysis is necessary to evaluate if potentially varying levels of the L1 protein machinery could play a role in any observed differences in *trans*-mobilization frequencies.
10. To investigate if potential discrepancies in mRNA production or stability could be affecting observed differences in *trans*-mobilization rates between the various passenger constructs, the amounts of spliced, *mneo*I-tagged passenger mRNAs in the differently cotransfected cells have to be quantified as reported [2].
11. 15-cm cell culture dishes are used in this assay because their lids can be removed to isolate single G418^R colonies after G418 selection is accomplished.
12. For each combination of cotransfected driver and passenger construct, three out of four cotransfection experiments are used to quantify retrotransposition frequencies resulting from coexpression of the SVA passenger and the L1 driver plasmid, or processed pseudogene formation frequencies resulting from coexpression of pCEPneo and the L1 driver plasmid.
13. Cell lysates and total RNA of the cotransfected cells have to be isolated directly before the initiation of the G418 selection procedure to evaluate if potentially varying levels of the L1 protein machinery could play a role in potential differences in *trans*-mobilization frequencies that might be observed later.
14. To calculate *trans*-mobilization frequencies after transient cotransfections, G418-resistant colonies are counted and adjusted for transfection efficiency (*see* Subheading 3.2, **step**

2c). Calculate the mean colony counts for the six flasks of the same transfection condition (six technical replicates). To calculate the adjusted retrotransposition mean, divide the mean colony counts and the standard deviation by the transfection efficiency. To express the adjusted retrotransposition values as a percentage of the L1 *cis* retrotransposition frequency, divide the retrotransposition mean of the passenger sample (e.g., the number of retrotransposition events generated from the passenger expression construct) by the retrotransposition mean of the L1 reporter element (e.g., pJM101/L1$_{RP}$ cotransfected with pCEP4) and then multiply by 100 to calculate the *trans*-mobilization frequency of the passenger.

Acknowledgements

The authors would like to thank Ulrike Held for helpful comments during the preparation of this manuscript. This work was supported in part by DFG grant SCHU1014/8-1 to G.G.S.

References

1. Dewannieux M, Esnault C, Heidmann T (2003) LINE-mediated retrotransposition of marked Alu sequences. Nat Genet 1:41–48
2. Raiz J, Damert A, Chira S, Held U, Klawitter S, Hamdorf M et al (2012) The non-autonomous retrotransposon SVA is trans-mobilized by the human LINE-1 protein machinery. Nucleic Acids Res 4:1666–1683
3. Hancks DC, Goodier JL, Mandal PK, Cheung LE, Kazazian HH (2011) Retrotransposition of marked SVA elements by human L1s in cultured cells. Hum Mol Genet 17:3386–3400
4. Hancks DC, Mandal PK, Cheung LE, Kazazian HH (2012) The minimal active human SVA retrotransposon requires only the 5′-hexamer and Alu-like domains. Mol Cell Biol 22:4718–4726
5. Esnault C, Maestre J, Heidmann T (2000) Human LINE retrotransposons generate processed pseudogenes. Nat Genet 4:363–367
6. Vanin EF (1985) Processed pseudogenes: characteristics and evolution. Annu Rev Genet 19:253–272
7. Weiner AM, Deininger PL, Efstratiadis A (1986) Nonviral retroposons: genes, pseudogenes, and transposable elements generated by the reverse flow of genetic information. Annu Rev Biochem 55:631–661
8. Wang H, Xing J, Grover D, Hedges DJ, Han K, Walker JA et al (2005) SVA elements: a hominid-specific retroposon family. J Mol Biol 4:994–1007
9. Hancks DC, Ewing AD, Chen JE, Tokunaga K, Kazazian HH (2009) Exon-trapping mediated by the human retrotransposon SVA. Genome Res 11:1983–1991
10. Ostertag EM, Goodier JL, Zhang Y, Kazazian HH (2003) SVA elements are nonautonomous retrotransposons that cause disease in humans. Am J Hum Genet 6:1444–1451
11. Ono M, Kawakami M, Takezawa T (1987) A novel human nonviral retroposon derived from an endogenous retrovirus. Nucleic Acids Res 21:8725–8737
12. Shen L, Wu LC, Sanlioglu S, Chen R, Mendoza AR, Dangel AW et al (1994) Structure and genetics of the partially duplicated gene RP located immediately upstream of the complement C4A and the C4B genes in the HLA class III region. Molecular cloning, exon-intron structure, composite retroposon, and breakpoint of gene duplication. J Biol Chem 11:8466–8476
13. Strichman-Almashanu LZ, Lee RS, Onyango PO, Perlman E, Flam F, Frieman MB et al (2002) A genome-wide screen for normally methylated human CpG islands that can identify novel imprinted genes. Genome Res 4:543–554
14. Zhu ZB, Hsieh SL, Bentley DR, Campbell RD, Volanakis JE (1992) A variable number of tandem repeats locus within the human complement C2 gene is associated with a retroposon derived from a human endogenous retrovirus. J Exp Med 6:1783–1787

15. Hancks DC, Kazazian HH (2012) Active human retrotransposons: variation and disease. Curr Opin Genet Dev 3:191–203
16. Ostertag EM, Kazazian HH (2001) Biology of mammalian L1 retrotransposons. Annu Rev Genet 35:501–538
17. Christensen SM, Eickbush TH (2005) R2 target-primed reverse transcription: ordered cleavage and polymerization steps by protein subunits asymmetrically bound to the target DNA. Mol Cell Biol 15:6617–6628
18. Cost GJ, Feng Q, Jacquier A, Boeke JD (2002) Human L1 element target-primed reverse transcription in vitro. EMBO J 21:5899–5910
19. Luan DD, Korman MH, Jakubczak JL, Eickbush TH (1993) Reverse transcription of R2Bm RNA is primed by a nick at the chromosomal target site: a mechanism for non-LTR retrotransposition. Cell 4:595–605
20. Damert A, Raiz J, Horn AV, Löwer J, Wang H, Xing J et al (2009) 5′-Transducing SVA retrotransposon groups spread efficiently throughout the human genome. Genome Res 11:1992–2008
21. Xing J, Wang H, Belancio VP, Cordaux R, Deininger PL, Batzer MA (2006) Emergence of primate genes by retrotransposon-mediated sequence transduction. Proc Natl Acad Sci U S A 47:17608–17613
22. Zabolotneva AA, Bantysh O, Suntsova MV, Efimova N, Malakhova GV, Schumann GG et al (2012) Transcriptional regulation of human-specific $SVAF_1$ retrotransposons by cis-regulatory MAST2 sequences. Gene 1:128–136
23. Savage AL, Wilm TP, Khursheed K, Shatunov A, Morrison KE, Shaw PJ et al (2014) An evaluation of a SVA retrotransposon in the FUS promoter as a transcriptional regulator and its association to ALS. PLoS One 6:e90833
24. Bantysh OB, Buzdin AA (2009) Novel family of human transposable elements formed due to fusion of the first exon of gene MAST2 with retrotransposon SVA. Biochemistry (Mosc) 12:1393–1399
25. Bennett EA, Coleman LE, Tsui C, Pittard WS, Devine SE (2004) Natural genetic variation caused by transposable elements in humans. Genetics 2:933–951
26. Chimpanzee Sequencing and Analysis Consortium (2005) Initial sequence of the chimpanzee genome and comparison with the human genome. Nature 7055:69–87
27. Locke DP, Hillier LW, Warren WC, Worley KC, Nazareth LV, Muzny DM et al (2011) Comparative and demographic analysis of orang-utan genomes. Nature 7331:529–533
28. Mills RE, Bennett EA, Iskow RC, Luttig CT, Tsui C, Pittard WS et al (2006) Recently mobilized transposons in the human and chimpanzee genomes. Am J Hum Genet 4:671–679
29. Ventura M, Catacchio CR, Alkan C, Marques-Bonet T, Sajjadian S, Graves TA et al (2011) Gorilla genome structural variation reveals evolutionary parallelisms with chimpanzee. Genome Res 10:1640–1649
30. Alisch RS, Garcia-Perez JL, Muotri AR, Gage FH, Moran JV (2006) Unconventional translation of mammalian LINE-1 retrotransposons. Genes Dev 2:210–224
31. Wei W, Gilbert N, Ooi SL, Lawler JF, Ostertag EM, Kazazian H et al (2001) Human L1 retrotransposition: cis preference versus trans complementation. Mol Cell Biol 4:1429–1439
32. Freeman JD, Goodchild NL, Mager DL (1994) A modified indicator gene for selection of retrotransposition events in mammalian cells. Biotechniques 17(1):46, 48–9, 52
33. Moran JV, Holmes SE, Naas TP, DeBerardinis RJ, Boeke JD, Kazazian HH (1996) High frequency retrotransposition in cultured mammalian cells. Cell 5:917–927
34. Ostertag EM, Prak ET, DeBerardinis RJ, Moran JV, Kazazian HH (2000) Determination of L1 retrotransposition kinetics in cultured cells. Nucleic Acids Res 6:1418–1423
35. Gilbert N, Lutz S, Morrish TA, Moran JV (2005) Multiple fates of L1 retrotransposition intermediates in cultured human cells. Mol Cell Biol 17:7780–7795
36. Sassaman DM, Dombroski BA, Moran JV, Kimberland ML, Naas TP, DeBerardinis RJ et al (1997) Many human L1 elements are capable of retrotransposition. Nat Genet 1:37–43
37. Doucet AJ, Hulme AE, Sahinovic E, Kulpa DA, Moldovan JB, Kopera HC et al (2010) Characterization of LINE-1 ribonucleoprotein particles. PLoS Genet 10:e1001150
38. Ianc B, Ochis C, Persch R, Popescu O, Damert A (2014) Hominoid composite non-LTR retrotransposons-variety, assembly, evolution, and structural determinants of mobilization. Mol Biol Evol 11:2847–2864
39. Baillie JK, Barnett MW, Upton KR, Gerhardt DJ, Richmond TA, de Sapio F et al (2011) Somatic retrotransposition alters the genetic landscape of the human brain. Nature 7374:534–537
40. Muotri AR, Chu VT, Marchetto MC, Deng W, Moran JV, Gage FH (2005) Somatic mosaicism in neuronal precursor cells mediated by L1 retrotransposition. Nature 7044:903–910
41. Iskow RC, McCabe MT, Mills RE, Torene S, Pittard WS, Neuwald AF et al (2010) Natural mutagenesis of human genomes by endogenous retrotransposons. Cell 7:1253–1261

42. Konkel MK, Batzer MA (2010) A mobile threat to genome stability: the impact of non-LTR retrotransposons upon the human genome. Semin Cancer Biol 4:211–221
43. Ewing AD, Kazazian HH (2010) High-throughput sequencing reveals extensive variation in human-specific L1 content in individual human genomes. Genome Res 9:1262–1270
44. Witherspoon DJ, Xing J, Zhang Y, Watkins WS, Batzer MA, Jorde LB (2010) Mobile element scanning (ME-Scan) by targeted high-throughput sequencing. BMC Genomics 11:410
45. Goodier JL, Ostertag EM, Engleka KA, Seleme MC, Kazazian HH (2004) A potential role for the nucleolus in L1 retrotransposition. Hum Mol Genet 10:1041–1048
46. Garcia-Perez JL, Doucet AJ, Bucheton A, Moran JV, Gilbert N (2007) Distinct mechanisms for trans-mediated mobilization of cellular RNAs by the LINE-1 reverse transcriptase. Genome Res 5:602–611
47. Hulme AE, Bogerd HP, Cullen BR, Moran JV (2007) Selective inhibition of Alu retrotransposition by APOBEC3G. Gene 1–2:199–205
48. Kimberland ML, Divoky V, Prchal J, Schwahn U, Berger W, Kazazian HH (1999) Full-length human L1 insertions retain the capacity for high frequency retrotransposition in cultured cells. Hum Mol Genet 8:1557–1560
49. Ullu E, Tschudi C (1984) Alu sequences are processed 7SL RNA genes. Nature 5990: 171–172
50. Batzer MA, Deininger PL (2002) Alu repeats and human genomic diversity. Nat Rev Genet 5:370–379
51. Comeaux MS, Roy-Engel AM, Hedges DJ, Deininger PL (2009) Diverse cis factors controlling Alu retrotransposition: what causes Alu elements to die? Genome Res 4:545–555
52. Esnault C, Casella J-F, Heidmann T (2002) A Tetrahymena thermophila ribozyme-based indicator gene to detect transposition of marked retroelements in mammalian cells. Nucleic Acids Res 11:e49

Chapter 15

Detection of LINE-1 RNAs by Northern Blot

Prescott Deininger and Victoria P. Belancio

Abstract

Repetitive genetic elements have had an unprecedented success populating phylogenetically diverse species making them a common feature of most genomes. Hundreds of thousands of copies of active and nonfunctional transposable elements representing different classes and families can reside within and outside of host genes. In addition to creating structural variations in genomic DNA, some of these loci are expressed to contribute to the continuing amplification cycle. Transposable elements, specifically Long Interspersed Element-1 (LINE-1) produce a spectrum of RNAs, some of which are important for their mobilization, while others are processed forms of LINE-1 transcription that may or may not play relevant functions. Additionally, many LINE-1 sequences integrated into cellular genes are included into cellular transcripts creating substantial background when L1-related RNA expression is detected by some conventional methods. This chapter provides an in-depth description of the complexity of L1-generated mRNAs as well as sources of cellular transcripts containing L1 sequences. It also highlights the strengths and weaknesses of conventional methods used to detect LINE-1 expression.

Key words LINE-1, Retrotransposon, L1 RNA expression, RNA analysis, Northern blot, RT-PCR, Next-generation sequencing

1 Introduction

There are over 500,000 copies of L1 elements making up 17 % of the mass of the human DNA. Only a few thousand of these copies are full length and contain the internal 5′ UTR promoter that can make the RNA species that are capable of retrotransposition. These retrotranspositionally competent, as well as defective, full-length and 5′-truncated L1s are spread throughout the genome, with many found included in introns and even 3′ noncoding regions of cellular genes. Thus, partial sequences of L1 elements are incorporated into the transcripts of other genes; therefore, it is essential to differentiate between "authentic" L1 transcripts (those that are transcribed from the L1 promoter and are the only ones likely to contribute to retrotransposition) and those "L1-related" sequences

Jose L. Garcia-Pérez (ed.), *Transposons and Retrotransposons: Methods and Protocols*, Methods in Molecular Biology, vol. 1400, DOI 10.1007/978-1-4939-3372-3_15, © Springer Science+Business Media New York 2016

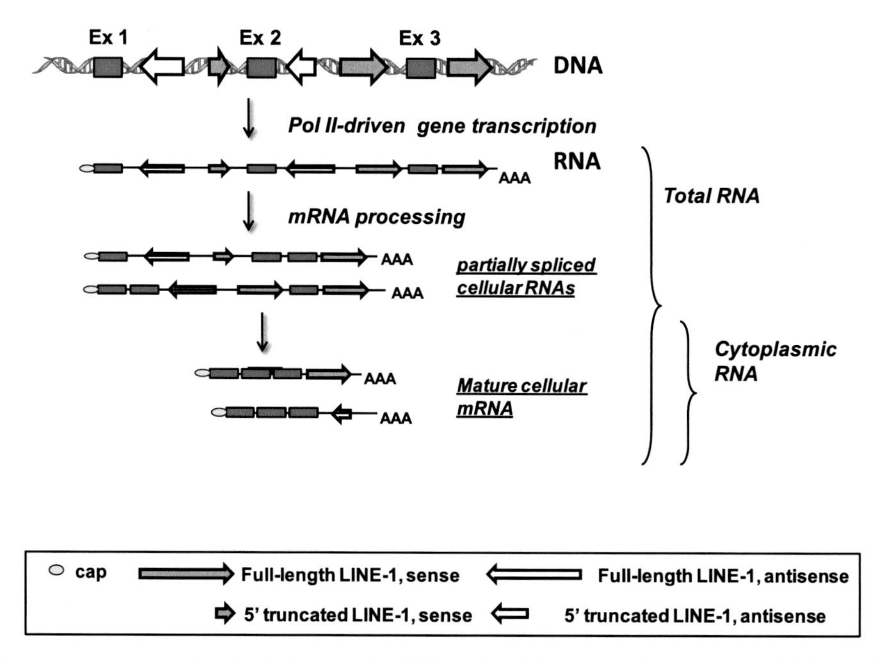

Fig. 1 L1 sequences present in the introns and 3′ UTRs of cellular genes. Truncated and full-length L1 loci, which are abundant in the sense and antisense orientations in the introns and 3′ UTRs of cellular genes, are included in pre-mRNAs generated during polymerase II (Pol II) transcription. These sequences are completely or partially removed during splicing with many mature cytoplasmic mRNAs still containing L1 sequences in their 3′ UTRs. Ex are exons, AAA are polyA tails. *Light grey rectangles* correspond to two open reading frames

that are simply parts of the L1 elements incorporated into other RNA molecules (Fig. 1).

This chapter provides a historic perspective on LINE mRNA detection and an extensive overview of sources of LINE-1 sequences in mammalian cells with the goal of providing an unambiguous guide to understanding the difference between authentic LINE-1 mRNAs and L1-related sequences incorporated into cellular transcripts. The majority of the points made in this chapter regarding LINE-1 expression are also applicable to SINE expression, even though expression of these repetitive elements is not specifically addressed. We also discuss pros and cons of different techniques used to study LINE expression with a detailed focus on northern blot analysis.

1.1 "Authentic" L1 RNAs Generated by L1 Promoters

LINE repeats, which are common features of many genomes, accumulated to over 500,000 copies in the primate and rodent genomes. Both genomes contain old, mostly retrotranspositionally inactive elements, some of which can be expressed, and young L1 subfamilies capable of expression and retrotransposition [1, 2].

Primate L1 loci express a complex variety of RNAs [3–6], some of which are able to retrotranspose and others representing "dead-end" transcripts with a yet unknown role in L1 amplification or cellular fate [3, 4]. Limited evidence suggests that a similarly complex set of RNAs is generated from the mouse L1 subfamilies [4]. The complexity of authentic primate L1 transcripts, which are defined as mRNAs generated through the activity of L1 promoters, begins with the presence of multiple transcription start sites. There are at least two regions of sense start sites [7–9] and an antisense start site [6] identified within the sense and antisense sequences of the human L1 5′ UTR. While the relationship between the activity of the promoters associated with these transcription start sites remains unknown, their strength is almost certainly cell-type and locus-specific [10–12]. This variation is most likely related to the availability of specific transcription factors and epigenetic markers in the region of each locus [13]. Due to their opposite orientation these promoters produce sense and antisense L1 transcripts, respectively (Fig. 2). The sense L1 transcripts are further classified based on their length as full-length or 5′- or 3′-truncated L1-related mRNAs (Fig. 2).

The full-length L1 mRNAs are the transcripts that include the entire length of the L1 locus, which is about 6 kb long in the case of the human L1. These are the conventional L1 mRNAs generated by the sense L1 promoter located within the first 100 base pairs of the L1 5′ UTR [8]. At least some (and perhaps all) of these transcripts are capped and are capable of retrotransposition when they encode and generate functional ORF1 and ORF2 proteins. There is a recent report showing that some human L1 transcripts are 5′-truncated, missing much of the 5′ UTR due to transcription initiation near the downstream end of the L1 5′ UTR [7]. It is possible that both the full-length and 5′ truncated transcripts are generated through mostly the same promoter sequences, but in one case the polymerase reaches upstream and the other downstream of the promoter. Even the 5′-truncated L1 mRNAs would be expected to be able to mobilize given that they encode both L1 proteins as demonstrated for the L1 lacking its 5′ UTR in cultured cells as well as endogenous L1s missing most of the 5′ UTR due to splicing [3, 14]. However, retrotransposed 5′ truncated RNAs would be unable of any subsequent mobilization because they do not include the L1 promoter.

The 3′-truncated L1 transcripts are generated when internally located polyadenylation sites are utilized during transcription initiated by the sense L1 promoters [4] (Fig. 2). Even though not confirmed experimentally, 5′- and 3′-truncated L1 mRNAs are expected to exist based on the position of L1 promoters and polyA sites. Further complexity of the L1-generated mRNAs originates from the presence of numerous splice donor and acceptor sites distributed throughout the L1 sequence with a dominant splice donor site located at position 97 of a reference L1.3 element [3, 15]. Examples of spliced L1 mRNAs missing the middle portion of the 5′ UTR are found in the human genomes demonstrating that these processed transcripts are expressed endogenously and can

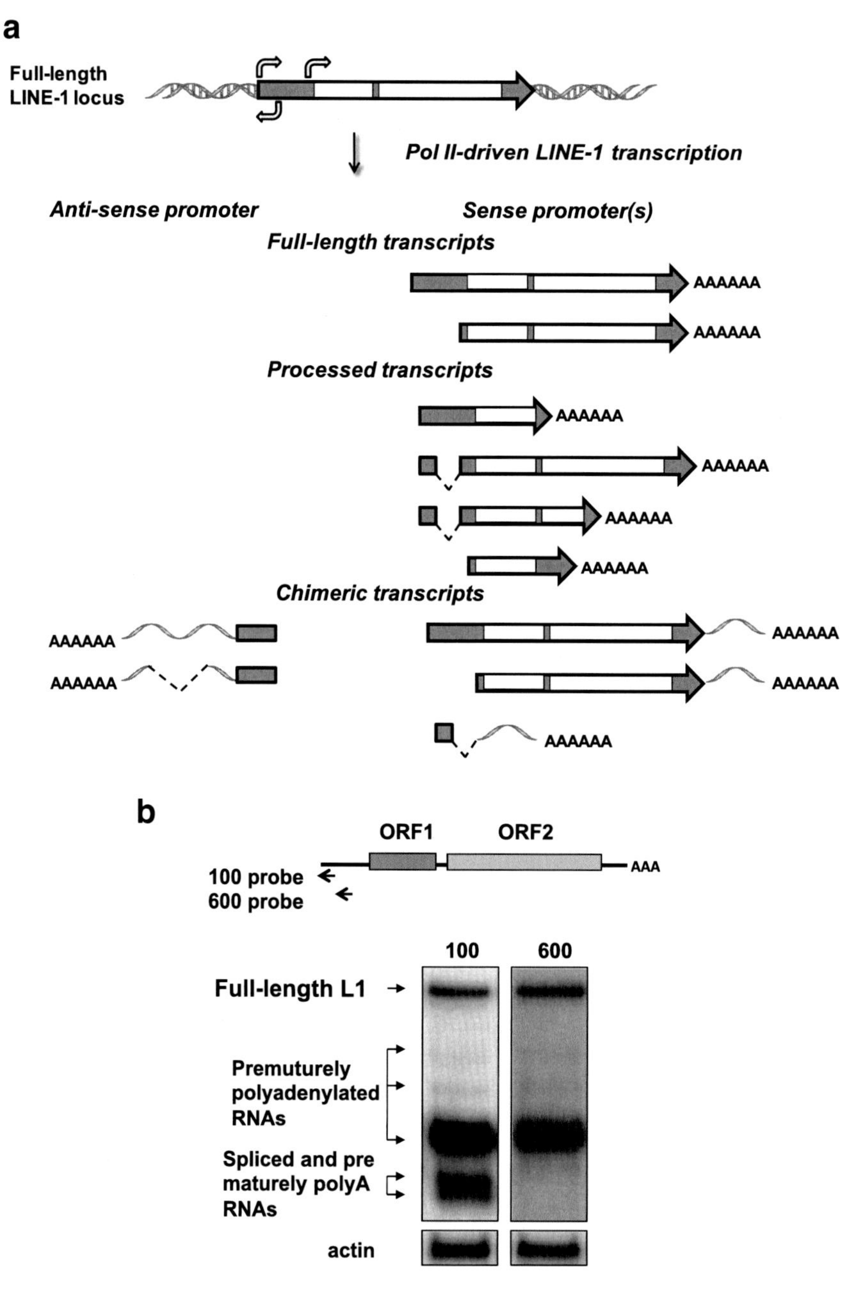

Fig. 2 (**a**) A schematic representation of various L1 transcripts generated by a full-length transcriptionally active L1 locus. *Horizontal arrows above* and *below* of the full-length L1 locus indicate relative positions of alternative sense and antisense transcription start sites, respectively. Pol II refers to the polymerase II driven transcription, AAAAAA indicate polyA tails, *dashed lines* indicate spliced mRNAs, *wavy lines* represent genomic sequences adjacent or distant to L1 locus. (**b**) Results of a northern blot analysis of sense L1 transcripts generated by a plasmid containing untagged, full-length, wild type, human L1 sequence transiently transfected in NIH 3T3 cells. *Horizontal arrows* indicate relative positions of the antisense RNA probes complementary to the first 100 bp and to the 600 region of the L1 5′ UTR. ORF1 and ORF2 indicated open reading frame 1 and 2 sequences within the L1 sequence

retrotranspose in the human germ line [3]. Another identified spliced mRNA with potential biological relevance is the one that lacks the entire ORF1 sequence but retains the ORF2 sequence [3]. This mRNA may produce functional ORF2p, which could drive Alu retrotransposition or introduce DNA double strand breaks (DSBs) [3, 16, 17]. Many other spliced L1 transcripts are expected to be incapable of retrotransposition because they cannot generate proteins required for de novo integration.

L1 loci containing functional promoters can also generate chimeric transcripts that read into adjoining genes [3, 5, 6, 18]. These chimeric mRNAs may contain full-length or partial L1 sequences joined with adjacent or distant genomic sequences. Authentic transcripts longer than the size of individual L1 locus are produced during transcription by the pol II reading through the polyA site at the end of a L1 locus. Transcription termination at the next available polyA signal present downstream of the L1 locus incorporates these genomic sequences into the L1 mRNA. In contrast to other hybrid transcripts described below, these mRNAs can retrotranspose leading to the duplication of genomic sequences incorporated into the L1 mRNAs. These RNAs represent 10–15 % of normal transcripts [19–21]. In addition to these simple read through chimeric transcripts, L1 sequences can also be joined with distant genomic sequences. This typically results from canonical splicing between splice donor sites in the sense or antisense L1 sequences and splice acceptor sites within genomic sequences [3, 5]. The antisense promoter of L1 can also make transcripts that include about 600 bases of the antisense portion of the 5′ UTR along with sequences present upstream of L1 loci. The exact nature and stability of these sequences is expected to vary between L1 loci.

1.2 L1 Sequences in Cellular Transcripts

Despite their relative depletion within genes the full-length and truncated, sense and antisense L1 inserts are still highly abundant in introns and 3′ UTRs of cellular genes [22]. Their presence as part of these cellular transcripts that are unrelated to the L1 life cycle creates background signal can mask or interfere with the detection of authentic L1 transcripts. The earliest attempts to detect L1 mRNAs through northern blots recognized that background expression far exceeds the expression of authentic L1 mRNAs [23, 24]. This was particularly true for whole cell RNAs and it was found that only in some cells could low levels of what appeared to be authentic L1 RNAs be detected in the cytoplasmic RNA. Similar findings were also reported for the expression of the mouse L1 [25].

Figures 1 and 2 summarize the various transcripts discussed above. The vast majority of L1-related RNA sequences are unrelated to the L1 retrotransposition process and represent the inclusion of L1-related sequences, in both orientations, in various celluar transcripts as shown in Fig. 1. Regardless of the technique,

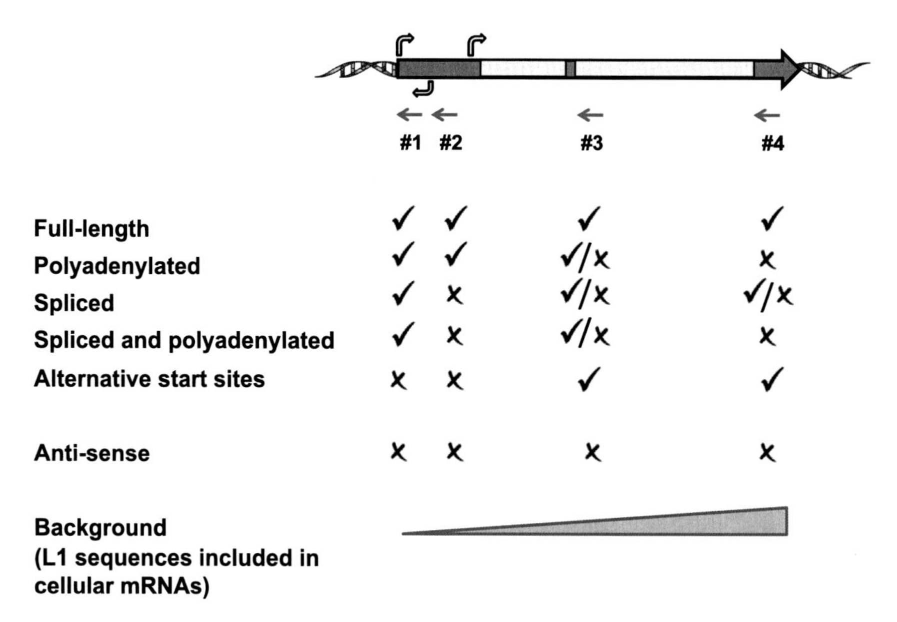

Fig. 3 Spectrum of L1 transcripts that could be identified by each of the antisense RNA probes complementary to the indicated regions of L1. The relative positions of the probes are indicated by *blue horizontal arrows*. *Check marks* indicate L1 transcripts that can be detected by specific probes. *Crosses* indicate L1 transcripts that cannot be detected by specific probes

when probing various regions of L1, there will be much more background from the L1-related sequence if the probe is designed to detect a region near the 3′ end of the L1 element because of the high level of truncated L1 elements in genic regions (Fig. 3). The low levels of transcripts from the L1 promoter will create the various transcripts shown in Fig. 2, with only the full-length transcripts clearly contributing to retrotransposition. Thus, the most enriched source of authentic L1 transcripts will be cytoplasmic, polyadenylated RNAs. Including the nuclear material, as in whole-cell RNAs will include some intronic L1 RNA-related sequences, as well as some aberrant polyadenylated and/or spliced RNAs that might not otherwise be fated to become mature mRNAs.

However, even polyadenylated cytoplasmic mRNAs will include a significant level of background from portions of L1-related sequences found in the 3′ UTR of mRNAs, as well as any background from nuclear contamination. The contribution to the signal from almost all of the background sources of L1-related transcripts will be biased towards the 3′ end of the elements because of the high copy number of truncated L1 elements in the genome (five to tenfold enrichment of L1 3′ UTR sequences relative to L1 5′ UTR sequences) and will be represented by both sense and antisense strands of the L1 sequence. One would expect the authentic L1 transcripts (those generated by its 5′ UTR promoter) to be almost exclusively sense transcripts, with only a portion of antisense transcripts corresponding to the 5′ UTR (those generated from the antisense promoter). All of these background issues make it imperative to utilize the cleanest RNA sources possible and to apply methods that can distinguish the authentic L1 transcripts from the various background L1-related transcripts. Thus, methods such as RT-PCR, S1- or ribonuclease-protection, and other approaches that do not provide any information on the 5′ end of the RNAs and whether they encompass the entire coding region cannot unambiguously distinguish authentic from background L1 RNA species.

1.3 Transiently Expressed L1 Elements

Most of the more conclusive studies of authentic L1 RNAs have been generated by overexpressing L1 elements from transiently transfected plasmids [3, 4, 14, 17, 20, 26]. Some of the existing L1 expression plasmids contain untagged L1 sequences, others include L1 sequences tagged with various intron-containing reporter cassettes expressing neomycin (Neo) or blasticidin (Blast) resistance genes or producing a green fluorescent protein (GFP) or a luciferase protein, which can be expressed in their active form upon L1 retrotransposition [14, 20, 26, 27]. Transcription of these tagged L1s produces the full spectrum of full-length, spliced and prematurely polyadenylated transcripts as well as spliced mRNAs that contain some L1 sequence coupled with the reporter cassette. The prematurely polyadenylated and spliced mRNAs are much more abundant than the retrotranspositionally competent full-length transcripts, at least in the case of the untagged human and mouse L1s and human L1 tagged with a Neo cassette [3, 4, 15]. The processed L1 transcripts expressed by plasmids containing wild type L1 sequences are almost completely eliminated upon codon optimization of L1 sequences corresponding to open reading frames. Codon-optimization efficiently removes the majority of the splice and polyadenylation sites present in the L1 genome and therefore greatly enhances production of full-length transcripts [4, 28–31].

1.4 Techniques for Measuring L1 Expression

Commonly used techniques to measure RNA levels can be, and have been, used to assess expression of endogenous and transiently transfected L1 elements as long as the experimental design and the interpretation of the results are performed taking into consideration the complexity of L1-generated transcripts.

RT-PCR and qRT-PCR have been frequently used to detect and measure L1 expression, mostly because of their convenience and sensitivity. Because PCR-based techniques only look at a small segment of mRNA at any one time, they are unable to unambiguously distinguish the authentic L1 mRNA transcripts from the L1-containing RNAs from other sources. Thus, these methods can be effectively utilized in some of the L1 overexpression experiments where the appropriate controls for background versus transfected L1 signal can be carried out. They are unlikely to produce reliable results from endogenous L1 expression unless the results are validated by other techniques measuring L1 expression. DNAase-treated, polyA-selected cytoplasmic RNA is the best starting material for RT-PCR analysis of authentic L1 sequences because it removes many sources of L1 sequences that generate the background signal (Fig. 1). The use of the L1-specific or oligo-dT primers for cDNA generation is always preferred over random primers, which will efficiently amplify intronic L1 sequences as well as those included in mature cellular transcripts. Figure 3 demonstrates which authentic L1 RNAs can be detected if the strand-specificity is included in the experimental design. The results of the RT-PCR analysis should be interpreted based on the L1 region selected for detection by this approach. PCR primers targeting the 3′ UTR of L1 are by far the worst choice for detection of authentic L1 transcripts by RT-PCR because of the high background generated from the 5′-truncated L1 sequences included into cellular mRNAs. This is mostly due to the lack of any other discriminating parameters (such as product size in the case of northern blot approach), which can aid in distinguishing between authentic L1 expression and L1 sequences incorporated into cellular mRNAs.

Less problems are encountered when overexpressed codon-optimized L1 elements are analyzed by this approach because the sequence of their mRNAs is completely different from the sequence of endogenously expressed authentic L1 transcripts or L1 sequences included in the cellular mRNAs [28, 31]. The codon-optimized L1 elements tagged with reporter cassettes also rarely produce processed products because they are lacking most of the splice and polyadenylation sites [3, 30]. These processed transcripts could be an issue when using primer pairs amplifying across the intron junction in the selectable marker of a tagged, wild type L1 for quantitative comparison of the presumed full-length L1 mRNAs between different L1 constructs or different cell types. The main reasons are that the splice products generated by splicing between splice sites within L1 and the selection cassette can be much more abundant than the full-length L1 transcripts and could be cell type-specific, easily masking a significant change in the levels of the full-length L1 mRNAs [3]. One strong attempt to differentiate authentic endogenous L1 transcripts from the background L1 sequences was made by identifying 3′ ends of the expected read-through transcripts generated by full-length L1 elements [32]. In this analysis inclusion of the portion of genomic sequences located downstream or upstream of the full-length L1 loci into L1 transcripts provided a unique tag for the individual L1 loci. While informative, these studies did not unambiguously define detected transcripts as those transcribed from the L1 promoter for the full-length L1 loci inserted in the sense orientation of genes.

Northern blot analysis is the most reliable technique for detection of authentic L1 transcripts, particularly the full-length L1 mRNAs [3, 4]. A major strength of this approach is that it allows identification of the size of the transcripts, which helps significantly in proper classification of the nature of these mRNAs. Another strength of the northern blot approach is its specificity achieved by a targeted selection of a region or regions within the L1 sequences to be used for probe generation as well as by making RNA strand-specific probes that can discriminate between sense and antisense L1 transcripts (Fig. 2). For example, a RNA probe complementary to the first 100 bp of the sense strand of human L1 5′ UTR is expected to detect most, if not all, L1 related transcripts generated by the full-length sense L1 promoter identified in this region. A probe complementary to the sense L1 sequence in the middle of the 5′ UTR will miss most of the spliced L1 mRNAs. Similarly, a probe complementary to the 3′ end of the L1 will not detect any prematurely polyadenylated transcripts (Fig. 3). In contrast to double-stranded, DNA-based probes, which detect both sense and antisense L1 sequences, strand-specific RNA probes are designed to detect only sense (or antisense) L1 transcripts. It is worth noting that probes complementary to the 3′ end of the L1 sequence have a higher likelihood of detecting cellular mRNAs containing 5′-truncated L1 sequences. We have generally found it beneficial to utilize polyA selection to enrich the mRNA component and it may even be beneficial to utilize cytoplasmic polyA mRNA fraction. Some of the limitations of the northern blot approach are the need for a relatively high amount of RNA to reach the threshold of sensitivity and its inability to discriminate between transcripts produced by functional and nonfunctional L1 loci.

The advent and wide usage of NGS (Next-Generation Sequencing) analysis of cellular transcriptome has resulted in accumulation of an impressive amount of data. The sensitivity and inclusive nature of NGS approaches are very appealing features that are irreplaceable when it comes to analysis of unique or rare cellular transcripts. However, the application of standard bioinformatic tools to analysis of repetitive sequences, such as LINEs and SINEs, using libraries generated by NGS approaches that are not strand-specific and performed on total cellular RNAs revealed serious problems with the interpretation of results. Most such studies have not clearly differentiated authentic L1-directed transcripts from those that represent background inclusion in other genes [33, 34]. The application of a nanoCAGE approach to enrich for transcript 5′ ends primarily confirmed that even with the enrichment, the vast majority of L1-related RNAs were not from the L1 promoter and therefore did not reflect on the authentic L1 transcripts [33, 35]. The situation for Alu transcription is much worse, with many more Alu elements included in genes creating extremely high backgrounds. Additionally, SINE RNAs are not capped therefore any SINE sequences identified in the nanoCAGE selected RNA pool represent sequences associated with cellular transcripts.

In this chapter, we provide a protocol designed to detect L1 RNAs by northern blot.

2 Materials

1. *Common reagents*: TRIzol (Invitrogen) or similar, formamide, formaldehyde, bromophenol blue, ethidium bromide (EtBr), MOPS, Sodium Acetate, EDTA, NaCl, sodium citrate tribasic Ficoll (type 400), bovine serum albumin, SDS, chloroform, isopropanol, herring sperm DNA, yeast tRNA, a polyA-selection kit, agarose, NaOH, nuclease-free water, formaldehyde, 100 % ethanol, nuclease-free water, Hybond-N membrane (Amersham) or similar, a T7 in vitro transcription kit, [α-^{32}P] UTP, NucAway spin columns or equivalent, and heat-resistant plastic bags can be obtained from the indicated source or from local suppliers.
2. *Special Equipment*: temperature controlled hybridization chamber, Gene Linker (Bio-Rad) or equivalent, a phosphorimager or X-ray film processor.

3 Buffers

(a) *10× MAE Buffer (pH 7.0)*: 200 mM MOPS, pH 7.0, 50 mM NaAce, 10 mM EDTA, pH 8.0. Prepare 1 L with ultrapure H_2O, adjust pH to 7 with NaOH, filter and store at room temperature (wrapped in aluminum foil).

(b) *Loading buffer*: combine 150 μl formamide, 100 μl formaldehyde, 50 μl of 10× MAE, 1.3 μl of EtBr (10 mg/ml), and 0.04 mg bromophenol blue (to approximately a 0.02 (w/v)% final). This buffer should be made fresh each time.

(c) *20× SSC (pH 7.0)*: dissolve 175.3 g NaCl and 88.2 g sodium citrate tribasic in 1 L of water.

(d) *Hybridization buffer*: combine 30 ml formamide, 2 ml Denhardt's solution (1 % (w/v) Ficoll (type 400)), 1 % (w/v) polyvinylpyrrolidone, 1 % (w/v) bovine serum albumin, 1 g SDS, 5.84 g NaCl, 1 ml herring sperm DNA at 1 μg/ml, sheared by extensive sonication. Complete to 100 ml using deionized water.

(e) *High stringency wash buffer (in 1 L of water)*: combine 5 ml 20× SSC with 1 g SDS and complete to 1000 ml using deionized water.

4 Methods

1. *RNA harvest.* A single T75 flask of Hela cells transfected with one of the L1 expression plasmids described above produces sufficient RNA to detect transient L1 expression. A single confluent T75 flask of Hela cells or other human cancer cells is sufficient to detect endogenous L1 expression. Cells are harvested in 7.5 ml of TRIzol by scraping. After the addition of 1.5 ml of chloroform and vigorous mixing, the samples are centrifuged at 1500 × *g* for 30 min at 4 °C. After collection, the aqueous (top) portion is combined with 4.5 ml of chloroform, vigorously mixed and the samples are centrifuged at 1500 × *g* for 10 min at 4 °C. After collection, the aqueous (top) portion is combined with 4 ml of isopropanol and precipitated overnight at −80 °C. The next day, samples are centrifuged at 1500 × *g* for 30 min at 4 °C. The pellet is washed with ethanol and resuspended in 500 μl of nuclease-free water for polyA selection. This is expected to generate approximately 100–150 μg of whole-cell RNA.
2. *polyA selection* is performed using a polyA-selection kit (for example Promega) following the manufacturer's protocol [3]. Store purified RNAs at −20 °C.
3. *RNA sample fractionation.* Add 1 g of agarose to 85 ml ultrapure H_2O. Dissolve the agarose by heating in a microwave, cool to 45 °C; add 10 ml of 10× MAE Buffer and 5 ml of formaldehyde, pour into a 10 cm × 10 cm slab gel box with a well comb generating approximately 1 cm wells and allow gel to solidify. Remove samples from −20 °C (from polyA selection, see protocol), centrifuge at 12,000 × *g* for 10 min at 4 °C, wash the pellets with 500 μl of cold 100 % ethanol, centrifuge at 12,000 × *g* for 10 min at 4 °C. Discard ethanol, drain excess liquid, use applicator (sterile q-tip) to remove liquid from sides of tube, and air-dry the samples for 10–15 min. Resuspend the dry pellet in 30 μl of nuclease-free water. Add 15 μl of loading buffer, and pipette up and down.

   Heat samples in loading buffer at 70 °C for 3 min. Place immediately on ice for 2–3 min. Load the samples on a gel and run in 1× MAE at 110 V until the dye has reached the desired length, typically near the bottom of the gel.
4. *RNA transfer.* Use Hybond-N membrane (GE Amersham) or a similar alternative and 6× SSC to transfer RNA overnight as previously described using a simple capillary blot procedure [36]. The next day, crosslink RNA samples to the membrane by 2× exposure to 250 mJ of UV light using a Gene Linker (Bio-Rad) or equivalent. Pre-hybridize the membrane in a

heat-sealed plastic bag with 4 ml of hybridization buffer by rocking for at least 2 h at 60 °C.

5. *Probe, hybridization, and detection.* Generate radioactively labeled RNA probe specific to the sense (or if needed to the antisense) strand of the L1 sequence using 100–150 bp PCR products containing a T7 promoter sequence (5′-TAATACGACTCACTATAGG-3′) upstream of the primer portion of the downstream oligonucleotide to serve as a template for in vitro transcription. In vitro transcription is performed using a T7 in vitro transcription kit following manufacturer instructions supplemented with 5 μl of [α-^{32}P] UTP (6000 Ci/mmol at concentration of 10 mCi/ml). The resulting radioactively labeled RNA probe is separated from the free NTPs using NucAway spin columns or equivalent. The region and the strand-complementarity of L1 sequence selected for probe generation is important because it dictates the spectrum of RNA species that could be detected by this approach (*see* northern blot under Subheading 1.4 and Figs. 2 and 3). A probe complementary to the first 100 bp and a probe complementary to the 600 bp region of the human L1 5′ UTR provide robust detection of all L1-related sense transcripts or full-length and prematurely polyadenylated transcripts, respectively [3] (Fig. 2b). Hybridize the membrane in a heat-sealed plastic bag with the probe in 3–4 ml of hybridization buffer ($2–4 \times 10^6$ cpm ^{32}P-labeled RNA probe/ml) by rocking overnight at 60 °C. The next day wash the membrane three times with high stringency wash buffer for 5 min at 60 °C. Use a phosphorimager or X-ray film to detect the signal.

5 Notes

This approach also works well for detection of Alu RNAs generated from expression plasmids transiently transfected into mammalian cells [37]. However, it is not suitable for detection of endogenously expressed Alu elements because of the abundance of Alu sequences in cellular RNAs, length heterogeneity of authentic Alu transcripts, and their sequence similarity to the 7SL RNA. A standard northern blot approach to detecting endogenously expressed Alu RNAs using an Alu-specific probe produces a smeared signal mostly originating from Alu containing cellular mRNAs and a much weaker smeared signal expected to be generated by authentic Alu transcripts [38, 39]. It also produces a strong discrete band corresponding to the 7SL RNA [38, 39].

Due to their abundance in host genomes and unique complexity of mRNA expression, transposable elements present a serious challenge when it comes to analysis of their expression and activity.

A better understanding of this complexity is expected to significantly improve the choice of methods applied to detection of L1 mRNA as well as the interpretation of the results obtained by these methods.

References

1. Lander ES, Linton LM, Birren B, Nusbaum C, Zody MC, Baldwin J, Devon K, Dewar K, Doyle M, FitzHugh W et al (2001) Initial sequencing and analysis of the human genome. Nature 409:860–921
2. Waterston RH, Lindblad-Toh K, Birney E, Rogers J, Abril JF, Agarwal P, Agarwala R, Ainscough R, Alexandersson M, An P et al (2002) Initial sequencing and comparative analysis of the mouse genome. Nature 420:520–562
3. Belancio VP, Hedges DJ, Deininger P (2006) LINE-1 RNA splicing and influences on mammalian gene expression. Nucleic Acids Res 34:1512–1521
4. Perepelitsa-Belancio V, Deininger P (2003) RNA truncation by premature polyadenylation attenuates human mobile element activity. Nat Genet 35:363–366
5. Nigumann P, Redik K, Matlik K, Speek M (2002) Many human genes are transcribed from the antisense promoter of L1 retrotransposon. Genomics 79:628–634
6. Speek M (2001) Antisense promoter of human L1 retrotransposon drives transcription of adjacent cellular genes. Mol Cell Biol 21: 1973–1985
7. Alexandrova EA, Olovnikov IA, Malakhova GV, Zabolotneva AA, Suntsova MV, Dmitriev SE, Buzdin AA (2012) Sense transcripts originated from an internal part of the human retrotransposon LINE-1 5′ UTR. Gene 511:46–53
8. Swergold GD (1990) Identification, characterization, and cell specificity of a human LINE-1 promoter. Mol Cell Biol 10:6718–6729
9. Athanikar JN, Badge RM, Moran JV (2004) A YY1-binding site is required for accurate human LINE-1 transcription initiation. Nucleic Acids Res 32:3846–3855
10. Lavie L, Maldener E, Brouha B, Meese EU, Mayer J (2004) The human L1 promoter: variable transcription initiation sites and a major impact of upstream flanking sequence on promoter activity. Genome Res 14:2253–2260
11. Yu F, Zingler N, Schumann G, Stratling WH (2001) Methyl-CpG-binding protein 2 represses LINE-1 expression and retrotransposition but not Alu transcription. Nucleic Acids Res 29:4493–4501
12. Yang N, Zhang L, Zhang Y, Kazazian HH (2003) An important role for RUNX3 in human L1 transcription and retrotransposition. Nucleic Acids Res 31:4929–4940
13. Hata K, Sakaki Y (1997) Identification of critical CpG sites for repression of L1 transcription by DNA methylation. Gene 189:227–234
14. Moran JV, Holmes SE, Naas TP, DeBerardinis RJ, Boeke JD, Kazazian HH Jr (1996) High frequency retrotransposition in cultured mammalian cells. Cell 87:917–927
15. Belancio VP, Roy-Engel AM, Deininger P (2008) The impact of multiple splice sites in human L1 elements. Gene 411:38–45
16. Belancio VP, Roy-Engel AM, Pochampally RR, Deininger P (2010) Somatic expression of LINE-1 elements in human tissues. Nucleic Acids Res 38:3909–3922
17. Gasior SL, Wakeman TP, Xu B, Deininger PL (2006) The human LINE-1 retrotransposon creates DNA double-strand breaks. J Mol Biol 357:1383–1393
18. Matlik K, Redik K, Speek M (2006) L1 antisense promoter drives tissue-specific transcription of human genes. J Biomed Biotechnol 2006:1–16
19. Goodier JL, Ostertag EM, Kazazian HH Jr (2000) Transduction of 3′-flanking sequences is common in L1 retrotransposition. Hum Mol Genet 9:653–657
20. Moran JV, DeBerardinis RJ, Kazazian HH Jr (1999) Exon shuffling by L1 retrotransposition. Science 283:1530–1534
21. Pickeral OK, Makalowski W, Boguski MS, Boeke JD (2000) Frequent human genomic DNA transduction driven by LINE-1 retrotransposition. Genome Res 10:411–415
22. Medstrand P, van de Lagemaat LN, Mager DL (2002) Retroelement distributions in the human genome: variations associated with age and proximity to genes. Genome Res 12:1483–1495
23. Skowronski J, Singer MF (1985) Expression of a cytoplasmic LINE-1 transcript is regulated in a human teratocarcinoma cell line. Proc Natl Acad Sci U S A 82:6050–6054

24. Skowronski J, Fanning TG, Singer MF (1988) Unit-length line-1 transcripts in human teratocarcinoma cells. Mol Cell Biol 8:1385–1397
25. Schichman SA, Severynse DM, Edgell MH, Hutchison CA III (1992) Strand-specific LINE-1 transcription in mouse F9 cells originates from the youngest phylogenetic subgroup of LINE-1 elements. J Mol Biol 224:559–574
26. Moran JV, Gilbert N, Boeke J, Kazazian H, Ostertag E, Loon S, Wei W (2000) Human L1s retrotransposition: cis-preference vs. trans-complementation. Am J Hum Genet 67:199
27. Xie Y, Mates L, Ivics Z, Izsvak Z, Martin SL, An W (2013) Cell division promotes efficient retrotransposition in a stable L1 reporter cell line. Mob DNA 4:10
28. Han JS, Boeke JD (2004) A highly active synthetic mammalian retrotransposon. Nature 429:314–318
29. Han JS, Szak ST, Boeke JD (2004) Transcriptional disruption by the L1 retrotransposon and implications for mammalian transcriptomes. Nature 429:268–274
30. Wagstaff BJ, Barnerssoi M, Roy-Engel AM (2011) Evolutionary conservation of the functional modularity of primate and murine LINE-1 elements. PLoS One 6:e19672
31. Wallace NA, Belancio VP, Deininger PL (2008) L1 mobile element expression causes multiple types of toxicity. Gene 419:75–81
32. Rangwala SH, Zhang L, Kazazian HH Jr (2009) Many LINE1 elements contribute to the transcriptome of human somatic cells. Genome Biol 10:R100
33. Fadloun A, Le Gras S, Jost B, Ziegler-Birling C, Takahashi H, Gorab E, Carninci P, Torres-Padilla ME (2013) Chromatin signatures and retrotransposon profiling in mouse embryos reveal regulation of LINE-1 by RNA. Nat Struct Mol Biol 20:332–338
34. Criscione SW, Zhang Y, Thompson W, Sedivy JM, Neretti N (2014) Transcriptional landscape of repetitive elements in normal and cancer human cells. BMC Genomics 15:583
35. Faulkner GJ, Kimura Y, Daub CO, Wani S, Plessy C, Irvine KM, Schroder K, Cloonan N, Steptoe AL, Lassmann T et al (2009) The regulated retrotransposon transcriptome of mammalian cells. Nat Genet 41:563–571
36. Maniatis T, Fritsch EF, Sambrook J (1982) Molecular cloning: a laboratory manual. Cold Spring Harbor Laboratory, Cold Spring Harbor
37. Kroutter EN, Belancio VP, Wagstaff BJ, Roy-Engel AM (2009) The RNA polymerase dictates ORF1 requirement and timing of LINE and SINE retrotransposition. PLoS Genet 5:e1000458
38. Deininger P (2011) Alu elements: know the SINEs. Genome Biol 12:236
39. Roy-Engel AM (2012) LINEs, SINEs and other retroelements: do birds of a feather flock together? Front Biosci (Landmark Ed) 17:1345–1361

Chapter 16

Monitoring Long Interspersed Nuclear Element 1 Expression During Mouse Embryonic Stem Cell Differentiation

Maxime Bodak and Constance Ciaudo

Abstract

Long Interspersed Elements-1 (LINE-1 or L1) are a class of transposable elements which account for almost 19 % of the mouse genome. This represents around 600,000 L1 fragments, among which it is estimated that 3000 intact copies still remain capable to retrotranspose and to generate deleterious mutation by insertion into genomic coding region. In differentiated cells, full length L1 are transcriptionally repressed by DNA methylation. However at the blastocyst stage, L1 elements are subject to a demethylation wave and able to be expressed and to be inserted into new genomic locations. Mouse Embryonic Stem Cells (mESCs) are pluripotent stem cells derived from the inner cell mass of blastocysts. Mouse ESCs can be maintained undifferentiated under controlled culture conditions or induced into the three primary germ layers, therefore they represent a suitable model to follow mechanisms involved in L1 repression during the process of differentiation of mESCs. This protocol presents how to maintain culture of undifferentiated mESCs, induce their differentiation, and monitor L1 expression at the transcriptional and translational levels. L1 transcriptional levels are assessed by real-time qRT-PCR performed on total RNA extracts using specific L1 primers and translation levels are measured by Western blot analysis of L1 protein ORF1 using a specific L1 antibody.

Key words LINE-1, mESCs, Embryonic body differentiation, Protein extraction and Western blot, RNA extraction and real-time qRT-PCR

1 Introduction

Transposable elements (TE) represents 39 % of the mouse genome [1] and Long Interspersed Elements-1 (LINE-1 or L1) which belong to the: retrotransposon non-LTR autonomous subclass of TE, account for 19 % [1]. Full length L1 are DNA mobile elements about 7 kb in length containing a 5′ untranslated region with an internal promoter activity, two open reading frames (ORF1 and ORF2) and a 3′ untranslated region ending with a polyA tail [2]. L1 encode the proteins necessary for their retrotransposition,

Jose L. Garcia-Pérez (ed.), *Transposons and Retrotransposons: Methods and Protocols*, Methods in Molecular Biology, vol. 1400, DOI 10.1007/978-1-4939-3372-3_16, © Springer Science+Business Media New York 2016

which requires an RNA intermediate involving a "copy and paste" mechanism [3]. To be considered as active, L1 elements must be able to perform a complete retrotransposition cycle that includes the following steps: transcription of L1 RNA, export into the cytoplasm, translation of ORF1 and ORF2, association of L1 RNA with ORF1 and ORF2 proteins, return to the nucleus, and integration into a new genomic location [2]. The 5′ UTR is frequently truncated [4] and the L1 sequences are prone to inversion [5] during this retrotransposition event, which result is an accumulation of almost 600,000 L1 fragments in the mouse genome [1]. Nonetheless, only 3000 full-length elements are still potentially able to retrotranspose [6, 7]. These actives L1 elements belongs to three different L1 subfamilies defined according to the variable numbers of monomer (tandem repeat units of 200 bp) contained in the 5′ untranslated region: A [8], Tf [6], and Gf [7]. Due to their retrotransposition ability, active L1 have a high mutagenic potential and must be regulated. A range of mechanisms have been described to be involved in L1 regulation, and L1 5′ UTR methylation is the main repression process by inhibiting L1 transcription [2]. However during early development, at the blastocyst stage, L1 elements undergo a wave of demethylation [9] taking off the inhibition, which result in their reactivation: L1 are transcribed, translated, and able to retrotranspose [2, 10–12]. Then during development, DNA methylation is restored as well as L1 repression [13].

Mouse embryonic stem cells (mESCs) are derived from the inner cell mass of the blastocyst and in appropriate culture conditions, they can retain their pluripotent state and can proliferate indefinitely. This pluripotent state is maintained by three major essentials factor: Pou5f1 (also called Oct4), Sox2, and Nanog [14, 15]. Moreover, mESCs can be differentiated into the three primary germ layers and mimic in vitro early development. These features bring mESCs as a useful model to follow L1 expression, as L1 are transcribed and translated at the pluripotent stage and then repressed by DNA methylation after differentiation. This chapter describes how to monitor L1 expression during mESCs differentiation at the transcriptional and translational level. As controls: Oct4, Nanog, and Sox2 pluripotency factors are assessed at the same levels to follow the differentiation status as their expression is shut down during differentiation [16].

This protocol has been subdivided into three parts in order to facilitate its understanding. The first part describes how to culture mESCs and how to set up differentiation culture suspension in low-adherent tissue culture dishes. Samples have been generated at three different time points: Day 0 of differentiation which corresponds to pluripotent cells, Day 6 and Day 10 of differentiation, which correspond to differentiated cells. In the second part, it is explained how to assess transcriptional expression level of L1 and pluripotency factors: total RNA extraction from cell pellets will be performed, followed by real-time qRT-PCR using specific primers. Primers sequences for: each pluripotency factor (Nanog, Sox2, and Oct4), L1 subfamily (L1_A, L1_Tf, and L1_Gf), all L1 elements (L1_ORF2, primers specific to ORF2, which is common to all subfamilies), and the reference gene *Rrm2* (for normalization) (*see* Table 1) will allow to determine the amount of corresponding messenger RNA. In the third part, we describe how to assess translational expression level of L1 and pluripotency factors: total protein extraction from cell pellets is performed, followed by Western blot (with SDS polyacrylamide gel electrophoresis). Specific antibodies described here: anti-L1_ORF1 (specific of ORF1p—ORF1 protein), anti-OCT-3/4 (specific of Oct4), and anti-α-tubulin (specific of the structural protein α-tubulin—a loading control) will allow the detection of the corresponding protein.

2 Materials

2.1 mESCs Culture: Proliferation and Differentiation

1. mESCs. In this protocol the E14TG2a mESC line has been used (obtained from ATCC).
2. mESC proliferative medium. Dulbecco's Modified Eagle Media (DMEM) containing: 15 % (v/v) of a special selected batch of fetal bovine serum (FBS; Life Technologies) tested for optimal growth of mESCs, 1000 U/mL of LIF (Millipore),

Table 1
Primers sequences

	Sequence (5′–3′)	
Name	**Forward**	**Reverse**
Rrm2	CCGAGCTGGAAAGTAAAGCG	ATGGGAAAGACAACGAAGCG
L1_Tf	CAGCGGTCGCCATCTTG	CACCCTCTCACCTGTTCAGACTAA
L1_Gf	CTCCTTGGCTCCGGGACT	CAGGAAGGTGGCCGGTTGT
L1_A	GGATTCCACACGTGATCCTAA	TCCTCTATGAGCAGACCTGGA
L1_ORF2	GGAGGGACATTTCATTCTCATCA	GCTGCTCTTGTATTTGGAGCATAGA
Nanog	CAGAAAAACCAGTGGTTGAAGA	GCAATGGATGCTGGGATACTC
Sox2	CACAGATGCAACCGATGCA	GGTGCCCTGCTGCGAGTA
Pou5f1 (Oct4)	CAACTCCCGAGGAGTCCCA	CTGGGTGTACCCCAAGGTGA

Primers have been specifically designed for qRT-PCR and all have a melting temperature of 60 °C

0.1 mM of 2-β-mercaptoethanol, 0.05 mg/mL of streptomycin, and 50 U/mL of penicillin. This medium must be kept at 4 °C.

3. mESC differentiation medium. DMEM (Invitrogen) containing: 10 % of FBS, 0.1 mM of 2-β-mercaptoethanol, 0.05 mg/mL of streptomycin, and 50 U/mL of penicillin. This medium is LIF-free and must be kept at 4 °C.
4. T75 tissue culture flasks.
5. 0.2 % gelatin solution and gelatin-coated flasks. Prepare 500 mL of 0.2 % (w/v) gelatin solution: dilute 1 g of gelatin from porcine skin with 500 mL of autoclaved water; mix well and then autoclave the solution. Mouse ESCs grow on gelatin-coated support in the absence of feeder cells. To gelatin-coat a flask or plate, pour a few milliliters of 0.2 % gelatin solution on the bottom of the flask, distribute homogeneously (the entire surface must be covered, e.g., 2 mL for T75 TPP flask), incubate for minimum 5 min at room temperature and finally remove the gelatin (*see* **Note 1**).
6. Cell culture incubator: mESCs grow at 37 °C in 8 % CO_2.
7. Cell culture media: 0.05 % Trypsin containing EDTA (Life Technologies) and PBS pH 7.4 (1×) without $CaCl_2$ and $MgCl_2$ (Life Technologies). This media must be kept at 4 °C.
8. Low-adherent tissue culture dish: 94×16 mm triple vented petri dish (Greiner Bio-One).
9. 15 mL Falcon tubes.
10. 1.5 mL Eppendorf tubes.

2.2 Total RNA Extraction, RNA Quality Check, and qRT-PCR

1. Cell/embryoid body pellets.
2. TRIzol Reagent (Life Technologies) (*see* **Note 2**).
3. Chloroform and isopropanol.
4. 1.5 mL Eppendorf tubes.
5. Ice-cold 70 % ethanol.
6. RNA concentration measurement device.
7. Agarose powder.
8. Ethidium bromide solution 10 mg/mL.
9. 6× loading dye.
10. Electrophoresis power supply and horizontal electrophoresis cell.
11. UV revelation device for electrophoresis gel.
12. RNase-Free DNase kit: RNase-Free DNase, Stop Solution, and RNase-Free DNase 10× reaction buffer.
13. RiboLock™ RNase Inhibitor (Thermo Scientific).

14. Random hexamer primers 100 μM.
15. GoScript Reverse Transcriptase Kit (Promega): GoScript Reverse Transcriptase, GoScript 5× Reaction buffer, and $MgCl_2$ 25 mM.
16. Compounds for PCR reactions: 5× Taq reaction buffer, dNTP mix 10 mM each, Taq Polymerase.
17. Autoclaved water.
18. Primers for PCR and qPCR 100 μM (*see* Table 1).
19. PCR tubes and thermo-cycler.
20. Appropriate DNA Ladder Marker.
21. SYBR Green.
22. Real-time PCR device (*see* **Note 3**).

2.3 Total Protein Extraction and Western Blot

1. Cell/embryoid body pellets.
2. NP40 lysis buffer: 137 mM NaCl, 1 % (v/v) IGEPAL®CA-630, 20 mM Tris–HCl, and 1 mM EDTA. To prepare 500 mL, mix and dissolve: 4 g of NaCl, 5 mL of IGEPAL®CA-630, 1.576 g of Tris–HCl, and 0.186 g of EDTA into 500 mL of autoclaved water. This buffer must be stored at 4 °C.
3. 10× Completed NP40 lysis buffer. To prepare 1 mL: dissolve one pellet of EDTA-free protease inhibitor cocktail tablets (Roche) into 1 mL of NP40 lysis buffer. This buffer must be stored at −20 °C.
4. Sonication device (*see* **Note 4**).
5. 1.5 mL Eppendorf tubes.
6. 4× Laemmli Buffer containing 2-mercaptoethanol. To prepare 1 mL: mix 900 μL of 4× Laemmli Buffer (Bio-Rad) and 100 μL 2-mercaptoethanol. This solution can be stored at −20 °C.
7. Acrylamide–bis-acrylamide 37.5:1 (40 % solution, AppliChem) (*see* **Note 5**).
8. Autoclaved water.
9. 1.5 M Tris–HCl pH 8.8. To prepare 500 mL: dissolve 90.9 g of Tris base into 500 mL of autoclaved and adjust pH to 8.8 using HCl (*see* **Note 6**).
10. 1 M Tris–HCl pH 6.8. To prepare 500 mL: dissolve 60.57 g of Tris base (Sigma) into 500 mL of autoclaved and adjust pH to 6.8 using HCl (*see* **Note 6**).
11. 20 % (w/v) SDS solution. To prepare 50 mL: dissolve 10 g of sodium dodecyl sulfate into 50 mL of autoclaved water.
12. 10 % (w/v) APS solution. To prepare 10 mL: dissolve 1 g of ammonium persulfate for electrophoresis into 10 mL of autoclaved water. Make aliquots of 1 mL and store them at −20 °C.

13. TEMED (from a local supplier).
14. Isopropanol and methanol.
15. Prestained protein ladder.
16. Necessary for Western blot running and blotting: 1.5 mm spacer plate and short plate (10.1 cm × 8.2 cm), 1.5 mm comb, tank, companion running module, trans-blot module, transfer cassette, sponge and a power supply.
17. PVDF (0.45 μm membrane) and Whatman cellulose chromatography blotting paper 3MM. Cut the PVDF membrane (*see* **Note 7**) and the Whatman paper to the same size as the casted acrylamide gel (10 cm × 8 cm).
18. 80 % ethanol.
19. 10× Tris–glycine–SDS (TGS) pH 8.3 buffer: To prepare 1 L: dissolve 30.3 g of Tris base, 144 g of glycine, and 10 g of sodium dodecyl sulfate into 1 L of autoclaved water. Then adjust pH to 8.3 (*see* **Note 6**).
20. Running buffer. To prepare 1 L: mix 100 mL of 10× TGS pH 8.3 buffer into 900 mL of autoclaved water.
21. Transfer buffer. To prepare 1 L: mix 100 mL of 10× TGS pH 8.3 buffer, 200 mL of 100 % ethanol, and 700 mL of autoclaved water.
22. Tris-buffered saline 10× (TBS 10×) pH 7.4 buffer. To prepare 1 L: dissolve 121.14 g of Tris base and 87.66 g of NaCl into 1 L of autoclaved water. Then adjust pH to 7.4 (*see* **Note 6**) and autoclave.
23. TBS 1× Tween (TBST 1×) buffer. To prepare 1 L: mix 100 mL of TBS 10× pH 7.4 buffer, 900 mL of autoclaved water and 1 mL of Tween 20.
24. 5 % Milk TBST 1×. To prepare 50 mL: dissolve 2.5 g of manufactured powdered milk into 50 mL of TBS 1× Tween buffer.
25. 3D agitator device.
26. Anti-L1_ORF1 antibody (gift of Dr. A. Bortvin, Carnegie Institution for Science, USA) (rabbit) 1:15,000 diluted. To prepare the dilution: put 0.7 μL of the antibody stock solution into 10 mL of 5 % Milk TBST 1×. This dilution can be stored at −20 °C.
27. OCT-3/4 antibody (H-134, Santa Cruz) (rabbit) 1:3000 diluted. To prepare the dilution: put 3.33 μL of the antibody stock solution into 10 mL of 5 % Milk TBST 1×. This dilution can be stored at −20 °C.
28. Anti-α-tubulin (GenScript) antibody (mouse) 1:10,000 diluted. To prepare the dilution: put 1 μL of the antibody

stock solution into 10 mL of 5 % Milk TBST 1×. This dilution can be stored at −20 °C (*see* **Note 8**).

29. Anti-rabbit IgG, HRP-linked antibody diluted as recommended by the manufacturer. To prepare the dilution: dilute the antibody stock solution into 10 mL of 5 % Milk TBST 1×. This dilution can be stored at −20 °C (*see* **Note 9**).
30. Anti-mouse IgG, HRP-linked antibody diluted as recommended by the manufacturer. To prepare the dilution: dilute the antibody stock solution into 10 mL of 5 % Milk TBST 1×. This dilution can be stored at −20 °C.
31. Membrane revelation kits (*see* **Note 10**).
32. Membrane revelation device.
33. Stripping membrane buffer (0.2 M NaOH). To prepare 500 mL: dissolve 4 g of NaOH into 500 mL of autoclaved water.

3 Methods

3.1 mESCs Culture: Proliferation and Differentiation

This part of the protocol concerns proliferation, passing and differentiation procedures for the E14TG2a mESC line. Before using any cell culture media, make sure that they are at 37 °C (using a water bath) or at least at room temperature (around 25 °C).

1. Proliferation and passing. During proliferation, mESC proliferative medium is changed on a daily basis (a T75 flask contains 10 mL medium). When mESCs reach of 70–80 % of confluence they must be passed (*see* **Note 11**). For one T75 flask: aspirate the medium and wash once with 4 mL of PBS 1× then add 2 mL of 0.05 % Trypsin containing EDTA and incubate for 5 min at 37 °C. Add 4 mL of mESC differentiation medium, pipet up and down and flush the flask several times to detach all the cells. Then collect 1 mL of this cell suspension in a 15 mL Falcon tube and spin the cells 5 min at 180×*g*. Aspirate the supernatant, resuspend the cell pellet in 10 mL of mESC proliferative medium and plate the cells in a new T75 gelatin coated flask (*see* **Note 12**). Put the remaining 3 mL of cell suspension into a 15 mL Falcon tube and spin the cells 5 min at 180×*g*. Aspirate the supernatant, resuspend the cell pellet in 2 mL of PBS 1×, distribute equally to two 1.5 mL Eppendorf tubes, and spin again for 5 min at 180×*g*. Finally aspirate the supernatant and keep the cell pellets. Cell pellets must be stored at −80 °C. Cell pellets generate during the passing can be considered as: Day 0 of differentiation (*see* Fig. 1), as cells are pluripotent (*see* **Note 13**).

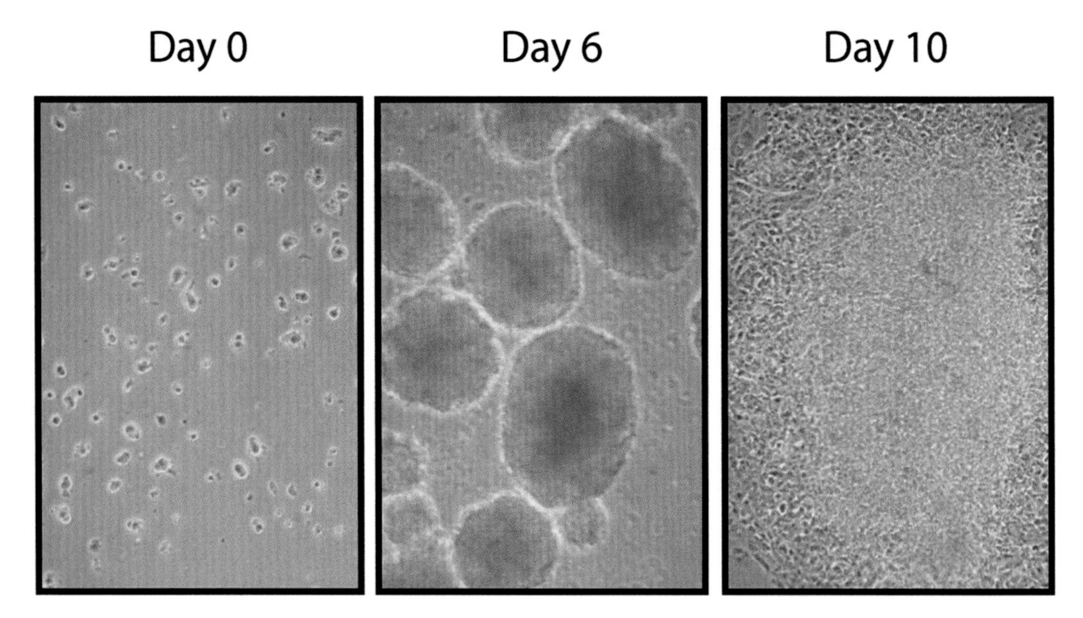

Fig. 1 Visualization of E14 mESCs during differentiation at Day 0, Day 6, and Day 10 of differentiation. During the proliferation phase—Day 0, mESCs remain pluripotent and are capable to propagate indefinitely in an undifferentiated state. At Day 6 of differentiation, embryoid bodies are easily observables as brownish spherical structures. At Day 10 of differentiation, embryoid bodies attach, spread and differentiated cells are observable at peripheries of EB

2. Differentiation set up—Day 0. When mESCs reach 70–80 % of confluence, differentiation can be set up. For one T75: aspirate the medium and wash once with 4 mL of PBS 1× then add 2 mL of 0.05 % Trypsin containing EDTA and incubate for 5 min at 37 °C. Add 4 mL of mESC differentiation medium, pipet up and down, and flush the flask several times to detach all the cells. Then collect everything in a 15 mL Falcon tube and spin the cells 5 min at 180 × *g*. Aspirate the supernatant, resuspend the cell pellet in 10 mL of mESC differentiation medium, plate the cells in two 94 × 16 mm triple vented petri dishes, and put them in the incubator. Do not manipulate the petri dishes until Day 3 of differentiation.
3. Differentiation—Day 3: changing medium. At Day 3 of differentiation, mESCs have started to aggregate to form embryoid bodies (EB). Put the EB in a 15 mL falcon tube using a 10 mL pipette. Add 5 mL of new differentiation medium in the petri dishes, which were containing the EB and put back the dishes in the incubator. Let the EB go down by gravity (3–5 min) in the 15 mL tubes, and aspirate the medium. Then resuspend them in 10 mL of mESC differentiation medium and replate them in the two original petri dishes. Put the dishes in the incubator (*see* **Note 14**).

4. Differentiation—Day 4 to Day 5: changing medium. If the color of the medium changes (from reddish to yellowish), mESC differentiation medium has to be changed. For one petri dish: put the EB in a 15 mL falcon tube using a 10 mL pipette. Let the EB sediment by gravity (3–5 min), and aspirate the medium. Then resuspend them in 10 mL of mESC differentiation medium and replate them in the original petri dish. Put the dish in the incubator (*see* **Note 14**).
5. Differentiation—Day 6: collect sample and attach the EB. At Day 6 of differentiation, EB are easily observable (Fig. 1). Put the EB in a 15 mL falcon tube using a 10 mL pipette from each petri dish, as previously explained (*see* **Note 14**). For one of the 15 mL Falcon tube: aspirate the supernatant, resuspend the EB in 10 mL of mESC differentiation medium, plate them in two new T75 gelatin-coated flasks, and put them in the incubator. For the other 15 mL Falcon tube: aspirate the supernatant, resuspend the EB in 2 mL of PBS 1×, distribute in two 1.5 mL Eppendorf tubes and spin again 5 min at 180 ×*g*. Finally aspirate the supernatant and keep the pellets at −80 °C. EB pellets generate at this step are considered as: Day 6 of differentiation.
6. Differentiation—Day 8. Change the mESC differentiation medium from the T75 flasks. Aspirate the medium and replace it with 10 mL of fresh medium.
7. Differentiation—Day 10. At Day 10 of differentiation, differentiated cells are visible around the attached EB (Fig. 1). For each T75 flask: aspirate the medium and wash once with 5 mL of PBS 1×, then add 2 mL of 0.05 % Trypsin containing EDTA and incubate for 5–10 min at 37 °C. Add 4 mL of mESC differentiation medium, pipette up and down and flush the flask several times to detach all the differentiated cells. Then transfer everything (both T75 contents) to one 15 mL Falcon tube and spin 5 min at 180 ×*g*. Aspirate the supernatant, resuspend the pellet in 4 mL of PBS 1×, distribute to four 1.5 mL Eppendorf tubes and spin again 5 min at 180 ×*g*. Finally aspirate the supernatant and keep the pellets at −80 °C. Cell pellets generate at this step are considered as: Day 10 of differentiation.

3.2 Total RNA Extraction, RNA Quality Check, and qRT-PCR

As RNA molecules are easily degraded, the samples must be kept on ice, while manipulated (unless otherwise indicated) and the centrifuge must be cooled down in advance to 4 °C. In order to avoid any kind of contamination by RNases, it is strongly advised for each total RNA extraction to use fresh aliquots of buffers or chemicals, to use RNase-free products (or at least autoclaved), and to clean the working area. In this part of the protocol, samples generated at Subheading 3.1 are used, meaning that total RNA is extracted from pellets corresponding to: a half of a T75 flask for

Day 0 of differentiation, a half of a 94 × 16 mm triple vented petri dish for Day 6 of differentiation and an half of a T75 flask for Day 10 of differentiation. To generate the qRT-PCR data presented here (Fig. 3), a specific real-time PCR device has been used allowing the utilization of 384 Multiwell Plate (*see* **Note 3**). In order to simplify the protocol and for a better comprehension, here we specify the composition for only one real-time PCR reaction. The gene *Rrm2* is used as a reporter gene [17].

1. Proceed directly with the pellet samples from Subheading 3.1. If the samples have been frozen at −80 °C, let them slowly thaw on ice for 30 min.
2. Total RNA extraction. Add 1 mL of TRIzol Reagent to each pellet, mix by pipetting until the pellet is completely dissolved and vortex strongly 10 s (*see* **Note 2**). Then add 200 μL of Chloroform, vortex for a minimum of 10 s and spin at 12,000 × *g* for 15 min at 4 °C. For each sample, transfer the supernatant to a new 1.5 mL Eppendorf tube (*see* **Notes 2** and **15**). Then add 600 μL of isopropanol, vortex each tube for a minimum of 10 s and spin at 12,000 × *g* for 30 min at 4 °C. Discard the supernatant, resuspend each pellet in 1 mL of ice-cold 70 % ethanol, and spin at 12,000 × *g* for 10 min at 4 °C. Finally, discard the supernatant (*see* **Note 16**), air-dry the pellet for 5 min, and resuspend the RNA pellet in 120 μL of autoclaved water. The samples must be stored at −80 °C.
3. Concentration measurement and quality assessment. Measure the total RNA extracts concentration of the samples and verify the quality by loading from each sample 1 μg of total RNA on a 1 % agarose gel with an ethidium bromide final concentration of 0.1 μg/mL. Do not forget to add the right amount of 6× loading dye (to reach a final concentration of 1×) to each 1 μg of total RNA extract before loading. Then run the gel 30 min at 110 V (*see* **Note 17**). If the total RNA are not degraded, two clear bands must appears on the gel, each for one of the ribosomal RNA species: 28S and 18S (Fig. 2) (*see* **Note 18**). If the quality is good, then proceed directly to the following step or the samples can be stored at −80 °C. If smears appear on the gel then the total RNAs are degraded and total RNA extraction (Subheading 3.2, **step 2**) has to be repeated on new pellet samples.
4. DNase treatment. If the samples have been frozen at −80 °C, let them slowly thaw on ice. For each sample, prepare two PCR tubes, each containing 1 μg of total RNA extract in a final volume of 6 μL of autoclaved water (*see* **Note 19**). Then add to each tube:

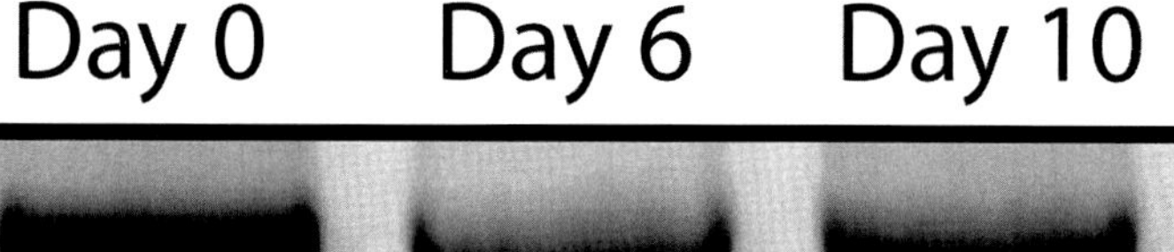

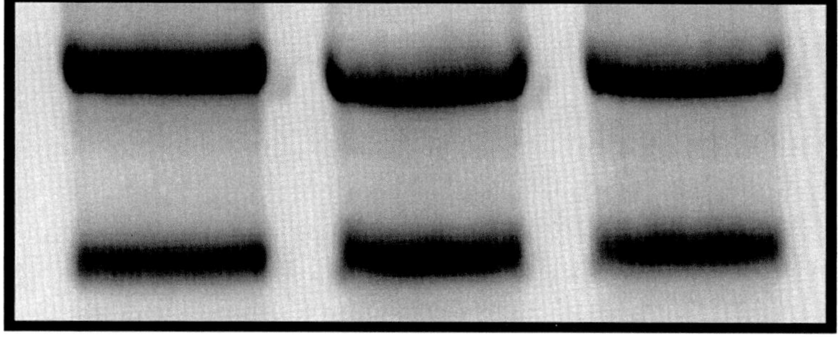

Fig. 2 RNA quality control. One microgram of each total RNA extract sample has been run on a 1 % agarose gel for 30 min at 110 V. Intact total RNA will have two clear bands corresponding to the 28S and 18S ribosomal RNA species, whereas partially degraded RNA will have a smeared appearance

Stock	For 1 reaction (μL)
Total RNA sample	6
RQ1 RNase-Free DNase buffer	2
RQ1 RNase 10× reaction buffer	0.9
RiboLock™ RNase Inhibitor	0.1
Total (μL)	9

Incubate at 37 °C for 1 h in a thermo-cycler. Then add 1 μL of Stop Solution to each tube and incubate for 10 min at 65 °C.

5. Reverse transcription. Add to each tube, 1 μL of Random Hexamer Primers (100 μM) and incubate for 5 min at 95 °C. For each sample, fill one PCR tube with RT mix and the other one with No_RT mix, which correspond to a negative control, according to the following composition:

Stock	RT mix—for 1 PCR reaction (μL)	No_RT mix—for 1 PCR reaction (μL)
GoScript Reverse Transcriptase buffer	1	0
GoScript 5× Reaction buffer	4	4
$MgCl_2$ 25 mM	2	2
dNTP mix 10 mM	2	2
RiboLock RNase Inhibitor	0.5	0.5
Autoclaved water	0.5	1.5
Total (μL)	10	10

Place the PCR tubes into a thermo-cycler and run the following program (*see* **Note 20**):

Step 1: 25 °C for 5 min
Step 2: 42 °C for 60 min
Step 3: 72 °C for 10 min
Step 4: 12 °C HOLD

Finally add 80 μL of autoclaved water to each sample (1:5 dilution). The samples must be stored at −20 °C.

6. Validation PCR. For each sample, prepare a PCR reaction as followed:

Stock	**For 1 PCR reaction (μL)**
Taq Polymerase (5 U/μL)	0.2
5× Taq reaction buffer (with Mg)	4
Rrm2 forward primer 100 μM	0.1
Rrm2 reverse primer 100 μM	0.1
dNTP mix 10 mM each	0.2
cDNA sample (1:5)	2
Autoclaved water	13.4
Total (μL)	20

Do not forget to perform a negative water control, containing autoclaved water instead of cDNA. Put the PCR tubes into a thermo-cycler and run the following PCR program:

Step 1: 94 °C for 5 min
Step 2: 94 °C for 30 s
Step 3: 60 °C for 30 s
Step 4: 72 °C for 30 s
Step 5: GOTO step 2, 34 times
Step 6: 72 °C for 5 min
Step 7: 12 °C HOLD

Subsequently prepare a 2 % agarose gel with an ethidium bromide final concentration of 0.1 μg/mL. For each sample, load the gel with 10 μL of the PCR and the loading dye. Store the rest at 4 °C. Reserve one slot to load a DNA Ladder and run the electrophoresis at 110 V for 1 h. A band at 100 bp must appear for the RT samples whereas nothing should appear for the No_RT samples and the negative water control. If a 100 bp band is observed for the No_RT samples, this means that the samples are contaminated and in that case, the steps from

Subheading 3.2, **step 2** until Subheading 3.2, **step 5** must be repeated for the contaminated samples. If the cDNA samples produced at Subheading 3.2, **step 5** are clean, the protocol can be continued.

7. Real-time PCR. Here the composition is specified for one reaction (*see* **Note 3**):

Stock	For 1 PCR reaction (μL)
SYBR Green mix	6.5
Specific forward primer 10 μM	0.5
Specific reverse primer 10 μM	0.5
cDNA sample	2
Autoclaved water	0.5
Total (μL)	10

The specific primers for qPCR are at a concentration of 10 μM and are described in Table 1.

Put the reactions into a real-time PCR device and run the following program:

Step 1: 95 °C for 10 min
Step 2: 99 °C for 15 s
Step 3: 60 °C for 30 s
Step 4: GOTO step 2, 39 times

8. Real-time qPCR results analysis. Differences between samples and controls were calculated based on the $2^{-\Delta CT}$ method. Calculation and graphics have been performed using Excel. *Rrm2* transcript levels were used for normalization. LINE-1 and pluripotency factors transcripts are highly expressed in undifferentiated cells (Day 0 of differentiation) (Fig. 3). On the other hand, less amount of these transcripts are detected at Day 6 and Day 10 of differentiation, which is coherent with the repression of LINE-1 by DNA methylation [2] and the lost of the mESCs pluripotent makers that occur during differentiation (Fig. 3).

3.3 Total Protein Extraction and Western Blot

As proteins are easily degraded, the samples must be kept on ice, while manipulated (unless otherwise indicated) and the centrifuge must be cooled down in advance to 4 °C. In this part of the protocol, samples generated at the Subheading 3.1 are used, meaning that total protein is extracted from pellets corresponding to: an half of a T75 flask for Day 0 of differentiation, an half of a 94 × 16 mm triple vented petri dish for Day 6 of differentiation and an half of a T75 flask for Day 10 of differentiation. The three

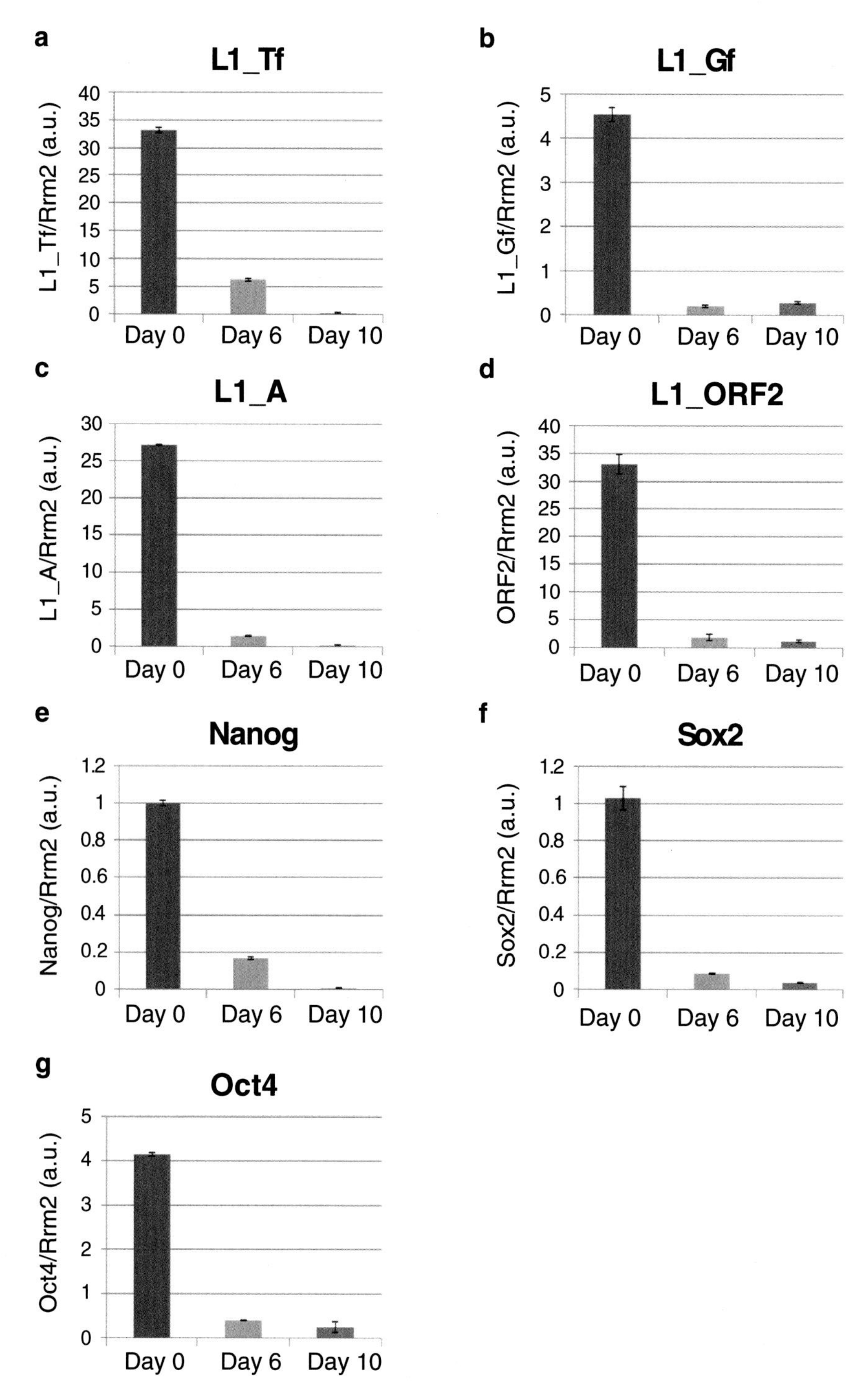

Fig. 3 qRT-PCR analysis of LINE-1 element and pluripotency factors messenger RNA levels during mESCs differentiation. (**a–d**) L1 messenger RNAs from all subfamilies are strongly expressed at Day 0, whereas less amounts are found at Day 6 and Day 10 of differentiation, which corresponds to the LINE-1 repression occurring during differentiation. (**e–g**) Same observations can be done for pluripotency factors: messenger RNA levels decreased significantly during differentiation, as expected

proteins detected in this protocol: ORF1p, OCT4, and α-tubulin have very close molecular weights, around 40 kDa, 42 kDa, and 55 kDa, respectively. It is therefore not possible to use the same PVDF membrane to detect the three proteins by cutting it. In order to overcome this problem, in this protocol the membrane is stripped between each revelation, which allows the detection of the three different proteins from the same Western blot. Alternatively, the experiment can be performed on independent gels from the same total protein extracts. Here, the structural protein α-tubulin is used as reference and loading control: as its expression does not differ between the samples, this protein must be detected in each sample in equal amount if the gel has been loaded correctly. All the membrane-washing steps are performed at room temperature under agitation using 3D gyratory rockers at 40 rpm (unless otherwise indicated).

1. Proceed directly with the pellet samples from Subheading 3.1. If the samples have been frozen at −80 °C, let them slowly thaw on ice.
2. Total protein extraction. Prepare 1 mL of 1× Completed NP40 lysis buffer by mixing 100 μL of 10× Completed NP40 lysis buffer with 900 μL of NP40 lysis buffer and keep it on ice. Add 50 μL of 1× Completed NP40 lysis buffer to each sample and mix by pipetting. Then sonicate each sample two times for 10 s at 10 % amplitude using a sonication device (*see* **Notes 4** and **21**). Centrifuge the samples 10 min at 9500 ×*g* at 4 °C. Finally, transfer the supernatant in new 1.5 mL Eppendorf tubes. Total protein extracts must be stored at −80 °C.
3. Determine protein concentration (*see* **Note 22**). For each sample, prepare aliquot of 10 μg of protein and complete to 15 μL with 1× Completed NP40 lysis buffer (*see* **Note 23**). Then add 5 μL of 4× Laemmli Buffer containing 2-mercaptoethanol. Finally, boil the samples for 5 min at 95 °C. The aliquots can be stored at −20 °C if not loaded subsequently on the acrylamide gel.
4. Cast a 1.5 mm thick 10 % acrylamide gel. Mix the reagents in that specific order to cast one small 1.5 mm gels (10.1 cm × 8.2 cm) (*see* **Note 24**):

	Resolving	**Stacking**
Acrylamide–bis-acrylamide 37.5:1 (40 % solution)	2.25 mL	625 μL
Autoclaved water	4.35 mL	3.67 mL
1.5 M Tris–HCl, pH 8.8 buffer	2.25 mL	0
1 M Tris–HCl, pH 6.8 buffer	0	625 μL

	Resolving	Stacking
20 % SDS solution	50 μL	25 μL
10 % APS solution	90 μL	50 μL
TEMED	10 μL	5 μL
Total (mL)	9 mL	10 mL

Pour first the resolving part of the gel but do not fill the cast chamber completely, stop 2 cm before the top of the cast chamber. Fill the 2 cm of the chamber left with 500 μL of isopropanol (*see* **Note 25**). Wait until the resolving gel is polymerized (*see* **Note 26**). Then remove the isopropanol, wash quickly with autoclaved water, pour the stacking gel and immediately insert the 1.5 mm comb. Wait until the stacking gel is polymerized (*see* **Note 26**).

5. Migrate the proteins. Place the gel in the companion running module, put it in the tank, and fully fill it with running buffer. Remove the 1.5 mm comb (*see* **Note 27**). For each sample, load the entire aliquot prepared at Subheading 3.2, **step 3** (10 μg of protein per well per sample). Keep one well free to load a Prestained Protein Ladder (*see* **Note 28**). Close the tank and run the gel at 90 V until the samples go through the stacking gel and reach the resolving gel (around 20 min). Then run it 2 h at 110 V.
6. Transfer of the proteins. For one gel: activate the PVDF membrane 5 min in 80 % ethanol (*see* **Notes 7** and **29**) and then rinse it in transfer buffer. Meanwhile, soak two sponges and two Whatman papers in transfer buffer. Prepare the transfer cassette, open it and on the part that will be in contact with the cathode lay down on this order: a sponge, a Whatman paper, the acrylamide gel (from Subheading 3.2, **step 5**), the activated membrane, a Whatman paper, and a sponge. When the activated membrane is put down on the acrylamide gel, be sure that there is no air bubbles between the membrane and the gel. Air bubbles can be removed using a roll. Close the transfer cassette, insert it in the trans-blot module and put this module in the tank. Fill the tank with transfer buffer and run the transfer at 110 V for 1 h 30 min (*see* **Note 30**).
7. Block the membrane. Remove the transfer cassette from the trans-blot module, and carefully open it. Verify that the Prestained Protein Ladder markers have been transferred to the membrane. Then incubate the membrane for 30 min in 5 % milk TBST 1× at room temperature under agitation (*see* **Note 31**). Then wash the membrane three times for 10 min in TBST 1× buffer.
8. Anti-L1_ORF1 antibody. Incubate the membrane in the anti-L1_ORF1 antibody dilution overnight at 4 °C under agita-

tion. Then wash the membrane three times for 10 min in TBST 1× buffer. Incubate the membrane in the corresponding secondary antibody solution: Anti-rabbit IgG, HRP-linked antibody at the indicated dilution, for 1 h at room temperature under agitation. Finally, wash the membrane three times for 10 min in TBST 1× buffer (*see* **Note 32**). Then reveal the membrane by following the revelation kit instructions (*see* **Note 10**). A band around 40 kDa appears if the ORF1 protein is detected (Fig. 4).

9. Membrane stripping. Wash the membrane one time 10 min in TBST 1× buffer. Incubate the membrane for 30 s in methanol, then wash it two times for 5 min and two times for 10 min (four washes in total) in stripping membrane buffer. Perform an additional wash of 5 min in autoclaved water. Then incubate the membrane for 30 min in 5 % milk TBST 1× and finally wash it one time 10 min in TBST 1× buffer. The membrane is ready to be reused (*see* **Note 33**).
10. Anti-OCT4 antibody. Incubate the membrane in the OCT-3/4 antibody (rabbit) dilution overnight at 4 °C under agitation. Then wash the membrane three times for 10 min in

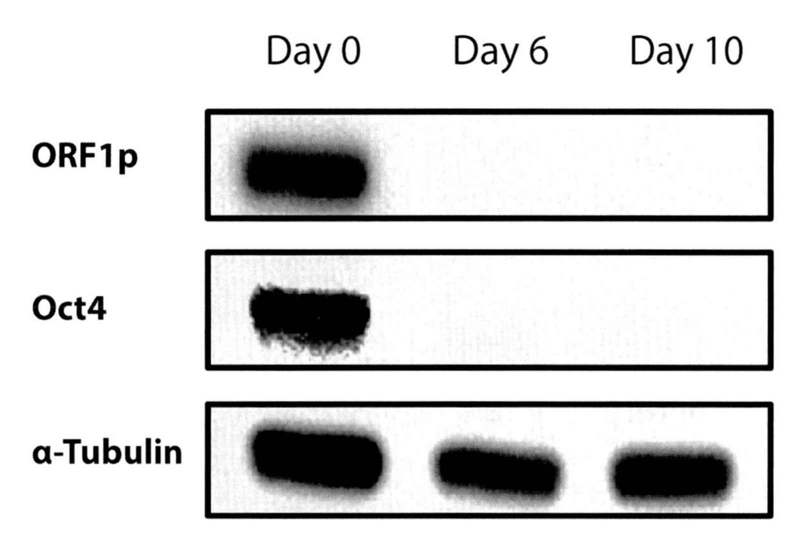

Fig. 4 Western analysis of ORF1p, Oct4, and α-tubulin protein levels during mESCs differentiation. Ten micrograms of total protein extract has been loaded per well per sample. For ORF1p a band appears around 40 kDa at Day 0 but does not appear at Day 6 and 10 of differentiation that corresponds to the LINE-1 repression occurring during differentiation. Similarly for Oct4, a band appears around 40 kDa at Day 0 corresponding to a strong expression of Oct4 protein that is consistent with the pluripotent features of mESCs at Day 0. And Oct4 protein is no more detectable at Day 6 and Day 10 of differentiation as this pluripotency factor is no more expressed in differentiated cells. For α-tubulin, a band appears around 55 kDa for each sample

TBST 1× buffer. Incubate the membrane in the corresponding secondary antibody solution: Anti-rabbit IgG, HRP-linked antibody, for 1 h at room temperature under agitation. Finally, wash the membrane three times for 10 min in TBST 1× buffer. Then reveal the membrane by following the revelation kit instructions (*see* **Note 10**). A band around 42 kDa appears if the Oct4 protein is detected (Fig. 4).

11. Membrane stripping. Wash the membrane one time for 10 min in TBST 1× buffer. Incubate the membrane for 30 s in methanol, then wash it two times for 5 min and two times for 10 min (four washes in total) in stripping membrane buffer. Perform an additional wash of 5 min in autoclaved water. Then incubate the membrane for 30 min in 5 % milk TBST 1× and finally wash it one time for 10 min in TBST 1× buffer. The membrane is ready to be reused (*see* **Note 33**).
12. α-tubulin antibody (*see* **Note 8**). Incubate the membrane in the anti-α-tubulin antibody (mouse) dilution overnight at 4 °C under agitation. Then wash the membrane three times for 10 min in TBST 1× buffer. Incubate the membrane in the corresponding secondary antibody solution: Anti-mouse IgG HRP-linked antibody, for 1 h at room temperature under agitation. Finally, wash the membrane three times for 10 min in TBST 1× buffer. Then reveal the membrane by following the revelation kit instructions (*see* **Note 10**). A band around 55 kDa appears if the α-tubulin protein is detected (*see* Fig. 4).

4 Notes

1. Gelatin-coated flasks or plates can be prepared in advance: pour few milliliters of 0.2 % gelatin on the bottom of the flask, distribute the gelatin homogeneously, and keep them in the incubator (no more than 2 days to avoid evaporation). Five minutes before usage take the plate out of the incubator and redistribute the gelatin homogeneously and do not forget to remove the gelatin solution before to add the cells.
2. TRIzol reagent is extremely toxic. This product must be manipulated under a chemical hood and the use of a lab-coat (forearms must be covered), gloves, and safety glasses is mandatory. Every solid product contaminated by TRIzol reagent (pipette tips, Eppendorf tubes, …) is considered as a special waste and must be disposed of in a specific container stored under the chemical hood.
3. To generate the qRT-PCR data presented in this protocol (Fig. 3), the LightCycler ® 480 II device (Roche) has been used. This device allows the use of 384 Multiwell. With this

kind of plate, each sample for each condition is processed in triplicates, which allows assessing and minimizing pipetting errors. It is advised to use an optimized qPCR kit for this real-time PCR device, like the KAPA SYBR ® FAST qPCR Kit Optimized for Light Cycler ® 480 (KAPA Biosystems).

4. If no sonication device is available, an alternative method consists of using a rotation wheel. Put the samples on a rotation wheel at 7 rpm at 4 °C (place the rotation wheel in a 4 °C room) during 15 min.
5. When not polymerized, acrylamide is toxic. Manipulate this product with caution.
6. Adjust the pH of the different buffers using: 1 M HCl or 1 M NaOH solutions depending on the needs.
7. Do no touch directly the PVDF membrane with fingers. Always manipulate it with tweezers.
8. In this protocol the anti-α-tubulin antibody (mouse) is used to detect the presence of the tubulin protein. It is a control performed to be sure that the same amount of total protein extract has been loaded in the acrylamide gel. An alternative method consists of performing a Coomassie Blue staining of the PVDF membrane. But after a Coomassie Blue staining, the PVDF membrane cannot be reused.
9. It is recommended to prepare an aliquot of secondary antibody for each primary antibody: anti-L1_ORF1 and OCT-3/4 antibody. Use the same aliquot of secondary antibody with the same primary antibody in order to avoid any kind of contamination. These aliquots have to be stored at −20 °C and can be used several times.
10. To generate the Western blot data presented in this protocol (Fig. 4), two membrane revelation kits have been used. For the anti-L1_ORF1 and the anti-α-tubulin antibodies are really efficient, with 10 μg of total proteins extract loaded, a strong signal is observed after 10 and 2 s respectively of exposure using the Clarify Western ECL substrate (Bio-Rad) kit. When revealing the membrane for the anti-α-tubulin antibody, do not incubate the membrane too long (no more than 2 min) in the kit reagents at the risk of burning the PVDF membrane. For the OCT-3/4 antibody, with 10 μg of total proteins extract loaded, a strong signal is observed after 120 s of exposure using the SuperSignal West Femto Maximum Sensitivity Substrate (Thermo Scientific) kit.
11. Mouse ESCs must never be confluent.
12. In this protocol, the mESCs are passed to the sixth (1/6). After passage, it should take 2 or 3 days to mESCs to reach

70–80 % of confluence again. The cells can be diluted more if needed.

13. Mouse ESCs in proliferation phase can be frozen for long-term storage and thawed to be cultured again. Here is the procedure for freezing and thawing:

    To freeze one T75 flask: aspirate the medium and wash once with 4 mL of PBS 1× then add 2 mL of 0.05 % trypsin containing EDTA and incubate for 5 min at 37 °C. Add 4 mL of mESC differentiation medium, pipet up and down and flush the flask several times to detach all the cells. Then collect everything in a 15 mL Falcon tube and spin the cells 5 min at 180 × *g*. Aspirate the supernatant, resuspend the cell pellet in 4 mL of freezing medium (special selected batch of fetal bovine serum tested for optimal growth of mESCs supplemented with 10 % of dimethyl sulfoxide (tissue culture grade)). Then split into four screw cap cryotubes, close the tubes, put them in a freezing container, close it, and put the container at −80 °C. Twenty four hours later, transfer the cryotubes to liquid nitrogen.

    To thaw one cryotube: put out the cryotube from liquid nitrogen and transfer it immediately to a 37 °C water bath. Incubate for 5 min, transfer the entire content in a 15 mL Falcon tube containing 3 mL of mESC differentiation medium, and spin cells 5 min at 180 × *g*. Then aspirate the supernatant, resuspend the cell pellet in 10 mL of mESC proliferative medium, and plate the cells in a new T75 gelatin coated flask. Finally put the cells in the incubator and change the medium 24 h after the thawing. Dimethyl sulfoxide is toxic to the cells, so the freezing and thawing procedures must be done as fast as possible.

14. Always manipulate the EB with 10 mL pipettes in order to reduce the risk of disrupting their structure. Handle embryoid bodies carefully and avoid any formation of air bubbles during the pipetting.

15. After centrifugation three phases are observable: one pinkish lower phase, one whitish middle phase, and one clear upper phase. Handle tubes carefully in order to not perturb the phases. Only the upper phase is required for total RNA extraction. When taking out the upper phase, pay attention to not touch any of the other phases. If the phases have been disturbed, spin the samples at 12,000 × *g* for 10 min at 4 °C and try again.

16. At this step, a pellet corresponding to the precipitated total RNA should be observable at the bottom of the tubes. Handle the tubes carefully in order not to displace the pellet. If the pellet is not visible, assume there is one. During the removal

of the supernatant avoid touching the pellet and try to remove as much supernatant as possible.

17. Horizontal electrophoresis cell and relative products can be contaminated by RNases. Before casting and running the gel, wash everything (fist wash with liquid soap, second wash with 70 % ethanol), otherwise total RNA can be degraded during the migration.
18. In this protocol, messenger RNA quality is assessed by electrophoresis. This method relies on the assumption that ribosomal RNA quality and quantity reflect the one of the messenger RNA population. Other methods can be used.
19. If the concentrations of total RNA extracts are too low and do not allow to have 1 μg of total RNA into 6 μL, then equilibrate all the samples to the lowest sample concentration. But do not get under 0.5 μg of total RNA into 6 μL. If it is still not possible, it means that the concentrations are too low, and the total RNA extraction from a new pellet have to be redone for the concerned samples.
20. The program can vary depending on the Reverse Transcriptase Kit used and it might be modified according to the needs.
21. Keep the samples on ice even during sonication.
22. To generate the Western blot data presented in this protocol (Fig. 4), the total protein extracts concentrations have been measured using the DC Protein Assay Reagents Package (Bio-Rad).
23. If the concentrations of total protein extracts are too low and do not allow to have 10 μg of total protein, then equilibrate all the samples to the lowest sample concentration. But do not get under 5 μg of total protein. If it is still not possible, it means that the concentrations are too low, and the total protein extraction from a new pellet have to be redone for the concerned samples.
24. TEMED is the reagent responsible of the acrylamide polymerization; it has to be added to the solution last. Once the TEMED has been added, act fast: mix the preparation by pipetting up and down using a 10 mL pipette and fill the cast chamber.
25. Adding isopropanol to the top of the resolving gel will allow to remove the air bubbles formed during the pouring of the resolving mix and to have a smooth surface to cast the stacking gel afterward.
26. It is advised to prepare the resolving and stacking mixes in 15 mL Falcon tubes. When one of the mix has been poured, keep the remaining mix in the Falcon tube. When the mix is

polymerized in the Falcon tube, it means that it is also polymerized in the cast chamber.

27. After having removed the 1.5 mm comb, some polymerized acrylamide residues can stuck or fall into the wells. That can result in problems during the migration. It is advised to wash each well with a syringe plus needle filled with running buffer.
28. If there are empty wells on the acrylamide gel, fill them with 5 μL of 4× Laemmli Buffer containing 2-mercaptoethanol in order to avoid to have a "smiling gel" during the migration.
29. PVDF 0.45 μm membranes can also be activated with methanol. Put the membrane in methanol for 30 s then wash it for 5 min in transfer buffer.
30. During the transfer, transfer buffer temperature can increase. It is advised to add an ice block into the tank and to change it for a new one after 45 min of transfer.
31. After having checked that proteins have been transferred to the membrane, it is advised to mark the membrane (cut on of the edge) in order to know on which side are the proteins. This will facilitate the handling of the membrane afterwards.
32. This protocol has also been applied with a commercial antibody targeting mouse ORF1p: LINE-1 (M-300, Santa Cruz) antibody (rabbit). It has been used 1:1000 diluted. The results obtained (data not shown) are not comparable with the Anti-L1_ORF1 antibody (gift of Dr. A. Bortvin, Carnegie Institution for Science, USA) (rabbit). The commercial LINE-1 antibody (M-300) is not specific, as several bands appear for all samples (Day 0, Day 6 and Day 10 of differentiation) and also able to bind to some Prestained Protein Ladder, which makes the analysis difficult.
33. Efficiency of stripping can be checked by revealing the membrane after the stripping, using a by following the revelation kit. If there is no signal detected, then the membrane have been correctly stripped. If a signal is still detected, the membrane has to be stripped again.

Acknowledgments

We thank Dr. Tobias Beyer and the Ciaudo laboratory for the critical reading of the manuscript and for fruitful discussions. We also thank Dr. A. Bortvin for the gift of the L1_ORF1 antibody. This work was supported by a core grant from ETH-Z (supported by Roche). M.B. is supported by a PhD fellowship from the ETH-Z foundation (ETH-21 13-1).

References

1. Waterston RH, Lindblad-Toh K, Birney E et al (2002) Initial sequencing and comparative analysis of the mouse genome. Nature 420:520–562. doi:10.1038/nature01262
2. Bodak M, Yu J, Ciaudo C (2014) Regulation of LINE-1 in mammals. Biomol Concepts 5:409–428. doi:10.1515/bmc-2014-0018
3. Engels WR, Johnson-schlitz DM, Eggleston WB, Svedt J (1990) High-frequency P element loss in Drosophila is homolog dependent. Cell 62:515–525
4. Lander ES, Linton LM, Birren B et al (2001) Initial sequencing and analysis of the human genome. Nature 409:860–921. doi:10.1038/35057062
5. Ostertag EM, Kazazian HH (2001) Twin priming: a proposed mechanism for the creation of inversions in L1 retrotransposition. Genome Res 11:2059–2065. doi:10.1101/gr.205701.then
6. Naas TP, DeBerardinis RJ, Moran JV et al (1998) An actively retrotransposing, novel subfamily of mouse L1 elements. EMBO J 17:590–597. doi:10.1093/emboj/17.2.590
7. Goodier JL, Ostertag EM, Du K, Kazazian HH (2001) A novel active L1 retrotransposon subfamily in the mouse. Genome Res 11:1677–1685. doi:10.1101/gr.198301
8. Loeb DD, Padgett RW, Hardies SC et al (1986) The sequence of a large LlMd element reveals a tandemly repeated 5′ end and several features found in retrotransposons. Mol Cell Biol 6:168. doi:10.1128/MCB.6.1.168.Updated
9. Howlett SK, Reik W (1991) Methylation levels of maternal and paternal genomes during preimplantation development. Development 113:119–127
10. Fadloun A, Le Gras S, Jost B et al (2013) Chromatin signatures and retrotransposon profiling in mouse embryos reveal regulation of LINE-1 by RNA. Nat Struct Mol Biol 20:332–338. doi:10.1038/nsmb.2495
11. Van den Hurk JA, Meij IC, Seleme MDC et al (2007) L1 retrotransposition can occur early in human embryonic development. Hum Mol Genet 16:1587–1592. doi:10.1093/hmg/ddm108
12. Kano H, Godoy I, Courtney C et al (2009) L1 retrotransposition occurs mainly in embryogenesis and creates somatic mosaicism. Genes Dev 23:1303–1312. doi:10.1101/gad.1803909
13. Lee HJ, Hore TA, Reik W (2014) Reprogramming the methylome: erasing memory and creating diversity. Cell Stem Cell 14:710–719. doi:10.1016/j.stem.2014.05.008
14. Boyer LA, Mathur D, Jaenisch R (2006) Molecular control of pluripotency. Curr Opin Genet Dev 16:455–462. doi:10.1016/j.gde.2006.08.009
15. Martello G, Smith A (2014) The nature of embryonic stem cells. Annu Rev Cell Dev Biol 30:647–675. doi:10.1146/annurev-cellbio-100913-013116
16. Dunn S-J, Martello G, Yordanov B et al (2014) Defining an essential transcription factor program for naïve pluripotency. Science 344:1156–1160. doi:10.1126/science.1248882
17. Ciaudo C, Jay F, Okamoto I et al (2013) RNAi-dependent and independent control of LINE1 accumulation and mobility in mouse embryonic stem cells. PLoS Genet 9:e1003791. doi:10.1371/journal.pgen.1003791

Chapter 17

Immunodetection of Human LINE-1 Expression in Cultured Cells and Human Tissues

Reema Sharma, Nemanja Rodić, Kathleen H. Burns, and Martin S. Taylor

Abstract

Long interspersed element-1 (LINE-1) is the only active protein-coding retrotransposon in humans. It is not expressed in somatic tissue but is aberrantly expressed in a wide variety of human cancers. ORF1p protein is the most robust indicator of LINE-1 expression; the protein accumulates in large quantities in cellular cytoplasm. Recently, monoclonal antibodies have allowed more complete characterizations of ORF1p expression and indicated potential for developing ORF1p as a clinical biomarker. Here, we describe a mouse monoclonal antibody specific for human LINE-1 ORF1p and its application in immunofluorescence and immunohistochemistry of both cells and human tissues. We also describe detection of tagged LINE-1 ORF2p via immunofluorescence. These general methods may be readily adapted to use with many other proteins and antibodies.

Key words LINE-1 ORF1p, p40, Immunofluorescence, Immunohistochemistry, Tissue microarray, Tumor marker, Biomarker

1 Introduction

Long interspersed element-1 (LINE-1) activity is a significant determinant of our genome sequence. LINE-1 and other sequences mobilized by LINE-1-encoded proteins together constitute about 35 % of our DNA [1]. Ongoing LINE-1-mediated retrotransposition makes these sequences important sources of genetic structural variation in humans as well as a source of instability in cancer genomes [2–6].

LINE-1 is the only active protein-coding family of transposable elements in humans. The first of its two open reading frames encodes ORF1p, a 40 kDa protein with RNA-binding activities required for LINE-1 retrotransposition [7–9]. The second, ORF2p, encodes the endonuclease and reverse transcriptase activities which are also essential for LINE-1 propagation [10, 11] and is expressed at extremely low levels [12].

Jose L. Garcia-Pérez (ed.), *Transposons and Retrotransposons: Methods and Protocols*, Methods in Molecular Biology, vol. 1400, DOI 10.1007/978-1-4939-3372-3_17,

ORF1p comprises an N-terminal domain, a coiled-coil domain (CCD), an RNA recognition motif (RRM), and a C-terminal domain. The protein associates with LINE-1 ribonucleoprotein (RNP) complexes as a trimer brought together through CCD interactions [13–15], and two to ten trimers associate with each LINE-1 RNP [16]. LINE-1 ORF1p is an abundant protein in cells expressing LINE-1. It can be readily seen on Coomassie gels of total protein when cells are transfected with LINE-1-expressing constructs under the native promoter; it may be a dominant band using CMV or CAG promoters. In terms of total protein expression, ratios of ORF1p:ORF2p are in the range of 1000–10,000:1. Using immunoprecipitation and two distinct quantitative staining methods as well as quantitative mass spectrometry, the ratio of ORF1p:ORF2p in RNPs in these cells is between 6:1 and 30:1. Together, these data show that a substantial portion of ORF1p is not engaged in the LINE-1 RNP [16].

To develop a mouse monoclonal antibody to detect LINE-1, 15 peptides from the human LINE-1 ORF1p sequence (AAB60344.1, 338 AA) were selected for immunization of BALB/c mice. The antibody referenced in this chapter recognizes amino acids 35–44 (MENDFDELRE) of ORF1p [17]. This is a region of the protein where mouse and human LINE-1 sequences diverge; it precedes the CCD and has not been solved in protein crystallography studies [15]. Using immunohistochemistry, we found nearly half of human cancers stain positively for ORF1p, with immunoreactivity in some common cancers approaching 100 % of cases. No staining was observed in the cognate normal tissues [17].

2 Materials

2.1 For Cell Culture IF

1. Cover slips, size 22 × 22 × 1.
2. Slides, size 25 × 75 × 1.
3. Forceps: Electron Microscopy Services Style 4A.
4. Parafilm M: 4″ width.
5. Fibronectin from human plasma.
6. Poly-L-lysine.
7. Hoechst 33342 trihydrochloride, trihydrate: 10 mg/mL Solution in water.
8. Bovine serum albumin (BSA), high avidity, 30 % solution.
9. *PBS/glycine*: We make standard 10× PBS and a 100× glycine-azide stock. Sodium-only or sodium-potassium PBS recipes may be used.

10. *100× Glycine*: 1 M glycine plus 2 % sodium azide, adjust pH to 7.4 with NaOH or HCl. This is stable at room temperature for months to years.
11. *Mounting media:* 0.1 M *N*-propyl-gallate in glycerol. Add *N*-propyl-gallate. Prepare and dissolve at 65 °C overnight. Stable at room temperature for months to years.
12. *Fixative*: 3 % paraformaldehyde (PFA) in PBS (makes 100 mL) (*see* **Note 1**).
    (a) Weigh 3 g PFA, and then add 80 mL ddH_2O and 250 μL 1 N NaOH.
    (b) Heat at 50 °C in a water bath to dissolve completely.
    (c) Add 10 mL 10× PBS and adjust volume to 100 mL with ddH_2O.
    (d) Check pH, and adjust to 7.4 with NaOH or HCl if needed.

2.2 For Tissue IHC and IF

1. Xylene.
2. Absolute ethanol.
3. TBS: 50 mM Tris–HCl, pH 7.4, 150 mM NaCl.
4. TBST: TBS plus 0.05 % (v/v) Tween 20.
5. BSA.
6. BSA-TBST: 1 % BSA (w/v) dissolved in TBST.
7. Citrate antigen retrieval buffer: 10 mM Sodium citrate, pH 6.0, 0.05 % Tween 20.
8. EDTA antigen retrieval buffer: 1 mM EDTA, pH 8.0, 0.05 % Tween 20.
9. 3 % (v/v) Hydrogen peroxide in dH_2O.
10. For signal amplification (option one): Histostain-SP Kit Broad Spectrum Kit (Life Technologies).
11. For signal amplification (option two): EnVision HRP Mouse (DAB+) or Rabbit (DAB+).
12. For HRP-catalyzed amplified immunofluorescence using tyramide: Perkin Elmer™ TSA Plus Fluorescein or Cy-5.

2.3 Antibodies

Mouse anti-LINE-1 ORF1p clone 4H1 (10 mg/mL), stored in 50 % glycerol at −20 °C.

Rabbit anti-LINE-1 ORF1p clone JH73 (gift of Jeff Han).

Alexa 488 GOAT Anti-Mouse.

Alexa 568 GOAT Anti-Rabbit.

Rabbit Anti-GFP polyclonal antibody.

Rabbit Anti-Flag polyclonal antibody.

Mouse Anti-Flag M2 (*see* **Note 2**).

3 Methods

3.1 Immunofluorescence (IF) Detection of LINE-1 ORF1p and ORF2p from Human Cell Lines

Immunofluorescent staining (IF) for LINE-1 allows sensitive detection of LINE-1 expression and demonstrates subcellular co-localization of both ORF proteins. LINE-1 proteins accumulate in the cytoplasm and are relatively excluded from cell nuclei. For reasons that are not understood, we find that expression of tagged ORF2p in HEK-293T and HeLa cell lines varies greatly from cell to cell in a seemingly stochastic manner [16]. IF was performed in transiently transfected or puromycin-selected HeLa and HEK-293T cells grown on glass cover slips using both constitutive and tetracycline-inducible promoters, but this standard IF protocol works well with many cultured cell lines and antigens. Many IF protocols use hundreds of μL of antibody solution per sample and often take days to complete. With this protocol [18], slides can be visualized in less than 2 h and, by using small droplets (50 μL) of antibody on Parafilm M, it saves limited and costly reagents. For ORF1p, untagged protein can be visualized using either of the two recently developed monoclonal antibodies, which were verified in comparison with Flag-tagged ORF1p. 3× Flag-tagged and GFP-tagged ORF2p can be detected readily; robust immunodetection of untagged ORF2p is currently a technical barrier in the field. Example images are shown in Fig. 1.

Many variations on the protocol below are effective. Here, we present our standard protocol for transient transfection, along with some options for customization.

3.1.1 Timeline

Day 1: Plating and Transfection

9 a.m.: Plating

5 p.m.: Transfection

Day 2: Permeabilization, Fixation, and Immunostaining

Considerations: For optimal results, plating should occur >8 h before transfection. For maximum transfection efficiency, cells can be plated 16–24 h beforehand, but because this results in fewer single cells under the scope, we prefer using the 8-h interval described. After 3 days on glass, cultured cells are often very clumpy and their borders may be difficult to distinguish. Therefore, for staining more than 18–24 h after transfection, plating on cover slips can be done after transfection. ORF2p expression peaks at 20–24 h post-transfection/induction and can be very difficult to detect at <16 h.

3.1.2 Protocol

Day 1: Plating and Transfection

1. *Plating considerations*

   For strongly adherent cells such as HeLa, uncoated glass cover slips can be used. For weakly adherent cells such as 293T or non-adherent hematopoietic lines, cover slips should be

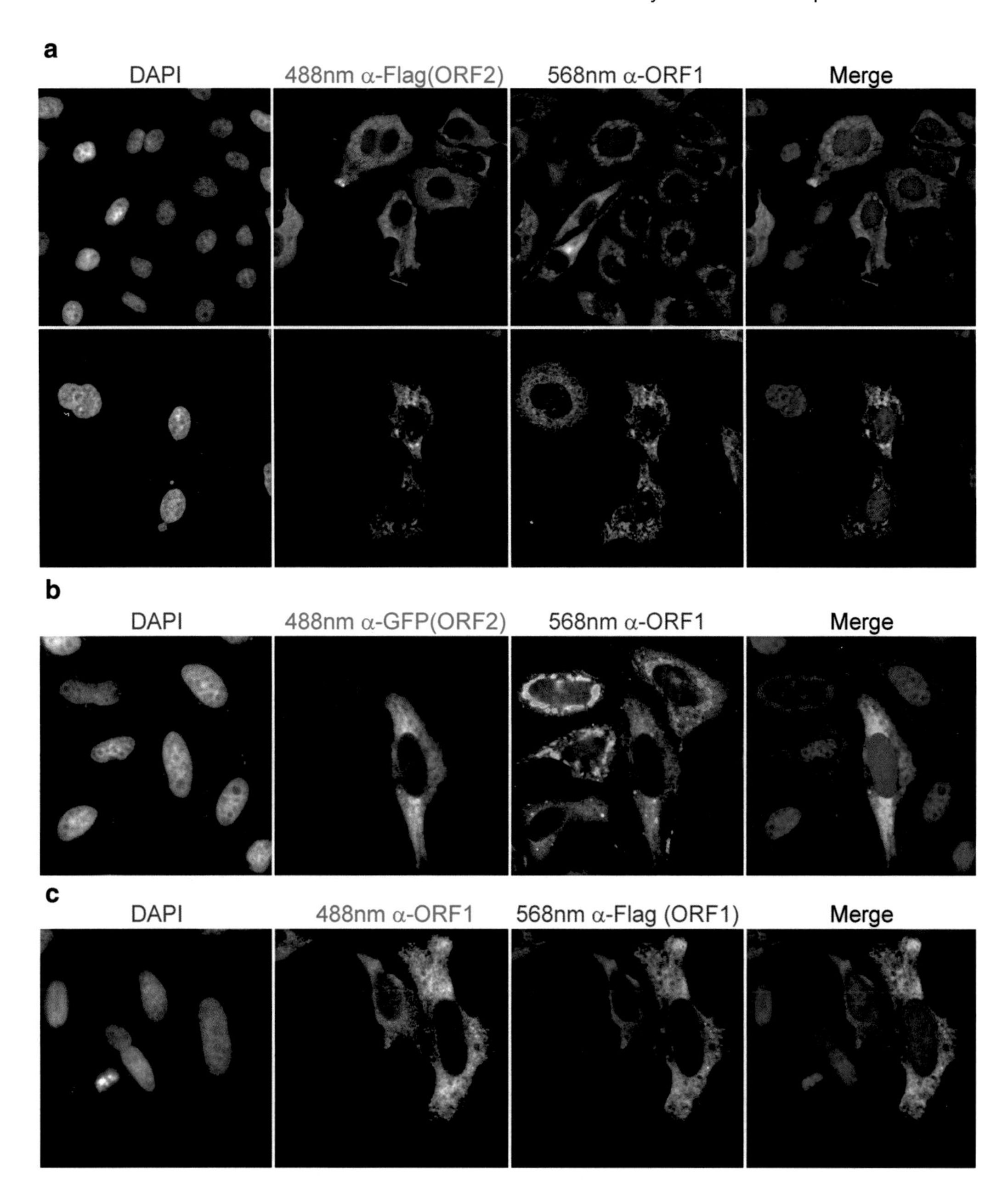

Fig. 1 Example immunofluorescent staining of LINE-1. (**a**) Staining of HeLa cells expressing pLD401 (untagged ORF1, ORF2-3×Flag) using mouse anti-Flag and rabbit anti-ORF1 JH73. *Top row*: epifluorescence, 40×. *Bottom row*, confocal, 40×. (**b**) Staining of HeLa cells expressing pLD402 (untagged ORF1, ORF2-GFP) using rabbit anti-GFP and mouse anti-ORF1 4H1; epifluorescence, 64×. (**c**) Control staining of HeLa cells expressing pLD288 (ORF1-Flag, untagged ORF2) with rabbit anti-flag and mouse anti-ORF1 4H1; epifluorescence, 64×

coated with a substrate to promote attachment of the cells. Fibronectin glycoprotein or poly-L-lysine, a positively charged synthetic amino acid chain, is widely used for this purpose. We autoclave cover slips in glass petri dishes; cover slips stuck together after autoclaving are discarded.

2. *Coating cover slips with fibronectin.*
   1. Fibronectin from human plasma is resuspended in sterile phosphate-buffered saline (PBS) and stored in 60 μL aliquots at −80 °C.
   2. Dilute a 60 μL aliquot 1:100 to 6 mL in sterile PBS (10 μg/mL final).
   3. With cover slips in 6-well wells or 35 mm dishes, add 1 mL to each cover slip, leaving it beaded up on top of the glass.
   4. Incubate for >2 h at room temperature in a tissue culture hood. Alternatively, >1 h at 37 °C, or overnight at 4 °C. For overnight incubation, seal with Parafilm.
   5. When ready to use, aspirate and wash with PBS or media.
3. *Coating cover slips with poly-L-lysine*

   Poly-L-lysine is less expensive than fibronectin and works well with HEK-293T cells.
   1. Poly-L-lysine (mol wt ≥300,000) is dissolved to 10 mg/mL in sterile ddH_2O and 400 μL aliquots are stored at −20 °C.
   2. Make 4 mL of a 1 mg/mL working dilution in ddH_2O (400 μL stock + 3.6 mL ddH_2O).
   3. With cover slips in 35 mm dishes or 6-well plates, apply the solution to each cover slip for ~30 s to 1 min.
   4. Pipet to sequential dishes/wells, coating ~5–10 per 1 mL solution. We routinely coat up to 40 cover slips with 4 mL.
   5. Wash each coated cover slip with 2× 1 mL ddH_2O, aspirating off all of the solution.
   6. After the second wash, aspirate all the water off and allow to air-dry completely in the hood (*see* **Note 3**).
   7. Store at room temperature for weeks to months before use, or use once dry (minimum ~30 min).
4. *Plating*

   Cultured cells can be subconfluent or confluent, but not overgrown.
   1. Insert an autoclaved cover slip or coated cover slip into each 35 mm dish or 6-well plate, discarding cover slips that are clumped together.
   2. If plating HEK-293T the day before staining as in the above timeline, plate 300,000–400,000 cells. Plate 200,000 if plating the day before transfection (2 days on the plates). If

cells are to be on the plates longer than 2 days, plate fewer cells.

5. *Transfection*

We prefer Fugene 6 or HD (Promega) over Lipofectamine (Life Technologies) and XtremeGene (Roche) reagents in HEK-293T and HeLa due to reduced toxicity and better reproducibility. Transfection is done following the manufacturer's standard protocol with a 3 μL:1 μg ratio of reagent:DNA. DNA should be midi- or maxi-prepped; minipreps are not suitable.

Day 2: Fixation, Permeabilization, and Staining (*See* **Note 4**)

1. *Paraformaldehyde (PFA) Fixation and Triton Permeabilization*

This is a standard fixation method for immobilizing antigens. Paraformaldehyde acts as a cross-linking reagent and is followed by permeabilization with Triton X-100 detergent. After PFA fixation, glycine is added to the buffers as both a quench and blocking agent. Saponin detergent is also popular and may expose different epitopes. A methanol treatment at −20 °C can be used as an alternative to cross-linking fixation and permeabilization. Organic solvent fixations remove lipids and dehydrate cells and may provide better access to antigens or better preservation of some organelles such as the endoplasmic reticulum.

1. Thaw an aliquot of 3 % PFA solution.
2. Make fresh working stock of 0.5 % Triton X-100 in PBS/glycine from 10 % stock (at least 1 mL per sample).
3. Aspirate media, and wash cells with PBS (without glycine).
4. Aspirate, wash, and fix cells: Add 1 mL 3 % PFA in PBS for 10 min.
5. Aspirate PFA, and wash with PBS/glycine.
6. Aspirate, wash: Permeabilize cells by adding 1 mL fresh 0.5 % Triton X-100 in PBS/glycine. Incubate for 3 min at room temperature.
7. Rinse in PBS/glycine, and let sit in PBS/glycine for at least 5 min to fully quench (or longer).

2. *Immunostaining*

To minimize the volume of antibody solution used, we take advantage of the hydrophobicity of Parafilm M, applying the cover slip cell-side down onto a beaded drop of antibody solution. The cell side of the cover slip is coated in the antibody solution via surface tension. The minimum volume required for this using a standard 22 mm cover slip is approximately 35 μL; the extra volume reduces edge drying and makes handling the cover slips easier. All antibodies are diluted in *PBS/glycine/1 % bovine serum albumin (BSA)*; BSA serves as both diluent and blocking agent.

Multichannel fluorescent staining and the species problem. Using standard UV-fluorescent microscopes and fluorophores, up to four orthogonal channels may be studied simultaneously; our common filter set is blue/DAPI (excite: 360 nm, emit: ~460 nm), green/FITC (excite: 488 nm, emit: ~525 nm), red/Texas Red (excite: 568 nm, emit: ~615 nm), and far-red/Cy5 (excite: 633 nm, emit: ~680 nm) (*see* **Note 5**). Thus, in principle, up to four different targets may be visualized simultaneously using immunofluorescence. In practice, the ultraviolet channel is usually reserved for DNA dyes (DAPI, Hoechst), and detecting three different targets requires primary antibodies produced in three different species with a panel of orthogonally reactive fluorescent-labeled secondary antibodies. In selecting the panel of secondary antibodies, using three antibodies raised in the same species can reduce background. Further, using highly cross-adsorbed secondary antibodies can eliminate cross-binding to off-target primary antibodies. Note that in the first use of any secondary antibody combination, it is important to include "no-primary" controls for each primary antibody individually (*see* **Note 6**).

For LINE-1, a typical staining is a rabbit anti-ORF1p monoclonal JH73 (0.25 μg/mL) and mouse anti-Flag M2 (2 μg/mL, ORF2p-3×Flag) along with Hoechst stain for DNA, and might include chicken anti-actin. In our hands, the signal-to-noise ratio tends to be better in the green (488 nm) channel, so we usually stain for ORF2p with this secondary antibody. Alternatively, mouse anti-ORF1 4H1 (1 μg/mL) can be used with rabbit polyclonal anti-flag or many other commercial DYKDDDK reagents; however the 3×Flag signal is weaker in this case because the 3×Flag tag is optimized for binding the M2 antibody.

1. Dilute primary and secondary antibodies. Note: primary antibody dilutions should be optimized. First make a solution of PBS/glycine/1 % BSA using the 30 % BSA stock (*see* **Note 7**).
2. Stretch and label Parafilm on the bench as in Fig. 2. Label each square with both the cover slip number and which stain that square receives.
3. Add 50 μL of the primary antibody mix to the center of the corresponding squares on the Parafilm. Avoid bubbling the drop.
4. Using a pair of sharp forceps, carefully remove the cover slip from the dish. (This is much easier with buffer present.) Dry the cell side of the cover slip by blotting and flip it *cell-side down* onto the antibody drop (*see* Fig. 3). *Incubate for 20 min at room temperature.*
   (a) Set a stack of two paper towels on the bench for blotting.

Fig. 2 Antibody droplets on Parafilm M. For each cover slip, one square of 4″ wide Parafilm M is used. Label each square with the antibody combination to be used. Pipet a 50 μL droplet of primary or secondary antibody, and then apply the cover slip *cell-side down* for 20 min at room temp. Note that 30–35 μL is the minimum needed, but higher volume increases consistency and ease of handling. Avoid bubbling the drop; remove bubbles with a pipet

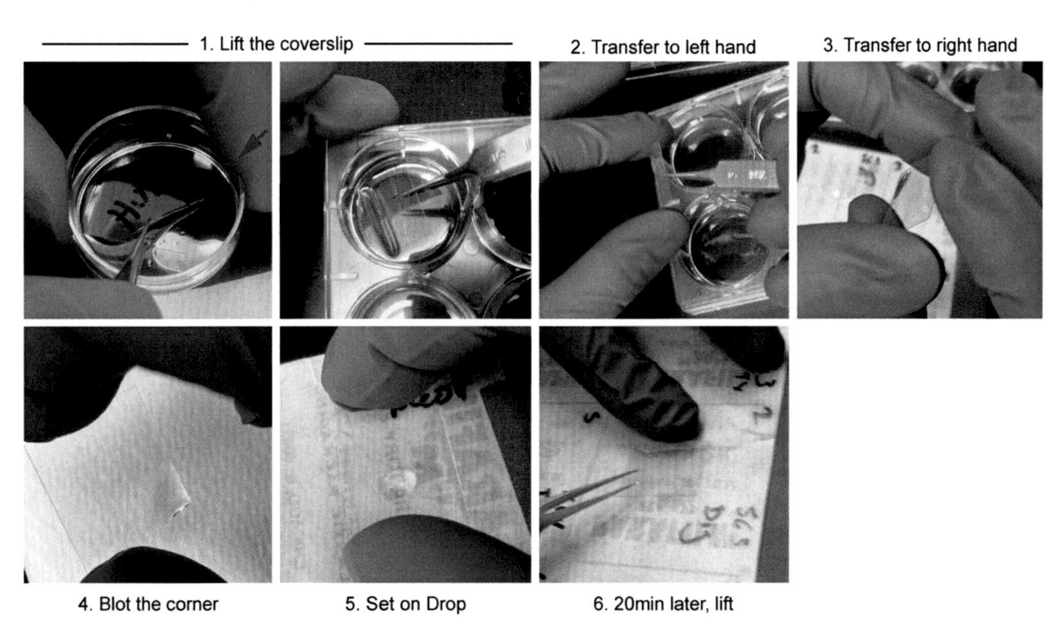

Fig. 3 Transferring the cover slip from dish or plate to droplets on Parafilm. Lift the cover slip from buffer-containing dish/plate with forceps (*1*). With dishes, pinching the dish firmly (*red arrows*) helps. Wear gloves. To avoid breaking or dropping the cover slip, transfer the cover slip carefully from the forceps to the left hand (*2*), then back to the right hand (*3*) as shown. Tilt the cover slip so that a drop forms in the corner (*4*) and blot the bulk off using a small stack of paper towels, touching only the edges of the slide to the paper towel. Set the blotted cover slip *cell-side down* on the drop (*5*). If any bubble forms, gently chase it out by pushing on the back or lift the cover slip and remove. Incubate for 20 min at room temperature. To lift cover slip, use a finger and gentle pressure as shown to keep the cover slip in place (*6*)

(b) Lift the cover slip with fine forceps (*see* **Note 8**).

(c) Very coordinated scientists can use forceps to flip, blot, and transfer the cover slip to the drop. For the rest of us, we recommend the following method (right-handedness is presumed):

- Lift the cover slip with forceps and transfer the cover slip into your left hand, holding it gently by the edges and paying careful attention to which side is the cell side. Set the forceps down and transfer the cover slip back to your right hand, grasping the free edges.
- Angle the cover slip downwards, cell-side down. A drop of liquid will accumulate in the corner. Gently blot the edge on the paper towel, and then set the cover slip cell-side down down on the antibody droplet.
- Notes:

  During this step it is easy to make a mistake and shatter the cover slip by dropping it or lose track of which side has the cells. Work carefully and focus; *it gets easier with practice.*

  If the cover slip falls and you are not sure which side the cells are on, the best way to check is to lift the slip and carefully scrape a corner. The cell side scrapes clean; nothing happens on the other side.

5. During the incubation, aspirate the PBS/glycine out of the dishes and replace with fresh buffer for the first wash.
6. Using forceps, carefully lift the cover slip off the Parafilm and flip it back CELL-SIDE UP into the dish (containing fresh PBS/glycine wash).

   (a) You can keep the cover slip from sliding away by putting a fingernail on the opposite side (*see* **Note 9**).
7. Shake gently to wash, and aspirate off the wash.
8. Wash again with PBS/glycine.
9. Clean the Parafilm with a paper towel and rinse, dry, and repeat for two washes total. (Alternatively, replace the Parafilm with a clean sheet.)
10. Stain with secondary antibody: Repeat **steps 2**–**7** using the secondary antibody, but modify the wash procedure as below to include Hoechst stain in the first wash.

    (a) During the secondary antibody incubation, make a working stock of Hoechst DNA Dye by diluting the 10 mg/mL stock 1:100,000 total in PBS/glycine in sequential 1:100 and then 1:1000 dilutions (final 100 ng/mL).

(b) Add 1 mL to each well for the first wash. Once the cover slips are added, incubate for 5 min at room temperature.

(c) Aspirate and wash with PBS/glycine. Let sit in PBS/glycine until ready to mount.

3. *Mounting*: Many different mounting media are available, and there are pros and cons to each. We typically use a simple and inexpensive mounting media using 0.1 M *N*-propyl gallate in glycerol [19]. Slides are stable for weeks to months if stored flat, and can be viewed on the scope the instant they are mounted. They can also be un-mounted easily if needed (*see* **Note 10**). Some mounting media also contain DAPI stain. We prefer not to seal cover slips with clear nail polish as the raised surface can damage microscope objectives, although this can be helpful for preserving slides with liquid mounting media over time.

   1. Pre-label the slides, and clean with Kimwipes and/or a compressed air can. Each slide should be marked with sample name, date, permeabilization, and stain, as appropriate. If two samples are to be compared directly, it is helpful to put both cover slips on the same slide.

   2. To mount each cover slip, we follow a protocol analogous to the staining procedure, followed by aspiration of excess mounting media.

      (a) Put a drop of mounting media for each cover slip on the slide. Remove the cover slip from the dish, blot to remove buffer from the cell side, and then flip cell-side down onto the drop. It may be helpful to dry the back of the cover slip before mounting.

      (b) Apply gentle downward pressure on each cover slip with index and middle fingers and a cut pipet-tip aspirator while aspirating excess mounting media from around the edges of the cover slip. Then, maintain firm downward pressure with your fingers. Push downward and sideways with the plastic pipet tip against the top of the glass cover slip, chase out large bubbles, and continue to remove excess mounting media accumulating at the edges of the glass. Push firmly enough to move air bubbles to the edge but not excessively, as sliding cover slips or excessive pressure can shear the cells.

      (c) For those inclined to remove *every single drop* of excess mounting media, allow the slides to sit for a few minutes and then make a second pass over the completed slides.

3.2 Immunohistochemical and Immunofluorescent (IF) Detection of LINE-1 ORF1p from Human Tissues

This protocol describes immunolabeling of formalin-fixed, paraffin-embedded (FFPE) tissue sections using the monoclonal antibody for LINE-1 ORF1p. This is a general staining protocol we have used in human pancreatic tissue, but it can be modified to accommodate staining of other tissues and cell lines. Two alternative approaches for detection are described here: immunohistochemistry (IHC), where the protein of choice is detected via binding of a chromogen such as 3,3′-diaminobenzidine (DAB) and visualized using a light microscope; or immunofluorescence (IF), where the protein of choice is detected via binding of fluorophores and visualized using a fluorescence microscope. Alternatives to DAB for immunohistochemistry include chromogens such as 3-amino-9-ethylcarbazole (AEC). Both the IHC and IF protocols described herein have been optimized and validated with proprietary signal amplification technologies as well as fluorescent-conjugated secondary antibodies. We routinely use 10 % neutral buffered formalin for tissue fixation, although B5, Bouin's, zinc formalin, or alcohol-based fixatives are also commonly used.

For IHC, in [17] we used Invitrogen's Histostain-Streptavidin-Peroxidase (SP) Broad Spectrum Kit for LINE-1 ORF1p detection, with a protocol modified from that provided by the manufacturer. The kit uses a biotinylated secondary antibody and horseradish peroxidase-labeled streptavidin for signal amplification. This complex is then visualized through the use of a DAB chromogen mixture and results in highly specific signal for LINE1 ORF1p. Example IHC images of LINE-1 ORF1p using the 4H1 monoclonal antibody are shown in Fig. 4.

As an alternative to the Histostain kit, we use the EnVision System for signal amplification. EnVision consists of a poly-horseradish peroxidase (HRP) chain secondary antibody. In our experience, this system gives more robust signal over background than the Histostain kit.

For immunofluorescence, detection can be achieved directly with fluorescent-conjugated secondary antibodies; we prefer Alexa-fluor-labeled antibodies raised in goat as described in cell culture immunofluorescence above (*see* **Note 11**).

A common protocol follows for deparaffinization through primary antibody incubation. The protocol splits at this point for IHC and IF.

3.2.1 Protocol

1. *Deparaffinization and rehydration*:
   1. Label slides with a pencil (*see* **Note 12**).
   2. Bake slides at 60 °C for 20–30 min to melt paraffin.
   3. Perform the following series of washes to remove paraffin:
      (a) Xylene washes for 5 min, 5 min, and 3 min (using the fume hood).
      (b) Wash in 100 % ethanol for 2 min. Repeat twice for a total of three washes.

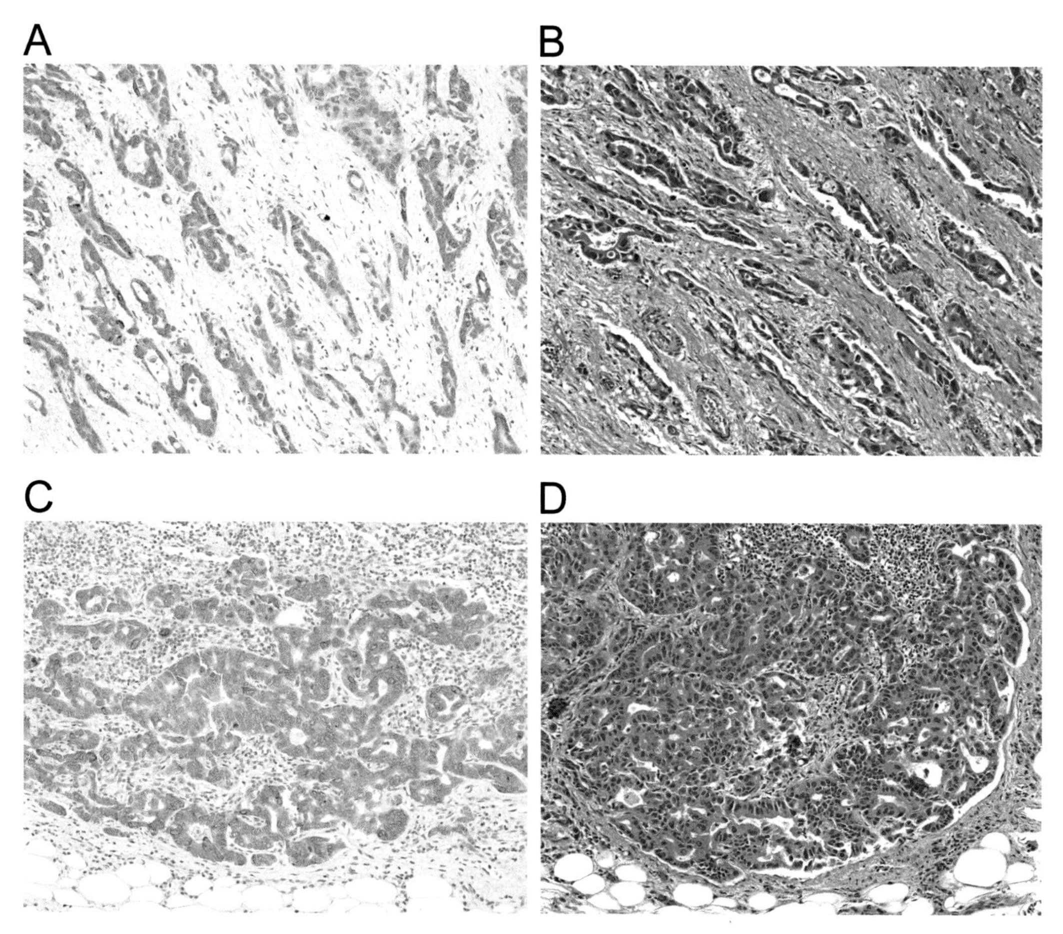

Fig. 4 IHC staining of LINE-1 ORF1p in pancreatic ductal adenocarcinoma. (**a**) and (**c**): Anti-ORF1p-stained sections with hematoxylin counterstain. (**b**) and (**d**): Hematoxylin and eosin (H&E) stain corresponding to (**a**) and (**c**), respectively

(c) Wash in 95 % ethanol for 2 min. Repeat twice for a total of three washes.

(d) Rinse under streaming tap water for 5 min.

2. *Antigen retrieval*:

Formalin and other aldehyde-based tissue fixations form extensive protein cross-links which require a subsequent antigen retrieval step for antibody penetration and antigen unmasking.

Our standard retrieval protocol uses a steamer as follows:

1. Immerse slides in citrate antigen retrieval buffer (pH 6).
2. Transfer the buffer and slides to a steamer and heat at 98 °C for 20 min.
3. Remove slides from steamer and allow them to cool to room temperature.
4. Place in TBST buffer.

An EDTA antigen retrieval buffer can be used in lieu of the citrate antigen retrieval solution.

In addition, the alternate hot plate approach below can be applied using either the citrate or the EDTA buffers. Thus, we provide four [4] different variations on conditions for antigen retrieval. Alternatively, several manufacturers make automated antigen retrieval systems using either heat or microwave energy; a review of these devices and their application is beyond the scope of this chapter. We suggest using the standard protocol above first and applying alternate protocols for optimization or based on the availability of heating equipment.

Alternate retrieval protocol: Hot plate approach

- Heat retrieval buffer to 55 °C.
- Immerse the rehydrated slides and increase the temperature to 95–98 °C.
- Incubate for 30 min.
- Let the slides cool to at least 45 °C in the antigen retrieval buffer. This step typically takes about 45 min.
- Rinse under streaming tap water for 5 min.

General note: The volume of each reagent used per slide from here on varies according to the surface area of the tissue. Reagent volumes can range between 100 μL and 1 mL per slide. It is recommended to add certain reagents (including blocking reagent, primary and secondary antibody solutions, chromogen and fluorophore mixes, and counterstains) dropwise to minimize their use.

3. *Permeabilization and blocking*:
   Two things are accomplished in this blocking protocol. First, activities of endogenous peroxidases are quenched. The protocol below uses a 3 % hydrogen peroxide solution (*see* **Note 13**). Secondly, blocking of nonspecific interactions is achieved using a 10 % non-immune serum blocking solution (Histostain kit) or BSA.
   1. Remove slides from TBST and wipe off any residual buffer on the slides using Kimwipes, gently drying the underside and edges. Take care to leave a small amount of buffer overlying the tissue, because drying the tissue during this step will interfere with antibody binding in subsequent steps.
   2. Incubate samples with 3 % hydrogen peroxide for 5 min at room temperature.
   3. Rinse slides with TBST (with shaking) for 5 min.
   4. Block:
      (a) Incubate samples with the blocking reagent (Reagent 1A, Histostain kit) at room temperature for 10 min.
      (b) Alternatively, overlay slides with 500–1000 μL of BSA-TBST blocking solution, and preincubate in a humid chamber for 30 min at room temperature (*see* **Note 14**).

4. *Primary Antibody Incubation*
   1. Remove slides from TBST and carefully wipe off any residual buffer on the slides.
   2. Dilute LINE-1 ORF1p antibody with BSA-TBST (TBST with 1 % BSA) to 1 μg/mL. The use of BSA reduces nonspecific binding of the primary antibody.
   3. Add enough diluted primary antibody to cover the tissue and incubate overnight at 4 °C in a humid chamber.
   4. Remove primary antibody solution and wash with TBST for 5 min. Repeat twice for a total of three washes.

Up to this point, the protocol has been identical for IF and IHC. We describe several approaches for subsequent staining:

- IF with fluorescent-conjugated secondary antibodies.
- IHC or IF with signal amplification using Histostain or EnVision reagents, followed by chromogenic or TSA-based IF detection.

Three starting points for these protocols follow:

A. Fluorescent secondary antibody incubation for IF
B. Secondary antibody incubation, chromogen incubation, and counterstain: Histostain Kit
C. Secondary antibody incubation, chromogen incubation, and counterstain: EnVision Kit

Departure points from each of the latter two methods to *amplified immunofluorescence using TSA* are noted. Finally, IHC detections conclude with *dehydration and coverslipping (for IHC only)*, and IF detections with *coverslipping (for IF only)* protocols.

A. Fluorescent secondary antibody incubation for IF
   1. Remove slides from TBST and carefully wipe off any residual buffer on the slides.
   2. Dilute fluorescent-conjugated secondary antibody or antibodies in BSA-TBST to 2 μg/mL (1:1000 for Alexa Fluor-conjugated goat antibodies but might change depending on the provider).
   3. Add enough diluted antibody to cover the tissue and incubate for 1 h at room temperature in a humid chamber.
   4. Remove the primary antibody solution and wash with TBST for 5 min. Repeat twice for a total of three washes.
   5. Proceed to *coverslipping (for IF only)*.

B. Secondary antibody incubation, chromogen incubation, and counterstain: Histostain Kit

1. Incubate slides with a biotinylated secondary antibody: Pipet Reagent 1B over tissue and incubate at room temperature for 10 min.
2. Remove the secondary antibody solution and wash with TBS (no tween) for 2 min. Repeat twice for a total of three washes.
3. Incubate slides with a streptavidin-peroxidase conjugate (Reagent 2) at room temperature for 10 min.
4. Remove the enzyme conjugate solution and wash with TBS (no tween) for 2 min. Repeat twice for a total of three washes.

If performing IF using the Histostain Kit, jump to the *amplified immunofluorescence using tsa (for IF only)* protocol below. Otherwise, for IHC, continue to **step 5** immediately below.

5. Incubate samples with DAB chromogen mix (one drop each of Reagents 3A, 3B, and 3C added to 1 mL water). The DAB chromogen mix should be protected from light and used immediately after mixing. The ideal development time for pancreatic tissue usually occurs at approximately 10 min but may vary across samples. Color development (brown) should be monitored using a light microscope.
6. Rinse with deionized water for 3 min with shaking. Repeat once for a total of two washes.
7. Incubate with the hematoxylin solution (Reagent 4). The ideal staining time for pancreatic tissue is approximately 7 min but may vary across samples.
8. Rinse with deionized water for 3 min with gentle shaking.
9. Proceed to *dehydration and coverslipping (for IHC only)*.

C. Secondary antibody incubation, chromogen incubation, and counterstain: EnVision™ Kit

1. Select the species-appropriate EnVision solution. Solutions are available with peroxidase-labeled polymers conjugated to anti-mouse or anti-rabbit antibodies.
2. Pipet the EnVision solution and incubate in a light-protected humid chamber for 1 h at room temperature (*see* **Note 15**).
3. Wash a total of three times with TBST.

If performing IF using the EnVisionKit, jump to the *amplified immunofluorescence using TSA (for IF only)* protocol below. Otherwise, for IHC, continue to **step 4** immediately below.

4. Prepare substrate-chromogen solution which consists of one drop (20 μL) of DAB+ chromogen added to 1 mL DAB+ substrate buffer. Use this solution immediately after mixing.
5. Wipe excess liquid off of the slides and pipette DAB mixture onto each slide.
6. Rinse with deionized H_2O for 3 min with shaking. Repeat once for a total of two washes.
7. Wipe excess liquid off of the slides and pipet on a small amount of hematoxylin, enough to cover the tissue surface, and incubate at room temperature for 1 min.
8. Rinse slides in deionized H_2O for 3 min with gentle shaking.
9. Proceed to *dehydration and coverslipping (for IHC only)*.

5. *Dehydration and coverslipping (for IHC only).*
   1. Sequetially dehydrate samples by immersing them in the following series of increasing ethanol concentrations and ending with xylene:
      (a) Water, 5 min.
      (b) 70 % ethanol, 2 min.
      (c) 95 % Ethanol, 2 min. Repeat once for a total of two 95 % EtOH soaks.
      (d) 100 % Ethanol, 2 min. Repeat twice for a total of three 100 % EtOH soaks.
      (e) Xylene for 3 min, 5 min, and 5 min (fume hood).
   2. Remove slides from xylene and wipe excess solution of the underside and sides of the slides; as before, make sure not to allow the tissue to dry entirely.
   3. Add a few drops of HistoMount, CytoSeal 60, or other mounting medium and cover slip immediately (*see* **Note 16**).
6. *Amplified immunofluorescence using TSA (for IF only).*
   1. Remove TBST, and wipe slides carefully with Kimwipes, making sure not to dry out the tissue.
   2. Dilute fluorophore-tyramide 1:50 (Cy-5) or 1:100 (fluorescein) in amplification diluent and pipet enough reagent to cover tissue.
   3. Incubate in the dark at room temperature for 10 min.
   4. Rinse slides in 1× TBST for 5 min with shaking. Repeat once.
   5. Proceed to *coverslipping (for IF only)*.
7. *Coverslipping (for IF only).*
   1. Aliquot 50 to 100 μL of ProLong Gold with 4′,6-diamidino-2-phenylindole (DAPI) solution onto slides.

2. Incubate in the dark at room temperature for 5 min (*see* **Note 17**).
3. Place a cover slip cleaned with a Kimwipe or an air can duster carefully onto slides (*see* **Note 18**).

4 Notes

1. PFA is toxic. Weigh and dissolve in a fume hood and wear appropriate protection.
2. Antibodies can be obtained from several suppliers. Different Anti-Flag antibodies can be obtained and Mouse Anti-Flag M2 Sigma F3165 is not affinity purified but works well in our hands for this application.
3. The drying process is important for surface stickiness.
4. *It is critical not to let the cover slips dry out.* They should be without buffer for the shortest time possible. When working with many samples, aspirate one 6-well plate at a time, wash that plate, and then move to the next samples.
5. Far-red is not visible to the eye and requires charge-coupled device (CCD) camera-based detection.
6. We have had good success using Alexa-fluor-labeled highly cross-adsorbed goat antibodies from Life Technologies, with combinations of anti-mouse-rabbit-chicken, anti-mouse-rabbit-rat, and anti-mouse-rabbit-sheep.
7. *Helpful trick*: If more than one combination of primary antibody is being used (or if any volume needed to pipet is less than 0.5 μL), it is helpful to first make 2× working stocks of each antibody (or 3× stocks for three-channel immunostaining).
8. If using plastic dishes, pinch firmly to free the cover slip.
9. It is helpful if the forceps have a very slight upward curve, to prevent stabbing into the Parafilm. If the tip of the instrument is getting stuck in the Parafilm, try using the other tip (flip them over), or gently bend the end inward.
10. Commercial mounting media such as VECTASHIELD (Vector labs) and ProLong (Life Technologies) are popular and more permanent, but some require solvents and longer drying times, and all are more expensive.
11. Alternatively, either of the Histostain or EnVision™ HRP-based amplification systems can be used with the Tyramide Signal Amplification™ (TSA™) System (PerkinElmer, Inc., Waltham, MA) for increased sensitivity. A variety of fluorescent substrates are available for TSA™; we mostly use TSA™-plus fluorescein and TSA™-plus Cyanine-5 (Cy-5). Multiplexed immunofluorescent detection using TSA and EnVision is described in [20].

12. Ink from many laboratory markers will wash off in ethanol.
13. An alternative is to use Peroxo-Block™ (Invitrogen Cat. No. 00-2015).
14. We make the humid chamber using a small sealable plastic container, lined on the bottom with damp paper towels and containing an elevated stage to hold the slides.
15. The binding time is a step that can be optimized for each primary antibody.
16. Use gloves during this step, because any fingerprint(s) on the cover slip may interfere with subsequent microscopic analysis. Allow slides to dry in hazardous chemical hood, label with permanent marker, and store slides at room temperature.
17. Keep slides protected from light from the beginning of this step and onwards. Other mounting media (as described in the cultured cell IF protocol above) can be used here as well.
18. Use gloves during this step, because any fingerprint(s) on the cover slip may interfere with subsequent microscopy. Label slides with permanent marker. Store slides shielded from ambient light at room temperature or 4 °C.

Acknowledgement

We thank Carolyn Machamer and Travis Ruch for their immunofluorescence expertise. We thank Jeff Han for the gift of rabbit anti-ORF1 JH73. We thank Norman Barker for photography of IHC slides. Funding for these projects has been provided in part by R01CA163705 (KHB) and the Burroughs Wellcome Fund Career Awards for Medical Scientists (KHB).

References

1. Lander ES et al (2001) Initial sequencing and analysis of the human genome. Nature 409:860
2. Burns KH, Boeke JD (2012) Human transposon tectonics. Cell 149:740
3. Shukla R et al (2013) Endogenous retrotransposition activates oncogenic pathways in hepatocellular carcinoma. Cell 153:101
4. Ostertag EM, Kazazian HH Jr (2001) Biology of mammalian L1 retrotransposons. Annu Rev Genet 35:501
5. Hancks DC, Kazazian HH Jr (2012) Active human retrotransposons: variation and disease. Curr Opin Genet Dev 22:191
6. Beck CR, Garcia-Perez JL, Badge RM, Moran JV (2011) LINE-1 elements in structural variation and disease. Annu Rev Genomics Hum Genet 12:187
7. Martin SL, Bushman FD (2001) Nucleic acid chaperone activity of the ORF1 protein from the mouse LINE-1 retrotransposon. Mol Cell Biol 21:467
8. Kolosha VO, Martin SL (2003) High-affinity, non-sequence-specific RNA binding by the open reading frame 1 (ORF1) protein from long interspersed nuclear element 1 (LINE-1). J Biol Chem 278:8112
9. Martin SL et al (2005) LINE-1 retrotransposition requires the nucleic acid chaperone activity of the ORF1 protein. J Mol Biol 348:549
10. Feng Q, Moran JV, Kazazian HH Jr, Boeke JD (1996) Human L1 retrotransposon encodes a

conserved endonuclease required for retrotransposition. Cell 87:905

11. Mathias SL, Scott AF, Kazazian HH Jr, Boeke JD, Gabriel A (1991) Reverse transcriptase encoded by a human transposable element. Science 254:1808
12. Alisch RS, Garcia-Perez JL, Muotri AR, Gage FH, Moran JV (2006) Unconventional translation of mammalian LINE-1 retrotransposons. Genes Dev 20:210
13. Martin SL, Branciforte D, Keller D, Bain DL (2003) Trimeric structure for an essential protein in L1 retrotransposition. Proc Natl Acad Sci U S A 100:13815
14. Callahan KE, Hickman AB, Jones CE, Ghirlando R, Furano AV (2012) Polymerization and nucleic acid-binding properties of human L1 ORF1 protein. Nucleic Acids Res 40:813
15. Khazina E et al (2011) Trimeric structure and flexibility of the L1ORF1 protein in human L1 retrotransposition. Nat Struct Mol Biol 18:1006
16. Taylor MS et al (2013) Affinity proteomics reveals human host factors implicated in discrete stages of LINE-1 retrotransposition. Cell 155:1034
17. Rodic N et al (2014) Long interspersed element-1 protein expression is a hallmark of many human cancers. Am J Pathol 184:1280
18. Hicks SW, Machamer CE (2002) The NH2-terminal domain of Golgin-160 contains both Golgi and nuclear targeting information. J Biol Chem 277:35833
19. Giloh H, Sedat JW (1982) Fluorescence microscopy: reduced photobleaching of rhodamine and fluorescein protein conjugates by n-propyl gallate. Science 217:1252
20. Brown JR et al (2014) Multiplexed quantitative analysis of CD3, CD8, and CD20 predicts response to neoadjuvant chemotherapy in breast cancer. Clin Cancer Res 20:5995

Chapter 18

Cellular Localization of Engineered Human LINE-1 RNA and Proteins

Aurélien J. Doucet, Eugénia Basyuk, and Nicolas Gilbert

Abstract

The human LINE-1 retrotransposon has the ability to mobilize into a new genomic location through an intracellular replication cycle. Immunofluorescence and in situ hybridization experiments have been developed to detect subcellular localization of retrotransposition intermediates (i.e., ORF1p, ORF2p, and L1 mRNA). Currently, these protocols are also used to validate the interaction between retrotransposition complex components and potential cellular partners involved in L1 replication. Here, we describe in details methods for the identification of LINE-1 proteins and/or RNA in cells transfected with vectors expressing engineered human LINE-1 elements.

Key words LINE-1 retrotransposon, Immunofluorescence, MS2 fluorescent in situ hybridization, ORF1p, ORF2p, L1 mRNA

1 Introduction

The human genome is constituted of more than ~45 % of transposon- and retrotransposon-derived sequences [1]. The long interspersed element-1 (LINE-1 or L1) is a retrotransposon and corresponds to 17 % of the human genome sequence. The vast majority of L1 copies are unable to mobilize due to the accumulation of mutations or rearrangements and can be considered as molecular fossils. However, it is estimated that, in average, ~80–100 active L1 copies are present per haploid genome [2]. Their ongoing activity is responsible for genetic diversity among the human population [3, 4].

Full-length active L1 elements are 6 kb long (for review [5]). An active L1 contains an internal RNA polymerase II promoter in the 5′UTR that serves to produce the bicistronic L1 mRNA. L1 contains two open reading frames (ORF1 and ORF2) that encode two proteins (ORF1p and ORF2p), both required for retrotransposition [6]. L1 ends with a 3′UTR that includes a polyadenylation signal

Jose L. Garcia-Pérez (ed.), *Transposons and Retrotransposons: Methods and Protocols*, Methods in Molecular Biology, vol. 1400, DOI 10.1007/978-1-4939-3372-3_18,

and a variable-length poly(A) tail [7, 8]. ORF1p is an RNA-binding protein [9, 10] that also includes putative nucleic acid chaperone activity [11]. ORF2p contains two enzymatic activities essential for retrotransposition: endonuclease (EN) and reverse transcriptase (RT) activities [12, 13]. A round of L1 retrotransposition initiates with the production of the L1 mRNA, and its export to the cytoplasm where translation of ORF1 and ORF2 occurs. There, an RNP complex forms in *cis* by the association of L1 proteins to their encoding RNA molecule [14, 15]. The complex then enters the nucleus where a new insertion occurs by a mechanism termed target-primed reverse transcription (TPRT). It first consists of an endonucleolytic cleavage of the DNA target site (EN activity of ORF2p) followed by L1 RNA reverse transcription using the free 3′-OH of the cleaved target as a primer (RT activity of ORF2p) (reviewed in [5]). The L1 insertion is then completed by mechanisms not yet established [5].

Identification of cellular retrotransposition intermediates has been the purpose of several studies in the last decade [16, 17]. However, protocols were only recently designed to follow the cellular localization of RNA and proteins from a retrotransposition-competent human L1 copy [14, 18]. From these latest works the human L1-encoded proteins and L1 RNA have been localized in cytoplasmic foci that often are associated with stress granules.

The following protocols provide details to conduct either concomitant or independent detection of L1 RNA and L1 proteins. To overcome the difficulty to generate efficient ORF2p antibodies, we developed an epitope-tagging strategy and facilitated the immunofluorescent detection of L1 proteins. We combined this approach with the enhanced detection of L1 mRNA using the bacteriophage MS2 detection system [19]. This strategy allowed us to successfully identify the localization of the L1 mRNA and proteins expressed from an active L1 element in transfected human cells. Previously, individual fluorescent detection of human L1 proteins was performed with retrotransposition competent L1 constructs [20], modified expression vectors including fusions with GFP [16, 17, 21], or in teratocarcinoma cells expressing endogenous ORF1p [17]. Our protocol and others have successfully been used since for human L1 proteins and RNA localization in transfected cells alongside with other cellular factors [18, 22–27].

2 Materials

1. Mammalian cell line (e.g., U-2 OS), cell culture media, and dishes (including 12-well dishes).
2. Glass cover slips, round (15 mm of diameter), Number 1.5 (0.17 mm of thickness).

3. MilliQ water.
4. Solution of nitric acid in milliQ water (1:1) (20 mL).
5. 4 mM EDTA (40 mL).
6. 70 % ethanol (40 mL).
7. Transfection-grade plasmid DNA purification of expression vector for active L1 with epitope-tagged ORF1 (T7 *gene 10*) and ORF2 (TAP), and MS2-binding sites subcloned in the 3′UTR (i.e., pAD3TE1; Fig. 1a) [14]; MS2-binding protein fused to fluorescent proteins (e.g., pMS2-GFP) [19, 28].
8. Opti-MEM reduced serum media.
9. FuGENE 6 transfection reagent (Promega).
10. PAGE or reverse-phase HPLC-purified MS2 DNA oligonucleotide with amino-allyl T (depicted by T*), resuspended in H_2O at a final concentration of 1 μg/μL. 5′-AT*GTCGACCT GCAGACAT*GGGTGATCCTCAT*GTTTTCTAGGCAAT T*A-3′ (Fig. 1b).
11. Dimethyl sulfoxide (DMSO).
12. Cy3 mono-reactive dye pack (GE Healthcare).
13. 0.1 M $Na_2CO_3/NaHCO_3$ buffer, pH 8.8. Prepare by adding Na_2CO_3 0.1 M to $NaHCO_3$ 0.1 M until reaching pH 8.8.
14. Bacterial tRNA.
15. 100 % Ethanol.
16. 3 M Sodium acetate, pH 5.2.
17. Tabletop centrifuge.
18. Tris-EDTA buffer, pH 7.5.
19. 1× PBS (dilution of 10× PBS stock solution in milliQ H_2O).
20. 40 % Paraformaldehyde (PFA).
21. 4 % PFA (dilution of 40 % stock in 1× PBS).
22. RNase-free H_2O.
23. RNase-free 70 % ethanol (dilution of RNase-free ethanol 100 % in RNase-free H_2O).
24. 20× SSC, pH 7.0.
25. 100 % formamide.
26. 1× SSC pH 7.0, 10 % formamide (dilution of stock solutions in RNase-free H_2O).
27. Bovine serum albumin (BSA, Sigma).
28. 1 % (w/v) Bovine serum albumin (BSA).
29. 200 mM Vanadyl ribonucleoside complexes (VRC, Sigma).
30. 40 % Dextran sulfate.
31. Plastic paraffin film (Parafilm).

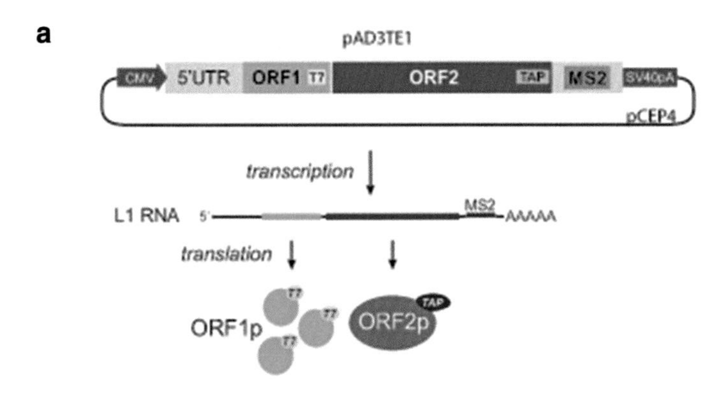

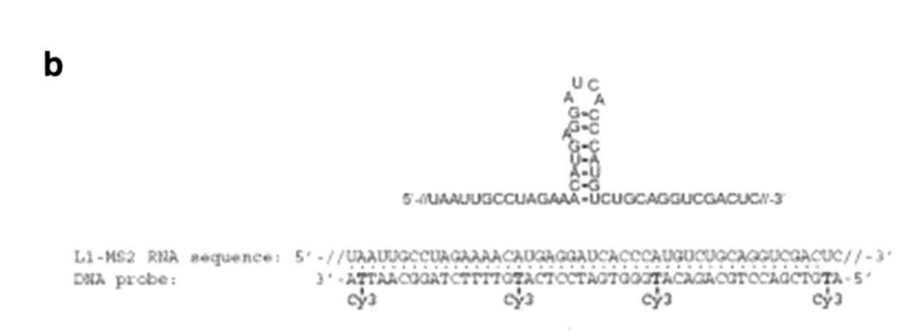

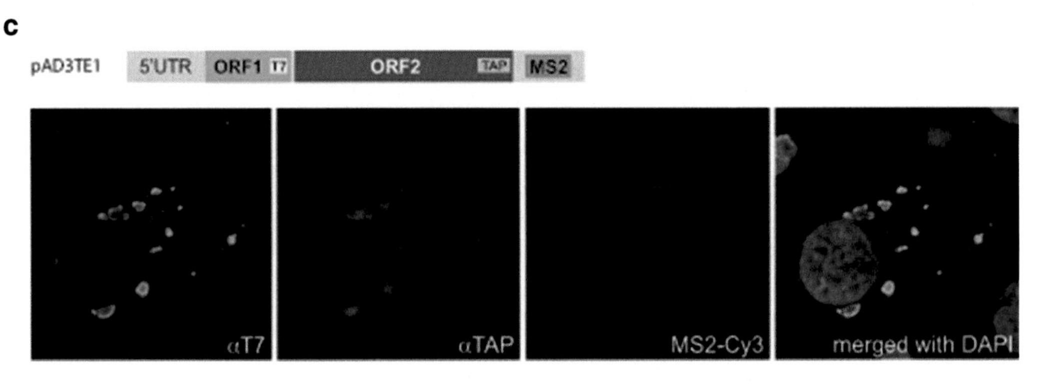

Fig. 1 Strategy and cellular localization of L1 RNA and proteins. (**a**) Schematic representation of the engineered L1 expression vector allowing the detection of L1 RNA with 24 repeats of MS2-binding site sequence (*red*) introduced in the 3′UTR, ORF1p with T7 tag (*green*), and ORF2p with TAP tag (*blue*). An active copy of L1 is subcloned in a mammalian episomal vector (pCEP4), flanked by CMV promoter and SV40 polyadenylation signal. (**b**) Details of the MS2-binding site sequence including the stem-loop structure recognized by the bacteriophage MS2 coat protein. The L1 RNA detection occurs by in situ hybridization of a Cy3-labeled oligonucleotide probe (with Cy3 and *bold T* indicating the modified Ts) with the MS2 repeat sequence present in the L1 RNA. (**c**) Cellular localization of ORF1p (*left picture* in *green*, with anti-T7 antibody), ORF2p (*center left picture* in *blue*, with anti-TAP antibody), and L1 RNA (*center right picture* in *red*, with MS2-Cy3 probe) obtained by FISH and IF 48 h after transfection of U-2 OS cells with pAD3TE1 vector. The *rightmost picture* corresponds to the merge of the first three pictures and the nuclear staining by DAPI. L1 RNA and proteins co-localize in cytoplasmic foci. Pictures of panel **c** were previously published in [14]

32. Humid chamber.
33. 37 °C Incubator.
34. Aluminum foil.
35. 1× SSC, pH 7.0, 10 % formamide, and 3 % BSA (dilution of stock solutions in RNase-free H_2O).
36. 100 % Anhydrous methanol.
37. 3 % BSA in 1× PBS.
38. Primary antibodies: Mouse anti-T7 Epitope Tag antibody (Millipore) and rabbit TAP Tag polyclonal antibody (Thermo Scientific).
39. Secondary antibodies: Alexa Fluor 488 goat anti-mouse IgG antibody (Life Technologies), Cy5 goat anti-mouse and anti-rabbit IgG antibodies (Jackson ImmunoResearch).
40. ProLong Gold Antifade Mountant with DAPI (Life Technologies).
41. Confocal microscope with adapted lasers.

3 Methods

The following methods describe the combination of fluorescent in situ hybridization (FISH) with immunofluorescence (IF) detections of L1 RNA and proteins on a single slide, as described in our previous publication [14]. However, it is not always required to conduct both techniques at the same time. We provide details in each section to allow independent detection of L1 RNA or L1 proteins. Finally, we provide an alternative detection of L1-MS2 RNA by co-expression of fluorescent MS2-binding proteins.

In our previous study, several L1 expression vectors have been built and are available upon request [14]. In this protocol, we use a single engineered L1 expression vector that combines epitope tags fused in C-terminus of ORF1p and ORF2p as well as 24 repeats of the MS2-binding motif in the L1 3′UTR (Fig. 1a, pAD3TE1) [14]. Other L1 expression vectors with different epitope tags on ORF1 and/or ORF2 or MS2-binding sites only have successfully been used [14, 22–25].

3.1 Preparation of Cover Slips for Cell Culture

To allow the manipulations required in the staining protocols, cells need to be grown on sterile cover slips suited for confocal microscopy (Number 1.5, 0.17 mm thickness). The following protocol describes the preparation of cover slips prior to optimal use for cell culture and subsequent microscopy observation.

1. Place ~250 glass cover slips inside of a 50 mL polyethylene conical tube.
2. Carefully prepare a 20 mL solution of nitric acid in milliQ water (1:1).

3. Pour the nitric acid solution on the cover slips.
4. Incubate for 2 h at room temperature with gentle shaking.
5. Remove the nitric acid solution and wash three to four times for 5 min with milliQ water.
6. Incubate the cover slips with a 4 mM EDTA solution for 3 h.
7. Remove the EDTA solution and wash three to four times for 5 min with milliQ water.
8. Incubate the cover slip stock in 70 % ethanol until ready to use.

3.2 Cell Culture and Transfection

For cellular localization in human cells, we routinely use U-2 OS cells in culture. However, multiple other cell lines can be used and protocols must be adapted to the desired cell type, including cell density plating, transfection conditions, time frame after transfection for staining, fixation, and permeabilization conditions (see below). In Subheading 3.7, we provide control conditions that can also serve as a starting point for troubleshooting with other cell lines.

Day 0:

1. Under sterile conditions, take one cover slip from the stock (see above, Subheading 3.1) and use a flame to burn the remaining drops of ethanol. Transfer the cover slip into one well of a 12-well culture plate. Prepare as many wells as experimental conditions, including untransfected cells for negative control condition.
2. Seed U-2 OS cells at a density of 5×10^4 cells per well in a total volume of 1 mL of growth media. Cell density can be adapted to optimize transfection efficiency, and to prevent over-confluent cells on the cover slips at the day of staining. The latter would affect the quality of microscopy observation.

Day 1:

3. Set up transfection reactions by mixing plasmid DNA and transfection reagent, following the manufacturer's recommendations. We generally use 3 μL of FuGENE-6 and 1 μg of plasmid DNA in 100 μL of Opti-MEM media. For co-transfection (see below), only 0.5 μg of each plasmid is used.
4. Add 100 μL of the transfection mix to the culture media in a well.

Day 2:

5. Change growth media in each well after transfection.

Day 3:

6. Take the culture plate outside of the incubator to follow up with FISH and/or IF protocols (see below, Subheadings 3.3.2 and 3.4.1). Time of incubation after transfection can be modified as desired. We generally start FISH or IF protocol 48 h post-transfection.

3.3 Fluorescent In Situ Hybridization to Detect L1-MS2 RNA

Fluorescent in situ hybridization (FISH) allows the detection of DNA or RNA molecules using nucleic acid probes. To detect L1 RNA, we took advantage of the MS2 RNA labeling strategy developed in Robert Singer laboratory [19]. This strategy consists of the addition of multiple repeats of the RNA-binding site of the bacteriophage MS2 coat protein in the 3′UTR of on active L1 copy. Here, we introduced 24 repeats of the RNA-binding site. The use of a fluorescent probe specific to a single repeat provides an enhanced signal allowing efficient L1 RNA localization (Fig. 1b). Below, we describe methods adapted from the Singer lab to generate MS2 fluorescent probes and to perform FISH to detect L1 RNA in transfected cells (for details, see http://www.singerlab.org/protocols). Alternately, L1 RNA can be detected after co-transfection of the L1 MS2-tagged construct and plasmids expressing the MS2 coat protein fused to fluorescent protein (see below, Subheading 3.5).

3.3.1 Preparation of MS2 Fluorescent Probe

The MS2 probe used for FISH is obtained by in vitro fluorescent labeling of an amino-allyl thymine-modified DNA oligonucleotide. The distance between two labeled thymines should not be less than ten nucleotides to avoid quenching of the fluorescent signal as well as high background issues.

1. Resuspend the MS2 DNA oligonucleotide dried oligonucleotide in H_2O to a final concentration of 1 μg/μL (*see* **Notes 1** and **2**).
2. Add 30 μL of DMSO into one tube of Cy3 mono-reactive dye (*see* **Note 3**).
3. In a clear 1.5 mL tube, mix the 30 μL of reactive dye with 5 μg of MS2 DNA oligonucleotide.
4. Add 65 μL of 0.1 M $Na_2CO_3/NaHCO_3$ buffer pH 8.8.
5. Incubate overnight at room temperature in the dark to allow chemical conjugation of the oligonucleotide with the activated Cy3 fluorophores.
6. The next day, perform ethanol precipitation of the labeled oligonucleotide by addition of 20 μg of carrier bacterial tRNA, 13 μL of 3 M sodium acetate pH 5.2 (1/10th V), and 390 μL of 100 % ethanol (3 V).
7. Incubate overnight at −20 °C.
8. The next day, centrifuge for 5 min at maximum speed at 4 °C. The pellet should be colored (in red when using Cy3 fluorophores).
9. Rinse the pellet with 1 mL of 100 % ethanol and centrifuge for 2 min at maximum speed at 4 °C.
10. Carefully remove the supernatant and resuspend the pellet in 100 μL of H_2O.
11. Repeat **steps 6–9**.

12. Carefully remove the supernatant, and let the pellet dry for 5 min at room temperature.
13. Resuspend the pellet in 500 μL of Tris-EDTA buffer, pH 7.5 to obtain a final concentration of 10 ng/μL. Aliquot the labeled oligonucleotide and store at −20 °C until ready to use.

3.3.2 Protocol for In Situ Hybridization with Fluorescent MS2 Probe

U-2 OS cells grown on cover slips and transfected with L1 expression plasmids need to be fixed to avoid loss of material and permeabilized to allow intracellular detection of L1 RNA using the Cy3-labeled MS2 DNA probe. The cover slips obtained in Subheading 3.2 are maintained in the original culture plate (unless otherwise indicated) and treated as follows with volumes corresponding to a 12-well dish format.

Day 1:

1. Take the culture plate containing transfected cells grown on cover slips out of the incubator. We usually perform FISH 48 h post-transfection.
2. Wash the cover slips three times for 5 min with 1 mL of 1× PBS directly in the wells.
3. Fix the cells with 0.5 mL of 4 % paraformaldehyde. Incubate for 15 min at room temperature.
4. Wash the cover slips twice for 5 min with 1 mL of 1× PBS.
5. Permeabilize the cells with RNase-free 70 % ethanol at 4 °C overnight (*see* **Note 4**).

Day 2:

6. Rince cells briefly in 1X PBS and incubate with 0.5 mL of 1× SSC pH 7.0 and 10 % formamide at least for 15 min at room temperature.
7. Prepare the hybridization mix as follows with amounts and volumes corresponding to one cover slip. Multiply by the number of samples to treat:

   Mix 1

   - 3 μL of 10 ng/μL bacterial tRNA (30 μg)
   - 3 μL of 100 % formamide (10 % final)
   - 1.5 μL of 20× SSC (1× final)
   - 0.75 μL of 10 ng/μL Cy3-labeled MS2 DNA probe (7.5 ng) (*see* Subheading 3.3.1)
   - 6.75 μL of RNase-free H_2O

   Incubate Mix 1 for 1 min at 90 °C. Briefly centrifuge to collect drops.

Mix 2

- 0.6 μL of 1 % BSA (0.02 % final)
- 0.3 μL of 200 mM VRC (2 mM final)
- 7.5 μL of 40 % dextran sulfate (10 % final)
- 6.6 μL of RNase-free H_2O

Add Mix 1 to Mix 2. Total volume for one cover slip should be 30 μL.

8. Spot 30 μL drops of the hybridization mix on parafilm placed in a humid chamber (*see* **Note 5**).
9. Place each cover slip on a drop, with cells facing the hybridization solution.
10. Incubate the humid chamber overnight at 37 °C in the dark.

Day 3:

11. Move the cover slips back into the 12-well culture dish. To avoid loss of fluorescent signal, protect cover slips from light in all the subsequent steps (e.g., wrap the 12-well dish with aluminum foil).
12. Wash the cells with 1 mL of 1× SSC and 10 % formamide for 30 min at room temperature (*see* **Note 6**).

13a. To finish the FISH protocol without coupling with the IF protocol, wash the cells a second time with 1 mL of 1× SSC and 10 % formamide for 30 min at room temperature. Rinse with 1× PBS and mount the cover slips on slides as described in **steps 22–27** of Subheading 3.4.2 (see below).

13b. To couple L1 RNA detection by FISH with protein detection by IF, wash a second time with 1× SSC, 10 % formamide, and 3 % BSA for 30 min at room temperature. The addition of BSA will allow reduction of background signal due to nonspecific binding of antibodies during the IF procedure. Then follow protocol directly in Subheading 3.4.2.

3.4 Immunofluorescence Detection of Epitope-Tagged L1 Proteins

The following steps describe the procedure for L1 protein detection by immunofluorescence (IF). The first paragraph (Subheading 3.4.1) includes steps that are required if IF is not coupled with the FISH protocol as described above. Otherwise, the main procedure is described in the second paragraph (Subheading 3.4.2).

3.4.1 Initial Steps of IF Protocol When not Coupled with FISH

1. Take the culture plate containing transfected cells grown on cover slips out of the incubator. We usually perform IF detection 48 h post-transfection.
2. Wash the cover slips three times for 5 min with 1 mL of 1× PBS directly in the wells.

3. Fix the cells with 0.5 mL of 4 % paraformaldehyde. Incubate for 15 min at room temperature (*see* **Note 7**).
4. Wash the cover slips three times for 5 min with 1 mL of 1× PBS.
5. Permeabilize the cells by incubation for 1 min precisely in 0.5 mL of anhydrous methanol.
6. Rehydrate the samples by progressive addition of 1× PBS onto the methanol-containing wells, few drops at a time, until adding 2 mL of 1× PBS per well.
7. Remove the methanol/PBS solution.
8. Wash the cover slips three times for 5 min with 1 mL of 1× PBS.
9. To saturate nonspecific antibody-binding sites, incubate the cover slips in 1 mL of 3 % BSA in 1× PBS solution for 30 min at room temperature. Then follow up with antibody incubation in the next paragraph.

3.4.2 Protocol for Immunofluorescence Detection of L1 Proteins

When coupled with FISH, the immunofluorescent detection of L1 proteins can be performed as follows after saturation of nonspecific antibody-binding sites with BSA (*see* **step 13b**, Subheading 3.3.2). It is necessary to perform all the following steps in the dark, to prevent loss of fluorescent signal. Commercially available secondary antibodies can be associated to various fluorophores. It is necessary to make sure that the fluorophores used in one preparation are compatible with each other and will not lead to nonspecific signal due to spectral cross talk.

10. Dilute the desired primary antibodies in 3 % BSA solution in 1× PBS: mouse anti-T7 antibody (1:1000) and rabbit anti-TAP antibody (1:200) (*see* **Note 8**). For each cover slip, prepare 30 μL of antibody dilution solution.
11. Spot 30 μL drops of the primary antibody dilution solution on parafilm, placed in a humid chamber.
12. Place each cover slip on a drop, with cells facing the antibody solution.
13. Incubate the humid chamber for 1 h at 37 °C.
14. Move the cover slips back into the 12-well culture dish.
15. Wash the cover slips three times for 5 min with 1 mL of 1× PBS.
16. Dilute secondary antibodies in 3 % BSA solution in 1× PBS: Alexa Fluor 488 anti-mouse antibody (1:1000), Cy5 anti-rabbit antibody (1:100), or Cy5 anti-mouse antibody (1:100) (*see* **Note 8**). For each cover slip, prepare 30 μL of secondary antibody dilution solution.
17. Spot 30 μL drops of the secondary antibody dilution solution on parafilm, placed in a humid chamber.
18. Place each cover slip on a drop, with cells facing the antibody solution.

19. Incubate the humid chamber for 30 min at 37 °C.
20. Move the cover slips back into the 12-well culture dish.
21. Wash the cover slips three times for 5 min with 1 mL of 1× PBS.
22. Briefly rinse the cover slips with H_2O.
23. Clean the required amount of microscopy slides with 70 % ethanol and wiper paper to remove dust (*see* **Note 9**).
24. Spot 10 μL drops of Prolong Gold Antifade with DAPI on microscopy slides (*see* **Notes 10** and **11**).
25. Quickly place the cover slips on the slide, with cells facing the mounting media solution.
26. Let the mounting media thicken by incubation for 2 h at room temperature in the dark.
27. Store the slides at 4 °C in the dark until ready for image acquisition of L1 RNA and L1-encoded protein cellular localization (Fig. 1c) (*see* **Note 12**).

3.5 Detection of L1-MS2 RNA with MS2-Binding Proteins

An alternative strategy to the detection of L1-MS2 RNA by FISH is to take advantage of the ability of the MS2 bacteriophage-binding protein to interact with MS2 repeats introduced in the 3′UTR of L1. Co-transfection of U-2 OS cells with plasmids expressing L1-MS2 RNA (e.g., pAD3TE1) and MS2-binding protein fused to a fluorescent protein (e.g., pMS2-GFP-NLS) allow cellular localization of L1-MS2 RNA. A collection of MS2-binding protein expression vectors is available through Addgene plasmid bank. This collection allows to choose between several fluorescent proteins (i.e., GFP, YFP, mCherry, CFP) as well as for the presence of a nuclear localization signal (NLS). Finally, the main advantage of this strategy is to allow for time-lapse capture of L1-RNA localization in live cells.

Here, we used pMS2-GFP-NLS and pMS2-CFP as tools to study cytoplasmic localization of L1-MS2 RNA (Fig. 2a, b).

1. U-2 OS cells are seeded on cover slips and transfected as described above (*see* Subheading 3.2) with pAD3TE1 and pMS2-GFP-NLS or pMS2-CFP.
2. Forty-eight hours post-transfection, take the culture plate containing transfected cells out of the incubator.
3. Wash the cover slips three times for 5 min with 1 mL of 1× PBS directly in the wells.
4. Fix the cells with 0.5 mL of 4 % paraformaldehyde. Incubate for 15 min at room temperature.
5. Wash the cover slips three times for 5 min with 1 mL of 1× PBS.
6. Briefly rinse the cover slips with H_2O.
7. Clean the required amount of microscopy slides with 70 % ethanol and wiper paper to remove dust (*see* **Note 9**).

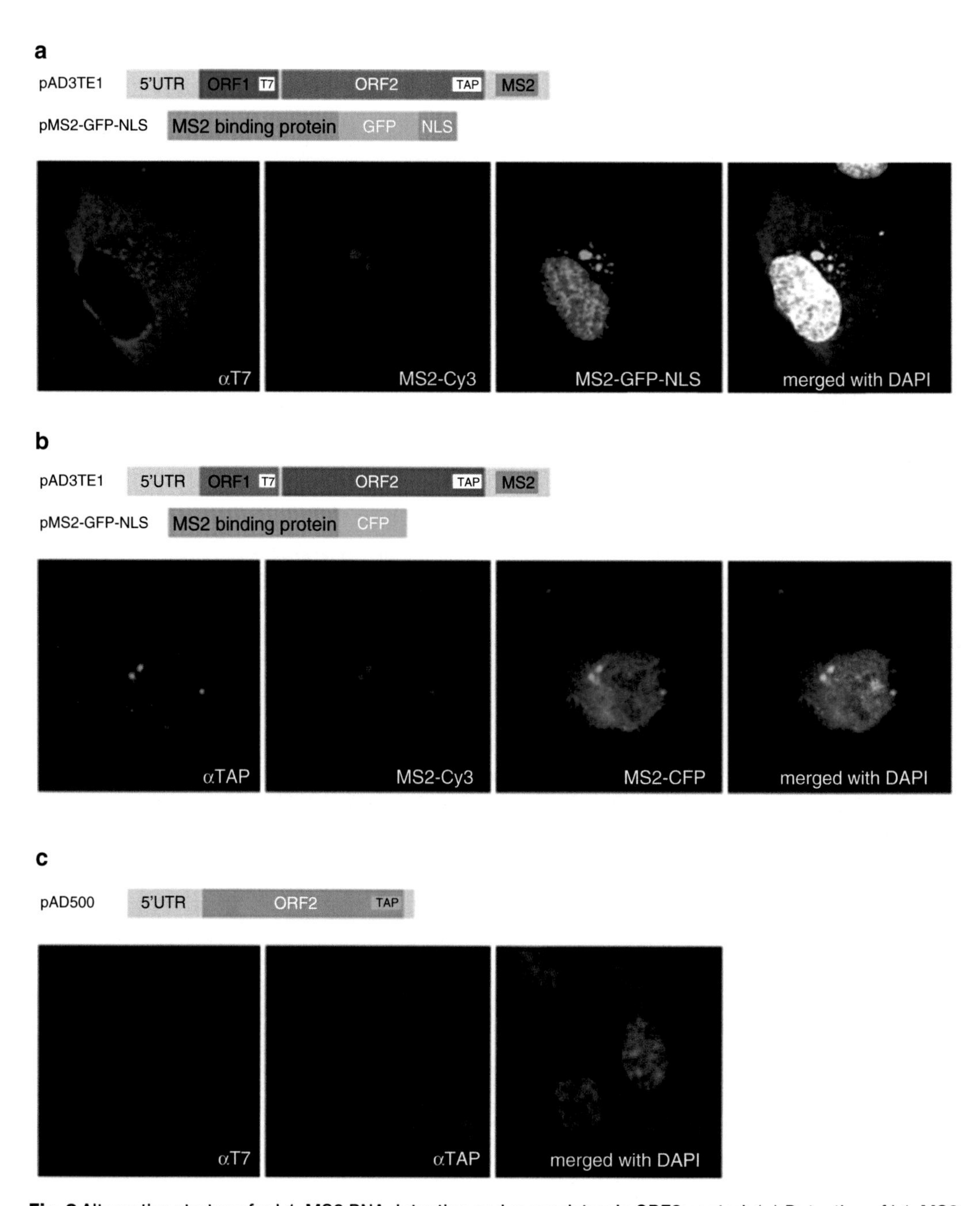

Fig. 2 Alternative strategy for L1-MS2 RNA detection and monocistronic ORF2 control. (**a**) Detection of L1-MS2 RNA by MS2 coat protein fused to green fluorescent protein (GFP) and a nuclear localization signal (NLS). The GFP signal is concentrated in the nucleus which correspond to free MS2-GFP, and cytoplasmic foci corresponds to MS2-GFP bound to RNA reporter. The colocalization with MS2-Cy3 probe (*center left picture* in *red*) confirms the specificity of the cytoplasmic GFP signal. ORF1p signal (*left picture* in *blue*, anti-T7 antibody) also co-localize with MS2-GFP and MS2-Cy3 signals in L1 RNP cytoplasmic foci. The *rightmost picture* corresponds to the merge of the first three pictures and the nuclear staining by DAPI. (**b**) Similar to the panel **a**, the L1-MS2 RNA accumulation can be detected in cytoplasmic foci through the CFP signal associated to MS2 coat protein accumulation (*center right panel* in *green*). The MS2-Cy3 FISH detection of L1-MS2 RNA (*center left panel* in *red*)

8. Spot 10 μL drops of Prolong Gold Antifade with DAPI on microscopy slides (*see* **Notes 10** and **11**).
9. Quickly place the cover slips on the slide, with cells facing the mounting media solution.
10. Let the mounting media thicken by incubation for 2 h at room temperature in the dark.
11. Store the slides at 4 °C in the dark until ready for image acquisition (*see* **Note 12**).

3.6 Acquisition by Confocal Microscopy

The identification of L1 RNA and protein cellular localization is then performed by image acquisition with a confocal microscope. It is important to make sure that the confocal microscope is equipped with lasers suited for the detection of fluorophores associated to MS2 DNA probe (Cy3 in our case, but could alternatively be Cy2, Cy5, or Oregon Green 488), secondary antibodies (Alexa Fluor 488 and Cy5 in this chapter, but could alternatively be other fluorescent dyes), co-transfected fluorescent proteins (if applicable), as well as nuclear DAPI staining.

3.7 Control Conditions and Adaptations to the Protocol

To adapt the described protocol to the use of other cell lines, we recommend to start with transfection of plasmid-expressing fluorescent proteins. This should allow the validation of the early steps of the protocol including cell density, transfection conditions, and interval after transfection for cell fixation and staining.

If other antibodies are to be used for specific detection of L1 proteins, it is required to test several antibody dilutions (for primary and secondary) as well as untransfected conditions. This will allow determining signal specificity and identifying cellular background signal.

For the detection of ORF2p, we suggest to first use a monocistronic construct (e.g., pAD500 in [14]). The translation mechanism of ORF2 is likely to be less efficient than the cap-scanning-dependent mechanism involved in ORF1 translation [29, 30]. We have demonstrated that high levels of ORF2 expression can be obtained with monocistronic expression plasmid, compared to a bicistronic construct [14]. The use of such vector should facilitate detection of ORF2p signals (Fig. 2c).

Fig. 2 (continued) confirms the signal obtained with MS2-CFP. The lack of NLS in the MS2-CFP fusion protein leads to a diffuse signal across the whole cell. The *rightmost picture* corresponds to the merge of the first three pictures and the nuclear staining by DAPI. (**c**) Cellular localization by immunofluorescence of ORF2p expressed from a monocistronic expression vector for which ORF1 is absent. The protocol was conducted by mixing the T7 and TAP primary antibodies and then a specific secondary antibody for each. The absence of signal with T7 antibody (*left picture*) corresponds to the absence of ORF1 in the expression vector. ORF2p signal is diffused in the cytoplasm in the absence of ORF1p (*center picture* in *red*, with anti-TAP). The *rightmost picture* corresponds to the merge of the first three pictures and the nuclear staining by DAPI. Pictures were previously published in [14]

4 Notes

1. Different probes can be designed to target other RNA sequences (e.g., L1 5′UTR, Alu SINE RNA). In that case, the oligonucleotide should be ~50 nucleotide long, with GC content around 50 %, and there should be ~10 nucleotides between each modified thymidine. If not targeting the MS2 repeat sequence, it is possible to enhance RNA detection signal by combining multiple individual probes against the same RNA molecule.
2. The use of Tris base-containing solution to resuspend the oligonucleotide is prohibited as it could lead to nonspecific amine reaction with the fluorophores.
3. Cy3 mono-reactive dye can be substituted with other amine-reactive dyes such as Cy2 or Cy5 mono-reactive dye (GE Healthcare), or Oregon Green 488 or Alexa 488 carboxylic acid succinimidyl ester (Life Technologies).
4. Cover slips can be stored for several weeks to months at this stage.
5. A humid chamber can easily be made of a 10 cm petri dish containing a water-soaked filter paper, with the parafilm placed on top of the wet paper.
6. To reduce background signal, 0.1 % SDS or NP40 can be added to the wash buffer.
7. Alternatively, permeabilization can be performed during fixation by adding 0.2 % Triton X-100 (Sigma) to the 4 % PFA solution. Then move directly to **step 9** to saturate nonspecific antibody-binding sites.
8. Antibody dilution can be adapted to reduce background or to increase specific signal.
9. We generally use one microscopy slide for three round cover slips.
10. To prevent the mounting media to dry before placing the cover slips, only add drops on one microscopy slide at a time.
11. Prolong Gold Antifade also exists without DAPI, if required (Life Technologies).
12. We generally wait until the next day for analysis of the slides under the microscope.

Acknowledgements

We are grateful to Edouard Bertrand (CNRS, Institut de Genetique Moleculaire de Montpellier, France) for the gift of plasmids, probes, and sharing protocols. We thank Robert Singer (Albert Einstein College of Medicine, Yeshiva University, USA) for sharing online protocols (available at http://www.singerlab.org/protocols) and

John V. Moran (Department of Human Genetics, University of Michigan, USA) for sharing plasmids. We are grateful to Nicole Lautredou at Montpellier RIO Imaging, France. This work was supported by the Institut National de la Santé Et de la Recherche Médicale (INSERM), the Centre National de la Recherche Scientifique (CNRS) and the Agence Nationale de la Recherche (ANR-12-BSV6-0003, RETROGENO) [to NG], Ministère de l'Enseignement Supérieur et de la Recherche, Association pour la Recherche contre le Cancer (ARC), and Fondation Recherche Médicale (FRM) [doctoral fellowships to AJD].

References

1. Lander ES, Linton LM, Birren B, Nusbaum C, Zody MC, Baldwin J, Devon K, Dewar K, Doyle M, FitzHugh W, Funke R, Gage D, Harris K, Heaford A, Howland J, Kann L, Lehoczky J, LeVine R, McEwan P, McKernan K, Meldrim J, Mesirov JP, Miranda C, Morris W, Naylor J, Raymond C, Rosetti M, Santos R, Sheridan A, Sougnez C, Stange-Thomann N, Stojanovic N, Subramanian A, Wyman D, Rogers J, Sulston J, Ainscough R, Beck S, Bentley D, Burton J, Clee C, Carter N, Coulson A, Deadman R, Deloukas P, Dunham A, Dunham I, Durbin R, French L, Grafham D, Gregory S, Hubbard T, Humphray S, Hunt A, Jones M, Lloyd C, McMurray A, Matthews L, Mercer S, Milne S, Mullikin JC, Mungall A, Plumb R, Ross M, Shownkeen R, Sims S, Waterston RH, Wilson RK, Hillier LW, McPherson JD, Marra MA, Mardis ER, Fulton LA, Chinwalla AT, Pepin KH, Gish WR, Chissoe SL, Wendl MC, Delehaunty KD, Miner TL, Delehaunty A, Kramer JB, Cook LL, Fulton RS, Johnson DL, Minx PJ, Clifton SW, Hawkins T, Branscomb E, Predki P, Richardson P, Wenning S, Slezak T, Doggett N, Cheng JF, Olsen A, Lucas S, Elkin C, Uberbacher E, Frazier M, Gibbs RA, Muzny DM, Scherer SE, Bouck JB, Sodergren EJ, Worley KC, Rives CM, Gorrell JH, Metzker ML, Naylor SL, Kucherlapati RS, Nelson DL, Weinstock GM, Sakaki Y, Fujiyama A, Hattori M, Yada T, Toyoda A, Itoh T, Kawagoe C, Watanabe H, Totoki Y, Taylor T, Weissenbach J, Heilig R, Saurin W, Artiguenave F, Brottier P, Bruls T, Pelletier E, Robert C, Wincker P, Smith DR, Doucette-Stamm L, Rubenfield M, Weinstock K, Lee HM, Dubois J, Rosenthal A, Platzer M, Nyakatura G, Taudien S, Rump A, Yang H, Yu J, Wang J, Huang G, Gu J, Hood L, Rowen L, Madan A, Qin S, Davis RW, Federspiel NA, Abola AP, Proctor MJ, Myers RM, Schmutz J, Dickson M, Grimwood J, Cox DR, Olson MV, Kaul R, Shimizu N, Kawasaki K, Minoshima S, Evans GA, Athanasiou M, Schultz R, Roe BA, Chen F, Pan H, Ramser J, Lehrach H, Reinhardt R, McCombie WR, de la Bastide M, Dedhia N, Blocker H, Hornischer K, Nordsiek G, Agarwala R, Aravind L, Bailey JA, Bateman A, Batzoglou S, Birney E, Bork P, Brown DG, Burge CB, Cerutti L, Chen HC, Church D, Clamp M, Copley RR, Doerks T, Eddy SR, Eichler EE, Furey TS, Galagan J, Gilbert JG, Harmon C, Hayashizaki Y, Haussler D, Hermjakob H, Hokamp K, Jang W, Johnson LS, Jones TA, Kasif S, Kaspryzk A, Kennedy S, Kent WJ, Kitts P, Koonin EV, Korf I, Kulp D, Lancet D, Lowe TM, McLysaght A, Mikkelsen T, Moran JV, Mulder N, Pollara VJ, Ponting CP, Schuler G, Schultz J, Slater G, Smit AF, Stupka E, Szustakowski J, Thierry-Mieg D, Thierry-Mieg J, Wagner L, Wallis J, Wheeler R, Williams A, Wolf YI, Wolfe KH, Yang SP, Yeh RF, Collins F, Guyer MS, Peterson J, Felsenfeld A, Wetterstrand KA, Patrinos A, Morgan MJ, Szustakowki J, de Jong P, Catanese JJ, Osoegawa K, Shizuya H, Choi S, Chen YJ (2001) Initial sequencing and analysis of the human genome. Nature 409(6822):860–921
2. Brouha B, Schustak J, Badge RM, Lutz-Prigge S, Farley AH, Moran JV, Kazazian HH Jr (2003) Hot L1s account for the bulk of retrotransposition in the human population. Proc Natl Acad Sci U S A 100(9):5280–5285
3. Beck CR, Collier P, Macfarlane C, Malig M, Kidd JM, Eichler EE, Badge RM, Moran JV (2010) LINE-1 retrotransposition activity in human genomes. Cell 141(7):1159–1170. doi:10.1016/j.cell.2010.05.021, S0092-8674(10)00557-X [pii]
4. Kaer K, Speek M (2013) Retroelements in human disease. Gene 518(2):231–241. doi:10.1016/j.gene.2013.01.008, S0378-1119(13)00046-2 [pii]
5. Beck CR, Garcia-Perez JL, Badge RM, Moran JV (2011) LINE-1 elements in structural variation and disease. Annu Rev Genomics

Hum Genet 12:187–215. doi:10.1146/annurev-genom-082509-141802
6. Moran JV, Holmes SE, Naas TP, DeBerardinis RJ, Boeke JD, Kazazian HH Jr (1996) High frequency retrotransposition in cultured mammalian cells. Cell 87(5):917–927
7. Dombroski BA, Mathias SL, Nanthakumar E, Scott AF, Kazazian HH Jr (1991) Isolation of an active human transposable element. Science 254(5039):1805–1808
8. Scott AF, Schmeckpeper BJ, Abdelrazik M, Comey CT, O'Hara B, Rossiter JP, Cooley T, Heath P, Smith KD, Margolet L (1987) Origin of the human L1 elements: proposed progenitor genes deduced from a consensus DNA sequence. Genomics 1(2):113–125
9. Hohjoh H, Singer MF (1996) Cytoplasmic ribonucleoprotein complexes containing human LINE-1 protein and RNA. EMBO J 15(3):630–639
10. Khazina E, Weichenrieder O (2009) Non-LTR retrotransposons encode noncanonical RRM domains in their first open reading frame. Proc Natl Acad Sci U S A 106(3):731–736. doi:10.1073/pnas.0809964106, 0809964106 [pii]
11. Martin SL, Bushman FD (2001) Nucleic acid chaperone activity of the ORF1 protein from the mouse LINE-1 retrotransposon. Mol Cell Biol 21(2):467–475
12. Feng Q, Moran JV, Kazazian HH Jr, Boeke JD (1996) Human L1 retrotransposon encodes a conserved endonuclease required for retrotransposition. Cell 87(5):905–916
13. Mathias SL, Scott AF, Kazazian HH Jr, Boeke JD, Gabriel A (1991) Reverse transcriptase encoded by a human transposable element. Science 254(5039):1808–1810
14. Doucet AJ, Hulme AE, Sahinovic E, Kulpa DA, Moldovan JB, Kopera HC, Athanikar JN, Hasnaoui M, Bucheton A, Moran JV, Gilbert N (2010) Characterization of LINE-1 ribonucleoprotein particles. PLoS Genet 6 (10). doi:10.1371/journal.pgen.1001150
15. Kulpa DA, Moran JV (2006) Cis-preferential LINE-1 reverse transcriptase activity in ribonucleoprotein particles. Nat Struct Mol Biol 13(7):655–660
16. Goodier JL, Ostertag EM, Engleka KA, Seleme MC, Kazazian HH Jr (2004) A potential role for the nucleolus in L1 retrotransposition. Hum Mol Genet 13(10):1041–1048
17. Goodier JL, Zhang L, Vetter MR, Kazazian HH Jr (2007) LINE-1 ORF1 protein localizes in stress granules with other RNA-binding proteins, including components of RNA interference RNA-induced silencing complex. Mol Cell Biol 27(18):6469–6483
18. Goodier JL, Mandal PK, Zhang L, Kazazian HH Jr (2010) Discrete subcellular partitioning of human retrotransposon RNAs despite a common mechanism of genome insertion. Hum Mol Genet 19(9):1712–1725. doi:10.1093/hmg/ddq048, ddq048 [pii]
19. Bertrand E, Chartrand P, Schaefer M, Shenoy SM, Singer RH, Long RM (1998) Localization of ASH1 mRNA particles in living yeast. Mol Cell 2(4):437–445
20. Ergun S, Buschmann C, Heukeshoven J, Dammann K, Schnieders F, Lauke H, Chalajour F, Kilic N, Stratling WH, Schumann GG (2004) Cell type-specific expression of LINE-1 open reading frames 1 and 2 in fetal and adult human tissues. J Biol Chem 279(26):27753–27763
21. Piskareva OA, Barron N, Clynes M, Shmatchenko VV (2004) In vivo cytoplasmic localization of the p40 protein of the L1 transposable element of human genome. Dokl Biochem Biophys 395:118–119
22. Guo H, Chitiprolu M, Gagnon D, Meng L, Perez-Iratxeta C, Lagace D, Gibbings D (2014) Autophagy supports genomic stability by degrading retrotransposon RNA. Nat Commun 5:5276. doi:10.1038/ncomms6276, ncomms6276 [pii]
23. Horn AV, Klawitter S, Held U, Berger A, Jaguva Vasudevan AA, Bock A, Hofmann H, Hanschmann KM, Trosemeier JH, Flory E, Jabulowsky RA, Han JS, Lower J, Lower R, Munk C, Schumann GG (2014) Human LINE-1 restriction by APOBEC3C is deaminase independent and mediated by an ORF1p interaction that affects LINE reverse transcriptase activity. Nucleic Acids Res 42(1):396–416. doi:10.1093/nar/gkt898, gkt898 [pii]
24. Taylor MS, Lacava J, Mita P, Molloy KR, Huang CR, Li D, Adney EM, Jiang H, Burns KH, Chait BT, Rout MP, Boeke JD, Dai L (2013) Affinity proteomics reveals human host factors implicated in discrete stages of LINE-1 retrotransposition. Cell 155(5):1034–1048. doi:10.1016/j.cell.2013.10.021, S0092-8674(13)01297-X [pii]
25. Zhang A, Dong B, Doucet AJ, Moldovan JB, Moran JV, Silverman RH (2014) RNase L restricts the mobility of engineered retrotransposons in cultured human cells. Nucleic Acids Res 42(6):3803–3820. doi:10.1093/nar/gkt1308, gkt1308 [pii]

26. Goodier JL, Cheung LE, Kazazian HH Jr (2012) MOV10 RNA helicase is a potent inhibitor of retrotransposition in cells. PLoS Genet 8(10):e1002941. doi:10.1371/journal.pgen.1002941, PGENETICS-D-12-00597 [pii]
27. Goodier JL, Cheung LE, Kazazian HH Jr (2013) Mapping the LINE1 ORF1 protein interactome reveals associated inhibitors of human retrotransposition. Nucleic Acids Res. doi:10.1093/nar/gkt512, gkt512 [pii]
28. Fusco D, Accornero N, Lavoie B, Shenoy SM, Blanchard JM, Singer RH, Bertrand E (2003) Single mRNA molecules demonstrate probabilistic movement in living mammalian cells. Curr Biol 13(2):161–167
29. Alisch RS, Garcia-Perez JL, Muotri AR, Gage FH, Moran JV (2006) Unconventional translation of mammalian LINE-1 retrotransposons. Genes Dev 20(2):210–224. doi:10.1101/gad.1380406, 20/2/210 [pii]
30. Dmitriev SE, Andreev DE, Terenin IM, Olovnikov IA, Prassolov VS, Merrick WC, Shatsky IN (2007) Efficient translation initiation directed by the 900-nucleotide-long and GC-rich 5′ untranslated region of the human retrotransposon LINE-1 mRNA is strictly cap dependent rather than internal ribosome entry site mediated. Mol Cell Biol 27(13):4685–4697. doi:10.1128/MCB.02138-06, MCB.02138-06 [pii]

Chapter 19

Purification of L1-Ribonucleoprotein Particles (L1-RNPs) from Cultured Human Cells

Prabhat K. Mandal and Haig H. Kazazian Jr.

Abstract

Almost two-thirds of the human genome is repetitive DNA, mostly derived from different kinds of transposon and retrotransposon sequences. Although most of these sequences are stable in the genome, one class called long interspersed element (LINE1 or L1) is actively jumping in the human genome, particularly in brain, germ cells, and certain types of cancer. Recent estimates predict that L1 activity combined with L1-mediated activity is responsible for a new insertion in 1 out of 25 newborns. In humans, more than 100 single-gene disease cases have been reported due to L1 activity. An active L1 encodes two proteins designated as ORF1p and ORF2p. L1 jumps by a target primed reverse transcription (TPRT) mechanism where L1 RNA forms L1-RNPs after binding with L1 proteins. L1-RNPs then enter into the nucleus where L1 RNA is converted to cDNA at the site of integration which subsequently integrates into the genome with the help of the L1 proteins (ORF1p and ORF2p) and other cellular factors. Although L1 is continuously jumping in the human genome the basic mechanism and requirement of other cellular factors in L1 retrotransposition are relatively unknown due to the difficulty in purifying intact L1-RNPs. Here we describe a detailed protocol for purification of L1-RNPs by an immunoaffinity method.

Key words Retrotransposons, L1 RNP, L1 RNA, L1ORF1p, L1ORF2p, LEAP

1 Introduction

Retrotransposons occupy the genomes of virtually all eukaryotes and are actively jumping in the genomes of human and mice [1, 2]. In humans, the master retrotransposon is L1, while other elements Alu, SVA, and processed pseudogenes depend upon L1 retrotransposon machinery for their mobility. An active L1 is 6.0 kb in length, and contains two open reading frames (ORFs), designated ORF1 and ORF2, separated by a small inter-ORF spacer sequence (Fig. 1). ORF2 encodes a protein with reverse transcriptase (RT) [3] and endonuclease (EN) [4] activity whereas ORF1 encodes a protein with demonstrated single-stranded RNA binding and nucleic acid chaperone activity [5]. Although the functions of the ORF-encoded proteins are poorly understood, both proteins are

Jose L. Garcia-Pérez (ed.), *Transposons and Retrotransposons: Methods and Protocols*, Methods in Molecular Biology, vol. 1400, DOI 10.1007/978-1-4939-3372-3_19, © Springer Science+Business Media New York 2016

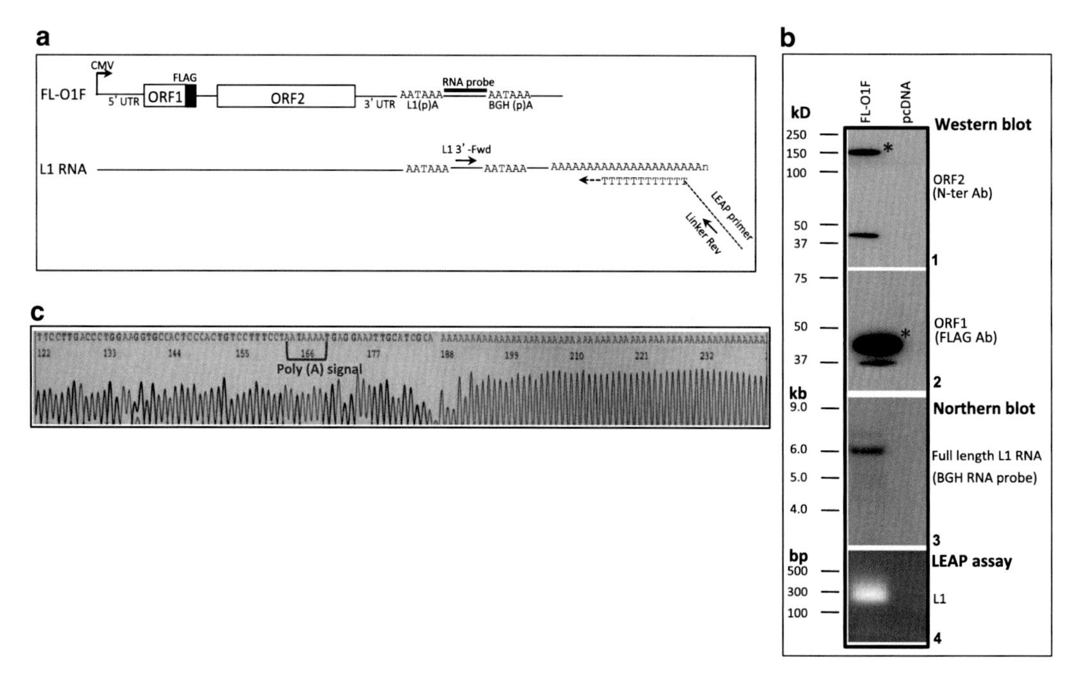

Fig. 1 Epitope tag L1 for L1-RNP purification. (**a**) Disease-causing full-length L1RP [24] containing a FLAG epitope at the C-terminus of ORF1p (FL-01F) was cloned between the CMV promoter and the BGH polyA signal sequence in pcDNA6 (Invitrogen). (**b**) Detection of ORF1p, ORF2p, L1RNA, and ORF2 RT activity in L1-RNPs. *Panel 1*: ORF2p was detected by anti-ORF2p N-terminal antibody [23]. *Panel 2*: ORF1p was detected with anti-FLAG antibody. *Panel 3*: L1 RNA detected by Northern blot analysis—lanes containing 600 ng and 150 ng total RNA isolated from RNPs after transfecting FL-01F and pcDNA6 into 293T cells, respectively. DIG-labeled BGH antisense RNA probe (200 bp) (marked thick line, **a**) detects 6.2 kb L1 RNA transcribed from the engineered L1 construct (FL-01F). *Panel 4*: L1 ORF2p reverse transcriptase activity on L1 RNA detected by the LEAP assay [19]. L1-RNPs were incubated with LEAP RT primer that contains a unique 20 nt linker sequence at the 5′-end followed by a 12 nt poly T sequence. The resultant cDNA was then amplified by an L1 3′-end-specific forward primer (L1 3′-Fwd) and a linker-specific reverse primer (Linker Rev) and resolved on 2.0 % agarose gel. (**c**) Sequence of a representative LEAP clone

critical for the process of L1 retrotransposition [6]. L1 retrotransposition is highly tissue specific, while retrotransposition is observed in germ cells, neurons, and certain types of cancers [7–10].

It is believed that the process of retrotransposition starts with transcription followed by transport of L1mRNA into the cytoplasm where L1mRNA binds with its own encoded proteins [11] along with certain cellular proteins to form an L1-ribonucleoprotein particle (L1-RNP), the retrotransposition intermediate. L1-RNPs then enter into the nucleus where at the genomic target site L1cDNA is synthesized by the RT activity of ORF2p. Subsequently, with the help of L1 proteins (ORF1p and ORF2p) and other unknown cellular proteins a new L1 copy is inserted at the genomic target site. The process of reverse transcription and integration into the genome is named target primed reverse transcription (TPRT) [12, 13], and is not well understood due to a lack of knowledge of the participating proteins that complete the process.

Insertional mutagenesis occurs when a retrotransposon inserts into or near a gene and thus alters the gene's expression or results in the production of mutated proteins. Kazazian et al. [14] first reported a human disease, hemophilia A, caused by de novo insertions of active L1s that disrupted the factor VIII gene. There are now ~100 cases in which L1 retrotransposon activity is responsible for single-gene disease in human [15]. Although L1 is continuously jumping in the human genome, many aspects of the basic mechanism of L1 retrotransposition remain unknown due to lack of knowledge concerning the components present in the L1-RNPs, the retrotransposition intermediate, and their action. Sucrose density gradient centrifugation was first used to purify mouse L1 RNPs from mouse cell line [16]. Subsequently, the same technique was used to purify L1-RNPs from human cultured cells after transfecting an engineered disease-causing active full-length L1 [17–19]. Using this method, it was first demonstrated that basal L1 RNPs contain ORF1p, ORF2p, and L1 RNA [19]. Although informative, this method has a high likelihood of co-purifying other cellular particles that co-sediment with the L1 RNP during the process of centrifugation. Recently, immunoaffinity purification has been employed by several groups to purify L1-RNPs followed by mass spectrometry and RNA sequence analysis to identify the protein [20, 21] and RNA components of the complex [22]. By using this method a few molecules with a role in L1 retrotransposition have been identified [20, 21]. Here we describe the detailed protocol for purification of L1-RNPs by an immunoaffinity method.

2 Materials

2.1 Cell Culture

1. High-glucose Dulbecco's modified Eagle medium (DMEM) without pyruvate and glutamine.
2. L-Glutamine (200 mM).
3. Penicillin-streptomycin (10,000 U/ml).

HEK 293T cells were maintained in a tissue culture incubator at 5 % CO_2, 37 °C in high-glucose Dulbecco's modified Eagle medium (DMEM) without pyruvate supplemented with 10 % fetal bovine calf serum, 2 mM L-glutamine, and 100 U/ml penicillin-streptomycin.

2.2 Components for Transfection

1. Fugene 6 (Promega).
2. Lipofectamine (Invitrogen).
3. Opti-MEM reduced serum medium (Gibco).
4. Plasmid maxiprep purification kit (Qiagen).

2.3 Reagents for Immunoprecipitation, Western Blotting, and Northern Blotting

1. Recombinant RNasin® Ribonuclease Inhibitor (20–40 U/μl) (Promega).
2. Proease inhibitor cocktail (Sigma).
3. Anti-FLAG agarose beads (Sigma).
4. 3× FLAG peptide (Sigma).
5. 1 M HEPES.
6. 2 M KCl.
7. 1 M $MgCl_2$.
8. NP-40.
9. Dithiothreitol (DTT).
10. Glycerol.
11. 4× Laemmli buffer: 200 mM Tris–HCl, pH 6.8, 400 mM DTT, 8 % (w/v) SDS, 40 % glycerol, 0.08 % (w/v) bromophenol blue.
12. NuPAGE® Novex® 4–12 % Bis–Tris Protein Gels, 1.0 mm, 12 well.
13. NuPAGE gel running buffer.
14. PVDF membrane.
15. Protein transfer buffer: 39 mM Glycine, 48 mM Tris base, 0.037 %(w/v) SDS, 20 % methanol.
16. Nonfat dry milk.
17. 1× Tris-buffered saline with Tween-20 (TBST): 20 mM Tris–HCl, pH 7.5, 150 mM NaCl, 0.1 %(v/v) Tween-20.
18. West Pico Chemiluminescent Substrate (Thermo Scientific).
19. X-ray film.
20. X-ray film-developing reagent.
21. Primary antibody to detect ORF2p [22].
22. Anti-FLAG antibody (Sigma).
23. Secondary antibody: ECL Rabbit IgG, HRP-Linked Whole Ab (from donkey) (GE Healthcare).
24. Secondary antibody: ECL Mouse IgG, HRP-Linked Whole Ab (from sheep) (GE Healthcare).
25. Anti-N terminus L1 ORF2p antibody [23].
26. SP6 RNA polymerase.
27. NorthernMax Gly Sample Loading Dye (Ambion).
28. NorthernMax Gly Kit (Ambion).
29. Nylon membrane, positively charged.
30. DIG Easy Hyb Granules (Roche).
31. DIG RNA labeling mix (Roche).
32. DIG wash and block buffer set (Roche).

33. Anti-DIG-AP, Fab fragments (Roche).
34. CDP-Star ready to use (Roche).
35. 1× Dulbecco's phosphate-buffered saline (DPBS): 2.67 mM KCl, 1.47 mM KH_2PO_4, 138 mM NaCl, 8.06 mM $NaH_2PO_4{\cdot}7H_2O$.
36. DNaseI recombinant, RNase-free (Roche).
37. Anti-FLAG agarose bead wash buffer: 100 mM KCl, 5 mM $MgCl_2$, 10 mM HEPES, pH 7.0.
38. Cell lysis buffer: 100 mM KCl, 5 mM $MgCl_2$, 10 mM HEPES, pH 7.0, 0.5 % NP-40, 0.5 mM DTT.
39. Wash buffer 1: 25 mM HEPES, pH 7.0, 250 mM KCl, 5 mM $MgCl_2$, 0.1 % NP-40.
40. Wash buffer 2: 25 mM HEPES, pH 7.0, 100 mM KCl, 5 mM $MgCl_2$, 0.1 % NP-40.
41. Go Taq green master mix (Promega).
42. DIG RNA labeling mix (Roche).
43. 1× Glyoxal RNA loading buffer (Ambion).
44. 10× DNA gel loading buffer.
45. Topo TA cloning Kit (Invitrogen).
46. RNP elution buffer (wash buffer 2 containing 150 μg/ml 3× FLAG peptide).

2.4 Primers for Synthesizing DIG-Labeled L1 RNA Probe and RNA Extraction

1. BGH-SP6rev: 5′-ATTTAGGTGACACTATAGACTCATTTT ATTAGGAAAGGACAG-3′.
2. BGHfwd: 5′-ATCTCAGAAGAGGATCTGAATATGC-3′.

   Bold sequences denote SP6 promoter primer sequence. Note that SP6 promoter primer sequence is attached with reverse primer which will make antisense RNA complementary to BGH poly A sense signal sequence if the PCR DNA template is used to transcribe by SP6 RNA polymerase.
3. Trizol reagent (Invitrogen).
4. Chloroform.
5. 100 % Isopropanol.
6. 20 mg/ml glycogen (Roche).
7. 70 % Ethanol (cold).

2.5 Primers for LEAP Reaction and Other Reagents Used in LEAP

1. LEAP primer: 5′-GCGAGCACAGAATTAATACGACTGGT TTTTTTTTTTT-3′.
2. L13′-Fwd: 5′-CCCGCTGATCAGCCTCGACTG-3′.
3. LinkerRev: 5′-GCGAGCACAGAATTAATACGACT-3′.
4. 5× LEAP buffer: 250 mM Tris–HCl, pH 8.3, 400 mM KCl, 20 mM $MgCl_2$.
5. 0.1 M DTT.

3 Methods

3.1 Transient Transfection of Epitope-Tagged L1 Construct and Preparation of Cell Lysate

The procedure described here is used to purify L1-RNPs by an immunoaffinity technique following transient transfection of an epitope-tagged disease-causing L1 (Fig. 1) into HEK293T cells. We used a single FLAG epitope which was cloned in the C-terminus of ORF1p in a full-length L1 (Fig. 1) [22]. We choose HEK293T cells because this cell line grows very quickly, is very efficient in transient transfection, and displays a high level of L1 retrotransposition. A 100 cm culture disc is sufficient to purify and detect the basal components in L1-RNPs, i.e., ORF1p and ORF2p, by Western blotting, transfected L1RNA by Northern hybridization, and ORF2p-mediated RT activity by the LINE Element Amplification Protocol (LEAP) assay [19]. To purify other RNA and protein components of the L1 RNP, the procedure can be scaled up by growing cells in large vessels with increased surface areas.

1. Eight to 12 h before transfection, plate 1.8×10^6 cells in a 100 mm disc containing 10 ml DMEM media to achieve ~50 % confluency at the time of transfection (*see* **Note 1**).
2. Both Fugene 6 and Lipofectamine are equally effective in transfecting DNA into HEK293T cells. To use Lipofectamine as the transfecting reagent, in one tube place 600 μl of optimem reduced serum media and mix with 10 μg of DNA. The plasmid DNA should be in supercoiled form and prepared using a midi or maxi prep kit (*see* **Note 2**). Incubate the tube for 5 min.
3. In another tube, place 600 μl of optimem reduced serum media and mix with 20 μl lipofectamine reagent. Incubate the tube for 5 min.
4. Combine the diluted DNA and diluted lipofectamine and incubate for 20 min at room temperature (21 °C).
5. Take the cells from the CO_2 incubator. Using a 1 ml pipette, add the DNA-transfection reagent mix to the cells drop by drop. (HEK293T cells are weakly attached to the surface; care should be taken to add the DNA-transfection reagent mixture gently.)
6. Swirl the disc to ensure distribution over the entire plate surface. Incubate cells in the presence of 5 % CO_2 for 40–48 h at 37 °C.

   ALTERNATIVE: To use Fugene 6 as transfection reagent, take 600 μl of serum-free media in a 1.5 ml tube, add 18 μl Fugene 6, mix by pipetting 2–3 times, and incubate for 5 min. After 5 min, add 8 μg DNA to the tube, pipette gently 2–3 times to mix all the components together, and then incubate for another 15 min. After the final incubation, remove the culture vessel from the incubator and add the mixture drop by drop to the cells. Swirl the disc to ensure distribution over the entire plate surface. Incubate the cells in the presence of 5 % CO_2 for 40–48 h at 37 °C.

7. After incubation, take the disc to the workbench, and decant the media into a 500 ml plastic beaker which will be used for liquid disposal.
8. Wash the cells with cold 10 ml 1× DPBS solution once. Scrape cells by adding 5 ml cold 1× DPBS and transfer them to a chilled 15 ml tube. Harvest the cells by centrifugation at 3000 × *g* at 4 °C for 5 min.
9. Remove 1× DPBS solution without disturbing the pellet and add 1 ml cold lysis buffer containing RNase inhibitor (1 U/μl) and protease inhibitor cocktail (10 μl for 1 ml). Mix cells with the lysis buffer by pipetting 6–8 times with a 1 ml pipette and transfer the entire volume to a chilled 1.5 ml tube.
10. Incubate the tube on ice for 15 min with intermittent mixing (by inverting the tube 2–3 times) for lysis of cells. Centrifuge the lysate at 10,000 × *g* for 10 min at 4 °C.
11. Collect the supernatant in a fresh chilled 1.5 ml tube without disturbing the pellet. For immediate use keep the tube on ice. If the supernatant is not to be used immediately, flash freeze with liquid nitrogen and store at −70 °C.

3.2 Preparation of Anti-FLAG Agarose Beads

1. We recommend using anti-FLAG agarose beads from Sigma for the purification of L1-RNPs. The anti-FLAG agarose beads are supplied in 50 % glycerol with buffer. Thoroughly resuspend the resin by gently inverting the container several times (*see* **Note 3**).
2. Transfer 40 μl anti-FLAG agarose gel suspension using a blunt cut 200 μl tip into a 1.5 ml tube. Forty microliter bead suspension is equivalent to 20 μl packed gel beads which will efficiently bind ~2 μg FLAG-tagged protein.
3. Add 1 ml cold wash buffer 1, mix with resin by inverting the tube several times, and then centrifuge at 1000 × *g* for 1 min at room temperature (RT). Keep the tube on ice for 2 min to settle the beads at the bottom of the tube. Remove the wash buffer without disturbing the resin. Repeat the washing steps twice more.

3.3 Immunoprecipitation of L1-RNPs

1. Transfer 1 ml cell lysate to the washed resin. Incubate the tube at 4 °C for 1 h with gentle shaking either in a platform shaker or in an overhead mixing device. Beads should not settle during the incubation (*see* **Note 4**).
2. After incubation, centrifuge the tube at 1000 × *g* for 1 min at 4 °C, and incubate the tube on ice for 2 min to settle the beads. Discard the supernatant carefully without disturbing the beads (*see* **Notes 5** and **6**).

3. Add 1 ml wash buffer 2, and mix with the beads gently by inverting several times. Repeat centrifugation to settle the beads and remove the supernatant (*see* **Notes 5** and **6**). Wash an additional four times with wash buffer 1 and one time with wash buffer 2.
4. Remove wash buffer 2, add 100 μl elution buffer, and incubate the tube for 20 min with gentle shaking on a platform shaker.
5. Collect the supernatant which contains the L1-RNPs by centrifugation at 1000 × *g* for 1 min and store in aliquots in 15 % glycerol at −70 °C if the RNPs are not to be used immediately. L1RNPs are stable for at least 2 months from the date of preparation.

3.4 Detection of ORF1p and ORF2p by Western Blot Analysis

1. To detect ORF2p, take 20 μl RNP elute, add 5 μl 4× Laemmli buffer, and heat at 95 °C for 5 min.
2. Centrifuge the tube at 10,000 × *g* in a microcentrifuge for 30 s.
3. Resolve samples on NuPAGE 4–12 % Bis–Tris polyacrylamide gel, 1 mm thick (Invitrogen) in an X-Cell Sure Lock mini cell following the manufacturer's instruction.
4. Following electrophoresis, transfer the proteins to a PVDF membrane using a Xcell II blot module (or a similar device) at 25 V for 1 h.
5. Block the membrane in 1× TBST containing 5 % w/v nonfat dry milk at room temperature for 1 h.
6. Dilute the primary antibody for ORF2p [23] in 1× TBST with 5 % w/v nonfat dry milk and incubate overnight at 4 °C.
7. Rinse the blot three times for 10 min with 1× TBST.
8. Dilute the secondary antibody in 1× TBST with 5 % w/v nonfat dry milk and incubate for 1 h at room temperature.
9. Rinse the blot three times for 10 min with 1× TBST.
10. Develop the blot using West Pico Chemiluminescent Substrate.
11. Expose the blot to X-ray film or imaging system (Fig. 1b; panel 1).
12. To detect ORF1p use 4 μl RNP elute. Use anti-FLAG antibody to detect ORF1p (Fig. 1b; panel 2). Use the RNPs purified after transfecting with pcDNA6 (vector control) as a control.

3.5 Preparation of DIG-Labeled L1 RNA Probe

1. PCR amplify the bovine growth hormone (BGH) poly A signal sequence present in pcDNA6/myc-His B (Invitrogen) using primer pair BGH-SP6 rev and BGHfwd.

   Set up the PCR reaction in a 25 μl reaction volume containing 1 μl pcDNA6 template (5–10 ng DNA), 12.5 μl 2× Go Taq green master mix, 1 μl 20 μm BGH fwd, 1 μl 20 μM BGH-SP6rev, and 9.5 μl water.

2. Use PCR conditions: one cycle at 94 °C for 30 s, followed by 30 cycles at 94 °C for 20 s, 58 °C for 20 s, and 72 °C for 20 s and finally one cycle at 72 °C for 2 min.
3. Resolve products in 2.0 % agarose electrophoresis, excise the band, and gel extract DNA using a gel extraction kit.
4. Take 100 ng of the purified PCR DNA template to generate digoxigenin (DIG)-11-UTP-labeled antisense BGH poly A signal sequence RNA probe. Perform a transcription reaction in 20 μl volume containing 2 μl template DNA (100 ng), 2 μl 10× DIG RNA labeling mixture, 2 μl 10× SP6 RNA polymerase buffer, 1 μl SP6 polymerase, and 13 μl nuclease-free water.
5. Incubate tube at 37 °C for 1 h.
6. Add 1 μl RNase-free DNase (10 U/μl) to remove template DNA.
7. Incubate the tube for 15 min at 37 °C.
8. Stop the reaction by adding 2 μl 0.2 M EDTA, pH 8.0.
9. Take 2 μl reaction and mix with 13 μl 1× DNA gel loading buffer. Incubate tube at 65 °C for 15 min to denature RNA.
10. Chill the tube on ice for 5 min. Run the sample in a 1.5 % neutral agarose gel to check the quality and quantity of the RNA probe. The yield of RNA should be around 200 ng/μl.
11. Store the DIG-labeled RNA probe at −70 °C until use. Use 1 μl probe per 10 ml pre-hybridization solution.

3.6 Detection of Engineered L1 RNA by Northern Blot Analysis

1. Take 50 μl of purified L1-RNP, and increase its volume to 250 μl by adding RNase-free water.
2. Add 250 μl Trizol reagent, vortex for 15 s, and incubate at room temperature for 3 min.
3. Add 100 μl chloroform, vortex for 15 s, and again incubate for 2 min at room temperature.
4. Centrifuge at 11,000 ×*g* for 5 min at 4 °C using a microfuge.
5. Transfer upper aqueous phase to another tube, and add 250 μl 100 % isopropanol and 1 μl 20 mg/ml glycogen. Incubate the tube at room temperature for 10 min.
6. Centrifuge at 11,000 ×*g* for 10 min at 4 °C to precipitate RNA.
7. Wash the RNA pellet with 500 μl 70 % ethanol (cold) by centrifugation at 11,000 ×*g* for 5 min at 4 °C.
8. Dry the RNA pellet in a 37 °C incubator for 3–4 min. Dissolve RNA in 20 μl 1× NorthernMax Gly Sample Loading Dye.

9. Denature the RNA by incubating at 65 °C for 15 min. Chill the tube on ice for 5 min before separating RNA on a 0.8 % denatured agarose gel. Use NorthernMax Gly Kit for running the gel and blotting the RNA onto a nylon membrane.
10. Place the wet nylon membrane RNA side up on a Whatman 3 MM paper and UV-cross-link. Transfer the membrane in a hybridization tube and add 10 ml preheated DIG Easy Hybe Granules. Rotate the tube in a hybridization oven at 68 °C for 1 h for pre-hybridization.
11. Hybridize overnight at 68 °C with a BGH antisense RNA probe (150 nucleotides) labeled with digoxigenin (DIG)-11-UTP.
12. Wash membrane using DIG Wash and Block Buffer Set as per the manufacturer's instruction.
13. Immunodetect RNA with anti-digoxigenin-AP, Fab fragments as per the manufacturer's instruction.
14. Apply the chemiluminescent substrate CDP-Star ready to use to the blot following the manufacturer's suggestions.
15. Capture the chemoluminescent signals using X-ray film or imaging system (Fig. 1b; panel 3).

3.7 Detection of ORF2p-Mediated Reverse Transcriptase (RT) Activity in Purified L1-RNP by LINE Element Amplification Protocol (See Notes 7 and 8)

1. Set up reaction in a 20 μl reaction volume containing 1 μl L1-RNP, 4 μl 5× LEAP buffer, 1 μl 10 mM dNTPs, 1 μl 50 μM LEAP primer, 0.5 μl RNasin (20 U/μl), 2 μl 0.1 M DTT, and 10.5 μl nuclease-free water.
2. Incubate tube at 37 °C for 1 h.
3. Perform PCR reaction in 25 μl total volume containing 1 μl LEAP reaction, 0.5 μl 10 μm linker-specific reverse primer (LinkerRev), 0.5 μl 10 μM transcript-specific forward primer (L13′-Fwd), 12.5 μl 2× Go Taq green master mix, and 10.5 μl water.
4. Use PCR conditions of one cycle at 94 °C for 30 s, followed by 35 cycles at 94 °C for 20 s, 56 °C for 20 s, and 72 °C for 20 s and finally one cycle at 72 °C for 2 min.
5. Resolve products in a 2.0 % agarose gel (Fig. 1b; panel 4).
6. Excise bands, gel extract, and clone in a TOPO TA cloning vector or similar.
7. Check clones by colony PCR using M13F and M13R.
8. Resolve colony PCR products in 1.2 % agarose gel, band excise, and sequence using M13R primer.
9. Analyze LEAP product sequences that show L1 3′ end with long poly A sequences (Fig. 1c).

4 Notes

1. HEK293T cells should be healthy and 30–50 % confluent before transfection.
2. The plasmid DNA used for transfection should be freshly prepared and in supercoiled form.
3. Anti-FLAG agarose beads need to be resuspended thoroughly. Twenty microliter packed beads is sufficient for cells obtained from one 100 mm disc. Use of more beads will show nonspecific interaction.
4. For immunoprecipitation of L1-RNPs 1-h incubation of cell lysate with anti-FLAG agarose is sufficient. Longer incubation leads to nonspecific interaction and also has chance of RNA degradation.
5. It is recommended to use flat gel loading tip to remove buffer from anti-FLAG agarose beads (pore size of the tips is smaller than the size of agarose beads and thus beads will not be lost during repeated washing steps).
6. Never dry anti-FLAG agarose beads during purification of L1-RNPs.
7. It is recommended to use RNase-free lab space and take extreme precaution to avoid any RNase contamination during the purification of L1-RNPs.
8. All chemicals and solution used to purify L1-RNPs must be RNase free.

Acknowledgements

This work was supported by NIH grant (RO1GM099875) to HHK. We thank John L. Goodier for his valuable input to develop the protocol. We thank all the members of the Kazazian lab for critical discussion and sharing valuable reagents and constructs. We thank members of the Mandal Lab for critical reading of the protocol.

References

1. Goodier JL, Kazazian HH Jr (2008) Retrotransposons revisited: the restraint and rehabilitation of parasites. Cell 135:23–35
2. Ostertag EM, Kazazian HH Jr (2001) Biology of mammalian L1 retrotransposons. Annu Rev Genet 35:501–538
3. Mathias SL, Scott AF, Kazazian HH Jr, Boeke JD et al (1991) Reverse transcriptase encoded by a human transposable element. Science 254:1800–1810
4. Feng Q, Moran JV, Kazazian HH Jr, Boeke JD (1996) Human L1 retrotransposon encodes a conserved endonuclease required for retrotransposition. Cell 87:905–916
5. Martin SL, Bushman FD (2001) Nucleic acid chaperone activity of the ORF1 protein from the mouse LINE-1 retrotransposon. Mol Cell Biol 21:467–475
6. Moran JV, Holmes SE, Naas TP, DeBerardinis RJ, Boeke JD, Kazazian HH Jr (1996) High

frequency retrotransposition in cultured mammalian cells. Cell 87:917–927
7. Kano H, Godoy I, Courtney C, Vetter MR et al (2009) L1 retrotransposition occurs mainly in embryogenesis and creates somatic mosaicism. Genes Dev 23:1303–1312
8. Muotri AR, Chu VT, Marchetto MC, Deng W et al (2005) Somatic mosaicism in neuronal precursor cells mediated by L1 retrotransposition. Nature 435:903–910
9. Solyom S, Ewing AD, Rahrmann EP, Doucet T, Nelson HH et al (2012) Extensive somatic L1 retrotransposition in colorectal tumors. Genome Res 22:2328–2338
10. Lee E, Iskow R, Yang L, Gokcumen O, Haseley P et al (2012) Landscape of somatic retrotransposition in human cancers. Science 337:967–971
11. Wei W, Gilbert N, Ooi SL, Lawler JF et al (2001) Human L1 retrotransposition: cis preference versus trans complementation. Mol Cell Biol 21:1429–1439
12. Luan DD, Eickbush TH (1995) RNA template requirements for target DNA-primed reverse transcription by the R2 retrotransposable element. Mol Cell Biol 15:3882–3891
13. Cost GJ, Feng Q, Jacquier A, Boeke JD (2002) Human L1 element target-primed reverse transcription in vitro. EMBO J 21:5899–5910
14. Kazazian HH Jr, Wong C, Youssoufian H, Scott AF et al (1988) Haemophilia A resulting from de novo insertion of L1 sequences represents a novel mechanism for mutation in man. Nature 332:164–166
15. Hancks DC, Kazazian HH Jr (2012) Active human retrotransposons: variation and disease. Curr Opin Genet Dev 22:191–203
16. Martin SL (1991) Ribonucleoprotein particles with LINE-1 RNA in mouse embryonal carcinoma cells. Mol Cell Biol 11:4804–4807
17. Hohjoh H, Singer MF (1996) Cytoplasmic ribonucleoprotein complexes containing human LINE-1 protein and RNA. EMBO J 15:630–639
18. Kulpa DA, Moran JV (2005) Ribonucleoprotein particle formation is necessary but not sufficient for LINE-1 retrotransposition. Hum Mol Genet 14:3237–3248
19. Kulpa DA, Moran JV (2006) Cis-preferential LINE-1 reverse transcriptase activity in ribonucleoprotein particles. Nat Struct Mol Biol 13:1655–1660
20. Goodier JL, Cheung LE, Kazazian HH Jr (2013) Mapping the LINE1 ORF1 protein interactome reveals associated inhibitors of human retrotransposition. Nucleic Acids Res 41:7401–7419
21. Taylor MS, Lacava J, Mita P, Molloy KR et al (2013) Affinity proteomics reveals human host factors implicated in discrete stages of LINE-1 retrotransposition. Cell 15:1034–1048
22. Mandal PK, Ewing AD, Hancks DC, Kazazian HH Jr (2013) Enrichment of processed pseudogene transcripts in L1-ribonucleoprotein particles. Hum Mol Genet 22:3730–3748
23. Goodier JL, Ostertag EM, Engleka KA, Seleme MC, Kazazian HH Jr (2004) A potential role for the nucleolus in L1 retrotransposition. Hum Mol Genet 13:1041–1048
24. Kimberland ML, Divoky V, Prchal J, Schwahn U, Berger W, Kazazian HH Jr (1999) Full-length human L1 insertions retain the capacity for high frequency retrotransposition in cultured cells. Hum Mol Genet 8:1557–1660

Chapter 20

Characterization of L1-Ribonucleoprotein Particles

Martin S. Taylor, John LaCava, Lixin Dai, Paolo Mita, Kathleen H. Burns, Michael P. Rout, and Jef D. Boeke

Abstract

The LINE-1 retrotransposon (L1) encodes two proteins, ORF1p and ORF2p, which bind to the L1 RNA in *cis*, forming a ribonucleoprotein (RNP) complex that is critical for retrotransposition. Interactions with both permissive and repressive host factors pervade every step of the L1 life cycle. Until recently, limitations in detection and production precluded in-depth characterization of L1 RNPs. Inducible expression and recombinant engineering of epitope tags have made detection of both L1 ORFs routine. Here, we describe large-scale production of L1-expressing HEK-293T cells in suspension cell culture, cryomilling and affinity capture of L1 RNP complexes, sample preparation for analysis by mass spectrometry, and assay using the L1 element amplification protocol (LEAP) and qRT-PCR.

Key words LINE-1, Ribonucleoprotein, Affinity purification, Protein complexes, Interactomics, Cryomilling, Mass spectrometry, Metabolic labeling

1 Introduction

LINE-1 is the only active protein-coding transposon in humans. Hundreds of thousands of copies of L1 make up ~20 % of the human genome and are a source of genetic structural variation and between humans and instability in cancer [1–4]. Although most L1s are truncated or mutated, ~90 copies per cell are capable of replication [5]. As a streamlined DNA parasite with which we have coevolved since early eukaryotic existence [1], L1 encodes only two proteins and coopts cellular machinery in order to replicate. To defend against L1-mediated mutagenesis, our cells possess multiple mechanisms to suppress its activity [6–18]. Although teleological roles of retrotransposition in evolution are somewhat controversial [19–21], critical interactions with the host functions are highly conserved; for example, the PIP box, a motif required for L1 ORF2p-PCNA binding and retrotransposition, is conserved from corn to humans [6].

Jose L. Garcia-Pérez (ed.), *Transposons and Retrotransposons: Methods and Protocols*, Methods in Molecular Biology, vol. 1400, DOI 10.1007/978-1-4939-3372-3_20,

L1 ORFs bind the L1 mRNA to form a ribonucleoprotein (RNP) complex [22, 23]. ORF1p is a trimeric nucleic acid binding protein that is highly expressed in human cancers and cell culture models and thought to have chaperone activity; it is required for retrotransposition but its precise role is unknown [24–29]. ORF2p possesses both endonuclease and reverse transcriptase activities [30, 31] and, in contrast to ORF1p, is expressed at such low levels even after overexpression (due to an unconventional translation mechanism) that detection was a technical barrier in the field until recently [6, 32, 33]. However, the combination of a tetracycline-inducible promoter and improved epitope tagging now makes detection and purification of both ORFs routine [6], reviewed in [34]. A synthetic codon-optimized L1, *ORFeus*-Hs [35, 36], produces ~40-fold more L1 RNA and ORF2p and was critical to establishing protocols for purification of ORF2p; interactors were similar to native L1RP [6].

Here, we outline our methods for suspension production, cryomilling, and affinity capture of L1 RNPs with subsequent characterization by mass spectrometry, the L1 element amplification protocol (LEAP) [23], and quantitative real-time reverse transcription PCR (qRT-PCR). Suspension production improves cell densities as compared to adherent cell culture, and allows for sufficient cell material for solid-phase lysis under liquid N_2 by cryomilling. This approach has a number of practical advantages over liquid-phase lysis including reduced background in coimmunoprecipitation and the ability to store the milled cell powder at −80 °C, allowing repeated experiments to be performed from the same sample without the need to produce more starting material [37, 38].

We express inducible L1 in Tet-On HEK-293$_{LD}$ cells using pCEP4-based episomal vectors and either large-scale transient transfection or quasi-stable episomal puromycin-selected cell pools [6]. Because no Tet-On HEK-293T cells were available commercially, we produced this line using a linearized pTet-On Advanced (Clontech), modified to contain blasticidin resistance instead of neomycin (pLD215). Cells are maintained in square bottles on an orbital shaker [39]. Alternative systems include GNTI-HEK293S cells, HEK-293F cells (Life Technologies), and the T-REx system (Life Technologies), and can be achieved in spinner flasks, conical flasks, and wave bags [40, 41].

For L1 affinity isolation, we couple antibody to functionalized micron-scale paramagnetic beads with relatively inert surfaces (Dynabeads, Life Technologies). After coupling, antibody is immobilized on the surface of the bead. For the isolation of protein complexes, these beads provide a number of advantages over agarose and porous synthetic resins (which both contain antibody bound within the pores) including the ability to bind and release larger complexes, faster binding and release, and reduced background [37, 38, 42]. Switching to this medium was critical for the successful characterization of L1 RNPs.

L1 RNPs purified by affinity capture provide the purest, most active L1 elements reported to date and are excellent starting material for assay by mass spectrometry, LEAP, and qRT-PCR [6]. We provide several of our working protocols for L1 sample preparation and analysis.

2 Materials

2.1 Suspension Cell Culture

1. Humidified CO_2-controlled tissue culture incubator.
2. Orbital shaker platform at 130 rpm fitted with racks.
3. 20 mm 40-place test tube racks.
4. Diagonal cutting pliers, flat wood file, bandsaw (for modifying racks).
5. Corning Pyrex 1 L glass bottles.
6. 7×® Cleaning solution (Bellco Glass).
7. Hybridoma SFM medium (Life Technologies).
8. Freestyle 293 Medium (Life Technologies).
9. Opti-MEM Reduced Serum Medium.
10. TrypLE Express (Life Technologies).
11. Certified tetracycline-Free FBS (Tet-free FBS).
12. DMEM medium.
13. Phosphate-buffered saline (PBS).
14. PEI Max (MW 40,000), Polysciences: Two grams is enough to transfect >600 L.
    (a) To prepare working 1 mg/mL PEI Max solution:
    (b) Dissolve 100 mg PEI Max in 90 mL ddH_2O.
    (c) Adjust pH to 7.0 using 1 M NaOH.
    (d) Adjust volume to 100 mL, filter sterilize, and store at 4 °C.
    (e) *NEVER FREEZE PEI working stock.* Working stocks can be used for up to 6 months if stored at 4 °C.

2.2 Cell Harvest

15. Large-volume floor centrifuge with appropriate rotor (e.g., 4×1 L, 6×500 mL).
16. 16ga needles.
17. Luer-lock syringes, 5 mL, 10 mL, or 30 mL.
18. Luer-lock syringe end caps (BioRad).
19. Liquid nitrogen and Dewar flask.
20. Gloves for handling liquid nitrogen.
21. Small Styrofoam box.

2.3 Cryomilling

22. Ice pan (Fisher).
23. RETSCH Planetary ball mill PM 100 or PM 100 CM.
24. RETSCH Stainless steel grinding balls 20 mm diameter.
25. RETSCH Stainless steel "comfort" grinding jars 50 mL and/or 125 mL.
26. Stainless steel measuring spoons for small amounts (e.g., "hint, pinch, dash").
27. Stainless steel spatulas.
28. Extra-large forceps.

2.4 Coupling of Magnetic Medium (Dynabeads)

29. Dynabeads M270 Epoxy (Life Technologies).
30. Anti-Flag M2 Antibody (Sigma). *See* **Note 1.**
31. Anti-ORF1 Antibody 4H1 (obtained from Kathleen H. Burns).
32. Magnetic separator for microcentrifuge tubes (Dynamag 2, Life Technologies).
33. Magnetic separator for 15 mL conical tubes (Dynamag 15, Life Technologies). *See* **Note 2**.
34. Zeba Spin Desalting Columns 7 K MWCO (Thermo) or chromatography system with preparatory-scale desalting column.
35. Nutating mixer or orbital shaker.
36. Rotating test tube wheel in a 37 °C environment.
37. 100 mM Sodium phosphate buffer pH 7.4 (makes 1 L):
    (a) 2.62 g Sodium phosphate monobasic monohydrate.
    (b) 14.42 g Sodium phosphate dibasic dihydrate.
    (c) Dissolve in 900 mL ddH_2O, adjust pH if necessary with HCl and NaOH, and adjust to 1 L.
38. 3 M Ammonium sulfate (in phosphate buffer) (makes 100 mL):
    (a) 39.6 g Ammonium sulfate.
    (b) Dissolve in 0.1 M sodium phosphate buffer (pH 7.4) and adjust to 100 mL.
39. 10× PBS—pH 7.4. *See* **Note 3** (makes 1 L):
    (a) 2.62 g Sodium phosphate monobasic monohydrate.
    (b) 14.42 g Sodium phosphate dibasic dihydrate.
    (c) 87.8 g Sodium chloride.
    (d) Dissolve in 900 mL with ddH_2O, adjust pH if necessary with HCl and NaOH, and adjust to 1 L.
40. Resuspension buffer (PBS, 50 % glycerol, 0.5 mg/mL BSA; makes 10 mL):
    (a) 1.0 mL 10× PBS.
    (b) 6.3 g glycerol (place tube + rack on a balance, tare, and pipet).

(c) 5 mg BSA.
(d) ddH_2O to 10 mL.
41. PBS + 0.5 % Triton X-100 (w/v) in 100 mL.
42. 100 mM Glycine-HCl, pH 2.5 (adjust pH with HCl).
43. Triethylamine.
44. 10 mM Tris–HCl, pH 8.8.
45. 10 % (w/v) Sodium azide (NaN_3).

2.5 Affinity Capture (See Note 4)

46. HEPES buffer, pH 7.4.
47. Sodium chloride.
48. Triton X-100.
49. Protease inhibitor cocktail: Complete EDTA-free (Roche).
50. Ultrasonic liquid processor with micro tip (Branson Sonifier or similar).

2.6 LEAP (L1 Element Amplification Protocol) and Real-Time Reverse Transcription PCR (qRT-PCR)

51. LEAP reaction mixture (Table 1), shown for ten reactions.
52. Primers for LEAP and qRT-PCR (*see* Table 2).
53. SuperScript III Reverse Transcriptase (Life Technologies) or similar.
54. FastStart Taq DNA Polymerase (Roche Applied Science) or similar.
55. TRIzol Reagent (Life Technologies).
56. RNA from GFP-transfected human cells, purified (*see* **Note 5**).

Table 1
LEAP reaction mixture for ten reactions

Sample	Final concentration	Stock concentration	µL each	×10
RNP prep			2	
Tris pH 7.5	50 mM	1 M	2.5	25
KCl	50 mM	1 M	2.5	25
$MgCl_2$	5 mM	1 M	0.25	2.5
DTT	10 mM	1 M	0.5	5
3′ Anchor primer	0.4 µM	10 µM	2	20
RNAsin	20 U/rxn	40 U/µL	0.5	5
dNTPs	0.2 mM	10 µM	0.5	5
Tween 20	0.05 %	10 %	0.25	2.5
ddH_2O			39	390

Table 2
Primers for LEAP and qRT-PCR

Name	Sequence
JB11560	5′-GCGAGCACAGAATTAATACGACTCACTATAGGTTTTTTTTTTTT-3′
JB11564	5′-GCGAGCACAGAATTAATACGACTC-3′
JB14067	5′-GGATCCAGACATGATAAGATACATTGATGA-3′
JB13415	5′-GCTGGATGGAGAACGACTTC-3′
JB13416	5′-TTCAGCTCCATCAGCTCCTT-3′
JB13417	5′-CTGATCAGCCGCATCTACAA-3′
JB13418	5′-TGGTCTTGATCTGCATCTCG-3′
JB13766	5′-ACGTAAACGGCCACAAGTTC-3′
JB13767	5′-AAGTCGTGCTGCTTCATGTG-3′

57. StepOne Plus Instrument or similar (Life Technologies) and appropriate reaction plates.
58. Fast SYBR Green Master Mix (Life Technologies) or similar.
59. RNAseZap (Life Technologies) or similar.

3 Methods

3.1 Suspension Culture of HEK-293T Cells

Suspension culture allows production of large amounts of human cell pellets with minimal work, waste, and cost when compared to monolayer adherent growth. We use a square bottle system in an orbital shaker, modified from [39] (Fig. 1a). For comparison, one confluent 15 cm plastic adherent culture dish provides about 100 mg wet cell weight (cell pellet, WCW), corresponding to 20–30 million cells, and uses 20 mL of media. To produce 1.5 g of cell pellet, 15 disposable dishes and 300 mL of media are required, and scale up, induction, and harvest are laborious. In suspension, cell densities of 4–5 million cells/mL are readily achieved in log phase using reusable 1 L glass bottles, yielding 4–5 g WCW from 333 mL media. Thus, each bottle is the equivalent of approximately 50 culture plates. The total yield from 48 culture bottles (Fig. 1a) was approximately 250 g WCW and was harvested in 90 min by one person; an entire incubator full of plates (Fig. 1b) produced 16 g WCW and was harvested in 4 h by two people.

We culture Tet-On HEK-293T_{LD} cells in Freestyle 293 medium supplemented with 1 % tetracycline-free FBS and 2 mM l-glutamine (suspension medium, *see* **Note 6**). Basic shaker setup and maintenance are described below. Initially, for each construct we

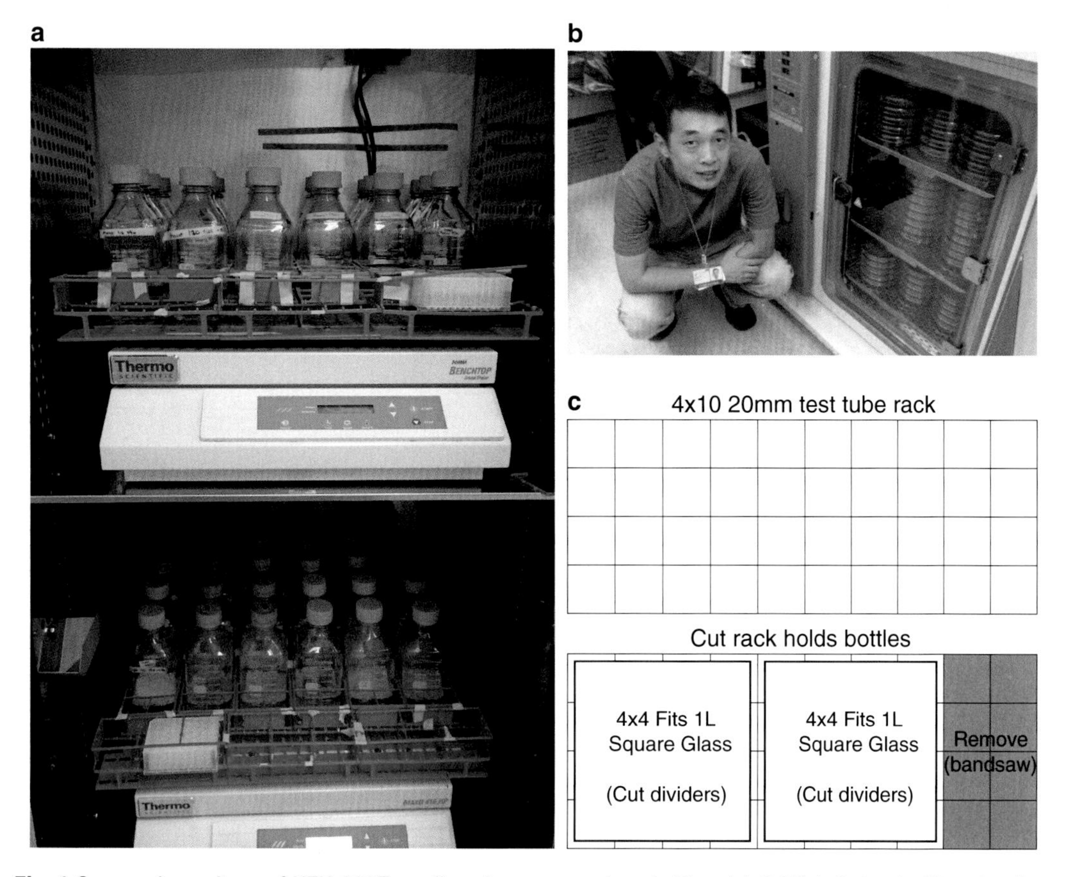

Fig. 1 Suspension culture of HEK-293T$_{LD}$ cells using square glass bottles. (**a**) Orbital shaker bottle setup in a CO_2-controlled humidified incubator cabinet. Total yield from 48 bottles was approximately 250 g wet cell weight (WCW), the equivalent of 2500 15 cm adherent culture plates. (**b**) Dr. Lixin Dai proudly showcases an incubator full of adherent culture plates for harvest. Harvesting this entire incubator full of cells yielded 16 g WCW, which can now be accomplished in three to four suspension bottles. (**c**) Diagram of modifications to Nalgene 4×10 configuration 20 mm test tube racks to fit square glass bottles. Dividers are cut with diagonal cutting pliers and smoothed with a file. Sections of rack can be removed with a bandsaw. The diversity of available rack sizes and shapes allows adaptation of this strategy for many size bottles

transfected, puromycin-selected, and scaled up adherent cultures to suspension. This protocol is effective but time consuming. Subsequently, we found that large-scale transient transfection using PEI-MAX (Polysciences) provides equivalent per-cell yields with much less work. Both protocols are described in separate sections below (*see* **Note 7**).

3.1.1 Square Bottle Orbital Shaker Setup

We use a Thermo Forma Model 416 orbital shaker in a 5 % CO_2 humidified incubator cabinet, shaking at 130 rpm. After repeated inconsistency issues with failed transfections and poor growth in plastic bottles made of polypropylene, polyethylene, and other plastics, we can only recommend culture in square Corning Pyrex glass bottles. Bottles are held in place using autoclavable test tube racks, cut to size. The racks last for a number of years in regular use before needing to be replaced.

1. Cut tube racks to fit bottle configurations (Fig. 1c). Use diagonal cutting pliers to remove undesired dividers and then a flat wood file to smooth sharp edges. To shorten racks, cut to length with a bandsaw. We start with 10 × 4 racks and cut most racks to fit two 1 L glass bottles, a configuration with two 4 × 4 openings as shown. Other configurations or other size starting racks would allow use of alternate or smaller bottles.
2. Clean and autoclave racks before installation.
3. Install racks to shaker platform using stainless steel flat-head machine screws and, if needed, washers. The bottom of the racks may need to be drilled.
4. Use paper tape between the racks to hold weak sections together.

3.1.2 Bottles and Volumes

1. Clean freshly gloved hands with 70 % ethanol before going into the shaker cabinet.
2. Media is maintained at up to 33 % of the indicated capacity for optimal mixing and gas exchange.
   (a) Minimum volume is 5 % of indicated capacity (50 mL for 1 L bottles).
   (b) Cell shearing is inversely proportional with bottle size; too little volume can result in excessive shearing.
3. Bottle caps are kept loose to allow gas exchange, but not loose enough to easily fall off.
4. The outsides of bottles are wiped clean before returning to the shaker.
5. After use, bottles must be immediately rinsed and filled with tap water. Bottles are left soaking until they are ready to clean. A small number of "beached" cells stuck to the side of the glass at the media-air interface is normal for bottles that have been used for a number of weeks.
6. When a batch of bottles is ready for sterilization, bottles are thoroughly scrubbed with a brush cleaned using 1 % solution of 7× cleaner, rinsed six times using tap water, and finally rinsed with deionized water.
7. Bottles are sterilized by autoclaving using a dry cycle at 121 °C, with 45-min sterilization and 15-min drying. *See* **Note 8**.
8. In the event of contamination, bottles are soaked in 10 % bleach for at least 2 h and then washed and autoclaved as above. *See* **Note 9**.

3.1.3 General Cell Line Maintenance

1. Most operations are done by pouring. Be careful to consider that only the threads of the bottle are sterile.
2. Do not allow non-sterile bottle sides to be positioned over the mouth of an open bottle. If a bottle is to be totally emptied, we pipet the last ~40 mL.
3. Caps are stored face-down on the hood. Remember that the hood surface and the rim of the cap are never sterile but the cap threads and inside are sterile.
4. This media has no pH indicator dye, but one can be added if desired.
5. Cells are generally maintained at a density of ~0.2–4 million/mL (*see* **Note 10**). Cells will grow to ~7 million/mL [39] but growth slows after ~5 million/mL.
6. We do not spin the cells down to fully exchange media. Passaging is done by dilution. Centrifugation is time consuming, risks contamination, and is generally not helpful.
7. A typical 1:5 split of a 333 mL culture is done as follows:
   (a) Count the cells.
   (b) Fill to 1 L (there is an indentation in the bottles at exactly this volume).
   (c) Cap tightly and gently shake to mix.
   (d) Pour 200 mL of diluted culture into each bottle.
   (e) Fill each bottle to 333 mL.
8. Antibiotics and/or antimycotics can only be used if cells are not to be transfected in suspension because the combination of PEI and Pen-strep is toxic. Most of our cultures are antibiotic free.

3.1.4 Counting Suspension HEK-293Ts with a Hemocytometer

HEK-293Ts grow in small clumps of 1–30 cells in suspension. Accurate counting requires dissociation of the clumps by gentle trituration using a pipet. Due to differences in light scattering by different clump sizes, optical density is not an accurate measure of cell number. We visualize the cells before and after dissociation because shearing in the dissociation protocol lyses a small fraction of the cells.

1. Using a 1 mL serological pipet, aliquot 200 μL (*see* **Note 11**) of culture to a clean microcentrifuge tube (*see* **Note 12**).
2. Mix by flicking, and pipet 10 μL onto one side of the hemocytometer.
3. With a 200 μL pipet, set the volume to ~150 μL and triturate 30 times to break up clumps. Try not to foam.
4. Pipet 10 μL onto the remaining half of the hemocytometer.

3.1.5 Suspension Transient Transfection Using PEI Max

Transfection is done using 1 μg DNA/mL media. Transfection complexes are prepared in 1/20th the culture volume hybridoma SFM with a 3:1 ratio of PEI Max to DNA (*see* **Note 13**). A transfection protocol for 1 L culture (three 1 L bottles, each containing 333 mL culture) is outlined below.

1. Prep DNA. High-quality endotoxin-free DNA is critical to success. We have had best results with PureLink HiPure Maxi and Giga prep kits (Life Technologies) using the manufacturer's protocol (*see* **Note 14**).
2. Day 1:
   (a) Grow cells to 2.5–3.5 million/mL. Count the cells.
   (b) Warm 50 mL of hybridoma SFM to room temperature (*see* **Note 15**).
   (c) Dilute 50 μg DNA in the hybridoma media. Mix well (*see* **Note 16**).
   (d) Add 150 μL 1 mg/mL PEI Max (pH 7.0). Mix well.
   (e) Incubate for 15 min at room temperature to allow DNA-PEI complex to form.
   (f) Pipet 16.7 mL into each 333 mL culture and return to the incubator.
3. Days 2–3 or 2–5:
   (a) Option 1: Induce cells by adding 1 μg/mL doxycycline. Harvest 24 h later on day 3 (*see* **Note 17**).
   (b) Option 2: Split cells 1:3 on day 2, induce on day 4, and harvest 24 h later on day 5 (*see* **Note 18**).

3.2 Adherent Transfection, Selection, and Adaptation to Suspension

Transfection is done on 6-well plates and followed by a puromycin selection done simultaneously with scale-up and transition to suspension medium. Adherent cells are maintained in DMEM supplemented with 10 % tetracycline-free FBS and penicillin-streptomycin (adherent medium).

1. Transfection (*see* **Note 19**):
   (a) Day 1: Plate 300,000 cells per well 6-well plates. Plate four wells per construct. Plate an additional four wells for a killing control.
   (b) Day 2: Prepare transfection mixtures of 400 μL Opti-Mem, 4 μg of DNA, and 12 μL Fugene HD (3:1 reagent:DNA ratio), following the manufacturer's protocol. After complex formation, add 100 μL of transfection mixture to each well and mix by rocking the plate.
   (c) Day 3:
      i. Dissociate the cells by banging or by using TrypLE Express (*see* **Note 20**).

ii. Pool the four wells used for each construct. Plate on 10 cm dishes or T-75 culture flasks in a 50:50 mixture of suspension medium (*see* **Note 21**) and adherent medium, supplemented with 1 μg/mL puromycin; higher concentrations may also be used [43] (*see* **Note 22**). For a good transfection, >80 % of cells will survive this harsh transition.

(d) Day 5–6: Control (untransfected) cells should be almost completely dead, with few cells adhered. Once transfected cells are >80 % confluent, split onto 2 × 15 cm dishes using an 80:20 mixture of suspension medium and adherent medium, supplemented with puromycin (*see* **Notes 23** and **24**).

(e) Days 7–9: Once the cells are confluent, adapt to suspension:

i. Prepare 50 mL of a 90:10 mixture of suspension medium and adherent medium (*see* **Note 25**).

ii. Dissociate and count the cells. Remove the media.

iii. Resuspend the cells in 10 mL media and triturate aggressively to break up clumps. Increase volume to 50 mL (typically ~1 million/mL), transfer to a 1 L bottle, and immediately put in the shaker.

(f) Days 9–11: Check the cells after 2 days in suspension. Typical density should be >2 million per mL. Cultures are typically very clumpy at this point, but healthy and doubling every 24–36 h. Sometimes the clumps are large enough to see by the naked eye. The goal of the following steps is to reduce the serum concentration and select non-clumping cells.

i. Transfer 25 mL of the culture (50 mL) into each of the two 50 mL conical tubes.

ii. Triturate five times.

iii. If the cells are very clumpy, vortex three pulses of 5 s (maximum intensity) to break up clumps.

iv. Transfer tubes to a rack in the hood. Count the cells. While counting, allow the cells to settle for a total of 5 min; large clumps will settle rapidly.

v. Split the cells to 0.5 million per mL, adding only suspension medium.

vi. Remaining cells may be used for a test induction at this point.

(g) Expand cells to desired volume and induce at 3.5–4.5 million/mL with 1 μg/mL doxycycline. Harvest 24 h later (*see* **Note 26**).

3.3 Harvest of Suspension Cells to Make Liquid Nitrogen "BB's" (See Note 27)

Cells are spun down in syringes and injected into tubes containing liquid nitrogen [38]. A video protocol of this process done in yeast cells is available at http://www.ncdir.org/public-resources/protocols/. The below protocol is modified for human cells.

1. Gather an ice bucket and clean centrifuge bottles.
2. Count cells. Counts are useful for normalization and blotting.
3. Transfer a 1 mL aliquot(s) of culture for western blotting to a microcentrifuge tube. Spin at 500 × *g* for 30 s, aspirate the media, and store on ice until freezing is convenient.
4. Spin the cultures at 1000 × *g* for 10 min at 4 °C to pellet. We use both 4 × 1 L and 6 × 500 mL centrifuge rotors (*see* **Note 28**). Pour off the media.
5. Resuspend the pellets in a minimal volume of PBS (approximately equal volume to the pellet). Pool resuspended cells.
6. Pellet cells inside syringes.
   (a) Select syringe(s) for cell pelleting. Five milliliter syringes are ideal for small samples. Use 10 mL or 30 mL syringes for large samples.
   (b) Remove the plungers and set aside.
   (c) Securely cap syringes with luer-lock end caps and place inside 50 mL conical tubes (*see* **Note 29**).
   (d) Spin at 200 × *g* for 10 min at 4 °C. Spinning harder risks breaking the syringes. If cells are not well pelleted, spin again. Transfer syringes to an ice bucket.
   (e) Aspirate the PBS, leaving wet cells in the syringe (*see* **Note 30**).
7. Insert a conical tube rack in a small Styrofoam box. Fill with liquid N_2 (LN_2) to the top of the rack (*see* **Note 31**). Pre-label 50 mL conical tubes, transfer to the rack, and fill with LN_2 (*see* **Note 32**) (Fig. 2).
8. Pre-chill a small stainless steel spatula, standing it vertically in the liquid nitrogen.
9. Punch a number of holes in the caps of the tubes using a 16ga needle. This allows drainage of the LN_2 without loss of cells.
10. Inject the cells gradually into the tube containing LN_2 (Fig. 2) (*see* **Note 33**). If injected too fast, they will form large clumps. Use the pre-chilled spatula as needed to break apart any clumps.
11. Cap the tube using the punched lid. Careful! Pressure will build up and liquid can shoot out. Decant into a sink. Replace the punched lid with a new, intact lid and store tubes at −80 °C until cryomilling (*see* **Note 34**).

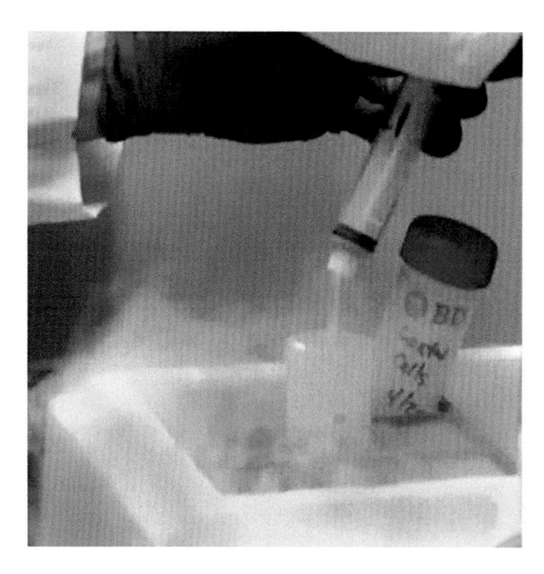

Fig. 2 Injection of pelleted human cells into liquid nitrogen to make "BB's." After centrifugation in a capped syringe, cells are injected into 50 mL conical tubes containing liquid nitrogen. Injection at a moderate rate prevents over-clumping of the cells. Use a pre-chilled spatula to break up any clumps

3.4 Cryomilling

We cryomill cells under liquid nitrogen in a Retsch PM 100 planetary ball mill; model PM 100 CM is also suitable. A newer model, the CryoMill, may be suitable for smaller samples. This protocol was initially developed for yeast, which are harder to break than mammalian cells due to the tough cell wall, and then adapted for mammalian cells with two stages of grinding [38]. Depending on the amount of cells to grind, we either use a 50 mL jar (~1–8 g cells) or a 125 mL jar (~5–30 g cells, Fig. 3a) (*see* **Note 35**). We present here a simplified protocol, after finding that the second stage of grinding was not necessary (Fig. 3b). Custom-made PTFE insulators (Fig. 3a) minimize warming of the sample during grinding and improve safety and performance (*see* **Note 36**). We use a homemade LN_2 decanter made using a spatula and 50 mL conical tube to pour LN_2 into and over the grinding jars [38].

1. Pre-clean grinding jar, lid, balls, two small steel spatulas, and large forceps using Windex glass cleaner or similar. Inspect the PTFE gasket for signs of damage. For a 50 mL chamber, use two 20 mm diameter balls. For a 125 mL chamber, use five 20 mm diameter balls.
2. Weigh the jar + insulators + balls and adjust PM 100 counterbalance accordingly.
3. Precool the jar, balls, spatulas, forceps, LN_2 decanter, and PTFE insulators in a clean Styrofoam box containing liquid LN_2 until the LN_2 stops boiling (*see* **Note 37**). Set up a working pan with LN_2 (*see* **Note 38**).

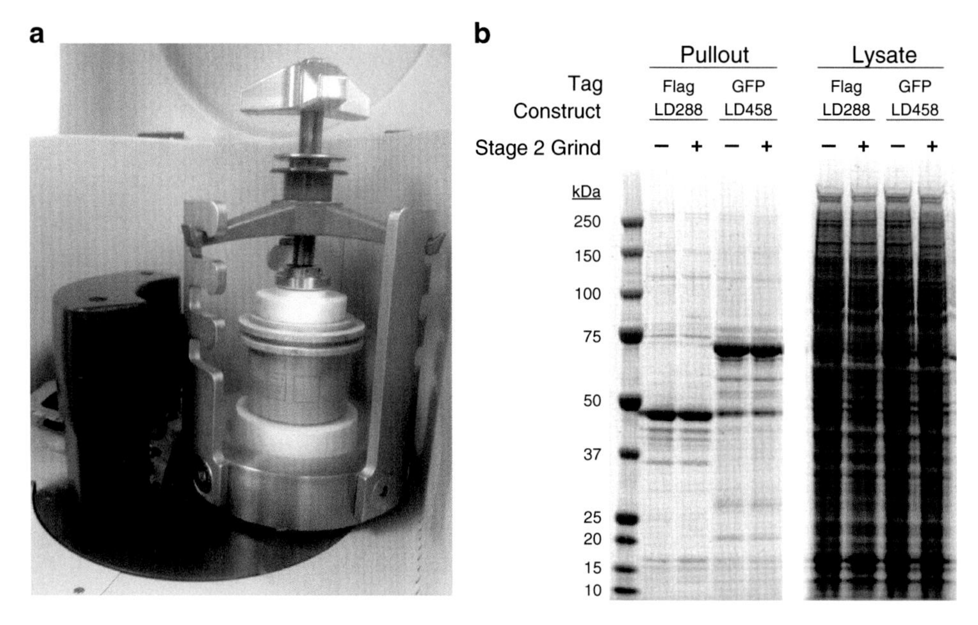

Fig. 3 Cryomilling setup for human cells. (**a**) Cryomilling apparatus in the Resch PM 100. 125 mL grinding jar is shown with custom PTFE insulators above and below. (**b**) Comparison of simplified one-stage and prior two-stage grinding protocols on protein extraction and affinity capture reveals equivalent results. HEK-$293T_{LD}$ cells expressing full-length L1 (ORFeus*HS* background) with ORF1p tagged with Flag (LD288) or GFP (LD258) were affinity captured (pullout) using respective epitope tag antibody-conjugated Dynabeads in our standard extraction solution and eluted under denaturing conditions. Total extracted lysate (lysate) before affinity capture is shown

4. Transfer the cold PTFE base to the grinder.
5. Transfer the frozen cell BB's into the grinding jar.
6. Fill with LN_2 to within ~0.5–1 cm of the top. Cover with the lid and Teflon top insulator, move jar plus lids *en bloc* onto the grinder, and clamp in place (*see* **Note 39**).
7. Pour LN_2 over the jar using the decanter.
8. Grind with three cycles of the following program: 400 rpm, 3 min, reverse rotation every 30 s, no interval breaks (*see* **Note 40**). Between grinding cycles, cool the jar as below:
   (a) Pour LN over the jar using the decanter to cool the lid and top.
   (b) Some pressure will have built up during the grinding. The jar may be gently hissing as pressure escapes: this is normal. Carefully remove the jar as in **step 9** below and transfer to the pan of LN_2 to recool.
   (c) Remove the lid and use a spatula to scrape any adhered powder back into the chamber. Do not scrape the gasket with a spatula: it will damage the PTFE seal. Submerse the lid to cool.

(d) Use a spatula to scrape around the lower corners of the jar, dislodging any packed cells.

(e) Refill the jar with LN_2 and reassemble as in **steps 6** and 7.

9. To remove the jar, slowly release clamping pressure (*see* **Note 41**). Transfer jar assembly to the pan of LN_2.
10. Put a pre-labeled 50 mL tube in a rack in the LN_2 pan. Remove steel balls with forceps, dislodging large chunks of grindate with a spatula.
11. Transfer grindate to conical tubes by pouring or with chilled spatulas or spoons. Once the sample is fully transferred, cap the tube loosely and move to a rack inside the Styrofoam box.
12. Store vertically at −80 °C overnight with the caps loose to allow LN_2 to evaporate, and then seal and store (*see* **Note 42**).

3.5 Conjugation of Dynabeads with Anti-Flag or Anti-ORF1 Antibody

This protocol was originally developed for bulk rabbit IgG [37] and has been adapted for use with precious/expensive antibodies, like anti-Flag M2. See also [44] for a general protocol. Nucleophilic side chains and N-termini on the antibody react with epoxide functional groups on the bead surface. It is critical that all other nucleophiles are absent from the buffer, or these will react with the beads and prevent antibody coupling. This includes tris, glycerol, azide, and other common antibody buffer components. It is safest to buffer exchange antibodies from commercial sources unless the absence of nucleophiles can be assured.

3.5.1 Antibody Buffer Exchange with Microcentrifuge Desalting Columns

We exchange the buffer twice to remove as much contaminant as possible. For large-scale couplings, we use a fast protein liquid chromatography (FPLC) system (ÄKTA, GE Healthcare) with a preparative-scale desalting column, but in the absence of this equipment, and for small, precious antibody samples, we use the below protocol. Alternative methods include ion exchange and dialysis.

1. Pre-equilibrate Zeba™ Spin Desalting columns in PBS three times according to the manufacturer's instructions.

   (a) Zeba™ columns have a maximum volume of 130 μL. Because we desalt twice, equilibrate two columns for every 130 μL. For example, for 250 μL antibody solution, equilibrate four columns.

2. Load, centrifuge, and recover exchanged antibody solution according to the manufacturer's instructions. Repeat once on fresh columns for a total of two exchanges (*see* **Note 43**).

3.5.2 Antibody Coupling

This protocol is for coupling of 300 mg Dynabeads. At the last step the beads are slurried by the addition of 2 mL buffer (*see* **Note 44**). We also routinely couple with 60 mg Dynabeads, slurried by the addition of 400 μL. For a 60 mg coupling, scale volumes in the protocol linearly. Smaller volumes can also be used [44]. A commercial coupling kit for Dynabeads is available from Life Technologies; this uses proprietary components but is effective.

Day 1: Coupling

1. Calculate how much antibody to use. One milligram of Dynabeads M-270 Epoxy has been estimated to immobilize 7–8 μg of antibody during coupling (*see* **Note 45**). Coupling is not 100 % efficient and excess antibody appears to help drive the reaction [37]. For commercially available antibodies we use 10 μg antibody/mg Dynabeads (3 mg α-Flag for a 300 mg Dynabead coupling, *see* **Note 46**). We use 5 μg/mg for precious custom antibodies, and 15 μg/mg when we have the hybridoma.
2. Resuspend an entire bottle of 300 mg Dynabeads M270 Epoxy with 10 mL of 100 mM phosphate buffer, pH 7.4. Directly add buffer to the bottle and vortex bottle.
3. Transfer to a 15 mL falcon tube and wash the bottle twice with 2 mL phosphate buffer (*see* **Note 47**).
4. Shake bead suspension for 10 min on a nutating mixer or orbital shaker.
5. While the beads mix, prepare the *antibody mixture* (20 μL/mg Dynabeads, 6 mL total).
   (a) 2 mL 3 M Ammonium sulfate.
   (b) 3 mg antibody, double buffer-exchanged or supplied in phosphate buffer or PBS, free of glycerol and interfering species (see the manufacturer's instructions).
   (c) 0.1 M Phosphate buffer to 6 mL.
6. Transfer beads to a magnetic separator (*see* **Note 48**). Aspirate the buffer.
7. Wash again with 10 mL 100 mM phosphate buffer. Add buffer, vortex for 15 s, apply magnet, and aspirate; no incubation is necessary.
8. Add the *antibody mixture* to the beads. Seal and Parafilm the tube, and mix well.
9. Incubate overnight (18–24 h) on a rotating wheel at 37 °C (*see* **Note 49**).

Day 2: Bead Washing

1. Separate beads from the antibody mixture with a magnet. Carefully remove the antibody mixture and set aside in a clean tube: it still contains 30–50 % of the antibody, unreacted, which can be recovered for reuse.
2. Wash beads once with 12 mL 100 mM glycine pH 2.5. Add buffer, vortex briefly, and take it off as fast as possible.

3. Wash once with 12 mL 10 mM Tris–HCl, pH 8.8.
4. Prepare fresh 100 mM triethylamine: Add 168 μL stock to 11.8 mL ddH_2O. Apply, mix, remove, and proceed to the next step as fast as possible.
5. Wash the beads with 12 mL 1× PBS, incubating for 5 min on the nutator. Repeat a total of four times.
6. Wash twice with 12 mL PBS + 0.5 % Triton X-100, incubating each wash for 10 min on the nutator.
7. Resuspend beads in 2 mL *resuspension buffer:* PBS, 50 % glycerol, and 0.5 mg/mL BSA (*see* **Note 50**).
8. Mix well and aliquot 100 μL each into Eppendorf tubes. Store at −20 °C (*see* **Note 50**).

Reuse and Optional Antibody Recovery Using Protein G affinity (*see* **Note 51**). Approximately 30–50 % of antibody is unbound after coupling. We routinely save antibody mixtures for use in other assays such as immunoblotting and immunofluorescence. Alternatively, recovered antibody can easily be re-captured using Protein G affinity. Recovery also allows concentration of antibody and transfer into a more permanent storage buffer. We use Protein G Sepaharose Fast Flow (GE Healthcare) (*see* **Note 52**) following the manufacturer's instructions. The antibody sample should be diluted with an equal volume of PBS for binding. Based on 50 % antibody-Dynabead binding, ~125 μL resin is needed for recovery after a 300 mg Dynabead coupling. After elution, add 50 % glycerol for storage or dialyze or desalt into the buffer of choice.

3.6 Affinity Capture of RNPs Using Conjugated Magnetic Medium

For a comprehensive review of considerations affecting expression systems, epitope tagging, and affinity medium choice see [45]. In brief, we find that Dynabeads, when conjugated to high-quality antibodies, provide for excellent quality recovery of endogenous protein complexes from human cells [37, 38]. We have successfully applied this approach to L1 RNPs [6]. For purifying 3× Flag-tagged constructs, α-Flag Dynabeads are necessary, prepared as above. When combined with neodymium magnet racks, antibody-conjugated magnetic medium (beads) is rapidly separated from the buffer and immobilized on the side of the tube. This allows near-complete aspiration of buffer without the risk of aspirating the beads.

1. Prepare clarified cell extracts:
   (a) Weigh out 200 mg of cell powder into a microcentrifuge tube—hold on LN_2 (*see* **Note 53**).
      i. Repeat (a) for as many purifications as you will carry out.
      ii. Multiple purifications can be pooled after elution if larger scale is required.
   (b) Move the tubes to room temperature for 1–2 min (*see* **Note 54**).

(c) Add 800 μL of extraction solution (*see* **Note 55**) (20 mM Na-HEPES pH 7.4, 500 mM NaCl, 1 % v/v Triton X-100; plus protease inhibitors) to each tube, vortex for ~30 s until powders are resuspended, and then place the crude extracts on ice. Some membrane aggregates may be observed (*see* **Note 56**).

(d) Sonicate each tube with a micro-tip probe on a low-power setting using 2 × 2 s pulses. Membrane aggregates should no longer be visible (*see* **Note 57**).

(e) Centrifuge for 10 min at full speed (20–30,000 RCF) in a refrigerated microcentrifuge at 4 °C.

   i. During this step the affinity medium can be pre-washed (**step 2a**).

(f) Remove supernatant—this is your clarified extract—and add to the tube containing α-Flag Dynabeads (**step 2b**).

   i. Set a fraction aside before combining with beads to compare pre- and post-bead binding (**step 2ci**) in order to assess the efficacy of the affinity capture.

2. Affinity capture:

(a) Prepare beads:

   i. Pipette 20 μL of α-Flag Dynabeads slurry into a 1.5 mL microcentrifuge tube (*see* **Note 58**).

   ii. Repeat for each affinity purification to be carried out.

   iii. Wash the beads twice with 500 μL of extraction solution.

   iv. Remove the supernatant, and hold the beads on ice until needed.

(b) Combine the clarified extract (**step 1f**) with the beads.

   i. Incubate at 4 °C for 30 min (*see* **Note 59**).

(c) Separate beads on a magnetic separator. Set a fraction aside to compare with input (**step 1fi**), and aspirate the remainder.

(d) Wash the beads with 1 mL of extraction solution and then remove the supernatant. Wash protocol for beads (used throughout):

   i. Add buffer, and vortex at full power for 2–3 s.

   ii. Pulse-spin in a benchtop microcentrifuge to remove any magnetic beads from the cap.

   iii. Separate beads on a magnetic separator, and remove buffer using a vacuum aspirator.

(e) Resuspend the beads in 1 mL of extraction solution, transfer to a fresh microcentrifuge tube, and then remove the supernatant (*see* **Note 60**).

(f) Wash again with 1 mL of extraction solution.

(g) Elute the L1 RNPs from the beads with 26 μL of 1 mg/mL 3× Flag peptide in extraction solution (native) or using 1× SDS-PAGE loading buffer without reducing agent (*see* **Note 61**).

i. Native elution: Incubate for 15–30 min at room temperature with gentle agitation (just enough to mix and suspend the beads).

ii. Denaturing elution: Incubate for 5–10 min at 70 °C with moderate agitation (*see* **Note 62**).

iii. If native elution is used, remove the eluate and then perform a second denaturing elution to assess the efficacy of the native elution.

3. Natively eluted samples can be carried forward to other assays (see LEAP and RT-PCR below) and/or subsequently prepared for SDS-PAGE (*see* **Notes 63–65**).

4. For samples to be analyzed by mass spectrometry, reduce and alkylate with iodoacetamide:

(a) Add SDS-PAGE loading dye to 1× and DTT to 20 mM (*see* **Note 66**).

(b) Incubate for 10 min at 70 °C.

(c) Cool to room temperature.

(d) Add iodoacetamide to 0.1 M and incubate in the dark at room temp for 30 min.

(e) Load directly on a gel.

3.7 Sample Preparation for Mass Spectrometry

Proper preparation of samples is critical for mass spectrometry as a number of interfering species can reduce sensitivity and compromise protein identification (for discussion and advice, see [45]). For identification of the most prominent species, readily observed by standard protein staining techniques (reviewed in [46, 47]), we excise regions of the gel containing stained protein bands and use MALDI-MS peptide mass fingerprinting (PMF, [48, 49]; example protocols that we use can be found in the supplement of [38]). For more sensitive detection and identification of the complement of proteins within the sample we use liquid chromatography-coupled tandem mass spectrometry (LC-MS/MS) approaches (reviewed in [50, 51]). If proteins are first metabolically labeled with stable isotopes prior to affinity capture, then isotopic differentiation of interactions as random or targeted (I-DIRT) can be implemented. This MS-based analysis provides statistical discrimination between interactors likely to have originated in vivo and those likely to be in vitro artifacts, and can be accomplished by modifying the procedures outlined in this chapter [6, 52, 53]. Whether MALDI-MS or LC-MS/MS will be used, we prefer gel-based peptide sample work-up [54]. Because different MS-based analytical approaches may require different sample work-up procedures, we recommend adopting appropriate procedures based on the preferences of your proteomics core facility or collaborator.

3.8 LEAP (L1 Element Amplification Protocol)

This protocol uses ORF2p reverse transcriptase and an anchor primer to reverse transcribe bound RNA [23]. cDNA is then amplified by PCR for visualization on a gel or quantification by real-time PCR. A boiled sample is used for a negative control. cDNA is made with commercial reverse transcriptase as a positive control.

1. Thaw RNPs on ice. Clean bench and work area using RNAseZap or similar. Open a clean box of pipet tips.
2. Prepare sufficient LEAP reaction mix for twice the number of samples (see recipe, Subheading 3).
3. Prepare LEAP + SuperScript reaction mixture:
   (a) Pipet half of the LEAP buffer into a clean tube.
   (b) Add SuperScript III Reverse Transcriptase (0.25 μL per reaction, 50 U), and mix well.
4. For the negative control, add 5 μL of one RNP sample to a clean tube. Boil for 5 min at 100 °C.
5. Aliquot LEAP and LEAP + SuperScript reaction mixtures into reaction tubes.
6. Add 2 μL of each RNP to both LEAP and LEAP + SuperScript mixture. Do not forget to add the boiled negative control.
7. Incubate LEAP reactions for 1 h at 37 °C. Incubate LEAP + SuperScript reactions for 1 h at 50 °C (*see* **Note 67**).
8. Optional pause point: Snap-freeze LEAP products.
9. Amplify LEAP products for visualization and/or sequencing:
   (a) Use 1 μL LEAP product in a 50 μL PCR reaction using FastStart Taq DNA Polymerase with primers JB11564 and JB14057.
10. Measure LEAP products by qRT-PCR using a StepOne Plus instrument or similar:
    (a) Use 0.5 μL of each LEAP product in triplicate 20 μL reactions with SYBR Green 2× master mix.
    (b) Measure the ORF1 region of the L1 RNA with primers JB13415 and JB13416 (*see* **Note 68**).
    (c) Measure the ORF2 region of the L1 RNA with primers JB13417 and JB13418.

3.9 RNA Isolation and qRT-PCR of L1 RNPs

Total RNA is isolated from L1 RNPs after spiking in purified GFP-transfected HEK-293T RNA (*see* **Note 4**). This serves as both a carrier and as an internal control for normalization. Alternatively, in vitro-transcribed GFP mRNA may be used along with glycogen as a carrier. A practical review of qRT-PCR methods can be found in [55].

1. Aliquot 2 μg purified GFP-transfected control RNA into each tube.
2. Add 10 μL of each RNP sample, and mix well.
3. Purify RNA using TRIzol Reagent according to the manufacturer's protocol.
4. Resuspend RNA in 20 μL RNase-free water.
5. Quantify by qRT-PCR as in Subheading 3.8, **step 10**, using 0.5 μL resuspended RNA in triplicate. Primers for GFP (JB13766 and JB13767) are used for normalization of each sample.

4 Notes

1. This affinity-purified version performs as well or better than the non-affinity-purified F3165, and at the time of writing is approximately half the cost.
2. The Dynamag 15 rack is useful for couplings starting with 300 mg Dynabeads. We often do smaller test couplings with 60 mg Dynabeads, which can be done with 2 mL tubes. To reduce cost, high-quality neodymium magnets from MAGCRAFT are widely distributed and available in a variety of shapes. These can be fitted into homemade racks or attached to tubes using rubber bands; arc-shaped magnets are particularly useful with rubber bands.
3. The pH of 10× PBS is between 6.6 and 6.7 at 10× but is 7.4 when diluted to 1×.
4. Buffer components for affinity capture and LEAP should be RNase free, and handled with precautions for cleanliness appropriate for RNA work. This includes buffers used for dissolution of 3× Flag peptide, etc..
5. We transfect HEK-293TLD cells with pCAG-eGFP (Addgene # 11150) and purify total RNA using an RNeasy mini kit (Qiagen).
6. Life Technologies reports that Freestyle 293 provides serum-free growth of their engineered HEK-293-F cells. In our experience with other HEK-293 and HEK-293T cells, 1 % tet-free FBS adds little cost to the medium and increases reliability. Similarly, supplemental l-glutamine, added fresh, provides more reliable growth than their GlutaMax alternative alone. We make a 50:50 mix of serum and 200 mM l-glutamine, filter, and add 20 mL per liter.

7. We have been unable to establish an effective protocol for puromycin selection after suspension transfection.
8. Always autoclave with loose bottle caps. The tops of the bottles can be covered in foil for additional protection. After the autoclave finishes, transfer bottles to a culture hood to cool, and then tighten caps for storage.
9. If any bottle is found to be contaminated, all bottles in the incubator must be checked for signs of contamination.
10. Until a suspension line is well established or after steps that can be toxic to cells, we find a minimum density of 0.5 million/mL is safer than 0.2 million/mL.
11. Do not directly transfer from the serological pipet to the hemocytometer.
12. Never use a micropipettor in suspension bottles: this risks contamination.
13. This ratio should be optimized for different cell lines and DNA preparation methods.
14. After column purification, precipitate with isopropanol and wash with ethanol. DNA prepared with their precipitator modules has been less effective in our experience.
15. To save time, hybridoma SFM aliquots can be left at room temperature overnight.
16. For small numbers of transfections, prepare DNA-PEI complex in conical tubes. For larger volumes, use a glass bottle.
17. ORF2p expression peaks 24 h after induction and falls thereafter. ORF1p expression peaks at ~18 h after induction and is constant for at least 4–5 days. Harvest time should be optimized for your protein of interest.
18. We have found that with episomal pCEP4-based plasmids, a 1:3 "split-back" minimally reduces per-cell yield but allows the use of threefold less DNA per gram WCW.
19. We find that transfection efficiency is higher in 6-well wells than larger scales, so we transfect a number of wells and then pool the cells later.
20. If protease is used, quench with serum-containing media and centrifuge for 5 min at 200 RCF to remove the protease.
21. Recipe for suspension medium is as described above, but when following the protocol for adherent transfection, selection, and adaptation to suspension it may include antibiotics.
22. Cells can be adapted into higher puromycin concentrations of 2–10 μg/mL, resulting in higher per-cell expression but slower growth and longer time needed for adaptation.
23. Suspension cells are more sensitive to puromycin than adherent cells.

24. It can be helpful to perform a test induction at this point. Plate 0.5 million cells in each of the two 6-well wells. Add 1 μg/mL doxycycline to induce one. Lyse 24 h later for western blotting and freeze lysate for later comparison.
25. Note that this 90:10 medium contains ~2 % FBS.
26. Once cells stably transfected with a construct of interest are growing in suspension, it is advisable to freeze aliquots for later use. We freeze 20 million cells per vial in suspension medium supplemented with 5 % DMSO and 20 % FBS.
27. A "BB" is a small round steel ball, like those used in toy guns. Cells frozen in this way form spheroid shapes (and globules of spheroids).
28. Pellets should appear approximately uniform. If centrifuged too hard, cells will be crushed, forming two different-colored layers.
29. Depending on the syringes used, it may be necessary to trim the finger grips so the syringe fits inside the tube. Thirty milliliter syringes are taller than 50 mL conical tubes. Make sure that these have enough clearance in your swinging bucket centrifuge before adding cells.
30. In order to thoroughly remove the liquid, we commonly suck off the very top layer of cells in order to remove the PBS. We use a vacuum aspirator and glass Pasteur pipet.
31. Use care and best practices in handling liquid nitrogen. Appropriate protective gear and goggles should be used to prevent injury.
32. Each tube will hold approximately 15–20 g BB's. For very large samples, we use polypropylene bottles.
33. A pair of pliers may be needed to remove the luer-lock syringe cap after centrifugation.
34. Storage with a punched cap is acceptable for a few days; however frost will form on cells over longer periods.
35. 1 g WCW is approximately the minimum amount of cells for grinding. A small amount of material (up to ~500 mg) is lost on the surface of the jar and balls, so practically we aim to produce at least 1.5–3 g WCW.
36. The custom PTFE insulators are not required for this protocol [38]. However, we find that grinding is greatly aided by the inclusion of LN within the jar ("wet grinding"). The insulators prevent warming of the jar, evaporation of the nitrogen, and pressure buildup, resulting in faster, more reliable grinds. To have insulators made at the Rockefeller University machine shop, contact the Rout Lab.
37. All tools for grinding should be chilled during use.

38. We use two containers: a Styrofoam box filled with enough N2 to completely submerge the jar, and a pan for sample prep and intermediate cooling steps in which the jar is not completely submerged. When grinding multiple samples, we put a tube rack in the Styrofoam box for storage of BB's and grindate and cover the box with its lid.
39. Clamping force should be firm, but not excessive such that removal of the jar becomes a problem.
40. Grinding should make a distinct clunking noise as the balls collide. This noise may stop at some point during a rotation, but should resume when rotation is reversed. If these sounds are not heard, inspect between cycles to make sure that grinding is occurring.
41. Releasing too quickly can cause rapid depressurization and loss of grindate. A controlled release allows gentle depressurization, which can be heard as a gentle hiss.
42. Grindate can be stored at −80 °C essentially indefinitely, without affecting performance.
43. Protein recovery can be verified by Bradford assay or SDS-PAGE with Coomassie, comparing input and output. For Bradford use gamma globulin as the standard to get an accurate concentration.
44. This results in ~2.2 mL final volume, and a slurry of approximately ~10–15 % by volume.
45. Measurements of coupling yield were done using rabbit polyclonal IgG. The apparent capacity of the beads may vary depending upon the coupling conditions, including the concentration of salt, ammonium sulfate, pH, and temperature of coupling.
46. With anti-Flag M2, beads conjugated at 8 μg antibody/mg Dynabeads do not perform as well as beads conjugated at 10 μg/mg. There is only a marginal improvement in going from 10 to 12.5 μg/mL; we use 10 μg/mg as a cost-effective compromise.
47. Sometimes resuspension requires more volume. Simply remove the buffer from the beads using a magnetic separator and wash the bottle with fresh buffer.
48. The addition of BSA increases the long-term stability of M2-Flag antibody-conjugated Dynabeads.
49. 30 °C is also effective.
50. With proper storage, coupled Dynabeads may be used without loss of performance for >1 year. Alternatively, if the beads will be completely consumed within ~8 weeks, storage at 4 °C is suitable—resuspend with PBS, 0.5 mg/mL BSA, and 0.02 % sodium azide and store at 4 °C.

51. Depending on the antibody species and subtype, protein A may be more appropriate than protein G.
52. A variety of Protein G preparations, including magnetic, are commercially available. At this scale, magnetic medium is much more expensive.
53. Use a small Styrofoam rack on a microbalance and pre-chill tubes and weighing instruments on LN2. We have found that this is easiest using inexpensive small stainless steel measuring spoons designed for culinary use.
54. Allowing the tube to briefly warm prevents extraction solution from flash freezing on the side of the tube.
55. Determination of the optimal extraction solution for each complex is critical but beyond the scope of this chapter. For more information see LaCava et al. (45).
56. Hold tubes on ice between each subsequent manipulation—working at room temperature is otherwise acceptable.
57. The power should be adjusted such that the minimum amount of energy is used that will disperse the aggregates. On a Branson Sonifier with Microtip, this is power setting 3.
58. The ratio of beads to lysate can be optimized. For α-ORF1p pullouts 50 μL Dynabeads were needed to deplete the ORF1p from extracts. For α-ORF2p pullouts, 10 μL was sufficient and background increased with larger amounts.
59. This time can be optimized. For α-ORF1p pullouts, 5 min was as effective as 30 min. For α-ORF2p pullouts using α-Flag Dynabeads, 1 h was more effective than 30 min.
60. We find that transfer to a fresh tube at this step reduces background because some protein nonspecifically sticks to the tube.
61. Reducing agent is omitted to reduce the release of IgG from the beads. It should be added after separation from the magnetic medium, before denaturation for SDS-PAGE.
62. We use a Thermomixer (Eppendorf) at full speed.
63. The advantage of native elution, even if SDS-PAGE is the next step, is its tendency to release only specific interactors of the tagged protein, reducing nonspecific contamination.
64. Regardless of the elution method, all samples should be reduced and alkylated prior to loading on the gel to maximize sensitive detection of cysteine-containing peptides during subsequent MS.
65. For storage of purified RNPs, for example for future LEAP, we dilute 1:1 with 50 % glycerol (25 % final), flash freeze with N2, and store at −80 °C.
66. This is 40 % of the typical 50 mM DTT concentration in reducing SDS-PAGE loading dye.

67. Incubation at 50 °C favors the SuperScript III reaction over the ORF2p reaction. At 37 °C both reactions occur simultaneously.
68. These primers are specific for ORFeusHS; appropriate primers should be chosen for other L1 elements.

Acknowledgements

We thank Dan Leahy, Jennifer Kavran, and Yana Li for help with suspension cell culture. This work was supported in part by NIH grant U54GM103511 to MPR, grant 5P50GM107632 to JDB, KHB, and MPR, U54GM103520 to JDB, R01CA163705 and R01GM103999 to KHB, and US DoD grant OC120390 to KHB.

References

1. Ostertag EM, Kazazian HH Jr (2001) Biology of mammalian L1 retrotransposons. Annu Rev Genet 35:501
2. Beck CR, Garcia-Perez JL, Badge RM, Moran JV (2011) LINE-1 elements in structural variation and disease. Annu Rev Genomics Hum Genet 12:187
3. Burns KH, Boeke JD (2012) Human transposon tectonics. Cell 149:740
4. Hancks DC, Kazazian HH Jr (2012) Active human retrotransposons: variation and disease. Curr Opin Genet Dev 22:191
5. Brouha B et al (2003) Hot L1s account for the bulk of retrotransposition in the human population. Proc Natl Acad Sci U S A 100:5280
6. Taylor MS et al (2013) Affinity proteomics reveals human host factors implicated in discrete stages of LINE-1 retrotransposition. Cell 155:1034
7. Arjan-Odedra S, Swanson CM, Sherer NM, Wolinsky SM, Malim MH (2012) Endogenous MOV10 inhibits the retrotransposition of endogenous retroelements but not the replication of exogenous retroviruses. Retrovirology 9:53. doi:10.1186/1742-4690-9-53
8. Dai L, Taylor MS, O'Donnell KA, Boeke JD (2012) Poly(A) binding protein C1 is essential for efficient L1 retrotransposition and affects L1 RNP formation. Mol Cell Biol 32(21):4323–4336
9. Goodier JL, Cheung LE, Kazazian HH Jr (2012) MOV10 RNA helicase is a potent inhibitor of retrotransposition in cells. PLoS Genet 8:e1002941
10. Goodier JL, Cheung LE, Kazazian HH Jr (2013) Mapping the LINE1 ORF1 protein interactome reveals associated inhibitors of human retrotransposition. Nucleic Acids Res 41:7401
11. Niewiadomska AM et al (2007) Differential inhibition of long interspersed element 1 by APOBEC3 does not correlate with high-molecular-mass-complex formation or P-body association. J Virol 81:9577
12. Suzuki J et al (2009) Genetic evidence that the non-homologous end-joining repair pathway is involved in LINE retrotransposition. PLoS Genet 5:e1000461
13. Peddigari S, Li PW, Rabe JL, Martin SL (2013) hnRNPL and nucleolin bind LINE-1 RNA and function as host factors to modulate retrotransposition. Nucleic Acids Res 41:575
14. Hata K, Sakaki Y (1997) Identification of critical CpG sites for repression of L1 transcription by DNA methylation. Gene 189:227
15. Soifer HS, Zaragoza A, Peyvan M, Behlke MA, Rossi JJ (2005) A potential role for RNA interference in controlling the activity of the human LINE-1 retrotransposon. Nucleic Acids Res 33:846
16. Yang N, Kazazian HH Jr (2006) L1 retrotransposition is suppressed by endogenously encoded small interfering RNAs in human cultured cells. Nat Struct Mol Biol 13:763
17. Mandal PK, Ewing AD, Hancks DC, Kazazian HH Jr (2013) Enrichment of processed pseudogene transcripts in L1-ribonucleoprotein particles. Hum Mol Genet 22:3730
18. Belancio VP, Whelton M, Deininger P (2007) Requirements for polyadenylation at the 3′ end of LINE-1 elements. Gene 390:98

19. Malik HS, Burke WD, Eickbush TH (1999) The age and evolution of non-LTR retrotransposable elements. Mol Biol Evol 16:793
20. Khan H, Smit A, Boissinot S (2006) Molecular evolution and tempo of amplification of human LINE-1 retrotransposons since the origin of primates. Genome Res 16:78
21. Muotri AR et al (2005) Somatic mosaicism in neuronal precursor cells mediated by L1 retrotransposition. Nature 435:903
22. Kulpa DA, Moran JV (2005) Ribonucleoprotein particle formation is necessary but not sufficient for LINE-1 retrotransposition. Hum Mol Genet 14:3237
23. Kulpa DA, Moran JV (2006) Cis-preferential LINE-1 reverse transcriptase activity in ribonucleoprotein particles. Nat Struct Mol Biol 13:655
24. Kolosha VO, Martin SL (2003) High-affinity, non-sequence-specific RNA binding by the open reading frame 1 (ORF1) protein from long interspersed nuclear element 1 (LINE-1). J Biol Chem 278:8112
25. Martin SL, Bushman FD (2001) Nucleic acid chaperone activity of the ORF1 protein from the mouse LINE-1 retrotransposon. Mol Cell Biol 21:467
26. Martin SL et al (2005) LINE-1 retrotransposition requires the nucleic acid chaperone activity of the ORF1 protein. J Mol Biol 348:549
27. Callahan KE, Hickman AB, Jones CE, Ghirlando R, Furano AV (2012) Polymerization and nucleic acid-binding properties of human L1 ORF1 protein. Nucleic Acids Res 40:813
28. Khazina E et al (2011) Trimeric structure and flexibility of the L1ORF1 protein in human L1 retrotransposition. Nat Struct Mol Biol 18:1006
29. Martin SL, Branciforte D, Keller D, Bain DL (2003) Trimeric structure for an essential protein in L1 retrotransposition. Proc Natl Acad Sci U S A 100:13815
30. Feng Q, Moran JV, Kazazian HH Jr, Boeke JD (1996) Human L1 retrotransposon encodes a conserved endonuclease required for retrotransposition. Cell 87:905
31. Moran JV et al (1996) High frequency retrotransposition in cultured mammalian cells. Cell 87:917
32. Alisch RS, Garcia-Perez JL, Muotri AR, Gage FH, Moran JV (2006) Unconventional translation of mammalian LINE-1 retrotransposons. Genes Dev 20:210
33. Doucet AJ et al (2010) Characterization of LINE-1 ribonucleoprotein particles. PLoS Genet 6. 10.1371/journal.pgen.1001150
34. Dai L, LaCava J, Taylor MS, Boeke JD (2014) Expression and detection of LINE-1 ORF-encoded proteins. Mob Genet Elements 4:e29319
35. An W et al (2011) Characterization of a synthetic human LINE-1 retrotransposon ORFeus-Hs. Mob DNA 2:2
36. Han JS, Boeke JD (2004) A highly active synthetic mammalian retrotransposon. Nature 429:314
37. Cristea IM, Williams R, Chait BT, Rout MP (2005) Fluorescent proteins as proteomic probes. Mol Cell Proteomics 4:1933
38. Domanski M et al (2012) Improved methodology for the affinity isolation of human protein complexes expressed at near endogenous levels. Biotechniques 0:1–6
39. Muller N, Girard P, Hacker DL, Jordan M, Wurm FM (2005) Orbital shaker technology for the cultivation of mammalian cells in suspension. Biotechnol Bioeng 89:400
40. Chaudhary S, Pak JE, Gruswitz F, Sharma V, Stroud RM (2012) Overexpressing human membrane proteins in stably transfected and clonal human embryonic kidney 293S cells. Nat Protoc 7:453
41. Reeves PJ, Callewaert N, Contreras R, Khorana HG (2002) Structure and function in rhodopsin: high-level expression of rhodopsin with restricted and homogeneous N-glycosylation by a tetracycline-inducible N-acetylglucosaminyltransferase I-negative HEK293S stable mammalian cell line. Proc Natl Acad Sci U S A 99:13419
42. Oeffinger M et al (2007) Comprehensive analysis of diverse ribonucleoprotein complexes. Nat Methods 4:951
43. Matentzoglu K, Scheffner M (2009) Ubiquitin-fusion protein system: a powerful tool for ectopic protein expression in mammalian cells. Biotechniques 46:21
44. Cristea IM, Chait BT (2011) Conjugation of magnetic beads for immunopurification of protein complexes. Cold Spring Harb Protoc. pdb prot5610
45. LaCava J et al (2015) Affinity proteomics for the study of endogenous protein complexes: pointers, pitfalls, preferences and perspectives. Biotechniques 58(3):103–119
46. Miller I, Crawford J, Gianazza E (2006) Protein stains for proteomic applications: which, when, why? Proteomics 6:5385
47. Gauci VJ, Wright EP, Coorssen JR (2011) Quantitative proteomics: assessing the spectrum of in-gel protein detection methods. J Chem Biol 4:3

48. Roepstorff P (2000) MALDI-TOF mass spectrometry in protein chemistry. EXS 88:81
49. Zhang W, Chait BT (2000) ProFound: an expert system for protein identification using mass spectrometric peptide mapping information. Anal Chem 72:2482
50. Dunham WH, Mullin M, Gingras AC (2012) Affinity-purification coupled to mass spectrometry: basic principles and strategies. Proteomics 12:1576
51. Oeffinger M (2012) Two steps forward – one step back: advances in affinity purification mass spectrometry of macromolecular complexes. Proteomics 12:1591
52. Tackett AJ et al (2005) I-DIRT, a general method for distinguishing between specific and nonspecific protein interactions. J Proteome Res 4:1752
53. Ong SE et al (2002) Stable isotope labeling by amino acids in cell culture, SILAC, as a simple and accurate approach to expression proteomics. Mol Cell Proteomics 1:376
54. Shevchenko A, Tomas H, Havlis J, Olsen JV, Mann M (2006) In-gel digestion for mass spectrometric characterization of proteins and proteomes. Nat Protoc 1:2856
55. Bustin SA (2002) Quantification of mRNA using real-time reverse transcription PCR (RT-PCR): trends and problems. J Mol Endocrinol 29:23

Chapter 21

LEAP: L1 Element Amplification Protocol

Huira C. Kopera, Diane A. Flasch, Mitsuhiro Nakamura, Tomoichiro Miyoshi, Aurélien J. Doucet, and John V. Moran

Abstract

Long INterspersed Element-1 (LINE-1 or L1) retrotransposons encode two proteins (ORF1p and ORF2p) that are required for retrotransposition. The L1 element amplification protocol (LEAP) assays the ability of L1 ORF2p to reverse transcribe L1 RNA in vitro. Ultracentrifugation or immunoprecipitation is used to isolate L1 ribonucleoprotein particle (RNP) complexes from cultured human cells transfected with an engineered L1 expression construct. The isolated RNPs are incubated with an oligonucleotide that contains a unique sequence at its 5′ end and a thymidine-rich sequence at its 3′ end. The addition of dNTPs to the reaction allows L1 ORF2p bound to L1 RNA to generate L1 cDNA. The resultant L1 cDNAs then are amplified using polymerase chain reaction (PCR) and the products are visualized by gel electrophoresis. Sequencing the resultant PCR products then allows product verification. The LEAP assay has been instrumental in determining how mutations in L1 ORF1p and ORF2p affect L1 reverse transcriptase (RT) activity. Furthermore, the LEAP assay has revealed that the L1 ORF2p RT can extend a DNA primer with mismatched 3′ terminal bases when it is annealed to an L1 RNA template. As the LINE-1 biology field gravitates toward studying cellular proteins that regulate LINE-1, molecular genetic and biochemical approaches such as LEAP, in conjunction with the LINE-1-cultured cell retrotransposition assay, are essential to dissect the molecular mechanism of L1 retrotransposition.

Key words LINE-1, Reverse transcriptase (RT), Ribonucleoprotein particle (RNP), L1 element amplification protocol (LEAP)

1 Introduction

Long INterspersed Element-1s (LINE-1s or L1s) are the only active, autonomous transposable elements in the human genome [1, 2]. L1s mobilize through an RNA intermediate by a "copy-and-paste" process termed retrotransposition (reviewed in [3]). An average individual human genome contains ~80–100 active (i.e., retrotransposition-competent L1s (RC-L1s)) [4, 5]. RC-L1s are ~6 kb in length and encode two proteins (ORF1p and ORF2p) that are required for their retrotransposition [1, 6–8]. L1 ORF1p is an approximately 40 kDa RNA-binding protein [9–15] with nucleic acid chaperone activity [16, 17]. L1 ORF2p is an

Jose L. Garcia-Pérez (ed.), *Transposons and Retrotransposons: Methods and Protocols*, Methods in Molecular Biology, vol. 1400, DOI 10.1007/978-1-4939-3372-3_21,

approximately 150 kDa protein [18–21] with both endonuclease (EN) [6] and reverse transcriptase (RT) activities [22].

Although L1 retrotransposition has had a dramatic impact on the structure of the human genome, the detailed molecular mechanism of L1 retrotransposition requires elucidation (reviewed in [3]). Assays using Ty1/L1 ORF2p expression constructs [22, 23], as well as L1 ORF2p produced in recombinant expression systems [24–26], have revealed that L1 ORF2p contains reverse transcriptase activity. Similarly, a recombinant L1 ORF2p EN domain produced in *E. coli* contains DNA endonuclease activity that can generate single-strand DNA breaks at thymidine-rich sequences in double-stranded DNA, leading to nicks that contain a 5′ phosphate and 3′ hydroxyl group [6].

The above experiments, in conjunction with the LINE-1-cultured cell retrotransposition assay [7] and molecular genetic approaches (reviewed in [3]), have demonstrated that L1 ORF1p and ORF2p preferentially associate with their encoding L1 RNA *in cis* [27–29], leading to the formation of a ribonucleoprotein particle (RNP) that is a necessary intermediate for retrotransposition [18, 21, 30]. Components of the L1 RNP are thought to gain access to the nucleus, where the L1 ORF2p EN activity generates a single-strand endonucleolytic nick in genomic DNA at a thymidine-rich consensus sequence (e.g., 5′-TTTT/A-3′, 5′-TTTC/A-3′, 5′-TTTA/A-3′, where "/" denotes the scissile bond) [31–35]. This endonucleolytic activity liberates a 3′-hydroxyl group, which can serve as a primer for the L1 ORF2p RT activity to initiate the reverse transcription of L1 RNA by a process termed target site-primed reverse transcription (TPRT) (Fig. 1) [6, 24, 36]. However, the difficulty in purifying high levels of recombinant L1 ORF2p in vitro, and the fact that L1 ORF2p is made in low quantities in vivo (as few as one molecule of L1 ORF2p is made per L1 RNA [37]), has hampered efforts to study L1 TPRT at the molecular level.

Epitope and RNA tagging strategies have been used to detect L1 ORF1p, L1 ORF2p, and L1 RNA from engineered L1 expression constructs [18, 20, 21, 28, 30]. Importantly, these technologies have also allowed the discrimination of the L1 proteins and RNA produced from transfected L1 expression constructs from the proteins and RNAs expressed from endogenous L1s. In 2006, we developed an assay, termed the L1 element amplification protocol (LEAP), that allows the detection of ORF2p reverse transcriptase activity in RNP preparations derived from cultured human cells transfected with engineered L1 expression constructs [28]. The LEAP assay uses a similar rationale that has been used to detect RT activity from mitochondrial plasmids of *Neurospora crassa* [38] and is similar to the strategy employed to detect telomerase activity using the telomere repeat amplification protocol (TRAP) assay [39].

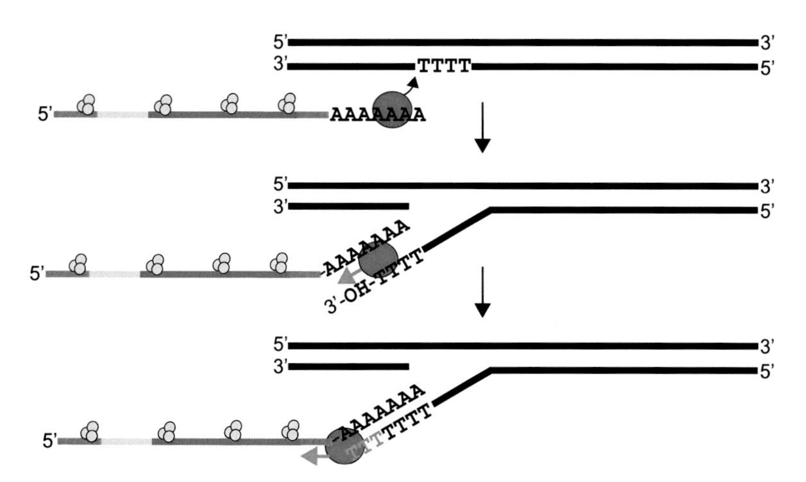

Fig. 1 Target site-primed reverse transcription (TPRT). The L1 RNA (*grey-yellow-blue-grey line*), ORF1p (*yellow circles*), and ORF2p (*blue circle*) minimally constitute the L1 RNP. The L1 RNP targets genomic DNA (*black bold lines*) and nicks one T-rich strand exposing a free 3′OH. L1 ORF2p RT uses the T-rich DNA strand as a primer for reverse transcription of the L1 RNA template. After initial priming, L1 ORF2p RT synthesizes a complementary L1 cDNA strand (*green arrow*) using L1 RNA as a template. Second-strand synthesis and completion of integration remain to be elucidated from L1 retrotransposition but may follow the *Bombyx mori* R2 retrotransposon model of integration [52]

For the LEAP assay, HeLa-JVM cells are transfected with engineered L1 constructs that express versions of ORF1p and ORF2p that contain different epitope tags at their respective carboxyl termini (Fig. 2) [18, 28, 30]. Hygromycin B is used to select for HeLa-JVM cells containing the engineered L1 constructs. Cellular RNP complexes then are isolated from hygromycin B-resistant HeLa-JVM cells using differential centrifugation through a sucrose cushion (Fig. 2b) [28, 30]. Alternatively, L1 RNPs can be isolated from hygromycin B-resistant HeLa-JVM cells by immunoprecipitation, using an antibody directed against the L1 ORF2p carboxyl terminal epitope tag (Fig. 2c) [18].

The resultant RNP then is incubated with an oligonucleotide (i.e., a LEAP adapter) to prime cDNA synthesis (Fig. 3a) [28]. The LEAP adapter contains a unique sequence at the 5′ end (RACE) followed by 12 thymidines (dT_{12}) and ends with VN nucleotides (where V represents adenosine (A), guanosine (G), or cytidine (C), and N represents any nucleotide). The L1 cDNAs are PCR-amplified using oligonucleotide primers to the RACE sequence and the engineered L1 construct. The LEAP PCR products then are visualized by gel electrophoresis and can be subsequently cloned and sequenced to characterize the products (Fig. 3b).

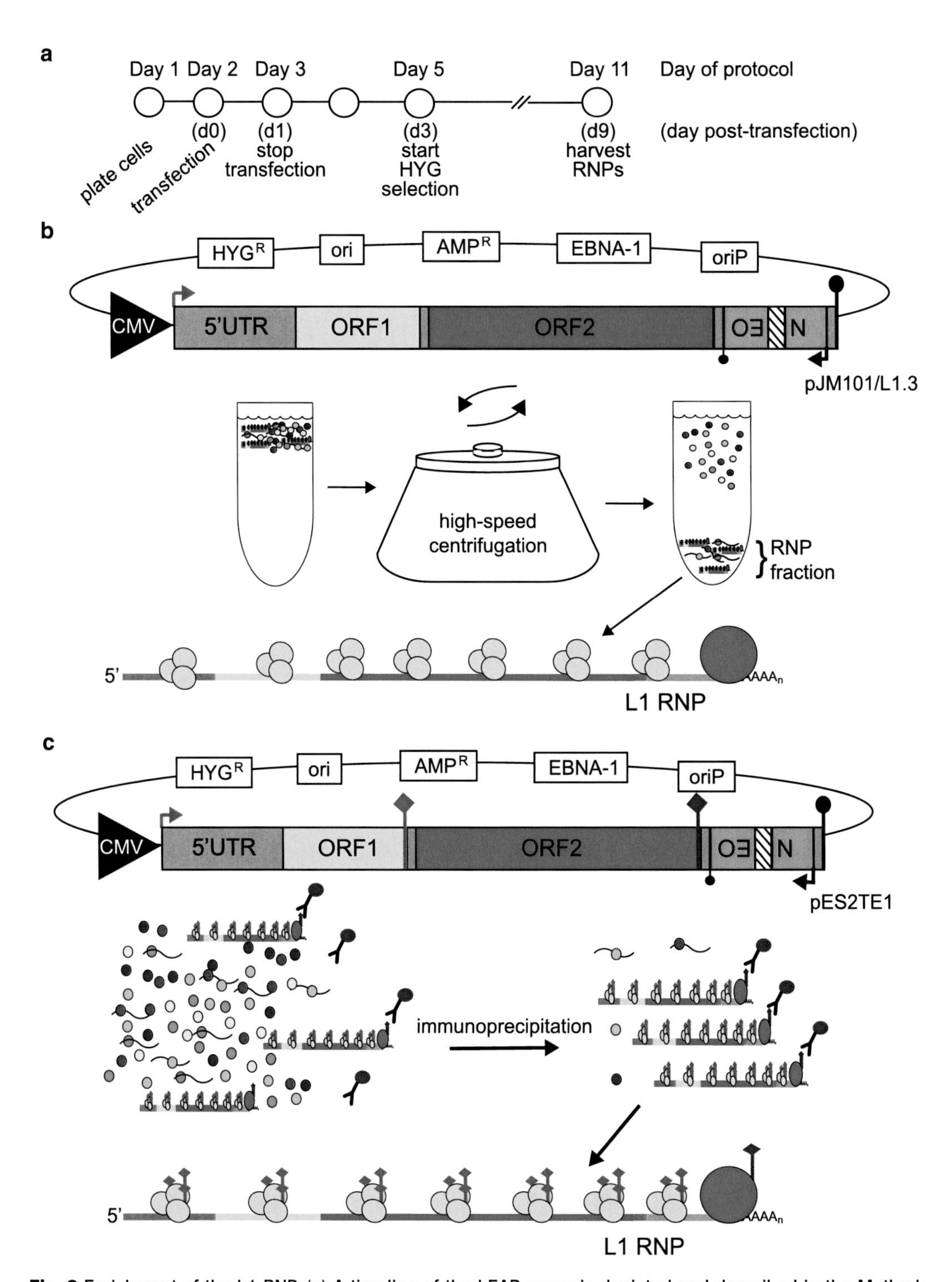

Fig. 2 Enrichment of the L1 RNP. (**a**) A timeline of the LEAP assay is depicted and described in the Methods. Days of the protocol are noted above and the corresponding days post-transfection (d0–9) are noted below. (**b**) The pJM101/L1.3 construct contains the L1.3 element (accession no. L19088). The pCEP4 plasmid (Life Technologies) backbone encodes for the EBNA-1 (EBNA-1) viral protein and contains an origin of viral replication (oriP), a hygromycin B-resistance gene (HYG^R), the cytomegalovirus (CMV) promoter (*large black triangle*), and an SV40 polyadenylation signal (*large black lollipop*) for plasmid replication, hygromycin selection, and transcription, respectively, in mammalian cultured cells. The pCEP4 backbone also has a bacterial origin of replication (ori) and an ampicillin-resistance gene (AMP^R) for replication and ampicillin-selection (respectively) in *E. coli*. (For details of the *mneoI* reporter cassette, please see the "LINE-1 Cultured Cell Retrotransposition Assay"

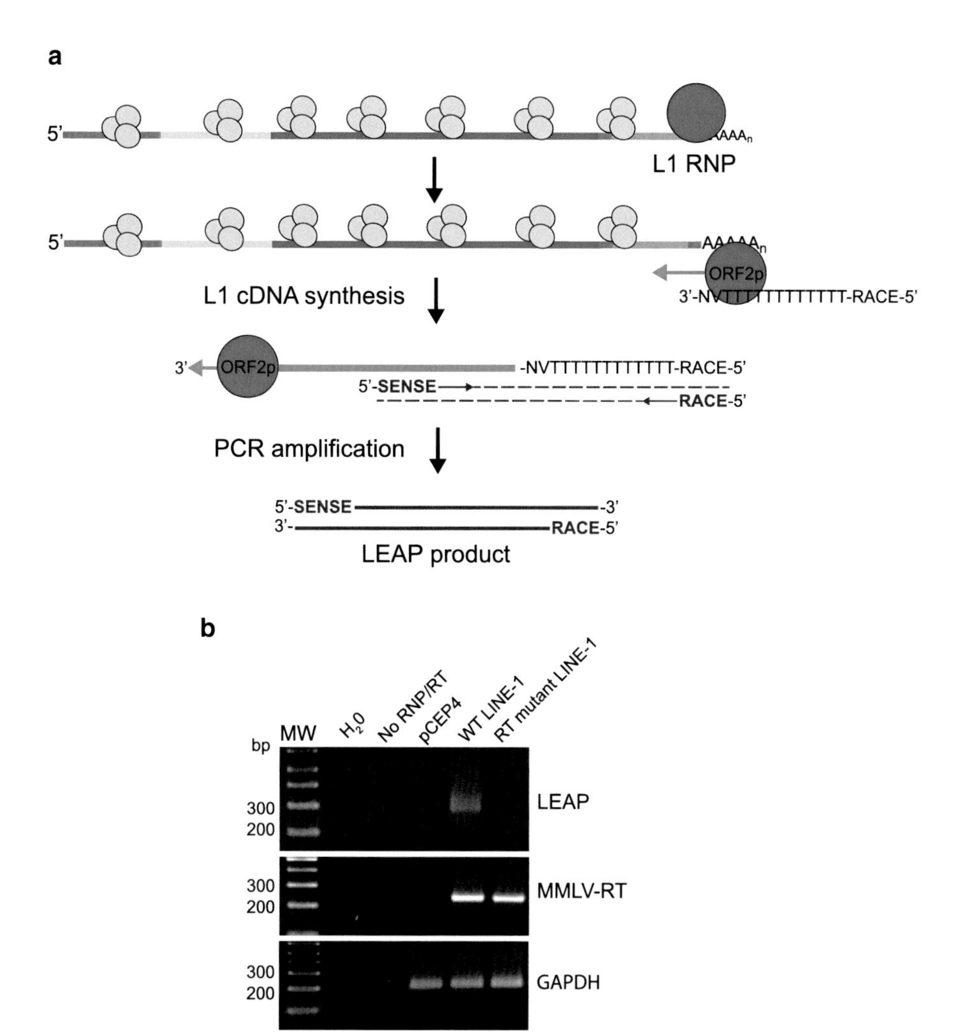

Fig. 3 The LEAP assay. (**a**) The L1 RNP minimally consists of L1 ORF1p (*yellow circles*) and ORF2p (*blue circle*) bound to the L1 RNA (*multicolored line*). The L1 RNP is incubated with a 5′-RACE-T_{12}VN-3′ primer and dNTPs. The L1 RT activity (*green arrow*) of ORF2p initiates L1 cDNA (*green line*) synthesis. Subsequently, the L1 cDNA is amplified using an engineered LINE-1 construct-specific primer (SENSE) and a RACE primer (*red arrows*), resulting in the LEAP product (*double red line*). (**b**) LEAP products can be resolved by electrophoresis and visualized by staining. LEAP products (*top panel*), L1 RNA present in RNPs (MMLV-RT; *middle panel*), and GAPDH RNA levels (GAPDH; *lower panel*) are shown. A standard LEAP assay includes a water control PCR reaction (H_2O); a control reaction without any RT (No RNP/RT); a reaction using RNPs from empty vector-transfected cells (pCEP4); a reaction with wild-type L1 RNPs (WT LINE-1); and a reaction with RT-mutant L1 RNPs (RT mutant LINE-1). The molecular weight (MW) ladder sizes are shown in base pairs (bp). L1 ORF1p and ORF2p protein levels can also be detected by western blot analyses (not shown) [18, 28]

Fig. 2 (continued) chapter in this volume.) After transfection and hygromycin B-selection in HeLa-JVM cells, the cells are lysed and subjected to high-speed centrifugation through a sucrose cushion. After centrifugation, cellular RNPs are enriched in the pellet fraction. This fraction contains the L1 RNA bound by L1 ORF1p and ORF2p, which minimally constitutes the L1 RNP (*see* Fig. 1). (**c**) The pES2TE1 L1 construct [18], like pJM101/L1.3, contains the L1.3 element in a pCEP4 backbone. Unlike pJM101/L1.3, the pES2TE1 construct encodes a T7-tagged (*orange diamond*) ORF1p and a FLAG-HA-tagged (*purple diamond*) ORF2p [18]. After transfection and hygromycin B-selection, HeLa cells are lysed and subjected to immunoprecipitation using an anti-FLAG antibody conjugated to beads (*red circle* with black "Y"). Immunoprecipitated complexes contain L1 ORF1p and ORF2p bound to its encoding RNA

Data obtained from the LEAP assay have revealed that, unlike retroviral reverse transcriptases (e.g., Moloney murine leukemia virus (MMLV)-RT), the L1 RT can initiate reverse transcription from a DNA primer with mismatched 3′ terminal bases when it is annealed to an L1 RNA template [28]. Indeed, the ability of L1 RT to extend mismatched DNA primer/RNA template duplexes explains how certain genomic DNA sequences (e.g., 5′-TTTA/A or 5′-TTTC/A, where "/" corresponds to the L1 EN nick) can serve as primers to initiate TPRT in vivo [31, 35, 40]. That being stated, increasing the length of DNA primer 3′ terminal mismatches to 4 mismatched nucleotides decreases the efficiency of reverse transcription [28, 41]. Notably, the presence of different VN dinucleotide pairs and poly (A) tail lengths in the resultant LEAP products allows independent products to be distinguished from one another in a single LEAP reaction [28]. Finally, the use of gene-specific primers has allowed the identification of cellular mRNAs that can be reverse transcribed at low levels by L1 RT *in trans* [28, 42–44].

LEAP adapters have also been designed to mimic genomic integration sites observed in cultured cell experiments. For example, the LEAP assay was used to examine how endonuclease-deficient L1s are able to integrate at dysfunctional telomeres in Chinese hamster ovary (CHO) cell lines that are defective in components required for the nonhomologous end-joining (NHEJ) pathway of DNA repair [45, 46]. Here, the LEAP adapter was modified to mimic potential telomere ends (e.g., 5′-RACE-$(TTAGGG)_3$-3′, 5′-RACE-$(TTAGGG)_3$TT-3′, 5′-RACE-$(TTAGGG)_2$TTAG-3′). A LEAP adapter with a telomeric repeat ending in 5′-$(TTAGGG)_3$TT-3′ was more efficient to prime first-strand L1 cDNA synthesis than an adapter ending in 5′-$(TTAGGG)_3$-3′ [45]. Characterization of the LINE-1 cDNA/primer junctions of these LEAP products revealed that they generally contained a perfect telomere repeat followed by a poly (T) sequence that resulted from the reverse transcription of the L1 poly (A) mRNA; hence, the LEAP products recapitulated the structure of endonuclease-deficient LINE-1 retrotransposition events observed in NHEJ-deficient CHO cells [33, 45, 46]. Indeed, these data further demonstrated that L1 RNP preparations are associated with a nuclease activity that can process the oligonucleotide adapter prior to its use as a primer in the LEAP reaction [45].

The LEAP assay has also been used to determine whether missense mutations in L1 ORF1p or ORF2p affect L1 RT activity [18, 28, 45]. For example, LEAP reactions revealed that missense mutations in ORF1p, that adversely affect L1 retrotransposition by decreasing the ability of ORF1p to bind L1 RNA, retain LEAP activity [18, 28]. Similarly, missense mutations in the EN or cysteine-rich domain of L1 ORF2p adversely affect L1 retrotransposition but retain LEAP activity [18, 28, 45].

A number of cellular proteins that interact with L1 RNPs recently have been identified and the overexpression of a subset of these proteins can adversely affect L1 retrotransposition in cultured human cells [21, 47, 48]. The LEAP assay can be used to determine if these cellular proteins affect L1 RT activity. For example, the LEAP assay revealed that the L1 mRNA template remains annealed to the L1 cDNA in an RNA/DNA hybrid after reverse transcription, protecting the L1 cDNA from APOBEC3A-mediated cytidine deamination [49]. However, the addition of RNase H (an enzyme that degrades RNA in an RNA/DNA hybrid) to the LEAP reaction renders the L1 cDNA susceptible to cytidine deamination [49]. These data, in conjunction with cultured cell-based experiments, revealed that APOBEC3A inhibits L1 retrotransposition, in part, by deaminating the transiently exposed single-strand L1 cDNA generated during TPRT [49].

In sum, the LEAP assay has been used to elucidate mechanistic details of TPRT. The following protocol has been adapted from previously published studies and is optimized for HeLa cells [18, 28]. Important experimental controls and technical tips are highlighted in Subheading 4 of this protocol.

2 Materials

Special care should be taken when preparing materials for L1 RNP isolation and RT-PCR reactions. Prepare all solutions in DNase/RNase-free water. All lab instruments, consumables (e.g., pipettor, pipets, conical tubes, and microcentrifuge tubes), and solutions should be RNase-free and used only for RNA work. Work areas should be cleaned regularly with RNase*Zap*® (Life Technologies) or other similar cleaning reagent.

2.1 Cell Culture Medium and Transfection Reagents

1. HeLa cells: We typically use HeLa-JVM cells for our assays [7, 28, 50].
2. Dulbecco's Modified Eagle medium (DMEM) (with 4.5 g/L d-glucose) containing 10 % fetal bovine serum (FBS), and 1× Pen Strep glutamine (100 U/mL penicillin, 100 μg/mL streptomycin, and 292 μg/mL glutamine). This is called HeLa-JVM DMEM growth medium in the protocol below.
3. 1× Phosphate-buffered saline (PBS), pH 7.4, sterilized.
4. L1 expression plasmid constructs (e.g., pJM101/L1.3 (Fig. 2b; [5]) or pES2TE1 (Fig. 2c; [18]).
5. A cell counter (e.g., Countess® Automated Cell Counter, Life Technologies) or hemocytometer.
6. Tissue culture T-25, T-75, or T-175 flasks.
7. FuGENE® 6 (Promega).
8. Opti-MEM® I (Life Technologies).
9. 50 μg/mL Hygromycin B.

2.2 Lysis Buffer

1. 1.5 mM KCl.
2. 2.5 mM $MgCl_2$.
3. 5 mM Tris-pH 7.5.
4. 1 % Deoxycholic acid (Calbiochem).
5. 1 % Triton X-100.
6. 1× Complete Protease Inhibitor Cocktail, EDTA-free (Roche, add fresh).
7. UltraPure™ DNase/RNase-free water.

2.3 Sucrose stock solution, 47 %

1. 80 mM NaCl.
2. 5 mM $MgCl_2$.
3. 20 mM Tris–HCl, pH 7.5.
4. UltraPure™ Sucrose (Life Technologies), 47 % weight to volume (w/vol) in UltraPure™ DNase/RNase-free water.
5. 1 mM Dithiothreitol (DTT).
6. 1× Complete Protease Inhibitor Cocktail, EDTA-free (add fresh).
7. UltraPure™ DNase/RNase-free water.
8. Sterile filter system, 0.22 μm CA 500 mL bottle.

2.4 Sucrose Dilution Buffer

1. 80 mM NaCl.
2. 5 mM $MgCl_2$.
3. 20 mM Tris-HCl, pH 7.5.
4. 1 mM DTT (add fresh).
5. 1× Complete Protease Inhibitor Cocktail, EDTA-free (add fresh).
6. UltraPure™ DNase/RNase-free water.

2.5 IP Flag Buffer

1. 100 mM KCl.
2. 20 mM Tris–HCl pH 8.0.
3. 1 mM DTT (add fresh).
4. 10 % UltraPure™ Glycerol (Life Technologies).
5. 0.1 % IGEPAL CA-630 (Sigma).
6. 1× Complete Protease Inhibitor Cocktail, EDTA-free (add fresh).
7. UltraPure™ DNase/RNase-free water.

2.6 Flag Elution Buffer

1. IP FLAG buffer (*see* Subheading 2.5 above).
2. 200 μg/mL 3× FLAG peptide (Sigma).

2.7 LEAP Reaction Buffer (L1 Reverse Transcriptase Reaction)

1. 50 mM Tris–HCl, pH 7.5.
2. 50 mM KCl.
3. 5 mM $MgCl_2$.
4. 10 mM DTT (add fresh).
5. 0.2 μM 3′ RACE adapter (*see* **Note 1**).
6. 20 U recombinant RNasin® ribonuclease inhibitor (Promega).
7. 0.2 mM dNTPs.
8. 0.05 % Tween®-20 (Fisherbrand).
9. UltraPure™ DNase/RNase-free water.

2.8 LEAP PCR Reaction Mix

1. 1× Pfu reaction buffer (Agilent).
2. 0.2 mM dNTPs.
3. 0.4 mM SENSE primer (*see* **Note 2**).
4. 0.4 mM OUTER primer (*see* **Note 2**).
5. 2.5 U PfuTurbo Hotstart DNA Polymerase (Agilent).
6. UltraPure™ DNase/RNase-free water.

2.9 Other

1. RNase*Zap*® RNase decontamination solution (Life Technologies).
2. Cell scrapers, small or large.
3. Conical tubes, 15 mL.
4. Barrier pipette tips.
5. M-MLV Reverse Transcriptase (MMLV-RT) (Promega).
6. RNeasy Mini Kit (Qiagen).
7. Agarose and low-melt agarose with 1× TAE.
8. Ethidium bromide, GelRed™ Nucleic Acid Stain (Biotium), or SYBR® Safe DNA gel stain (Life Technologies).
9. EZview™ Red ANTI-FLAG® M2 Affinity Gel (Sigma) (*see* **Note 3**).
10. Bradford reagent.
11. PCR machine.
12. Thin-walled PCR 96-well plate and adhesive PCR film or 0.2 mL tubes.
13. Refrigerated centrifuge that can accommodate conical tubes.
14. Thermo Scientific Sorvall MTX 150 Micro-Ultracentrifuge.
15. Micro-ultracentrifuge tubes (Thermo Scientific).
16. Nucleic acid electrophoresis system.
17. QIAquick Gel Extraction Kit (Qiagen).
18. PCR product cloning kit.

3 Methods

Detailed below is a scaled-down version of the original LEAP assay [28] as well as an alternative protocol using immunoprecipitation [18] instead of ultracentrifugation to isolate L1 RNPs. This protocol is optimized for HeLa cells but has also been used for other human cell lines [21, 42–44]. Hygromycin B concentrations and transfection protocols should be optimized for any different cell line used for this assay.

3.1 LINE-1 RNP Isolation Using Ultracentrifugation (Fig. 2a, b)

1. Day 1—Plate cells: Seed 2×10^6 HeLa-JVM cells in HeLa-JVM DMEM growth medium in a T-75 tissue culture flask. Cells are grown in a humidified incubator at 37 °C with 7 % CO_2 (*see* **Note 4**).
2. Day 2—Transfect cells: Cells typically are transfected 14–16 h post-plating, day zero (d0) (Fig. 2a), using the FuGENE® 6 transfection reagent following the manufacturer's instructions. The LEAP assay should include the following transfection conditions: (a) an empty vector (e.g., pCEP4); (b) a wild-type LINE-1 expression plasmid (e.g., pJM101/L1.3 [5]); and (c) an RT mutant LINE-1 plasmid (e.g., pJM105/L1.3, which has a D702A mutation in the ORF2 RT domain [29]) (*see* **Note 5**). Prepare a transfection mix in a 1.5 mL microcentrifuge tube containing 8 μg of the pCEP4 or LINE-1 expression plasmid and 32 μL FuGENE® 6 and 500 μL Opti-MEM® I. Incubate the solution at room temperature for 20 min. Add the transfection mix to the growth medium of one flask of cells.
3. Day 3—Stop the transfection: Approximately 16–24 h post-transfection, 1 day post-transfection (d1) (Fig. 2a), aspirate medium from the cells and add fresh DMEM growth medium to the cells.
4. Days 5–11—Select for transfected cells: Begin drug selection 3 days post-transfection (d3) and continue until 9 days post-transfection (d9). Grow cells in DMEM growth medium containing 200 μg/mL hygromycin B. Change the hygromycin B-containing media every day until 9 days post-transfection (d9) (Fig. 2a).
5. Day 11—Harvest the cells for ultracentrifugation: Rinse the cells once with cold 1× PBS. Aspirate the 1× PBS. Add 5 mL cold 1× PBS and scrape the cells from the flask using a cell scraper. Pellet the cells by centrifugation at 3000 × g at 4 °C for 5 min. Transfected cells from one T-75 flask can be divided into two aliquots and can be stored at −80 °C for at least 1 month without affecting reproducibility (*see* **Note 6**).
6. Lyse cells: Add 250 μL lysis buffer (Subheading 2.2) to the cell pellets. Pipette up and down to resuspend the pellets. Let the cells sit on ice for 15 min. Centrifuge the cell lysates at 3000 × g

at 4 °C for 10 min. Transfer the supernatants to new clean tubes and keep on ice. A small aliquot of the whole-cell lysates should be flash frozen in an ethanol/dry ice bath and stored at −80 °C (*see* **Note 7**).

7. Prepare the sucrose cushion: Using the sucrose dilution buffer (Subheading 2.4), dilute the 47 % sucrose stock solution (Subheading 2.3) to 8.5 and 17 % sucrose solutions. Prepare 500 μL of the 17 % sucrose solution and 250 μL of the 8 % sucrose solution per sample. Carefully layer the sucrose solutions such that 500 μL of the 17 % sucrose solution is at the bottom and 250 μL of the 8 % sucrose solution is on top in a 1 mL micro-ultracentrifuge tube. On top of the 8 % sucrose, gently layer 150–200 μL of the cell lysate. If using a fixed-angle rotor, leave an empty space between samples or use sealable tubes to minimize possible cross-contamination during the spin.
8. Run the micro-ultracentrifuge: Spin the cell lysates at 168,000 × *g* at 4 °C for 2 h.
9. Resuspend the RNP: After centrifugation, a pellet should be visible at the bottom of the tube. Aspirate off the sucrose solution, being careful not to disturb the pellet. Resuspend the pellet in 50–100 μL 1× Complete Protease Inhibitor Cocktail, EDTA-free (in DNase/RNase-free water) by gently pipetting up and down. Quantitate protein concentrations of RNPs using the Bradford protein assay reagent. If needed, dilute RNPs with 1× Complete Protease Inhibitor Cocktail, EDTA-free, to a concentration of 1.5–2 μg/μL of total protein. To preserve the enzymatic activity upon storage, dilute the RNPs to a concentration of 0.75–1 μg/μL of total protein with glycerol (final 50 % volume/volume glycerol concentration). Divide the RNPs into 25–50 μL aliquots, flash freeze in an ethanol/dry ice bath, and store the RNPs at −80 °C (*see* **Note 7**).

3.2 LINE-1 RNP Isolation Using Immunoprecipitation (Fig. 2a, c)

1. Day 1—Plate cells: Seed 6 × 10⁶ HeLa-JVM cells in HeLa-JVM DMEM growth medium in a T-175 tissue culture flask.
2. Day 2—Transfect cells: Cells typically are transfected 14–16 h post-plating, day zero (d0) (Fig. 2a), using the FuGENE® 6 transfection reagent following the manufacturer's instructions. The LEAP assay using RNPs isolated by immunoprecipitation should include the following transfection conditions: (a) an empty vector (e.g., pCEP4); (b) an epitope-tagged wild-type LINE-1 plasmid (e.g., pES2TE1 [18]); and (c) a non-tagged LINE-1 plasmid (e.g., pJM101/L1.3 [5], which does not have any epitope tag on ORF1p or ORF2p) (*see* **Note 5**). Prepare a transfection mix in a 1.5 mL microcentrifuge tube containing 20–30 μg of the pCEP4 or LINE-1 expression plasmid, 80–120 μL FuGENE® 6, and 1 mL Opti-MEM® I. Incubate

the solution at room temperature for 20 min. Add the transfection mix to the growth medium of one flask of cells.

3. Day 3—Stop the transfection: Approximately 16–24 h post-transfection, 1 day post-transfection (d1) (Fig. 2a), aspirate the medium from the cells and add fresh DMEM growth medium to the cells.
4. Days 5–11—Select for transfected cells: Begin drug selection 3 days post-transfection (d3) and continue until 9 days post-transfection (d9). Grow cells in DMEM growth medium containing 200 μg/mL hygromycin B. Change the hygromycin B-containing media every day until 9 days post-transfection (d9) (Fig. 2a).
5. Day 11—Harvest the cells for immunoprecipitation: Rinse the cells once with cold 1× PBS. Aspirate the 1× PBS. Add 10 mL cold 1× PBS and scrape the cells from the flask using a cell scraper. Pellet the cells by centrifugation at 3000 × *g* at 4 °C for 5 min. Transfected cells from one T-175 flask can be divided into two aliquots and can be stored at −80 °C for at least 1 month without affecting reproducibility.
6. Lyse the cells: Add IP Flag buffer (Subheading 2.5) three times the volume of the cell pellet. For example, if the volume of the cell pellet is approximately 100 μL, add 300 μL of the buffer. Pipette up and down to resuspend the pellets. Let the cells sit on ice for 15 min. Centrifuge the cell lysates at 3000 × *g* at 4 °C for 10 min. Transfer the lysates to new clean tubes and keep on ice. A small aliquot of the whole-cell lysates should be flash frozen in an ethanol/dry ice bath and stored at −80 °C (*see* **Note** 7).
7. Determine the cell lysate protein concentration using the Bradford protein assay reagent.
8. Prepare immunoprecipitation: For each immunoprecipitation, incubate 3 mg of protein from the cell lysate with 20 μL of equilibrated EZview™ Red ANTI-FLAG® M2 Affinity Gel (*see* **Note 3**) at 4 °C overnight on a rotating wheel or nutator (*see* **Note 8**). Collect the beads by centrifugation at 3000 × *g* at 4 °C for 10 min. Resuspend the beads in 1 mL cold IP Flag buffer. Wash the beads four more times with 1 mL cold IP Flag buffer.
9. Elute the RNPs: Incubate the beads with 50 μL IP Flag buffer containing 200 μg/μL of 3× FLAG peptide at 4 °C for 1 h on a rotating wheel or nutator. Spin down the beads at 6000 × *g* at 4 °C for 5 min and transfer the eluates to new tubes (*see* **Note** 7).

3.3 LEAP Reactions and PCR

The following reactions are performed the same for RNPs prepared by ultracentrifugation or immunoprecipitation. Reactions are prepared in a PCR workstation equipped with HEPA-filtered air circulation and UV-sterilization.

1. L1 reverse transcription reaction: Incubate 1 μL L1 RNP (0.75–1.0 μg) or IP sample with the LEAP reaction buffer (Subheading 2.7) in a total volume of 50 μL at 37 °C for 1 h. LEAP cDNAs can be stored at −20 °C.
2. PCR amplification of L1 RT-synthesized cDNA: Add 1 μL of LEAP cDNA to the LEAP PCR reaction mix (Subheading 2.8) in a total volume of 50 μL (*see* **Note 9**).
3. The LEAP PCR program is as follows:

1 Cycle	94 °C for 3 min
35 Cycles	94 °C for 30 s
	58 °C for 30 s
	72 °C for 30 s
1 Cycle	72 °C for 7 min
	4 °C hold

4. LEAP products are visualized by electrophoresis on a 2 % agarose gel (equal parts agarose to low-melt agarose) in 1× TAE stained with ethidium bromide, GelRed™, or SYBR® Safe nucleic acid stains (Fig. 3b). LEAP bands can be excised, extracted using the QIAquick Gel Extraction Kit, and cloned into a commercial sequencing vector (e.g., Zero Blunt PCR Cloning Kit). These plasmids can then be sequenced for characterization of individual LEAP products within one reaction.

4 Notes

1. The LEAP adapters for the RT reaction work best if they are HPLC-purified. LEAP adapter sequence: 5′-GCGAGCACAGAATTAATACGACTCACTATAGGTTTTTTTTTTTTVN-3′, where V is A, G, or C and N is any nucleotide [28].
2. PCR primers of standard purity are obtained from local suppliers.
   (a) SENSE: 5′-GGGTCCGAAATCGATAAGCTTGGATCCAGAC-3′. This “SENSE” primer is specific to the transfected engineered L1 construct and does not amplify endogenous L1s [28].
   (b) OUTER: 5′-GCGAGCACAGAATTAATACGACT-3′. This “OUTER” primer is depicted as the RACE primer in Fig. 3a [28].
3. EZview™ Red ANTI-FLAG® M2 Affinity Gel is the anti-FLAG M2 antibody covalently linked to agarose beads. Anti-FLAG

beads should be equilibrated according to the manufacturer's instructions. To equilibrate the anti-FLAG beads, spin down the affinity gel solution at 8200 × *g* at 4 °C for 30 s and aspirate the supernatant being careful not to disturb the beads. Resuspend the beads in at least ten times the packed bead volume in IP Flag buffer. Repeat this wash four more times. Spin the beads a final time and resuspend the beads in an equal volume of the packed bead volume in IP Flag buffer. For example, if your packed bead volume is 200 μL, resuspend the beads in 200 μL IP Flag buffer. This 50 % slurry of anti-FLAG beads is the working stock of equilibrated EZview™ Red ANTI-FLAG® M2 Affinity Gel.

4. HeLa-JVM cells can also be seeded at 2×10^5 cells per T-25 cell culture flask or 6×10^6 cells per T-175 cell culture flask. For cells in a T-25 flask, prepare a transfection mix in a 1.5 mL microcentrifuge tube containing 1 μg of the pCEP4 or LINE-1 expression plasmid, 4 μL FuGENE® 6, and 100 μL Opti-MEM® I. For cells in a T-175 flask, prepare a transfection mix in a 1.5 mL microcentrifuge tube containing 20–30 μg of the pCEP4 or LINE-1 expression plasmid, 80–120 μL FuGENE® 6, and 1 mL Opti-MEM® I. Incubate the solution at room temperature for 20 min. Add the transfection mix to the growth medium of one flask of cells.
5. Transcription of L1 elements being tested in the LEAP assay should be driven by the CMV promoter and have the pCEP4 backbone, which allows for episomal replication of the plasmid and hygromycin B-selection of cells containing the L1 expression construct. It is difficult to detect LEAP activity from L1 expression constructs where L1 transcription is driven by the promoter activity in the L1 5′UTR.
6. For ultracentrifugation, the entire cell pellet from a T-25 flask should be used for ultracentrifugation. The cell pellet from one T-175 flask can be divided into four equal aliquots and stored at −80 °C.
7. MMLV-RT and western blot analyses should be done to show comparable RNA and protein levels, respectively, of the isolated L1 RNP. Similarly, RNA and L1 proteins can be analyzed from whole-cell lysates and compared to the isolated L1 RNP fractions. RNA is purified from RNPs using the Qiagen RNeasy kit. The purified, DNA-free RNA is reverse transcribed and PCR amplified using the same adapters in the LEAP reactions but with MMLV-RT instead of L1 RNPs. The L1 cDNAs are PCR amplified using the same primers and conditions as listed for the LEAP PCR but for 25–30 cycles. Control cellular RNAs (e.g., GAPDH) are PCR amplified using gene-specific primers [28]. Western blots of RNPs expressed from pES2TE1

are performed using anti-T7 antibody (Millipore) to detect L1 ORF1p or anti-HA (Roche) to detect L1 ORF2p [18, 30]. Untagged L1 ORF1p can be detected from L1 RNPs and cell lysates using an anti-ORF1p antibody [30, 48, 51].

8. Cell lysates should be pre-cleared with agarose beads if immunoprecipitation is done with anti-FLAG-coupled agarose beads. Other resins, such as Dynabeads (Life Technologies), may have less nonspecific protein binding and may not require pre-clearing.
9. LEAP cDNAs may also be subjected to qPCR and quantified as detailed in Kulpa and Moran, 2006, and Doucet et al. [18, 28].

Acknowledgements

The authors would like to thank Nancy Leff for helpful comments during the preparation of this manuscript. This work was supported in part by NIH grant GM060518 to J.V.M. Authors were supported in part by fellowships from the American Cancer Society #PF-07-059-01GMC (H.C.K.), the NHGRI #T32-HG00040 (D.A.F.), the Japan Society for the Promotion of Science, the Uehara Memorial Foundation and the Kanae Foundation (T.M.), and an International postdoctoral fellowship from the Fondation pour la Recherche Medicale (A.J.D.). J.V.M. is an Investigator of the Howard Hughes Medical Institute.

Conflict of Interest

J.V.M. is an inventor on the patent: "Kazazian, H.H., Boeke, J.D., Moran, J.V., and Dombrowski, B.A. Compositions and methods of use of mammalian retrotransposons. Application No. 60/006,831; Patent No. 6,150,160; Issued November 21, 2000." J.V.M. has not made any money from this patent and voluntarily discloses this information.

References

1. Dombroski BA, Mathias SL, Nanthakumar E, Scott AF, Kazazian HH Jr (1991) Isolation of an active human transposable element. Science 254:1805–1808
2. Lander ES, Linton LM, Birren B et al (2001) Initial sequencing and analysis of the human genome. Nature 409:860–921
3. Beck CR, Garcia-Perez JL, Badge RM, Moran JV (2011) LINE-1 elements in structural variation and disease. Annu Rev Genomics Hum Genet 12:187–215
4. Brouha B, Schustak J, Badge RM, Lutz-Prigge S, Farley AH, Moran JV, Kazazian HH Jr (2003) Hot L1s account for the bulk of retrotransposition in the human population. Proc Natl Acad Sci U S A 100:5280–5285
5. Sassaman DM, Dombroski BA, Moran JV, Kimberland ML, Naas TP, DeBerardinis RJ, Gabriel A, Swergold GD, Kazazian HH Jr (1997) Many human L1 elements are capable of retrotransposition. Nat Genet 16:37–43
6. Feng Q, Moran JV, Kazazian HH Jr, Boeke JD (1996) Human L1 retrotransposon encodes a conserved endonuclease required for retrotransposition. Cell 87:905–916
7. Moran JV, Holmes SE, Naas TP, DeBerardinis RJ, Boeke JD, Kazazian HH Jr (1996) High

frequency retrotransposition in cultured mammalian cells. Cell 87:917–927

8. Scott AF, Schmeckpeper BJ, Abdelrazik M, Comey CT, O'Hara B, Rossiter JP, Cooley T, Heath P, Smith KD, Margolet L (1987) Origin of the human L1 elements: proposed progenitor genes deduced from a consensus DNA sequence. Genomics 1:113–125
9. Hohjoh H, Singer MF (1996) Cytoplasmic ribonucleoprotein complexes containing human LINE-1 protein and RNA. EMBO J 15:630–639
10. Hohjoh H, Singer MF (1997) Ribonuclease and high salt sensitivity of the ribonucleoprotein complex formed by the human LINE-1 retrotransposon. J Mol Biol 271:7–12
11. Holmes SE, Singer MF, Swergold GD (1992) Studies on p40, the leucine zipper motif-containing protein encoded by the first open reading frame of an active human LINE-1 transposable element. J Biol Chem 267:19765–19768
12. Khazina E, Weichenrieder O (2009) Non-LTR retrotransposons encode noncanonical RRM domains in their first open reading frame. Proc Natl Acad Sci U S A 106:731–736
13. Kolosha VO, Martin SL (1997) In vitro properties of the first ORF protein from mouse LINE-1 support its role in ribonucleoprotein particle formation during retrotransposition. Proc Natl Acad Sci U S A 94:10155–10160
14. Kolosha VO, Martin SL (2003) High-affinity, non-sequence-specific RNA binding by the open reading frame 1 (ORF1) protein from long interspersed nuclear element 1 (LINE-1). J Biol Chem 278:8112–8117
15. Martin SL (1991) Ribonucleoprotein particles with LINE-1 RNA in mouse embryonal carcinoma cells. Mol Cell Biol 11:4804–4807
16. Callahan KE, Hickman AB, Jones CE, Ghirlando R, Furano AV (2012) Polymerization and nucleic acid-binding properties of human L1 ORF1 protein. Nucleic Acids Res 40:813–827
17. Martin SL, Bushman FD (2001) Nucleic acid chaperone activity of the ORF1 protein from the mouse LINE-1 retrotransposon. Mol Cell Biol 21:467–475
18. Doucet AJ, Hulme AE, Sahinovic E, Kulpa DA, Moldovan JB, Kopera HC, Athanikar JN, Hasnaoui M, Bucheton A, Moran JV, Gilbert N (2010) Characterization of LINE-1 ribonucleoprotein particles. PLoS Genet 6(10), pii: e1001150
19. Ergun S, Buschmann C, Heukeshoven J, Dammann K, Schnieders F, Lauke H, Chalajour F, Kilic N, Stratling WH, Schumann GG (2004) Cell type-specific expression of LINE-1 open reading frames 1 and 2 in fetal and adult human tissues. J Biol Chem 279:27753–27763
20. Goodier JL, Ostertag EM, Engleka KA, Seleme MC, Kazazian HH Jr (2004) A potential role for the nucleolus in L1 retrotransposition. Hum Mol Genet 13:1041–1048
21. Taylor MS, Lacava J, Mita P, Molloy KR, Huang CR, Li D, Adney EM, Jiang H, Burns KH, Chait BT, Rout MP, Boeke JD, Dai L (2013) Affinity proteomics reveals human host factors implicated in discrete stages of LINE-1 retrotransposition. Cell 155:1034–1048
22. Mathias SL, Scott AF, Kazazian HH Jr, Boeke JD, Gabriel A (1991) Reverse transcriptase encoded by a human transposable element. Science 254:1808–1810
23. Dombroski BA, Feng Q, Mathias SL, Sassaman DM, Scott AF, Kazazian HH Jr, Boeke JD (1994) An in vivo assay for the reverse transcriptase of human retrotransposon L1 in Saccharomyces cerevisiae. Mol Cell Biol 14:4485–4492
24. Cost GJ, Feng Q, Jacquier A, Boeke JD (2002) Human L1 element target-primed reverse transcription in vitro. EMBO J 21:5899–5910
25. Piskareva O, Denmukhametova S, Schmatchenko V (2003) Functional reverse transcriptase encoded by the human LINE-1 from baculovirus-infected insect cells. Protein Expr Purif 28:125–130
26. Piskareva O, Schmatchenko V (2006) DNA polymerization by the reverse transcriptase of the human L1 retrotransposon on its own template in vitro. FEBS Lett 580:661–668
27. Esnault C, Maestre J, Heidmann T (2000) Human LINE retrotransposons generate processed pseudogenes. Nat Genet 24:363–367
28. Kulpa DA, Moran JV (2006) Cis-preferential LINE-1 reverse transcriptase activity in ribonucleoprotein particles. Nat Struct Mol Biol 13:655–660
29. Wei W, Gilbert N, Ooi SL, Lawler JF, Ostertag EM, Kazazian HH, Boeke JD, Moran JV (2001) Human L1 retrotransposition: cis preference versus trans complementation. Mol Cell Biol 21:1429–1439
30. Kulpa DA, Moran JV (2005) Ribonucleoprotein particle formation is necessary but not sufficient for LINE-1 retrotransposition. Hum Mol Genet 14:3237–3248
31. Gilbert N, Lutz-Prigge S, Moran JV (2002) Genomic deletions created upon LINE-1 retrotransposition. Cell 110:315–325
32. Jurka J (1997) Sequence patterns indicate an enzymatic involvement in integration of mam-

malian retroposons. Proc Natl Acad Sci U S A 94:1872–1877
33. Morrish TA, Gilbert N, Myers JS, Vincent BJ, Stamato TD, Taccioli GE, Batzer MA, Moran JV (2002) DNA repair mediated by endonuclease-independent LINE-1 retrotransposition. Nat Genet 31:159–165
34. Myers JS, Vincent BJ, Udall H, Watkins WS, Morrish TA, Kilroy GE, Swergold GD, Henke J, Henke L, Moran JV, Jorde LB, Batzer MA (2002) A comprehensive analysis of recently integrated human Ta L1 elements. Am J Hum Genet 71:312–326
35. Symer DE, Connelly C, Szak ST, Caputo EM, Cost GJ, Parmigiani G, Boeke JD (2002) Human l1 retrotransposition is associated with genetic instability in vivo. Cell 110:327–338
36. Luan DD, Korman MH, Jakubczak JL, Eickbush TH (1993) Reverse transcription of R2Bm RNA is primed by a nick at the chromosomal target site: a mechanism for non-LTR retrotransposition. Cell 72:595–605
37. Alisch RS, Garcia-Perez JL, Muotri AR, Gage FH, Moran JV (2006) Unconventional translation of mammalian LINE-1 retrotransposons. Genes Dev 20:210–224
38. Kuiper MT, Akins RA, Holtrop M, de Vries H, Lambowitz AM (1988) Isolation and analysis of the Neurospora crassa Cyt-21 gene. A nuclear gene encoding a mitochondrial ribosomal protein. J Biol Chem 263:2840–2847
39. Kim NW, Piatyszek MA, Prowse KR, Harley CB, West MD, Ho PL, Coviello GM, Wright WE, Weinrich SL, Shay JW (1994) Specific association of human telomerase activity with immortal cells and cancer. Science 266: 2011–2015
40. Gilbert N, Lutz S, Morrish TA, Moran JV (2005) Multiple fates of L1 retrotransposition intermediates in cultured human cells. Mol Cell Biol 25:7780–7795
41. Monot C, Kuciak M, Viollet S, Mir AA, Gabus C, Darlix JL, Cristofari G (2013) The specificity and flexibility of l1 reverse transcription priming at imperfect T-tracts. PLoS Genet 9:e1003499
42. An W, Dai L, Niewiadomska AM, Yetil A, O'Donnell KA, Han JS, Boeke JD (2011) Characterization of a synthetic human LINE-1 retrotransposon ORFeus-Hs. Mob DNA 2:2
43. Dai L, Taylor MS, O'Donnell KA, Boeke JD (2012) Poly(A) binding protein C1 is essential for efficient L1 retrotransposition and affects L1 RNP formation. Mol Cell Biol 32:4323–4336
44. Mandal PK, Ewing AD, Hancks DC, Kazazian HH Jr (2013) Enrichment of processed pseudogene transcripts in L1-ribonucleoprotein particles. Hum Mol Genet 22:3730–3748
45. Kopera HC, Moldovan JB, Morrish TA, Garcia-Perez JL, Moran JV (2011) Similarities between long interspersed element-1 (LINE-1) reverse transcriptase and telomerase. Proc Natl Acad Sci U S A 108:20345–20350
46. Morrish TA, Garcia-Perez JL, Stamato TD, Taccioli GE, Sekiguchi J, Moran JV (2007) Endonuclease-independent LINE-1 retrotransposition at mammalian telomeres. Nature 446:208–212
47. Goodier JL, Cheung LE, Kazazian HH Jr (2013) Mapping the LINE1 ORF1 protein interactome reveals associated inhibitors of human retrotransposition. Nucleic Acids Res 41:7401–7419
48. Moldovan JB, Moran JV (2015) The zinc-finger antiviral protein ZAP inhibits LINE and Alu retrotransposition. PLoS Genet 11 (5):e1005121
49. Richardson SR, Narvaiza I, Planegger RA, Weitzman MD, Moran JV (2014) APOBEC3A deaminates transiently exposed single-strand DNA during LINE-1 retrotransposition. Elife 3:e02008
50. Hulme AE, Bogerd HP, Cullen BR, Moran JV (2007) Selective inhibition of Alu retrotransposition by APOBEC3G. Gene 390:199–205
51. Leibold DM, Swergold GD, Singer MF, Thayer RE, Dombroski BA, Fanning TG (1990) Translation of LINE-1 DNA elements in vitro and in human cells. Proc Natl Acad Sci U S A 87:6990–6994
52. Christensen SM, Eickbush TH (2005) R2 target-primed reverse transcription: ordered cleavage and polymerization steps by protein subunits asymmetrically bound to the target DNA. Mol Cell Biol 25:6617–6628

Chapter 22

Biochemical Approaches to Study LINE-1 Reverse Transcriptase Activity In Vitro

Sébastien Viollet, Aurélien J. Doucet, and Gaël Cristofari

Abstract

In vitro reverse transcriptase assays have been developed to monitor the presence and activity of ORF2p, an essential protein product of the LINE-1 retrotransposon (L1), in cellular fractions. We describe methods for expression and isolation of L1 ribonucleoprotein particles, and identification of ORF2p reverse transcriptase activity. Two independent methods are described: L1 element amplification protocol (LEAP) and direct L1 extension assay (DLEA). The first method involves cDNA synthesis by primer extension using dNTPs followed by a step of PCR amplification. The second method involves primer extension by incorporation of radiolabeled dTMPs followed by dot-blot or gel separation detection. Finally, we discuss the output and benefits of the two methods.

Key words Reverse transcriptase, Non-LTR retrotransposon, Retroelement, RNA-dependent DNA-polymerase, ORF2p, Ribonucleoprotein particle, In vitro assay, LINE-1, L1

1 Introduction

Long interspersed element-1 (LINE-1 or L1) is an autonomous retrotransposon of the human genome [1]. Active L1 copies express a bicistronic RNA that encodes two distinct proteins, ORF1p and ORF2p, both required for retrotransposition (for review [2]). ORF1p is an abundant ~40 kDa RNA-binding protein [3, 4] that also contains nucleic acid chaperone activity [5]. ORF2p is an ~150 kDa protein that contains endonuclease (EN) and reverse transcriptase (RT) activities [6, 7]. Retrotransposition of an active L1 copy begins with transcription of the L1 RNA, expression of ORF1p and ORF2p in the cytoplasm, and then formation of a ribonucleoprotein particle (RNP) complex by the binding of L1 proteins to their encoding RNA (*cis*-preference model). After the nuclear import and chromatin binding, the L1 RNP can initiate reverse transcription directly at the target site, a process termed target-primed reverse transcription (TPRT). Briefly, the EN

Jose L. Garcia-Pérez (ed.), *Transposons and Retrotransposons: Methods and Protocols*, Methods in Molecular Biology, vol. 1400, DOI 10.1007/978-1-4939-3372-3_22,

domain of ORF2p cleaves the genomic DNA target and frees a 3′ hydroxyl that is subsequently used by the RT domain of ORF2p to prime L1 first-strand cDNA synthesis, with the L1 RNA as a template (for review [2]). The remaining steps for the completion of the insertion are not described yet, but likely involve cellular factors.

ORF2p is characterized by a low level of expression, likely caused by its unconventional translation mechanism [8, 9]. As a consequence, classical methods (e.g., protein immunoblotting or immunofluorescence) are difficult to apply for ORF2p detection. One alternative is to use epitope-tagging strategies [9, 10]. Another alternative is to benefit from the enzymatic activities encoded in ORF2p for its indirect detection.

To date, two different biochemical assays have been set up based on ORF2p reverse transcriptase activity detection: (1) the L1 element amplification protocol (LEAP) [11] and (2) the direct L1 extension assay (DLEA) [12]. Both are in vitro assays used to detect the presence of ORF2p and characterize its reverse transcriptase activity [6, 13, 14].

LEAP and DLEA are derivations of techniques developed previously to measure the activity of retroviral reverse transcriptases or telomerase [15–17]. They were engineered to mimic the early steps of the TPRT where the L1 RNP extends the DNA extremity generated upon cleavage of the target DNA by L1 EN activity. Both approaches use L1 RNP at varying degrees of purification as a source of L1 RT and RNA template. They mostly differ by the nature of DNA substrate that can be used, and by the method of detection of the product. In LEAP, ORF2p reverse transcription of the L1 RNA occurs when cellular extracts are incubated with dNTPs and a single-stranded DNA oligonucleotide (a poly(T) and a linker sequence). The produced cDNA is then amplified by PCR to allow its visualization on agarose gel (Fig. 1, left panel) [11]. In DLEA, reverse transcription is measured by the direct incorporation of radiolabeled dTMP to a linear DNA substrate, which can be single or double stranded. RT products are subsequently either resolved on sequencing polyacrylamide gels and visualized by autoradiography or quantified by spotting on DE81 paper to measure the incorporation of radiolabeled nucleotides [12]. Since DLEA reactions only contain ^{32}P-dTTP as a nucleotide, only the initial step of L1 poly(A) tail reverse transcription is monitored (Fig. 1, right panel).

LEAP has been implemented in several studies. It was originally used to demonstrate the stable and preferred association of ORF2p with the L1 RNA [11]. Since then, publications described its use to study various aspects of the L1 retrotransposition mechanisms: the formation of the L1 RNP complex [9], a potential

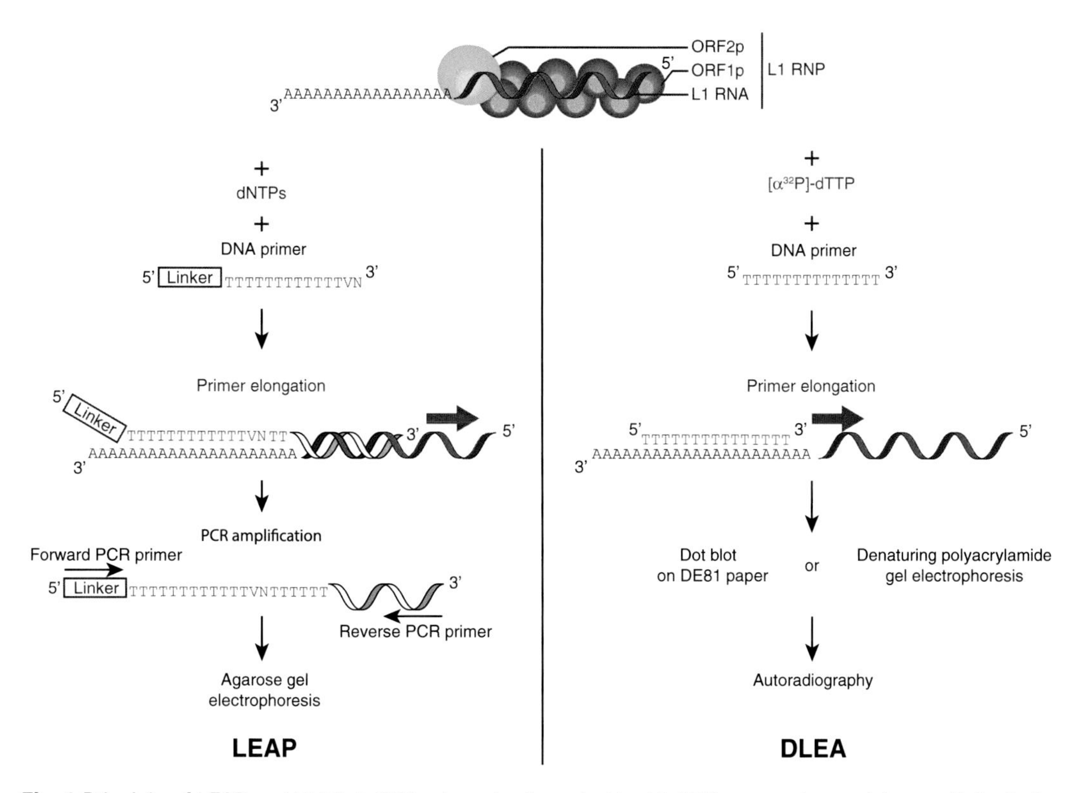

Fig. 1 Principle of LEAP and DLEA. A DNA primer is elongated by L1 RNP reverse transcriptase activity, in the presence of dNTPs (LEAP—*left*) or [α-^{32}P]-dTTP (DLEA—*right*). The reverse transcription products are amplified by PCR and visualized on agarose gel (LEAP), or either directly spotted on DE81 ion-exchange paper or resolved by denaturing polyacrylamide gel electrophoresis (DLEA)

endonuclease processing of DNA oligonucleotide primers [18], to test the impact of cellular inhibitors on ORF2p RT activity [19–21], or on the sequence of produced cDNAs [22]. Several other studies took advantage of the LEAP assay to simply test the RT activity of ORF2p in various L1 RNP preparations [10, 19, 23–26]. In contrast, DLEA was described more recently and used to evaluate the ability of L1 RT to extend DNA substrates of varying sequence and structure, as well as to show that efficient reverse transcription initiation requires a primer with a 3′ single-stranded extremity and is greatly enhanced by base complementarity with the L1 RNA poly(A) tail [12, 27].

The following protocols describe the methods used by Monot, Kuciak, and coworkers [12], a study for which both LEAP and DLEA were used. We first present the methods for L1 RNP expression in human cells and enrichment using adapted method for ultracentrifugation. We then detail protocols for the LEAP and DLEA assays and highlight their benefits and limitations.

2 Materials

1. L1 expression vectors: pJM101/L1.3 (active L1) and pJM105/L1.3 (reverse transcriptase deficient L1) [28].
2. Control empty vector: pCEP4 (Invitrogen).
3. GFP expression vector: phrGFP-C (Stratagene). It is used to verify transfection efficiency and can be substituted by any fluorescent protein expression vector.
4. $CaCl_2$ solution: 250 mM $CaCl_2$.
5. 2× Hanks' Balanced Salt (HBS) solution: 273.8 mM NaCl, 9.4 mM KCl, 2.8 mM Na_2HPO_4, 11.1 mM glucose, and 42 mM HEPES pH 7.05 (*see* **Note 1**).
6. 1× Phosphate-buffered saline (PBS): 138 mM NaCl, 2.7 mM KCl, 8.1 mM Na_2HPO_4, and 1.5 mM KH_2PO_4.
7. Growth medium: DMEM (supplemented with 4 mM Glutamax + 4.5 g/L D-glucose + 110 mg/mL pyruvate) (Life Technologies), 10 % fetal calf serum (v/v) (Life Technologies), and 100 U/mL penicillin/streptomycin (Life Technologies).
8. 0.25 % w/v Trypsin-EDTA (Life Technologies).
9. 50 mg/mL Hygromycin B (Life Technologies).
10. cOmplete, Mini, EDTA-free Protease Inhibitor cocktail tablets (Roche).
11. CHAPS buffer (should be filtered at 0.22 μm): 10 mM Tris pH 7.5, 0.5 % CHAPS (w/v), 1 mM $MgCl_2$, 1 mM EGTA, and 10 % glycerol (v/v).
12. CHAPS extract (ChE) buffer (prepare freshly from CHAPS buffer just before use): CHAPS buffer completed with 1× protease inhibitor (Roche) and 1 mM DTT.
13. 13.2 mL Ultra Clear tubes (Beckman Coulter).
14. SW 41 Ti swinging rotor (Beckman Coulter).
15. Optima LE-80K Ultracentrifuge (Beckman Coulter).
16. 47 % Sucrose stock solution filtered at 0.22 μm: 80 mM NaCl, 5 mM $MgCl_2$, 20 mM Tris pH 7.5, and 47 % sucrose (w/v).
17. Sucrose dilution buffer: 80 mM NaCl, 5 mM $MgCl_2$, 2 mM Tris pH 7.5, and 1 mM DTT (add freshly).
18. Working 47 % solution (prepare freshly): 47 % sucrose (w/v) and 1 mM DTT complemented with 1× protease inhibitor.
19. 17 and 8.5 % sucrose solution: For each 10 mL solution, sucrose dilution solutions should be prepared freshly by diluting 3.6 and 1.8 mL of 47 % working sucrose solution, respectively, with 6.4 and 8.2 mL of sucrose dilution buffer, respectively.

20. Bio-Rad Protein Assay Dye Reagent Concentrate.
21. 2× Loading buffer: 100 mM Tris–HCl pH 6.8, 4 % SDS (v/v), 20 % glycerol (v/v), and 0.4 % bromophenol blue (w/v).
22. Page ruler prestained protein ladder (Thermo Scientific).
23. 1× Running buffer: 25 mM Tris pH 8.3, 192 mM glycine, and 0.1 % SDS (v/v).
24. Vertical electrophoresis apparatus.
25. Transfer buffer: 25 mM Tris pH 8.3, 192 mM glycine, and 20 % methanol (v/v).
26. Whatman Grade 3MM Chr Blotting Paper (GE Healthcare).
27. Immobilon–FL PVDF membrane (Millipore).
28. Electrophoretic wet-transfer apparatus.
29. PBST: 1× PBS completed with 0.1 % Tween-20 (v/v).
30. Blocking buffer: 5 % nonfat dry milk (w/v) diluted in 1× PBST.
31. Human ORF1p polyclonal antibody SE-6798 (directed against peptide CERNNRYQPLQNHAKM) [12].
32. S6 Ribosomal Protein (5G10) rabbit monoclonal antibody (Cell Signaling Technology).
33. Odyssey blocking buffer (LI-COR Biosciences).
34. Antibody stripping buffer 1× (Gene Bio Application).
35. IRDye 800CW Goat anti-Rabbit IgG (H+L) (LI-COR Biosciences).
36. Odyssey infrared imaging system (LI-COR Biosciences).
37. LEAP buffer: 50 mM Tris pH 7.5, 50 mM KCl, 5 mM $MgCl_2$, 10 mM DTT, and 0.05 % Tween (v/v).
38. dNTP (Promega).
39. [α-^{32}P]-dTTP (Perkin Elmer).
40. Recombinant RNasin Ribonuclease Inhibitor (Promega).
41. Platinum Taq DNA Polymerase (Invitrogen).
42. DNA oligonucleotides (5′ to 3′): RACE primer: GCGAGCA CAGAATTAATACGACTCACTATAGGTTTTTTTTT TTVN; LOU312: GCGAGCACAGAATTAATACGACT; LO U851: GGGTTCGAAATCGATAAGCTTGGATCCAGAC; LOU852: GACCCTCACTGCTGGGGAGTCC, oligo(dT): $d(T)_{18}$.
43. Conventional thermocycler.
44. 0.5× TBE buffer: 44.5 mM Tris–HCl pH 8.0–9.0, 44.5 mM boric acid, and 1 mM EDTA.
45. TRI Reagent (Molecular Research Center).
46. RQ1 RNase-free DNase (Promega).

47. DNase stop solution: 20 mM EGTA pH 8.0.
48. RNaseOUT Recombinant Ribonuclease Inhibitor (Life Technologies).
49. 5× Superscript reverse transcriptase buffer: 250 mM Tris pH 8.0, 375 mM KCl (Life Technologies).
50. Superscript III Reverse Transcriptase (Life Technologies).
51. Milli-Q water (Millipore).
52. Grade DE81 ion exchange paper (GE Healthcare).
53. Stop buffer: 50 mM EDTA and 5 % SDS (w/v).
54. Saran wrap (Dow).
55. 2× SSC buffer: 300 mM NaCl and 30 mM sodium citrate.
56. Storage Phosphor Screen (GE Healthcare).
57. Typhoon FLA 9000 (GE Healthcare).
58. Recovery control DNA oligonucleotide (for example, RBD3, 5′ to 3′: TAC-GTTC-TATGCTA).
59. DNA ladder: ΦX174 DNA/HinfI Dephosphorylated Markers (Promega).
60. T4 Polynucleotide Kinase (New England Biolabs).
61. 10× T4 Polynucleotide Kinase Buffer (New England Biolabs).
62. Illustra MicroSpin G-25 Column (GE Heathcare).
63. [γ-^{32}P]-dATP (Perkin Elmer).
64. Molecular biology-grade phenol (Biosolve).
65. Chloroform.
66. 3 M Na-acetate solution pH 5.2.
67. Glycogen (Sigma Aldrich).
68. 100 % Ethanol.
69. Formamide loading buffer: 98 % deionized formamide (Biosolve), 10 mM EDTA, 0.02 % bromophenol blue (w/v) (Biosolve).
70. Sequagel-Urea Gel System (National Diagnostics).
71. Sequencing gel apparatus.
72. Gel dryer.

3 Methods

The following methods describe the production of native L1 RNPs in human cells, their enrichment by sucrose cushion, and the characterization of their reverse transcriptase activity by LEAP and DLEA. Although LEAP on crude cellular extracts is not robust and less reliable, likely due to PCR inhibitors present in this type of fraction, we successfully used DLEA on crude CHAPS-extract lysates. However, it is generally more desirable to first enrich L1 RNPs on sucrose cushion and/or by affinity chromatography to partly eliminate unrelated enzymatic activities.

3.1 Preparation of Native L1 RNPs from Cultured Cells

The RNP preparation protocol was originally described by Kulpa et al. [11]. We describe here the protocol used by Monot, Kuciak, and coworkers to obtain the high quantities of L1 RNPs required for a direct detection of reverse transcription activity [12]. L1 RNPs from endogenous or transfected vector expression can be extracted following this procedure. L1 RNPs are overexpressed from mammalian expression vectors such as pJM101/L1.3 and derivatives [28], or codon-optimized L1 elements [23, 29]. L1 expression is under the control of a CMV promoter and an SV40 poly(A) signal. The addition of a strong poly(A) signal compensates for the weakness of the L1 poly(A) signal located in its 3′ UTR. Here, all constructs are based on the pCEP4 vector containing a hygromycin resistance gene that allows the selection of transfected cells.

3.1.1 Transfection and Expression of L1 RNPs

Human cell lines commonly used for L1 RNP production are HeLa [11] or HEK-293T [23]. Here, we focus on the transfection of HEK-293T and present a complete expression/preparation procedure in 5 days to obtain high amounts of L1 RNPs. HEK-293T are transfected using a calcium phosphate-mediated method with high efficiency and very low cost, [30]. Common plasmids used for transfection are phrGFP-C (transfection efficiency control), pCEP4 (empty vector control), pJM101/L1.3 (active L1), and pJM105/L1.3 (reverse transcriptase-deficient L1).

Day 0

1. Seed 3×10^6 HEK-293T cells per 10 cm diameter Petri dish in 10 mL of growth medium.

Day 1

2. For each Petri dish, dilute 24 μg of plasmid into 600 μL of the $CaCl_2$ solution.
3. Mix ten times by pipetting the DNA-$CaCl_2$ solution. Add into a tube containing 600 μL of 2× HBS buffer. Gently tap the tube while adding the DNA-$CaCl_2$ solution. The final mixture should be cloudy.
4. Wait for 15 min at room temperature.
5. Mix ten times by pipetting the solution and add directly onto cells in the culture medium.
6. Change medium after 5 h.

Day 2

7. Check the GFP-transfected cells with a fluorescent microscope or by FACS to confirm the efficiency of transfection. We typically obtain 90 % of transfected cells. If transfection efficiency is below 50 %, we advise to repeat the transfection.
8. Wash (very) gently each dish with 5 mL of 1× PBS. Remove PBS. HEK-293T cells tend to detach easily from their plastic substrate.
9. Add 1 mL of trypsin-EDTA (0.25 %) to detach the cells. Incubate for 5 min in the 37 °C incubator. Proper cell dissociation can be monitored under the microscope. Inhibit trypsin with 9 mL of growth medium. Pipet up and down to break remaining cell aggregates.
10. Replate cells from one 10 cm dish into two new 10 cm dishes with selection medium (growth medium supplemented with 100 μg/mL hygromycin B).

Day 4

11. Replace the selection medium with fresh one.

3.1.2 Cell Pellet Collection and Storage

Day 5: At this time, cells have generally reached confluency.

1. Wash the cells with 1× PBS (5 mL) and detach them with trypsin-EDTA (1 mL) as above.
2. Harvest the cells with 10 mL of growth medium.
3. Count cells with a hemocytometer.
4. Prepare tubes containing 10^7 cells each.
5. Centrifuge for 5 min at 1000 ×*g* at room temperature.
6. Wash pellets with 2 mL of cold 1× PBS and centrifuge again for 5 min at 1000 ×*g*. From this stage, keep cells/extracts on ice or at 4 °C and use cold solutions.
7. Remove supernatant (*see* **Note 2**).

3.1.3 Cell Extraction and Sucrose Cushion Enrichment of Native L1 RNPs

1. Resuspend each cell pellet in 500 μL of ChE lysis buffer (*see* **Note 3**).
2. Vortex tubes for 10 s.
3. Incubate the cell suspensions at 4 °C on an end-over-end wheel for 15 min.
4. Centrifuge lysates at 4 °C at 20,000 ×*g* for 15 min in a benchtop centrifuge.

   During incubation and centrifugation:
5. Prepare the 17 and 8.5 % sucrose solutions (6 and 4 mL per tube, respectively).

6. Cool ultra clear tubes on ice and place a Pasteur pipette into each of them.
7. Load carefully 4 mL of the 8.5 % sucrose solution through the Pasteur pipette.
8. Repeat this step slowly with 6 mL of the 17 % sucrose solution, which should stay at the bottom, without mixing with the upper layer. The presence of the two independent sucrose layers, the higher density layer below the lower one, can be observed by eye. Verify the formation of two sucrose layers, with 17 % at the bottom and 8.5 % on top.
9. When lysates are cleared by centrifugation, take the ~500 μL supernatant and dilute it into 500 μL of ChE buffer (1 mL final volume).
10. Load gently this 1 mL of diluted cleared lysate at the top of the sucrose cushion tube.
11. Ultracentrifuge at 4 °C at 178,000 × *g* for 2 h (*see* **Note 4**).
12. Carefully remove the supernatant by inverting the tube and look for the presence of a colorless pellet at the bottom of the tube.
13. Resuspend pellets in 50 μL of RNase-free water (supplemented with 1× protease inhibitors) or with a buffer suitable for the desired downstream applications.
14. Vortex the tubes and incubate overnight at 4 °C on an orbital shaker (*see* **Note 5**).
15. Quantify the resuspended L1 RNPs with a protein dosage method (e.g., Bio-Rad Bradford protein assay, *see* **Note 6**) and divide into aliquots convenient for the subsequent experiments (e.g., 15 μL). We recommend adjusting aliquot size for a small number of assays. Multiple thawing/freezing cycles will reduce L1 RNP activity. We generally obtain 50 μg of RNP per 10^7 transfected cells. It should be stressed that many—if not all—cellular RNPs are enriched upon this sucrose cushion ultracentrifugation step. Thus this number does not specifically reflect L1 RNPs but all cellular RNPs. Consequently, empty-vector-transfected cells give similar yields.
16. Flash-freeze the aliquots in liquid nitrogen and store them at −80 °C.

3.2 Validation of L1 Protein Expression by ORF1p Detection

The following steps describe the procedure for the detection of ORF1p expression by immunoblotting. Here, we use a rabbit antibody targeting the C-terminus of ORF1p (Fig. 2a, top panel) [12]. Alternatively, the addition of epitope tags on ORF1p and/or ORF2p allows detection by commercial antibodies [9].

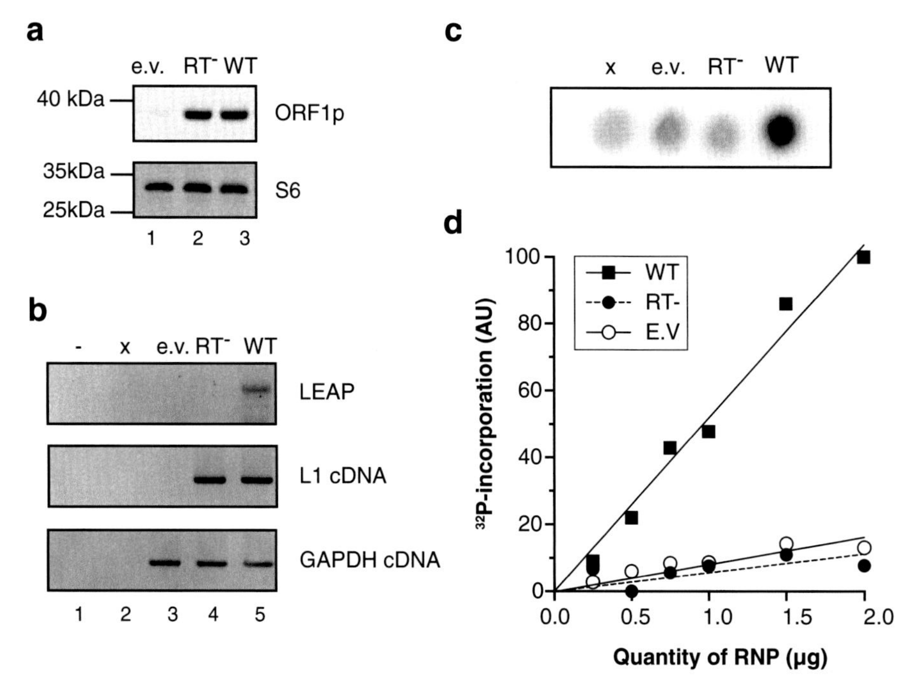

Fig. 2 Quality control of native L1 RNP preparations and reverse transcriptase assays. (**a**) Immunoblotting of human ORF1p (*top panel*) or S6 protein (*bottom panel*, loading control) in RNP prepared from cells transfected with different plasmid DNA: pCEP4 empty vector (e.v.), pJM105/L1.3 reverse transcriptase-deficient mutant (RT⁻) or pJM101/L1.3 wild-type L1 (WT). (**b**) Detection of L1 RT activity by LEAP (*top panel*), of L1 RNA by standard RT-PCR (*middle panel*, control of L1 RNA expression), and of cellular GAPDH RNA (*bottom panel*, control of RNA extraction). (–) and (x) correspond to PCR (no cDNA) and RT (no reverse transcriptase) controls, respectively. (**c**) Dot blot of DLEA products obtained with RNP prepared from cells transfected with pCEP4 empty vector (e.v.), pJM105/L1.3 reverse transcriptase-deficient mutant (RT⁻), or pJM101/L1.3 wild-type L1 (WT). A negative control was performed with water instead of RNP sample (x). (**d**) Dose-dependent incorporation of ^{32}P-dTMP by DLEA reactions performed with varying amounts of RNPs prepared from cells transfected with pJM101/L1.3 (WT), pJM105/L.3 (RT⁻), and pCEP4 (e.v.)

3.2.1 SDS-PAGE and Protein Transfer

1. Pour an SDS-PAGE gel (10 % polyacrylamide separating gel, 4 % polyacrylamide stacking gel) or use precast gels adapted to the available electrophoresis equipment. The range of separation should allow the detection of ~40 kDa (ORF1p) and ~32 kDa (S6 ribosomal protein control) proteins.
2. Prepare samples for loading by diluting 2 µg of L1 RNPs in 1× loading buffer on ice.
3. Heat the samples at 98 °C for 5 min directly from ice.
4. Let the tubes cool down at room temperature and centrifuge briefly.
5. Install the gel in the electrophoresis chamber, and fill with 1× running buffer.
6. Load the samples on the SDS-PAGE gel alongside with an appropriate prestained protein molecular weight marker.

7. Run the electrophoresis at 100 V for 2 h.
8. During electrophoresis:
   - Prepare 1× transfer buffer, cut four Whatman papers, and one Immobilon–FL PVDF membrane at the size of the gel.
   - Soak Whatman papers and transfer sponges in 1× transfer buffer.
   - Activate the Immobilon–FL PVDF membrane in 100 % methanol for 1 min.
   - Remove the membrane from methanol and wash with 1× transfer buffer for 5 min.
   - When the migration front reaches the bottom of the gel, remove from the electrophoresis chamber and rinse the acrylamide gel with 1× transfer buffer.
   - Prepare the wet transfer sandwich by combining in the following order: one sponge, two pieces of Whatman paper, the Immobilon-FL PVDF membrane, the polyacrylamide gel, two Whatman papers, and one sponge. Orientate the sandwich in the transfer apparatus to allow protein transfer from the gel to the membrane (gel closer from the minus pole electrode, membrane closer from the plus pole electrode).
   - Electrotransfer proteins at 30 V overnight at room temperature or at 80 V for 1.5 h. To avoid overheating, place the transfer apparatus in an ice bucket.

3.2.2 Immunoblotting

Before moving to the next step, briefly control the transfer efficiency by looking at the presence of the prestained protein ladder on the transfer membrane. Do not let the PVDF membrane dry. If this happens, re-soak the membrane in methanol and wash with 1× PBS.

1. Wash the membrane with 1× PBS for 2 min.
2. Incubate the membrane in blocking buffer for 1 h at room temperature.
3. Discard the blocking buffer and incubate the membrane with the primary antibody (ORF1p antibody at 1:1000) diluted in blocking buffer for 1 h or more at room temperature (or overnight at 4 °C) on a rocker. The volume should be sufficient to completely cover the membrane.
4. Wash the membrane four times 5 min with 1× PBST.
   Here, we use secondary antibodies cross-linked with infrared fluorophores and the LI-COR infrared detection system. To avoid signal loss, the membrane should be protected from light, from this step until the end of the immunoblotting procedure. Alternatively, it is possible to use secondary antibodies coupled with horseradish peroxidase (HRP) and to apply enhanced chemiluminescence (ECL) detection.

5. Incubate the membrane with IRDye 800CW goat anti-rabbit secondary antibody diluted 1:20,000 in Odyssey blocking buffer supplemented with 0.1 % Tween-20 (v/v), for 1 h at room temperature.
6. Wash the membrane four times 5 min with 1× PBST.
7. Rinse with 1× PBS for 2 min.
8. Scan the membrane on an Odyssey infrared imaging system. After immunoblotting against ORF1p, perform a second immunoblotting against Ribosomal S6 protein as a loading control (Fig. 2a, bottom panel). S6 protein is abundant in cellular RNP preparations.
9. Incubate the membrane in 1× antibody stripping buffer for 1 h at room temperature.
10. Rinse the membrane five times 5 min with Milli-Q water.
11. Repeat immunoblotting **steps 1–8** with S6 Antibody (1:2000) as primary antibody and IRDye 800CW goat anti-rabbit antibody (1:20,000) as secondary antibody.

3.3 LEAP Assay

In parallel with the detection of ORF1p in the L1 RNP fraction by immunoblotting, ORF2p and L1 RNA can be indirectly detected in L1 RNP preparations through ORF2p reverse transcriptase activity using L1 mRNA as a template. Two independent methods have been described to detect such activity: the L1 element amplification protocol (LEAP) [11] and the direct L1 extension assay (DLEA) [12]. Briefly, a DNA oligonucleotide serving as a primer is extended by ORF2p RT activity to generate an L1 cDNA.

In the LEAP, the DNA primer is combined with a set of the four dNTPs in order to get L1 cDNA synthesis from L1 RNP fraction that would subsequently be detected through a PCR amplification step (LEAP products) (Fig. 2b, top panel).

In the DLEA, the DNA primer is combined with radiolabeled dTTP and L1 RNP fractions in order to only detect reverse transcription priming in the L1 poly(A) tail.

3.3.1 L1 cDNA Synthesis

1. In a 1.5 mL microtube, prepare a 50 μL reaction mix containing 0.75 μg of RNPs, 0.2 mM dNTPs, 400 nM RACE primer, and 0.5 μL of Recombinant RNasin Ribonuclease Inhibitor (20 U, Promega) diluted in 1× LEAP buffer.
2. Mix tubes by pipetting up and down.
3. Incubate at 37 °C for 1 h.
4. Store on ice and proceed to the PCR amplification step.

Additional controls should include LEAP reactions performed: (1) with RNPs prepared from empty-vector-transfected cells (pCEP4), and (2) with RNPs prepared from RT-defective L1 elements (e.g., pJM105/L1.3, RT mutant).

3.3.2 PCR Amplification

1. In a PCR tube, prepare a 50 μL reaction containing 1 μL of the L1 cDNA synthesis reaction, 0.2 mM dNTPs, 3 mM $MgCl_2$, 10 pmol of LOU312, 10 pmol of LOU851, and 2 U of Platinum Taq Polymerase in 1× PCR buffer.
2. Cycle the reaction 35 times at 94 °C for 30 s, 60 °C for 30 s, and 72 °C for 30 s. Start the PCR reaction by an initial incubation at 94 °C for 2 min and finish it by an elongation step at 72 °C for 5 min.
3. Pour a 2 % agarose gel (w/v) in 0.5× TBE and 1× SYBR-safe.
4. Resolve the PCR products by 2 % agarose gel electrophoresis in 0.5× TBE. We generally use 1× SYBR-safe, directly poured in the gel, as a nucleic acid dye.

PCR conditions should include (1) a water control (no cDNA—PCR-negative control), (2) cDNA obtained from cells transfected with an empty vector, and (3) an RT-defective version of L1 (controls for DNA and cDNA synthesis reactions, respectively).

3.3.3 RNA Quality Control

LEAP assay should be combined with an RNA quality control assessment. First, RNA is extracted from the tested RNP fractions. Then, an RT-PCR reaction, using a commercial reverse transcriptase, will allow assessing the presence and verifying the quality of the L1 RNA for each tested fraction, including the one with ORF2p RT-deficient mutant. Similarly, we use RT-PCR amplification of the GAPDH RNA as a loading control (Fig. 2b, middle and bottom panels).

3.3.4 RNA Extraction

1. Mix 15 μg of L1 RNPs with RNase-free water to reach a final volume of 25 μL.
2. Add 250 μL of TRI reagent. Vortex for 15 s and incubate at room temperature for 5 min.
3. Add 50 μL of chloroform. Vortex for 15 s and store at room temperature for 15 min.
4. Centrifuge at 12,000 × *g* at 4 °C for 15 min.
5. Transfer the upper phase into a new tube and add 125 μL of isopropanol. Incubate at room temperature for 10 min.
6. Centrifuge at 12,000 × *g* at 4 °C for 8 min.
7. Remove the supernatant and wash the pellet with 125 μL of 75 % ethanol (v/v).
8. Centrifuge at 7500 × *g* at 4 °C for 5 min.
9. Remove the supernatant and air-dry the pellet for 5 min.
10. Resuspend in 20 μL RNase-free water.

3.3.5 RT-PCR of RNA Extracted from L1 RNPs

1. Perform a DNase treatment in a 10 μL reaction mixture containing 1 μg of extracted RNA and 1 U of RQ1 RNase-free DNase in 40 mM Tris–HCl pH 8.0, 10 mM $MgSO_4$, and 1 mM $CaCl_2$.

2. Incubate at 37 °C for 30 min.
3. Add 1 μL of RT stop solution.
4. Incubate at 65 °C for 10 min.
5. Mix 6 μL of treated RNA with 50 pmol of RACE primer and 10 nmol of dNTP in 13 μL reaction.
6. Incubate at 65 °C for 5 min and then chill on ice for 1 min.
7. Add 4 μL of 5× Superscript reverse transcriptase buffer, 1 μL of 0.1 M DTT, 40 U of RNaseOUT Recombinant Ribonuclease Inhibitor, and 200 U of SuperScript III Reverse Transcriptase to the previous reaction mix and adjust the volume to 20 μL with Milli-Q water.
8. Incubate at 50 °C for 1 h and then at 65 °C for 10 min.
9. Inactivate the reaction by heating the samples at 70 °C for 15 min.
10. After the first-strand cDNA synthesis, PCR amplification is performed using primers LOU312 and LOU851 following the protocol described in Subheading 3.3.1, **step 2**, with only 30 cycles of amplification. A parallel PCR reaction is performed to amplify GAPDH from the same cDNA samples. For this purpose, identical PCR conditions are used with LOU312 and LOU852 as primers.
11. Resolve the PCR products by 2 % agarose gel electrophoresis in 0.5× TBE. We generally use 1× SYBR-safe, directly poured in the gel, as a nucleic acid dye.

3.4 Direct LINE-1 Extension Assay

DLEA assay is a fast method to detect ORF2p RT activity from purified RNP fractions (Fig. 2c). Reactions are generally performed starting with 2 μg of L1 RNPs. However, reduction of RNP amounts will lead to a linear decrease of the detected ^{32}P incorporation (Fig. 2d, *see* **Note 7**). Thus, DLEA allows for quantitative measurement of primer extensions by ORF2p RT activity. This assay can accommodate a large diversity of primer, varying in sequence and structure [12], depending on experimenter needs. A single-stranded oligo$(dT)_{18}$ primer is robustly extended by L1 RNPs.

1. Add 2 μg of L1 RNP into a 25 μL reaction mixture containing 1× LEAP buffer, 400 nM of oligo$(dT)_{18}$ primer, and 10 μCi of [α-^{32}P]-dTTP (*see* **Notes 7** and **8**).
2. Incubate at 37 °C for 4 min (*see* **Note 9**).
3. Spot 5 μL of the reaction on DE81 ion exchange paper.
4. Incubate the paper at room temperature for 20 min.
5. Wash five times with 100 mL of 2× SSC. Proper washing can be controlled by measuring the radioactive signal in an unused region of the paper (no signal expected).

6. Seal the piece of DE81 paper in a plastic bag or with Saran wrap.
7. Place the sealed paper in an exposure cassette with a PhosphorImager screen for 1–16 h.
8. Scan the screen with a PhosphorImager.

Alternatively, DLEA products, obtained after **step 2** of the above procedure, can be resolved on denaturing polyacrylamide sequencing gels (Fig. 3), using the following procedure. A recovery control (DNA primer) is added before the phenol/chloroform extraction to monitor the loss of DLEA products during the purification process.

1. Add 5 μL of stop buffer and 1 μL of radiolabeled recovery control (*see* **Note 10**).
2. Mix 100 μL of phenol and 50 μL of chloroform into the reaction mixture.
3. Vortex the tubes for 1 min. Wait for 1 min and vortex again for 1 min.
4. Centrifuge at the maximum speed for 20 min at room temperature in a bench-top centrifuge.
5. Transfer the upper phase into a new tube.
6. Add 10 μg glycogen, 30 μL 3 M Na-acetate pH 5.2, and 250 μL of cold 100 % ethanol into each tube. Vortex for 1 min.
7. Precipitate at −20 °C for 1 h (or longer). If products are resolved by gel electrophoresis the same day, **step 12** can be done during precipitation.
8. Centrifuge for 30 min at 4 °C at 20,000 × *g*.
9. Remove carefully ethanol and dry the pellet at 37 °C for 5 min (or longer if required; no more traces of liquid should be seen).
10. Pour a 13 % denaturing polyacrylamide sequencing gel in 0.5× TBE using Sequa Gel-Urea gel system according to the manufacturer's protocol. Ensure that one of the two plates is siliconized to facilitate uncasting.
11. Pre-run the gel in 0.5× TBE for 30 min at 800 V (this can be modulated depending on the gel system). The plates should reach a temperature of 50–55 °C but not exceed 65 °C to avoid glass breaking.
12. Resuspend the pellet in 15 μL of formamide loading buffer. Heat samples at 95 °C for 5 min and put immediately on ice.
13. Just before loading, rinse the wells with running buffer using a needle and syringe.
14. Load 5 μL of each sample per well, including a well with the labeled size marker (*see* **Note 11**).

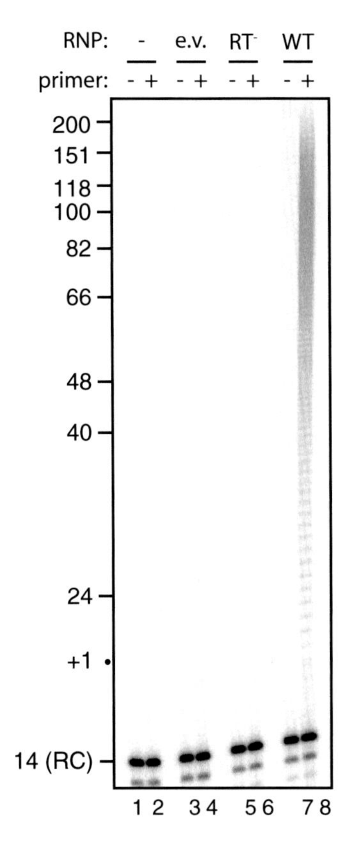

Fig. 3 Direct visualization of primer extension by L1 RNPs at nucleotide resolution. Upon DLEA reaction completion, products can be resolved on denaturing polyacrylamide sequencing gels. Direct L1 extension assay (DLEA) with or without a $(dT)_{18}$ primer in the presence of [α-^{32}P]-dTTP (even and odd lanes, respectively). Sucrose cushion RNP fractions prepared from L1-transfected cells (RT^-, RT-defective L1 RNP; WT, wild-type L1 RNP) or empty vector-transfected cells (e.v.) prepared in parallel were used as a source of RNPs. Trace amounts of a 14-nt 5′ end-labeled oligonucleotide was added after the reaction as a recovery control (denoted RC). The resulting signal is a ladder corresponding to L1 poly(A) tail reverse transcription. Oligo(dT) primers can anneal anywhere within the L1 poly(A) tail; therefore the product is heterogeneous in length. Modified from [12], doi: 10.1371/journal.pgen.1003499.g001, © 2013 Monot et al. licence CC-BY

15. Run the gel at 1500 V (this can be modulated depending on the gel system). Reduce the voltage if the temperature of the plates reaches 65 °C. Stop before the migration front runs out of the gel.
16. Uncast the gel. When separating the two glass plates, the gel should stick to the plate that was not siliconized.
17. Cut with a spacer or with a blade the front of the gel, which contains unincorporated radioactivity and discard it (following appropriate procedures for radioactive solid waste).

18. Place the glass plate on the bench with the gel up. Place a Whatman paper sheet on the gel, press softly, and peel the paper from a corner. The gel should stick to it. If not, use a plastic wash bottle with distilled water to detach slightly the gel from the glass plate, and repeat the procedure.
19. Place the gel (with the Whatman papers and Saran wrap) in a gel dryer at 80 °C for 2 h and 30 min.
20. Expose the covered gel overnight in an exposure cassette with a PhosphorImager screen.
21. Scan the screen with a PhosphorImager.

3.5 Comparison between LEAP and DLEA Assays

Although both LEAP and DLEA detect L1 RT activity through the extension of a DNA oligonucleotide primer, the output and benefits of the two methods are different.

LEAP concomitantly reveals ORF2p RT activity and confirms the use of L1 RNA as reverse transcription template. LEAP products can be cloned and sequenced, which can be used to monitor potential primer processing events before extension and cDNA editing during elongation or to measure RT error rates in vitro. Lower-than-expected molecular weight LEAP products have also been attributed to internal priming subsequent to ORF1p dysfunction. It is very sensitive but poorly quantitative due to the PCR step. The latter imposes also constraints on the type of primers that can be used.

The DLEA procedure has been more recently developed and is a quick and quantitative approach to detect ORF2p reverse transcriptase activity. Primers of varying structure or sequence may be used to test substrate specificity at the reverse transcription initiation step. Sequencing gels may also be used instead of dot blots to reveal the patterns of primer elongation or processing.

Effects of cellular factors or chemicals may be tested with both assays with different objectives: DLEA will provide quantitative information about RT inhibition, whereas LEAP will supply valuable data on sequence editing.

4 Notes

1. The efficiency of calcium phosphate-mediated transfection reaction is extremely dependent on pH. We recommend to test the transfection efficiency of different 2× HBS solutions with pH ranging from 7.00 to 7.10 and to select the solution with the best efficiency.
2. At this step, cell pellets can be snap-frozen in liquid nitrogen and stored for several weeks at −80 °C before preparing L1 RNPs. When needed, thaw cell pellets on ice for 10 min before use.

3. Fresh ChE buffer, 47 % working sucrose solution, and dilution buffer are recommended to avoid oxidization of DTT and inactivation of protease inhibitor. All buffers should be stored at 4 °C and used cold. The stock 47 % sucrose solution can be stored at 4 °C for 1 month.
4. Before ultracentrifugation, cool down rotor and buckets and centrifuge at 4 °C. Perform ultracentrifugation run with high acceleration and deceleration.
5. After ultracentrifugation, pellets may be difficult to resuspend. Start the resuspension by pipetting, then vortex, and let incubate overnight at 4 °C under agitation in the ultracentrifugation tube. The next day, resuspend by pipetting one more time before transferring into a microtube.
6. We routinely use the Bradford protein dosage method (Bio-Rad) with a bovine serum albumin standard, but other methods work as well. Caution should be taken to ensure that the standard (e.g., BSA) is diluted in the same solution/buffer as L1 RNPs. Indeed, detergents frequently react with the Bradford dye (or other reagents used for protein dosage).
7. Use radiolabeled element according to the legal specifications of your country and institution. Particular care should be taken to follow appropriate safety and waste disposal rules.
8. Linearity of the reaction should be controlled by using a range of RNP quantities (e.g., 2, 1, and 0.5 μg).
9. After 4 min with 2 μg of RNPs, the reaction is no longer in its initial velocity phase (not linear).
10. DLEA reaction may be stopped and stored at −20 °C for downstream application after adding 5 μL of stop buffer.
11. Recovery control and ladder for polyacrylamide gel electrophoresis are prepared in advance by end-labeling with [γ-^{32}P]-ATP and T4 polynucleotide kinase, followed by gel filtration (e.g., with Illustra MicroSpin G-25 column). For the recovery control oligonucleotide, incubate 100 pmol of primer RBD3, with 4 μCi of [γ-^{32}P]-ATP and 40 U of T4 polynucleotide kinase in a 50 μL reaction containing 70 mM Tris–HCl pH 7.6, 10 mM $MgCl_2$, and 5 mM DTT. Incubate at 37 °C for 30 min. For the DNA ladder, incubate 500 ng of the dephosphorylated ΦX174 DNA/HinfI DNA Ladder, 10 μCi [γ-^{32}P]-ATP, and 10 U of T4 polynucleotide kinase in a 25 μL reaction containing 70 mM Tris–HCl pH 7.6, 10 mM $MgCl_2$, and 5 mM DTT. Incubate at 37 °C for 30 min. Remove unincorporated labeled ATP by gel filtration. Dilute the labeled marker in formamide loading buffer and heat-denature before loading. A pilot electrophoresis can be useful to define the optimal dilution under defined experimental conditions.

Acknowledgements

We are grateful to John V. Moran (Univ. of Michigan, USA) and to Nicolas Gilbert (Institut de Génétique Humaine, France) for sharing plasmids. This work was supported by a joint Avenir grant from the Institut National de la Santé Et de la Recherche Medicale and the Institut National du Cancer [2009-340 to G.C.]; the European Research Council [243312 to G.C.]; and by Agence Nationale pour la Recherche [ANR-11-LABX-0028-01 to G.C.].

References

1. Lander ES, Linton LM, Birren B, Nusbaum C, Zody MC, Baldwin J et al (2001) Initial sequencing and analysis of the human genome. Nature 409:860–921
2. Beck CR, Garcia-Perez JL, Badge RM, Moran JV (2011) LINE-1 elements in structural variation and disease. Annu Rev Genomics Hum Genet 12:187–215
3. Hohjoh H, Singer MF (1996) Cytoplasmic ribonucleoprotein complexes containing human LINE-1 protein and RNA. EMBO J 15:630–639
4. Khazina E, Weichenrieder O (2009) Non-LTR retrotransposons encode noncanonical RRM domains in their first open reading frame. Proc Natl Acad Sci U S A 106:731–736
5. Martin SL, Bushman FD (2001) Nucleic acid chaperone activity of the ORF1 protein from the mouse LINE-1 retrotransposon. Mol Cell Biol 21:467–475
6. Mathias SL, Scott AF, Kazazian HH, Boeke JD, Gabriel A (1991) Reverse transcriptase encoded by a human transposable element. Science 254:1808–1810
7. Feng Q, Moran JV, Kazazian HH, Boeke JD (1996) Human L1 retrotransposon encodes a conserved endonuclease required for retrotransposition. Cell 87:905–916
8. Alisch RS, Garcia-Perez JL, Muotri AR, Gage FH, Moran JV (2006) Unconventional translation of mammalian LINE-1 retrotransposons. Genes Dev 20:210–224
9. Doucet AJ, Hulme AE, Sahinovic E, Kulpa DA, Moldovan JB, Kopera HC et al (2010) Characterization of LINE-1 ribonucleoprotein particles. PLoS Genet 6
10. Taylor MS, Lacava J, Mita P, Molloy KR, Huang CR, Li D et al (2013) Affinity proteomics reveals human host factors implicated in discrete stages of LINE-1 retrotransposition. Cell 155:1034–1048
11. Kulpa DA, Moran JV (2006) Cis-preferential LINE-1 reverse transcriptase activity in ribonucleoprotein particles. Nat Struct Mol Biol 13:655–660
12. Monot C, Kuciak M, Viollet S, Mir AA, Gabus C, Darlix JL et al (2013) The specificity and flexibility of L1 reverse transcription priming at imperfect T-tracts. PLoS Genet 9:e1003499
13. Cost GJ, Feng Q, Jacquier A, Boeke JD (2002) Human L1 element target-primed reverse transcription in vitro. EMBO J 21:5899–5910
14. Piskareva O, Denmukhametova S, Schmatchenko V (2003) Functional reverse transcriptase encoded by the human LINE-1 from baculovirus-infected insect cells. Protein Expr Purif 28:125–130
15. Baltimore D (1970) RNA-dependent DNA polymerase in virions of RNA tumour viruses. Nature 226:1209–1211
16. Temin HM, Mizutani S (1970) RNA-dependent DNA polymerase in virions of Rous sarcoma virus. Nature 226:1211–1213
17. Kim NW, Piatyszek MA, Prowse KR, Harley CB, West MD, Ho PL et al (1994) Specific association of human telomerase activity with immortal cells and cancer. Science 266:2011–2015
18. Kopera HC, Moldovan JB, Morrish TA, Garcia-Perez JL, Moran JV (2011) Similarities between long interspersed element-1 (LINE-1) reverse transcriptase and telomerase. Proc Natl Acad Sci U S A 108:20345–20350
19. Goodier JL, Cheung LE, Kazazian HH (2012) MOV10 RNA helicase is a potent inhibitor of retrotransposition in cells. PLoS Genet 8:e1002941
20. Zhao K, Du J, Han X, Goodier JL, Li P, Zhou X et al (2013) Modulation of LINE-1 and Alu/SVA retrotransposition by Aicardi-Goutières syndrome-related SAMHD1. Cell Rep 4:1108–1115

21. Horn AV, Klawitter S, Held U, Berger A, Vasudevan AA, Bock A et al (2014) Human LINE-1 restriction by APOBEC3C is deaminase independent and mediated by an ORF1p interaction that affects LINE reverse transcriptase activity. Nucleic Acids Res 42:396–416
22. Richardson SR, Morell S, Faulkner GJ (2014) L1 retrotransposons and somatic mosaicism in the brain. Annu Rev Genet 48:1–27
23. An W, Dai L, Niewiadomska AM, Yetil A, O'Donnell KA, Han JS et al (2011) Characterization of a synthetic human LINE-1 retrotransposon ORFeus-Hs. Mob DNA 2:2
24. Wagstaff BJ, Barnerssoi M, Roy-Engel AM (2011) Evolutionary conservation of the functional modularity of primate and murine LINE-1 elements. PLoS One 6:e19672
25. Dai L, Taylor MS, O'Donnell KA, Boeke JD (2012) Poly(A) binding protein C1 is essential for efficient L1 retrotransposition and affects L1 RNP formation. Mol Cell Biol 32:4323–4336
26. Mandal PK, Ewing AD, Hancks DC, Kazazian HH (2013) Enrichment of processed pseudogene transcripts in L1-ribonucleoprotein particles. Hum Mol Genet 22:3730–3748
27. Viollet S, Monot C, Cristofari G (2014) L1 retrotransposition: the snap-velcro model and its consequences. Mob Genet Elements 4:e28907
28. Moran JV, DeBerardinis RJ, Kazazian HHJ (1999) Exon shuffling by L1 retrotransposition. Science 283:1530–1534
29. Han JS, Boeke JD (2004) A highly active synthetic mammalian retrotransposon. Nature 429:314–318
30. Sambrook J, and Russell DW (2006) Calcium-phosphate-mediated Transfection of Eukaryotic Cells with Plasmid DNAs. CSH Protoc. doi is: 10.1101/pdb.prot387

Chapter 23

Methylated DNA Immunoprecipitation Analysis of Mammalian Endogenous Retroviruses

Rita Rebollo and Dixie L. Mager

Abstract

Endogenous retroviruses are repetitive sequences found abundantly in mammalian genomes which are capable of modulating host gene expression. Nevertheless, most endogenous retrovirus copies are under tight epigenetic control via histone-repressive modifications and DNA methylation. Here we describe a common method used in our laboratory to detect, quantify, and compare mammalian endogenous retrovirus DNA methylation. More specifically we describe methylated DNA immunoprecipitation (MeDIP) followed by quantitative PCR.

Key words DNA methylation, MeDIP, In vitro methylation, Endogenous retroviruses

1 Introduction

Mammalian genomes are littered with repetitive sequences among which are the remnants of millions of transposable element (TE) copies [1]. Endogenous retroviruses (ERVs), a major class of TEs, are responsible for 10 % of mouse spontaneous mutations [2] and contribute to thousands of human transcripts [3]. While some copies are able to escape host control, most ERV sequences are the target of epigenetic regulation such as DNA methylation [4, 5]. In mammals, the addition of a methyl group to a cytosine nucleotide (5-methylcytosine—5mC) in a CpG context is observed in ERV promoters and associated with their silencing. Indeed, DNA methyl transferase mutants, i.e., mutants lacking the machinery to deposit DNA methylation, show upregulation of several ERV families [6]. Different methods were developed in order to study general DNA methylation, but the study of TEs always poses difficulties related to their intrinsic multi-copy nature. Here we describe a 2-day method based on immunoprecipitation of methylated DNA followed by quantitative PCR (MeDIP-qPCR) [7, 8]. There are several advantages in using MeDIP-qPCR instead of other methods (Table 1). For instance, bisulfite treatment, which converts

Jose L. Garcia-Pérez (ed.), *Transposons and Retrotransposons: Methods and Protocols*, Methods in Molecular Biology, vol. 1400, DOI 10.1007/978-1-4939-3372-3_23, © Springer Science+Business Media New York 2016

Table 1
DNA methylation methods

	DNA methylation information on:					
	Single CpG	Single cell	Single ERV copy	5mC w/o 5hmC	Limitations	Suitable for:
Methylation-sensitive restriction enzyme	✓[a]	✗	✗	✗	Restriction enzyme recognition sites, one CpG at a time, average DNA methylation	Acquiring fast results on one CpG site
Bisulfite sequencing	✓	✓	✓	✗	Labor intensive	Search for bisulfite patterns across sequences; studying intra-sample variation
Combined bisulfite restriction analysis	✓[a]	✗	✗	✗	Restriction enzyme recognition sites, one CpG at a time, average DNA methylation	Acquiring fast results on one CpG site
Bisulfite-pyrosequencing	✓	✗	✓	✗	Average DNA methylation	Multiple sample comparison of one genomic region
Bisulfite-seq	✓	✗[b]	✓	✗	Average DNA methylation, bioinformatics labor intensive	Genome-wide DNA methylation analysis
MeDIP (qPCR or sequencing)	✗	✗	✓	✓	Average DNA methylation for one region. Antibody-based approach, more than a single CpG methylated is necessary	Cost-efficiently, genome-wide DNA methylation analysis (sequencing) and inter-sample comparison of multiple loci (qPCR). Suitable for samples with 5-hydroxymethylation (i.e., mouse ES cells [11])

[a]Restriction enzyme dependent

[b]Possibility to look at the sequenced reads but due to their small size it is difficult to find one that contains a nucleotide variation along with a CpG

nonmethylated cytosines into uracils while 5mC remains unchanged, does not differentiate between 5mC and 5-hydroxymethylcytosine (5hmC). Hence, when studying tissues where 5hmC is present, we suggest the usage of antibody-based assays like MeDIP. Furthermore, bisulfite-based methods can be labor intensive if multiple samples and loci are to be compared. For multiple comparisons including several ERV copies in the genome, MeDIP followed by quantitative PCR is a more suitable method [9]. Nevertheless, it is important to also confirm a subset of MeDIP results with another DNA methylation assay (Table 1) [9].

MeDIP is based on immunoprecipitation of methylated DNA [10]. Briefly, sonicated genomic DNA fragments are incubated overnight with a monoclonal antibody recognizing 5mC. The following day pre-blocked beads are added to create antibody-methylated DNA-bead complexes (Fig. 1). The bead complexes are washed and DNA is precipitated for qPCR detection. Real-time amplification can be carried out using primers recognizing single ERV copies (by taking advantage of unique flanking

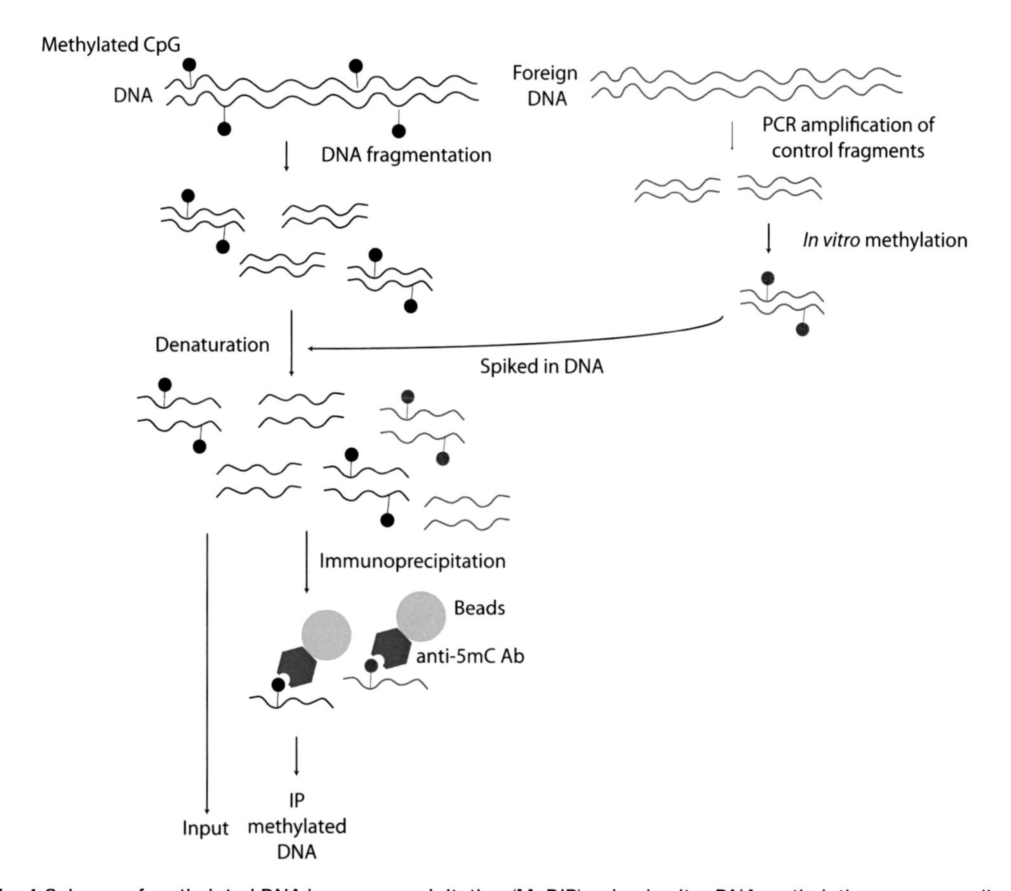

Fig. 1 Scheme of methylated DNA immunoprecipitation (MeDIP) using in vitro DNA methylation as a normalization control. Further details are included in the text

sequences) or multiple copies (by designing primers within the ERV sequences). In order to control immunoprecipitation and user errors, we include a step where in vitro-methylated DNA from another species (*Drosophila* is suitable when studying mammals) is spiked in the genomic DNA after sonication [9]. Alternatively, users can buy in vitro-methylated DNA from other species than the one studied. Amplification of the spiked methylated DNA is then used to normalize across samples, along with the input control.

2 Materials

2.1 In Vitro Methylation of Spiked-In DNA

1. 1 μg of purified PCR fragments (300–600 bp) from *Drosophila* (*see* **Note 1**).
2. CpG methyltransferase (M. SssI), S-adenosylmethionine (SAM), and 10× NEBuffer 2 (New England Biolabs).
3. Methylation-sensitive restriction enzyme and buffer (*see* **Note 2**).
4. Agarose, 1× TBE for gel electrophoresis, DNA loading dye, 100 bp DNA ladder.
5. Heat block and electrophoresis apparatus.

2.2 MeDIP

1. 4–6 μg of RNA-free high-quality genomic DNA (purified using standard phenol-chloroform extraction and RNase A treatment).
2. IP buffer 10× (to be made fresh and kept on ice): 100 mM NaPhosphate pH 7, 1.4 M NaCl, and 0.5 % (v/v) Triton X100 (*see* **Note 3**).
3. Digestion buffer: 50 mM Tris pH 8, 10 mM EDTA, and 0.5 % (w/v) SDS.
4. Agarose, 1× TBE for gel electrophoresis, DNA loading dye, and 100 bp DNA ladder.
5. 5-mC monoclonal antibody and anti-mouse IgG (Diagenode): 5–10 μl aliquots, stored at −20 °C.
6. PMSF 100×, 50 μl aliquots, stored at −20 °C.
7. 50 % Solution of pre-blocked protein A/G agarose beads (*see* **Note 4**).
8. 3 M NaCl.
9. 10× TE buffer.
10. Proteinase K.
11. UltraPure™ phenol:chloroform:isoamyl alcohol (25:24:1, v/v).
12. 3 M Sodium acetate.
13. Glycogen.
14. 100 and 70 % ice-cold ethanol.

15. 10 mM Tris–HCl pH 7.5.
16. Water bath sonicator (*see* **Note 5**), gel electrophoresis apparatus, heat block shaker, rotating wheel, and microcentrifuge.

2.3 qPCR

1. Primers for the regions to be studied (*see* **Note 6**) and spiked methylated DNA. Verify primer sequences by running an in silico PCR.
2. qPCR reagents according to the real-time instrument used.
3. Real-time instrument.

3 Methods

3.1 In Vitro DNA Methylation

1. On ice, mix 1 μg of purified PCR fragments with 320 μM SAM, 8 units of M.SssI, and 1× NEB buffer 2 for a 20 μl final volume. Mix thoroughly by pipetting up and down, pulse spin, and incubate at 37 °C for 4 h.
2. Digest 200 ng of in vitro-methylated DNA with methylation-sensitive restriction enzymes, i.e., enzymes digesting only unmethylated DNA. Follow the manufacturer's conditions.
3. Mix 1× loading dye to 80 ng of digested DNA and load into a 1.2 % agarose gel in 1× TBE, with a 100 bp DNA ladder. Run gel electrophoresis.
4. If all fragments are correctly methylated, proceed with MeDIP assay (*see* **Note 7**).

3.2 MeDIP

All procedures to be done on ice if not otherwise stated.

Day 1

1. Clean all working surfaces with 10 % bleach, prepare 10× IP buffer, and store it on ice.
2. Dilute 4–6 μg of genomic DNA in 10 mM $NaPO_4$ pH 7 for a final volume of 300 μl in a 1.5 ml Eppendorf tube. Mix well by pipetting up and down.
3. Sonicate the material using a water bath sonicator until DNA fragments range from 300 to 600 bp (*see* **Note 5**).
4. Check the size of fragmented DNA on a 1.2 % agarose gel by loading 5 μl of sonicated material mixed with 1× loading dye, along with a 100 bp DNA ladder, and run electrophoresis in 1× TBE.
5. Add 0.25 ng of in vitro-methylated DNA, mix thoroughly by pipetting up and down, and pulse spin (*see* **Note 8**).
6. Denature DNA for 10 min at 94 °C in a heat block. Immediately cool down on ice for at least 5 min (*see* **Note 9**).

7. Store 20 μl of fragmented material at 4 °C (input). Divide the remaining denatured DNA into two 1.5 ml Eppendorf tubes (140 μl each—IP).
8. Into each IP fraction add 50 μl of cold 10× IP buffer, 4 μl of anti-5mC and 8 μl anti-IgG (8 μg of antibody), 5 μl of PMSF 100×, and water for q.s. 500 μl. Mix thoroughly by pipetting up and down, pulse spin, and incubate overnight at 4 °C in a rotating wheel.

Day 2

1. Add 80 μl of 50 % pre-blocked beads solution and incubate for 2 h at 4 °C in a rotating wheel (*see* **Note 10**).
2. Prepare 1× IP buffer and 1× buffer-300 mM final NaCl, and store it on ice.
3. Centrifuge bead complexes for 2 min at 500 ×*g*. Carefully remove the supernatant without disrupting the bead pellets.
4. All washes are carried out at 4 °C for 5 min in a rotating wheel. After washing, centrifuge bead complexes for 2 min at 500 ×*g*. Carefully remove the supernatant without disrupting the bead pellets:
   (a) Wash bead complexes by adding 1 ml of cold IP buffer 1×. Repeat this step.
   (b) Wash bead complexes by adding 1 ml of cold IP buffer 1× supplemented with 300 mM final NaCl.
   (c) Wash bead complexes by adding 1 ml of 2× TE buffer.
5. Resuspend beads in 250 μl of digestion buffer with 70 μg of Proteinase K and incubate for 3 h (or overnight) at 55 °C in a heat block shaker (500–800 rpm).
6. Recover the input sample from 4 °C and bring the volume up to match the IP samples using 2× TE buffer.
7. Extract DNA from both the input and the IP samples by adding 1 volume of ultrapure phenol:chloroform:isoamyl alcohol in a fume hood. Vigorously vortex the samples for 10 s and spin at maximum speed for 10 min.
8. In the fume hood, transfer the top fraction (aqueous phase) into another 1.5 ml Eppendorf tube and dispose properly of the phenol waste.
9. Add 1/10th volume of 3 M sodium acetate and 1 μl of glycogen. Mix well by pipetting up and down.
10. Add 2.5 volumes of cold 100 % ethanol and mix well by pipetting up and down. Incubate at −80 °C for 30 min to overnight.
11. Spin at maximum speed for 30 min at 4 °C. Carefully discard the supernatant; a clear pellet should be visible.

12. Wash the pellet with freshly made 1 ml of cold 70 % ethanol, and spin at maximum speed for 15 min at 4 °C. Carefully discard the supernatant. Use a P10 to remove supernatant from around the pellet.
13. Air-dry the pellet and resuspend in 30 μl of 10 mM Tris–HCl, pH 7.5. Measure DNA concentration.

3.3 qPCR

General Guidelines

1. Test qPCR primers before hand and verify efficiencies are comprised between 1.9 and 2.1; the coefficient of determination is above 0.98 (linear standard curve) and the melt-curve only shows one peak (very important for TEs, as this shows that just one specific sequence is being amplified). For troubleshooting qPCR primer efficiencies, change temperature and primer concentration.
2. Use technical duplicates (run the sample twice in the same qPCR plate; less than 0.5 cycle differences should be seen between replicates) and biological duplicates (two independent MeDIP experiments). Use 8 ng of DNA per reaction.
3. To calculate enrichment simply follow the equation below using the efficiencies calculated for each primer (modified $\Delta\Delta C_T$ method):

$$\left(\frac{\text{Efficiency}^{-C_T(\text{IP})}}{\text{Efficiency}^{-C_T(\text{Input})}}\right) \Big/ \left(\frac{\text{Efficiency}^{-C_T(\text{IP spiked DNA})}}{\text{Efficiency}^{-C_T(\text{Input spiked DNA})}}\right)$$

4. We suggest confirming a set of MeDIP amplicons by using bisulfite sequencing [9].
5. Use the spiked unmethylated and IgG control values to plot noise (*see* **Note 11**).

4 Notes

1. Amplify regions in the *Drosophila* genome that are of the same range of sizes as your MeDIP fragments (300–600 bp) so no bias is observed during immunoprecipitation. Also, if carrying out this experiment for the first time, amplify fragments that contain one to four CpG sites so MeDIP sensitivity can be tested. qPCR primers should be designed within the PCR-purified fragments.
2. Methylation-sensitive restriction enzymes should be able to identify all methylated CpGs in the fragments that are in vitro methylated. Hence, if a fragment is designed to have four CpGs methylated, you should be able to verify that all four are

indeed in vitro methylated. For this, check the restriction digestion sizes of the fragments before amplifying them.

3. To make a 0.2 M NaPhosphate pH 7 solution just mix 6.1 ml of 0.2 M Na_2HPO_4 with 2.9 ml of 0.2 M NaH_2PO_4.
4. Verify protein A and G affinity before choosing the best beads. Alternatively, magnetic beads could be used to decrease the amount of sample loss during washes.
5. Sonication is an important step for MeDIP success. Use a water bath instead of probe sonicator in order to shear as many samples at the same time as possible. This decreases the fragmentation variation between samples. If your sonicator does not have a cooling system, add a cup of ice cubes to the water bath so the samples stay cool but be aware of the amount as this can change the fragmentation efficiency. Always check the DNA is well fragmented before continuing with the experiment even if the sonication conditions have worked previously. For information, using a Bioruptor sonicator from Diagenode to shear mouse DNA our settings were as follows: four cycles of 30 s on and 30 s off on high setting.
6. In order to study single copies, design one primer matching the unique flanking sequence and another primer matching the ERV copy. If the goal of the study is to quantify global ERV methylation, simply design the primers within the ERV copies such that they will amplify most copies of a particular family.
7. In vitro methylation fragments can be stored at −20 °C for at least 6 months without altering their detection through MeDIP-qPCR. For samples stored over 6 months, always check their methylation through restriction digest before spiking in an MeDIP assay.
8. If carrying out this experiment for the first time we suggest adding 0.25 ng of the four in vitro-methylated fragments (1–4 CpGs methylated) and one unmethylated fragment. This will help you to determine the sensitivity of your MeDIP assay. We have observed that a minimum of three methylated CpGs in a fragment is necessary to detect MeDIP enrichment. Once the assay sensitivity is determined, the user can simply add an ummethylated fragment along with the fragment showing the least number of methylated CpGs detected by MeDIP. If using commercially purchased in vitro-methylated DNA, the user will not test MeDIP sensitivity and can simply add a methylated and unmethylated control. Unmethylated controls are important to check the MeDIP noise, as it might be different from the IgG-negative control antibody used.
9. Denaturation of DNA is mandatory, as the anti-5mC antibody will only recognize single-stranded DNA. Keep samples on ice and use precooled buffers. If no enrichment is observed, one should change the denaturation conditions by adding water to

the heat block, verifying the heat block temperature, or simply boiling the 1.5 ml Eppendorf tubes in water for 10 min (make a small hole in the Eppendorf cap with a syringe to avoid pressure buildup inside the tube).

10. Significant variation can stem from this step, where the amount of beads added to each sample could be different. To ensure that the same amount is added to each tube simply lightly vortex the beads before pipetting each time and ensure that the solution is mixed. Cut the end off the pipette tips so the viscous bead slurry is easily transferred. Do not submerge the entire tip to avoid transferring beads on the tip's outside surface and pipette up and down to thoroughly flush the inside of the tip into the DNA tube.
11. We routinely design primers in regions where we know DNA methylation is absent in the samples studied to check the level of background noise.

References

1. Lander ES, Linton LM, Birren B et al (2001) Initial sequencing and analysis of the human genome. Nature 409(6822):860–921
2. Maksakova IA, Romanish MT, Gagnier L et al (2006) Retroviral elements and their hosts: insertional mutagenesis in the mouse germ line. PLoS Genet 2(1):e2
3. Conley AB, Piriyapongsa J, Jordan IK (2008) Retroviral promoters in the human genome. Bioinformatics 24(14):1563–1567
4. Walsh CP, Chaillet JR, Bestor TH (1998) Transcription of IAP endogenous retroviruses is constrained by cytosine methylation. Nat Genet 20(2):116–117
5. Maksakova IA, Mager DL, Reiss D (2008) Keeping active endogenous retroviral-like elements in check: the epigenetic perspective. Cell Mol Life Sci 65(21):3329–3347
6. Karimi MM, Goyal P, Maksakova IA et al (2011) DNA methylation and SETDB1/H3K9me3 regulate predominantly distinct sets of genes, retroelements, and chimeric transcripts in mESCs. Cell Stem Cell 8(6):676–687
7. Horard B, Eymery A, Fourel G et al (2009) Global analysis of DNA methylation and transcription of human repetitive sequences. Epigenetics 4(5):339–350
8. Gilson E, Horard B (2012) Comprehensive DNA methylation profiling of human repetitive DNA elements using an MeDIP-on-RepArray assay. Methods Mol Biol 859:267–291
9. Rebollo R, Miceli-Royer K, Zhang Y et al (2012) Epigenetic interplay between mouse endogenous retroviruses and host genes. Genome Biol 13(10):R89
10. Weber M, Davies JJ, Wittig D et al (2005) Chromosome-wide and promoter-specific analyses identify sites of differential DNA methylation in normal and transformed human cells. Nat Genet 37(8):853–862
11. Ficz G, Branco MR, Seisenberger S et al (2011) Dynamic regulation of 5-hydroxymethylcytosine in mouse ES cells and during differentiation. Nature 473(7347):398–402

Chapter 24

Profiling DNA Methylation and Hydroxymethylation at Retrotransposable Elements

Lorenzo de la Rica, Jatinder S. Stanley, and Miguel R. Branco

Abstract

DNA methylation is a key epigenetic modification controlling the transcriptional activity of mammalian retrotransposable elements. Its oxidation to DNA hydroxymethylation has been linked to DNA demethylation and reactivation of retrotransposons. Here we describe in detail protocols for three methods to measure DNA methylation and hydroxymethylation at specific genomic targets: glucMS-qPCR, and two sequencing approaches (pyrosequencing and high-throughput sequencing) for analyzing bisulfite- and oxidative bisulfite-modified DNA. All three techniques provide absolute measurements of methylation and hydroxymethylation levels at single-base resolution. Differences between the methods are discussed, mainly with respect to throughput and target coverage. These constitute the core techniques that are used in our laboratory for accurately surveying the epigenetics of retrotransposable elements.

Key words DNA methylation, DNA hydroxymethylation, Methylation-sensitive restriction enzyme, Quantitative PCR, Bisulfite, Oxidative bisulfite, Pyrosequencing, High-throughput sequencing

1 Introduction

1.1 Epigenetic Control of Retrotransposon Activity

Retrotransposon activity can be controlled at the transcriptional level via epigenetic mechanisms that primarily aim to repress retrotransposons. Loss of epigenetic control of retrotransposons in conditions such as cancer leads to their activation, with potentially detrimental consequences [1]. Interestingly, retroelements can epigenetically affect the activity of nearby genes, and have arguably made a major evolutionary contribution to the gene regulatory portfolio of our genome [2]. DNA methylation is one of the epigenetic defense mechanisms that have been extensively shown to play a major role in retrotransposon silencing. However, methylated cytosines (5mC) can be enzymatically oxidized to produce 5-hydroxymethylcytosine (5hmC, i.e., DNA hydroxymethylation) as part of a demethylation pathway [3], which has been associated with long interspersed element class-1 (LINE-1 or L1) activation in embryonic stem cells [4]. Notably, most classic techniques used

Jose L. Garcia-Pérez (ed.), *Transposons and Retrotransposons: Methods and Protocols*, Methods in Molecular Biology, vol. 1400, DOI 10.1007/978-1-4939-3372-3_24,

Table 1
Comparison of the methods described herein for quantification of 5mC/5hmC

	glucMS-qPCR	BS/oxBS pyroseq.	BS/oxBS amp. HTS
Total time	8 h	2 days	4 days[a]
Hands-on time	<2 h	3–4 h	4–5 h
Coverage	1 CpG per amplicon	CpGs within a 70 bp amplicon	CpGs within the HTS reads (up to 2× 300 bp)
Throughput (*n* of target/ sample combinations)	Up to 32 (per 384-well qPCR run)	Up to 96 (per pyrosequencing run)	1000+ (per HTS run)
Cost per run (excl. primers)	$120	$400	$1.5–2k
Cost per target/sample[b]	$4	$4	<$2

[a]The time for BS/oxBS amplicon high-throughput sequencing (HTS) will critically depend on the turnover of the HTS facility

[b]Costs are estimates and depend on several factors, including the number of samples processed for BS/oxBS

for DNA methylation profiling (e.g., bisulfite sequencing) cannot distinguish between 5mC and 5hmC [5]. Here we describe a range of techniques currently used in our lab to measure 5mC and 5hmC levels, which differ in assay time, cost, throughput, and data complexity (Table 1). These are effective tools for characterizing retroelement epigenetics and its dynamics.

1.2 GlucMS-qPCR

This assay is based on the digestion of genomic DNA by restriction enzymes that are sensitive to particular DNA modifications [6]. Firstly, T4 β-glucosyltransferase (BGT) is used to selectively add a glucose moiety to the 5hmC in genomic DNA. The DNA is then digested with either *Msp*I or *Hpa*II, both of which recognize CCGG sites, thus probing modifications in the internal CpG. *Hpa*II digestion is blocked by 5mC and by glucosylated 5hmC, whereas *Msp*I digestion is only blocked by glucosylated 5hmC, thus allowing discrimination of 5mC and 5hmC (Fig. 1).

Quantification of each undigested fraction by qPCR against a mock digestion control yields a percentage of each of the modifications at the queried CpG site. Importantly, an *Msp*I digestion without BGT is needed to control for genetic variability of the restriction site between retrotransposon copies. Advantages to using this technique include its low cost and short assay time, as results are obtainable within a day. However, it is a relatively low-throughput approach and it only queries one CpG site at a time. It is further limited by the availability of a restriction site at the region of interest.

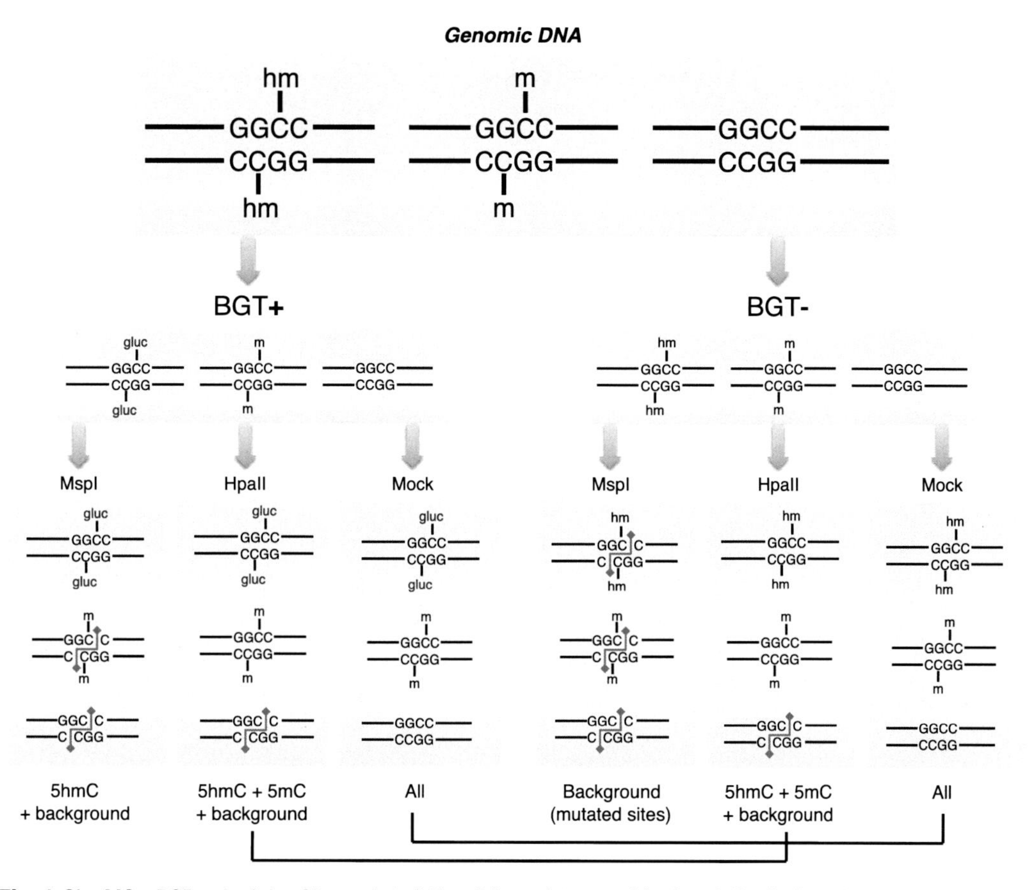

Fig. 1 GlucMS-qPCR principle: Glucosylated 5hmC is resistant to MspI and HpaII digestion, whereas 5mC is only resistant to HpaII digestion. Two of the BGT– samples are redundant to their BGT+ counterparts. The BGT–/MspI sample should have all sites digested; in the context of a repetitive target, undigested DNA in this sample is mainly indicative of mutation of the restriction site in a fraction of the copies

1.3 BS/oxBS Pyrosequencing

Bisulfite (BS) treatment of DNA converts unmodified cytosines to uracil, whereas 5mC and 5hmC are largely inert to this chemical. However, oxidative bisulfite (oxBS) treatment results in conversion of 5hmC as well; the combination of the two techniques therefore allows distinguishing between 5mC and 5hmC (Fig. 2) [7, 8].

Analysis of PCR products from BS/oxBS-treated DNA can be done via various sequencing platforms in a way that provides information on the frequency of 5mC/5hmC across a cell population. Pyrosequencing is a convenient and quick platform for producing such an output, where a series of enzymatic reactions yield a light signal that is proportional to the amount of incorporated nucleotide [9, 10]. The technology is commonly used for single-nucleotide polymorphism (SNP) detection, and in the case of BS/oxBS-treated DNA each CpG is effectively seen as a putative C/T SNP. The pyrosequencing reaction will measure the percentage of

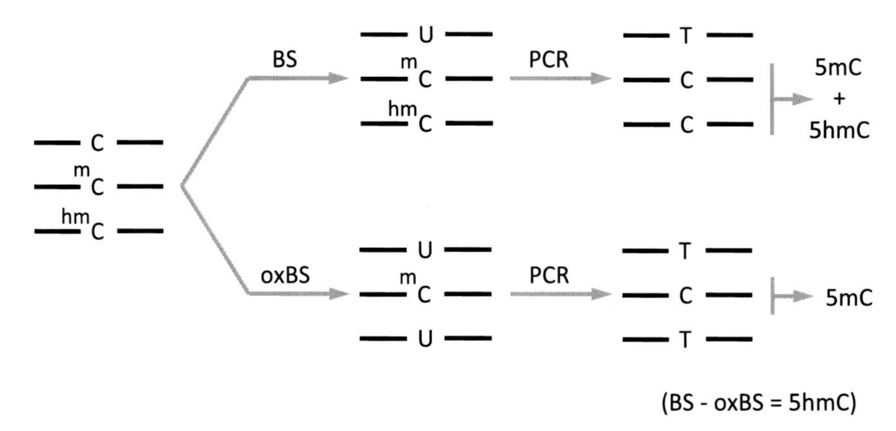

Fig. 2 Oxidative bisulfite principle: Bisulfite treatment of gDNA converts unmodified cytosines to uracil, which are detected as thymines after PCR, whereas oxidative bisulfite also converts 5hmC. Thus, oxidative bisulfite provides a direct readout of 5mC; 5hmC can be inferred from the combination of the two outcomes

C incorporated, which is a measure of the amount of modified cytosine (5mC + 5hmC in the case of BS and 5mC in the case of oxBS) at a given position. Importantly, this approach provides information on all CpGs within a PCR product (although it is limited to amplicons of about 70 bp) and is not limited by restriction sites as in the case of glucMS-qPCR. The cost is higher, but this is partially offset by the increased throughput, making it cost effective. The time scale for the complete assay (starting from unconverted genomic DNA) is around 2 days.

1.4 BS/oxBS Amplicon HTS

Another option for the sequencing of PCR products from BS/oxBS-converted DNA is high-throughput sequencing (HTS), which is most suitable when a very large number of samples and/or amplicons are to be analyzed. After BS/oxBS conversion of genomic DNA, a first PCR amplifies the region(s) of interest and adds part of the HTS adaptors, and a second PCR completes the adaptors (Fig. 3a). The second PCR also adds barcodes that allow for pooling of all the samples into a single sequencing run. HTS ensures that high sequencing depth is achieved, which is key for the accurate measurement of 5hmC.

As an example, we have obtained a median depth of 1000+ reads from sequencing 800 amplicon/sample combinations in a MiSeq run (2× 75 bp, v3 chemistry). The cost of HTS is relatively high, but the cost per amplicon/sample drops with increasing study size and quickly becomes the cheapest out of the three technologies described herein. Amplicons can be longer than those in pyrosequencing (current MiSeq chemistry gives up to 2× 300 bp read lengths), and the data are more detailed, providing clonal information on the distribution of DNA modifications across a region. For this same reason, processing of the data is less straightforward, requiring the use of a number of bioinformatic tools.

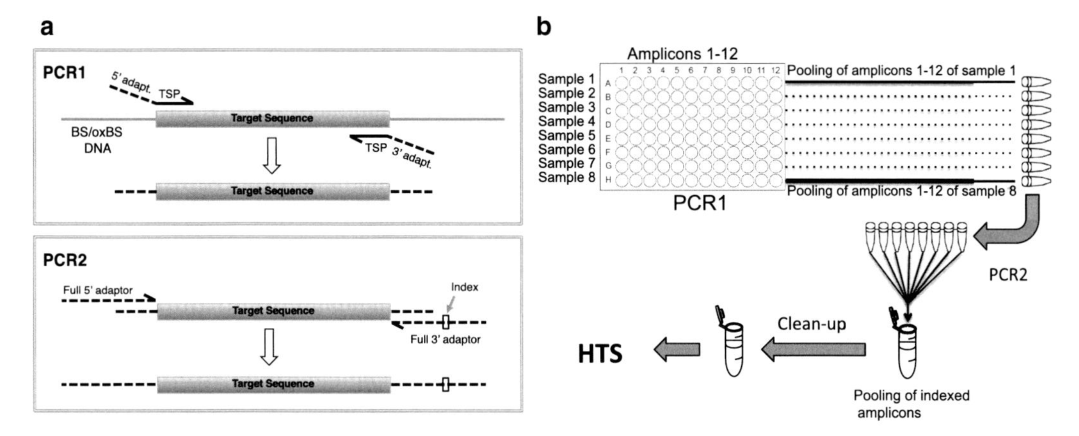

Fig. 3 BS/oxBS amplicon HTS design and workflow: (**a**) PCR1 uses primers with a target-specific portion (TSP) as well as part of the adaptors needed for HTS; PCR2, which is target independent, completes the adaptors and adds an index that allows sample pooling. (**b**) After PCR1 all amplicons from the same sample can be pooled, reducing the scale of the sample-indexing PCR2; after indexing all samples can be pooled into a single tube and sequenced

2 Materials

Use nuclease-free water throughout, unless otherwise stated.

2.1 glucMS-qPCR

1. Primers targeting the locus of interest (*see* **Note 1**).
2. T4 Phage β-glucosyltransferase (BGT) (New England Biolabs): This also includes NEBuffer 4 and 2 mM UDP-glucose, which should be diluted in water to a 0.4 mM working solution.
3. BGT storage buffer: 20 mM KPO_4, 200 mM NaCl, 0.25 mM DTT, 0.1 mM EDTA, 50 % glycerol pH 7.0.
4. *Hpa*II-HF (New England Biolabs), which includes CutSmart® buffer.
5. *Msp*I-HF (New England Biolabs), which includes CutSmart® buffer.
6. Heating block or incubator.
7. 2× SYBR green qPCR mix.
8. 96- or 384-well plates and adhesive seals.
9. Real-time PCR instrument.

2.2 BS/oxBS Pyrosequencing

1. MilliQ water (*see* comment in Subheading 3).
2. 10× oxidant: 150 mM $KRuO_4$ (Alpha Aesar) in 0.5 M NaOH. Store at −20 °C.
3. Micro Bio-Spin columns (Bio-Rad).
4. 1 M NaOH solution.

5. EpiTect Bisulfite Kit (Qiagen).
6. Assay-specific primers: Biotin-tagged forward primer (HPLC purified), reverse primer, and sequencing primer (*see* **Note 2**).
7. HotStarTaq DNA Polymerase including 10× PCR buffer and 25 mM $MgCl_2$.
8. 10 mM dNTP mix.
9. Thermal cycler.
10. 96-Well PCR plates.
11. Adhesive seals.
12. PyroMark Q96 ID system and associated reagents.

2.3 BS/oxBS Amplicon HTS

1. **Items 1–5** from Subheading 2.2 above, for BS/oxBS conversion.
2. Adaptor-tagged target-specific primers for PCR1, and barcode-containing adaptor primers for PCR2 (*see* **Note 3**).
3. **Items 7–11** from Subheading 2.2 above, for PCR.
4. AMPure XP magnetic beads (Beckman Coulter).
5. Magnetic rack.
6. MiSeq system (Illumina) and associated reagents.

3 Methods

When working with enzymes, carry out all the pipetting steps on ice, unless otherwise specified.

3.1 glucMS-qPCR

The amount of starting DNA material suggested has been optimized for the profiling of mouse LINE-1 elements; retroelements with substantially fewer copies may require larger amounts of starting material. Pipetting errors should be minimized by the use of master mixes and good technique.

1. To set up the glucosylation reaction: Dilute 200–300 ng of gDNA in water, up to a total of 15.56 μl. Add 2.22 μl of NEBuffer 4 and 2.22 μl of 0.4 mM UDP-glucose.
2. Split the starting mix into two tubes with 9 μl each ("BGT+" and "BGT−"). Add 1 μl of BGT to the "BGT+" tube and 1 μl of BGT storage buffer to the "BGT−" tube. Mix the content of each tube by pipetting.
3. Incubate for ≥3 h at 37 °C in a heating block or incubator (the reaction can be run overnight if using an incubator). After incubation, spin down the contents of the tube.
4. To set up the digestion reactions, first prepare three master mixes (*Msp*I, *Hpa*II, and mock). For each, add 6.6 μl of

CutSmart buffer to 51.7 μl of water. Then add 0.5 μl of *Msp*I, *Hpa*II, or BGT storage buffer.

5. Aliquot 3 μl of each of the glucosylation reactions (from **step 3**) into three tubes; to each of these tubes add 27 μl of one of the digestion master mixes described in the previous step. This should yield six separate reactions as follows (*see* also Fig. 1): (a) BGT+/HpaII, (b) BGT+/MspI, (c) BGT+/mock, (d) BGT−/HpaII, (e) BGT−/MspI, and (f) BGT−/mock.
6. Mix the contents of each tube by pipetting.
7. Incubate for ≥3 h at 37 °C in a heating block or incubator (the reaction can be run overnight if using an incubator). After incubation, spin down the contents of the tube.
8. Quantify by qPCR in a 96- or 384-well format, using 10 μl reactions as follows: 5 μl 2× SYBR green mix, 0.5 μl primer mix (forward and reverse primers at 5 μM each), and 4.5 μl of digestion reaction diluted 1:200 in water. Run each qPCR reaction in triplicate. Note that tubes (d) and (f) should be redundant to tubes (a) and (c), respectively, as BGT should not interfere with the outcome in these cases; after a recommended check that this is indeed the case, they can be excluded from the qPCR.
9. Normalize the qPCR data. After extracting Ct values, normalize each target to the loading control amplicon: $N = 2^{(Ct^{control} - Ct^{target})}$. Then calculate the % of undigested product in each digestion reaction by normalizing the N values to the mock digestion: $Q = N^{digestion} / N^{mock} \times 100$.
10. Calculate the amount of 5mC and 5hmC. The Q value for the BGT−/MspI reaction represents the % of copies lacking the target restriction site; if this value is too high the assay will have little sensitivity and yield highly variable data. Thus, to obtain the % of 5hmC, $Q^{BGT-/MspI}$ needs to be subtracted away and the data rescaled, as such $\%5hmC = (Q^{BGT+/MspI} - Q^{BGT-/MspI})/(100 - Q^{BGT-/MspI})$. The %5hmC is then used to calculate the % of 5mC: $\%5mC = (Q^{BGT+/HpaII} - Q^{BGT-/MspI})/(100 - Q^{BGT-/MspI}) - \%5hmC$.

3.2 BS/oxBS Pyrosequencing

PCR conditions should be optimized for each primer pair ahead of the experiment (*see* **Note 4**). We recommend using control samples that, amongst other things, assess the genetic variability across retroelement copies at each CpG position (*see* **Note 5**). Certain types of water interfere with the oxidation reaction in oxBS; we recommend using MilliQ water, which should be tested beforehand. Many buffers and ethanol also interfere with the oxidation; genomic DNA (around 500 ng) should be in MilliQ water (no more than 40 μl total volume) and free of any traces of ethanol. We recommend including a control template to test the success of the oxidation (described in detail in [8]), although the color of the oxidation reaction can be used as a rough indicator (*see* **Note 6**).

1. Replace buffer in Micro Bio-Spin columns: Spin columns for 1 min at 1000×*g* into a 2 ml collection tube and discard the eluate; wash three times with 500 μl MilliQ water, spinning for 1 min at 1000×*g* between washes. After a fourth wash, spin for 2 min at 1000×*g*. Do not allow the column to dry out.
2. Add the genomic DNA to the column and spin for 4 min at 1000×*g* into a fresh tube.
3. Split the eluate into two tubes: one will be oxidized (oxBS fraction) and the other run through a mock reaction (BS fraction). Top up each fraction with MilliQ water up to 21.75 μl.
4. Denature the DNA: Add 1.25 μl 1 M NaOH to both fractions and incubate for 30 min at 37 °C.
5. In the meantime, thaw the 10× oxidant on ice and dilute in MilliQ water to make a 1× solution.
6. Chill the denatured DNA on ice for 3 min.
7. Add 2 μl of 1× oxidant to the oxBS fraction. Incubate in an ice/water bath for 1 h, mixing sample every 20 min and checking for color (*see* **Note 6**).
8. Clean up reactions with Micro Bio-Spin columns: Spin columns for 2 min at 1000×*g* into a 2 ml collection tube, and discard supernatant; add each reaction to a column and spin for 4 min at 1000×*g* into fresh tubes.
9. Perform bisulfite conversion of the samples using the EpiTect kit according to the manufacturer's instructions for "Sodium Bisulfite Conversion of Unmethylated Cytosines in DNA Isolated from FFPE Tissue Samples," with the following modifications (*see* **Note 7**): to each sample (~25 μl) add 85 μl bisulfite mix and 30 μl DNA protect buffer; run the thermal profile twice over, i.e., 2× (95 °C 5 min, 60 °C 25 min, 95 °C 5 min, 60 °C 85 min, 95 °C 5 min, 60 °C 175 min), and hold at 20 °C.
10. Clean up the bisulfite reaction according to the manufacturer's instructions; elute the DNA with 2× 20 μl EB buffer.
11. PCR the BS/oxBS-modified DNA to amplify the region of interest: Prepare the PCR reactions on a 96-well plate according to the mix described in Table 2 and run the thermal profile outlined in Table 3 (with 35–45 cycles). Exact PCR conditions are primer pair dependent (*see* **Note 4**). Include a no-template control.
12. Run part of the PCR products (5–10 μl) on a 1.5 % agarose gel to check for amplification and specificity.
13. Perform pyrosequencing on the PyroMark Q96 ID system according to the manufacturer's instructions (*see* **Note 8**).

Table 2
PCR reaction mixes for pyrosequencing and HTS

Component	Volume (μl)		
	Pyroseq.	HTS PCR1	HTS PCR2
10× PCR buffer	2.5	0.5	1
25 mM $MgCl_2$	1.5	0.3	0.6
dNTP mix (10 mM of each)	0.5	0.1	0.2
Primer mix (F + R 5 μM each)	2	0.4	0.8
HotStartTaq DNA Polymerase	0.1	0.03	0.06
Template DNA	Variable[a]	Variable[a]	1
Nuclease-free water	Variable	Variable	6.34
TOTAL VOLUME	25	5	10

[a]The amount of DNA used depends on the copy number of the target region (*see* **Note 4**)

Table 3
PCR thermal profile for pyrosequencing and HTS PCRs

Step	Temperature (°C)	Duration (min)
Initial activation step	95	15
3-Step cycling[a]:		
1. Denaturation	95	0.5
2. Annealing	52–58[b]	0.5
3. Extension	72	1
Final extension	72	10
Store	4	Forever

[a]Guidelines on the number of cycles to use are given in the main text
[b]Annealing temperature should be optimized for each amplicon. For HTS PCR2 use 58 °C

14. Analyze the data using the PyroMark CpG SW package. The pyrogram peaks have to be analyzed carefully to confirm that there is no signal in the dispensation controls, in the BS conversion controls, or in the no-template PCR (Fig. 4). The DNA methylation levels of the 0 and 100 % methylation controls should be accurate (*see* **Note 5**). No peaks should be detected in between the nucleotide peaks. Values from oxBS samples represent 5mC levels, whereas 5hmC levels are obtained by subtracting the oxBS signal from the BS signal (Fig. 2).

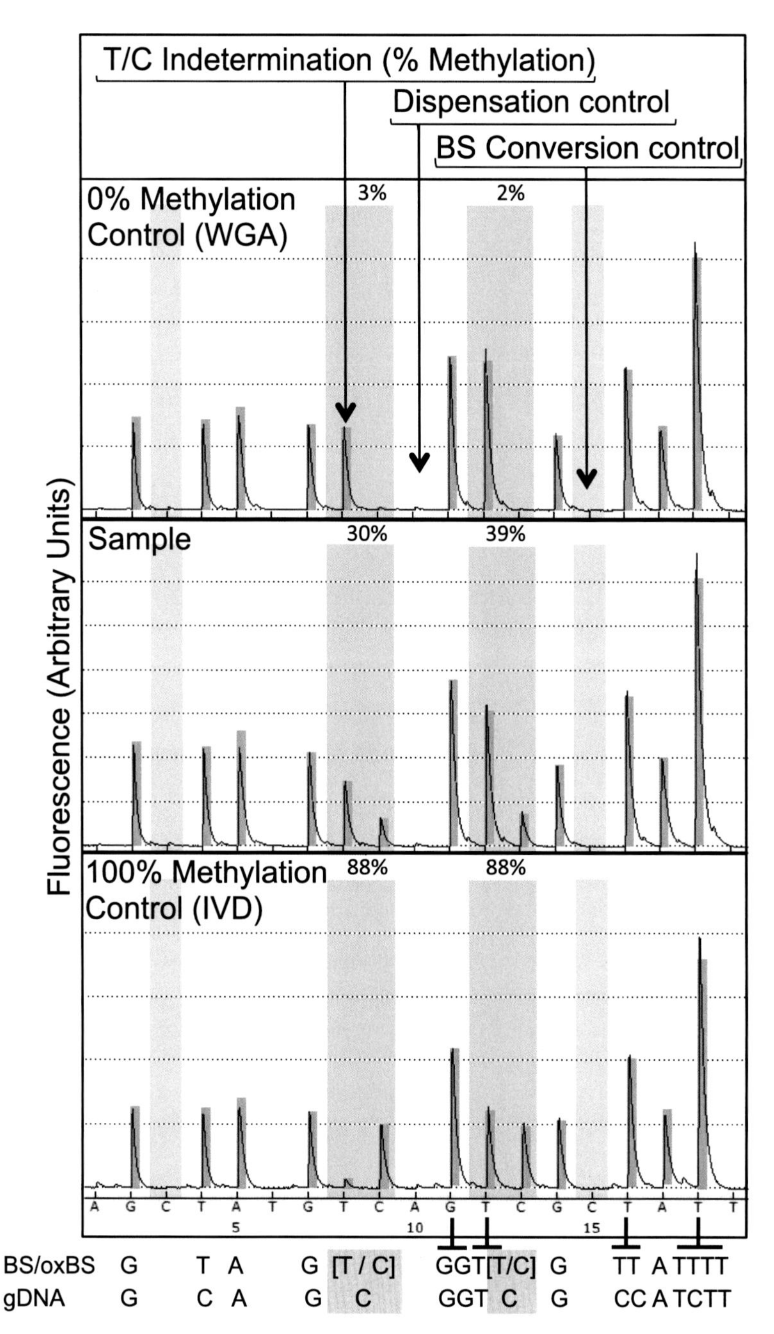

Fig. 4 Example pyrogram from BS/oxBS pyrosequencing: The nucleotide dispensation order (shown immediately below the pyrogram) is based on the target sequence but also includes bisulfite conversion and dispensation controls. CpG sites in the original gDNA sequence are treated as T/C SNPs and queried for the percentage of methylation. Samples from whole-genome amplification (WGA) or in vitro methylation of DNA (IVD) provide a control for the measurements. Note that poly-N stretches are detected by a single dispensation of that nucleotide

3.3 BS/oxBS Amplicon HTS

We have used this protocol with primers that are compatible with Illumina platforms (MiSeq, NextSeq, HiSeq; *see* **Note 3**). The protocol should be adaptable to other HTS platforms, but primers have to be restructured to ensure compatibility. A scheme depicting all the PCR and pooling steps in this protocol can be found in Fig. 3b. PCR conditions should be optimized for each target-specific primer pair ahead of the experiment (*see* **Note 4**). As for pyrosequencing, control samples with 0 or 100 % DNA methylation may be used (*see* **Note 5**). For large experiments, the use of multichannel and repetitive pipettes is recommended.

1. Perform BS/oxBS conversion of the genomic DNA according to **steps 1–10** of Subheading 3.2.
2. Perform PCR1 (target-specific PCR): Prepare the PCR reactions on 96- or 384-well plates according to the mix described in Table 2 and run the thermal profile outlined in Table 3 (with 35–45 cycles). Exact PCR conditions are primer pair dependent (*see* **Note 4**).
3. For each sample, pool together 1 μl of all the PCR1 amplicons. This is to allow for all the amplicons from a given sample to be simultaneously labeled with a barcode that identifies the sample. Note that the BS- and oxBS-modified fractions from the same genomic DNA should be treated as different samples and thus pooled separately.
4. Perform PCR2 (barcoding PCR). This PCR is target sequence independent, and thus its conditions are fixed. Nonetheless, before a large-scale experiment it is recommended to confirm that these conditions yield robust PCR products that are 56 bp longer than the corresponding PCR1 products (Fig. 5a). Prepare the PCR reactions on a 96-well plate according to the mix described in Table 2 and run the thermal profile outlined in Table 3 (with 8 cycles).
5. Pool 1 μl of each PCR2 reaction into a single tube. This constitutes the complete amplicon library, and should contain an even representation of all the samples (Fig. 5b).
6. To clean up the library, add a volume of AMPure XP magnetic beads that is 1.8× the volume of the library. Mix well and incubate for 5 min.
7. Place tube on magnetic rack and wait for 5 min for the beads to separate from the solution. Discard the supernatant.
8. Wash beads twice with 200 μl of 80 % ethanol, leaving the beads undisturbed on the magnetic rack.
9. Allow beads to dry well until there are no traces of ethanol.

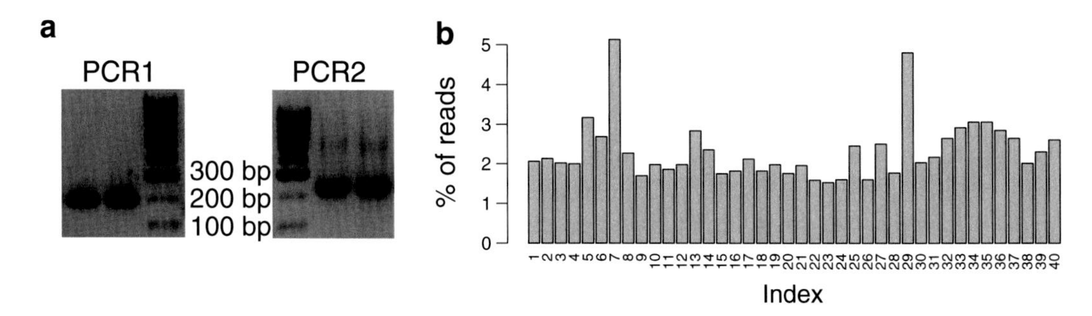

Fig. 5 Example of optimized PCRs for HTS: (**a**) PCR1 should be optimized to yield a robust and saturated signal that allows even amplicon pooling; after PCR2, products should be marginally longer (56 bp) due to the addition of the rest of the adaptors. (**b**) Proportion of reads assigned to each index in an HTS run, showing even representation of all pooled samples

10. Remove tube from magnetic rack and resuspend beads in 40 μl of water or Tris buffer.
11. Place tube back on the magnetic rack. After separation collect the supernatant to a new tube.
12. Perform HTS according to the manufacturer's instructions.
13. For data analysis, a basic knowledge of HTS data processing is required. Different tools and strategies can be used, and users are advised to set up their own analysis pipeline according to their preferences. Briefly, we use Bismark [11] to map the HTS reads to a custom genome that contains all the amplicon sequences, and to extract the methylation values. We further process these data using custom R scripts. Values from oxBS samples represent 5mC levels, whereas 5hmC levels are obtained by subtracting the oxBS signal from the BS signal (Fig. 2).

4 Notes

1. Primers for glucMS-qPCR should be designed around a single restriction site (CCGG); including additional sites in an amplicon will render the quantification meaningless. An additional pair of primers should be included that amplify a region without any restriction sites (ideally in the same retroelement to be targeted), which are used as a "loading control" for each digestion reaction. Perform quality control checks to ensure that the primers are specific and efficient.
2. The sequencing primer has to be in the opposite direction to the biotinylated primer. The region that will be pyrosequenced lies between the sequencing primer and the biotinylated one, so the CpGs of interest have to be in this region. We recommend using the PyroMark Assay Design SW 2.0 software to obtain

high-quality primers. We suggest targeting a PCR amplicon region of around 200 bp, with the pyrosequenced region not exceeding 70 bp, as the data quality will drop with longer reads. One amplicon can be pyrosequenced with several sequencing primers to increase the number of CpGs surveyed.

3. The primers described here are compatible with Illumina platforms (MiSeq, NextSeq, HiSeq, etc.). Target-specific primers have to be designed against the BS-modified sequence of interest and avoid overlap with CpGs. Part of the adaptor sequence has to be included as a 5′ overhang in each primer, to enable PCR2 (*see* Fig. 3a). In the example below (used for mouse L1Tf elements), the target-specific portion of the primer is underlined.

   Fwd: ACACTCTTTCCCTACACGACGCTCTTCCGATCTGAGGTAGTATTTTGTGTGGG Rev: TGACTGGAGTTCAGACGTGTGCTCTTCCGATCTCACCTATTCAAACTAATTTCC

   The second primer pair targets the portion of the adaptors added at PCR1, and will add the rest of the adaptors, including a 6 bp index (shown underlined below) in the reverse primer to allow sample multiplexing. Universal forward primer: AATGATACGGCGACCACCGAGATCTACACTCTTTCCCTACACGACGCTCTTCCGATCT

   Indexed reverse primer: CAAGCAGAAGACGGCATACGAGATCGTGATGTGACTGGAGTTCAGACGTGTGCTCTTCCGATC

   We recommend users follow Illumina's guidelines for index generation and pooling (https://support.illumina.com/downloads/truseqlibrary-prep-pooling-guide-15042173.html).

4. PCR conditions should be optimized for each primer set ahead of the experiment (template amount, cycle number, annealing temperature). The amount of template DNA necessary is inversely proportional to the copy number of the target repeat element. As an indication, for mouse LINE-1 elements we use 200 pg of BS/oxBS-converted DNA. PCR products should be checked on an agarose gel. For PCR1 of the HTS approach, it is also important to verify that saturated signals are obtained (Fig. 5a). Saturated signals will ensure that equal amounts of each PCR1 amplicon can be pooled together without the need for quantification while still achieving roughly even representation across all amplicons. It is key that the PCR conditions are robust and reliable, as it would be unfeasible to check all PCR products when doing a large-scale experiment.

5. We recommend using control samples in each pyrosequencing run, namely a "100 % methylation" control, which can be generated by in vitro methylation of genomic DNA (IVD) with M.SssI. This is particularly important in the context of repeti-

tive elements, as it will control for genetic variability at CpG positions—i.e., if the C is frequently mutated (commonly to a T), this control will show a methylation value significantly below 100 %. A "0 % methylation" control sample may also be included, which may be generated by whole-genome amplification.

6. The color of the oxidation reaction can be used as an indicator for the success of the oxidation. Ideally, the reaction should remain orange for the whole duration of the oxidation. An early shift to green, brown, or black indicates that contaminants were present and that the oxidation has failed. If the color slowly starts to change towards the end of the oxidation (after about 40 min), it is likely that it was still successful.
7. The bisulfite conditions outlined in this protocol (including the use of the Epitect kit) have been optimized for oxBS conversion, which has specific requirements. It is highly recommended that these conditions be strictly reproduced. Using a different kit or changing these conditions may render conversion of oxidized 5hmC inefficient.
8. The nucleotide dispensation order for the pyrosequencing platform is set up by the PyroMark Assay Design SW 2.0 software based on the target sequence. Importantly, it will include some dispensation controls, where a nucleotide that should not be incorporated is dispensed (Fig. 4). We recommend including bisulfite conversion controls, which are easily included in the dispensation order by testing for non-CpG methylation (which is extremely low in most tissues and therefore non-CpG positions should be fully converted).

References

1. Lee E, Iskow R, Yang L et al (2012) Landscape of somatic retrotransposition in human cancers. Science 337:967–971
2. Sundaram V, Cheng Y, Ma Z et al (2014) Widespread contribution of transposable elements to the innovation of gene regulatory networks. Genome Res 24:1963–1976
3. Branco MR, Ficz G, Reik W (2012) Uncovering the role of 5-hydroxymethylcytosine in the epigenome. Nat Rev Genet 13:7–13
4. Ficz G, Branco MR, Seisenberger S et al (2011) Dynamic regulation of 5-hydroxymethylcytosine in mouse ES cells and during differentiation. Nature 473:398–402
5. Nestor C, Ruzov A, Meehan R, Dunican D (2010) Enzymatic approaches and bisulfite sequencing cannot distinguish between 5-methylcytosine and 5-hydroxymethylcytosine in DNA. Biotechniques 48:317–319
6. Kinney SM, Chin HG, Vaisvila R et al (2011) Tissue-specific distribution and dynamic changes of 5-hydroxymethylcytosine in mammalian genomes. J Biol Chem 286:24685–24693
7. Booth MJ, Branco MR, Ficz G et al (2012) Quantitative sequencing of 5-methylcytosine and 5-hydroxymethylcytosine at single-base resolution. Science 336:934–937
8. Booth MJ, Ost TWB, Beraldi D et al (2013) Oxidative bisulfite sequencing of 5-methylcytosine and 5-hydroxymethylcytosine. Nat Protoc 8:1841–1851

9. Colella S, Shen L, Baggerly KA et al (2003) Sensitive and quantitative universal Pyrosequencing methylation analysis of CpG sites. Biotechniques 35:146–150
10. Tost J, Dunker J, Gut IG (2003) Analysis and quantification of multiple methylation variable positions in CpG islands by Pyrosequencing. Biotechniques 35:152–156
11. Krueger F, Andrews SR (2011) Bismark: a flexible aligner and methylation caller for Bisulfite-Seq applications. Bioinformatics 27:1571–1572

Chapter 25

A Large-Scale Functional Screen to Identify Epigenetic Repressors of Retrotransposon Expression

Gabriela Ecco, Helen M. Rowe, and Didier Trono

Abstract

Deposition of epigenetic marks is an important layer of the transcriptional control of retrotransposons, especially during early embryogenesis. Krüppel-associated box domain zinc finger proteins (KRAB-ZFPs) are one of the largest families of transcription factors, and collectively partake in this process by tethering to thousands of retroelement-containing genomic loci their cofactor KAP1, which acts as a scaffold for a heterochromatin-inducing machinery. However, while the sequence-specific DNA binding potential of the poly-zinc finger-containing KRAB-ZFPs is recognized, very few members of the family have been assigned specific targets. In this chapter, we describe a large-scale functional screen to identify the retroelements bound by individual murine KRAB-ZFPs. Our method is based on the automated transfection of a library of mouse KRAB-ZFP-containing vectors into 293T cells modified to express GFP from a PGK promoter harboring in its immediate vicinity a KAP1-recruiting retroelement-derived sequence. Analysis is then performed by plate reader and flow cytometry fluorescence readout. Such large-scale DNA-centered functional approach can not only help to identify the trans-acting factors responsible for silencing retrotransposons, but also serve as a model for dissecting the transcriptional networks influenced by retroelement-derived *cis*-acting sequences.

Key words Retroelement control, KRAB-ZFP, Protein-DNA interaction, Epigenetic regulation, Functional screen, Mammalian one-hybrid assay, GFP repression assay

1 Introduction

The sequence-specific deposition of chromatin marks has a profound influence on transcription. DNA methylation and histone modifications shape chromatin structure, guide the genomic recruitment of transcription factors, and dictate expression patterns. Retroelement-derived sequences account for more than half of the human and mouse genome, and epigenetic modifications play a paramount role in their transcriptional control. DNA methylation is a well-established mechanism of retroelement silencing, particularly in differentiated cells. Furthermore, during early embryonic development—when there is a wave of erasure of DNA methylation—covalent histone modifications, such as trimethylation

Jose L. Garcia-Pérez (ed.), *Transposons and Retrotransposons: Methods and Protocols*, Methods in Molecular Biology, vol. 1400, DOI 10.1007/978-1-4939-3372-3_25, © Springer Science+Business Media New York 2016

of lysine 9 on histone H3 (H3K9me3), take over and keep retroelements under control (reviewed in [1, 2]).

Some important mediators in the epigenetic silencing of retroelements are Krüppel-associated box domain zinc finger proteins (KRAB-ZFPs). These represent one of the largest families of transcriptional regulators with 300–400 members in mice and humans [3, 4]. KRAB-ZFPs bind to the DNA and recruit the universal co-repressor KAP1 (also known as Trim28, TIF1β, and KRIP-1) [5], which then acts as a scaffold for chromatin-modifying proteins, such as histone methyltransferases and histone deacetylases that alter the chromatin structure and silence the genomic region [6–9]. KAP1 and other proteins of the KAP1-nucleated silencing complex repress a large array of retroelements in mouse and human stem cells, including ERVs, LINEs, and SVAs [10–14]. However, while a few matches between a given KRAB-ZFP and its target retrotransposon have been established, the sequence-specific mediator of KAP1 recruitment to most retroelements remains unidentified [14, 15].

This gap in knowledge largely stems from the difficulty in identifying the proteins binding to specific DNA targets, even more so when these reside in repetitive genomic sequences. Methods to determine protein-DNA complexes from the protein perspective (protein-centered methods, such as ChIP-on-chip, ChIP-seq, SELEX, and MITOMI) have greatly improved in the last years, thanks to the advances in high-throughput DNA sequencing technologies. Techniques taking DNA as the starting point, in contrast, have evolved at a lower pace and rarely allow a large-scale approach.

The DNA-centric techniques most frequently used to characterize DNA-protein complexes are electrophoretic mobility shift assay (EMSA), yeast one-hybrid assay, and in vitro methods of affinity purification with DNA as bait usually followed by mass spectrometry (reviewed in [16, 17]). These approaches have the disadvantage of studying the DNA-protein interactions in an artificial environment, without a functional readout. In the case of the yeast one-hybrid assay, the interaction occurs in a eukaryotic cell environment, closer to mammalian systems, but it still lacks a functional component.

In order to identify KRAB-ZFPs responsible for the sequence-specific recognition of retrotransposon-derived elements, we have developed a transcriptional repression-based assay reminiscent of a mammalian one-hybrid system. A scheme of the screen is depicted in Fig. 1. A library of KRAB-ZFPs, for instance all mouse family members, is built into a gateway-compatible vector. In parallel, a DNA target of interest, such as a retrotransposon-derived KAP1-recruiting repressive sequence, is cloned upstream of a PGK-GFP reporter and the resulting vector is used to establish a stable cell line. The cell line is then transfected in a large-scale fashion together with the KRAB-ZFP library. The readout is done by the assessment of fluorescence

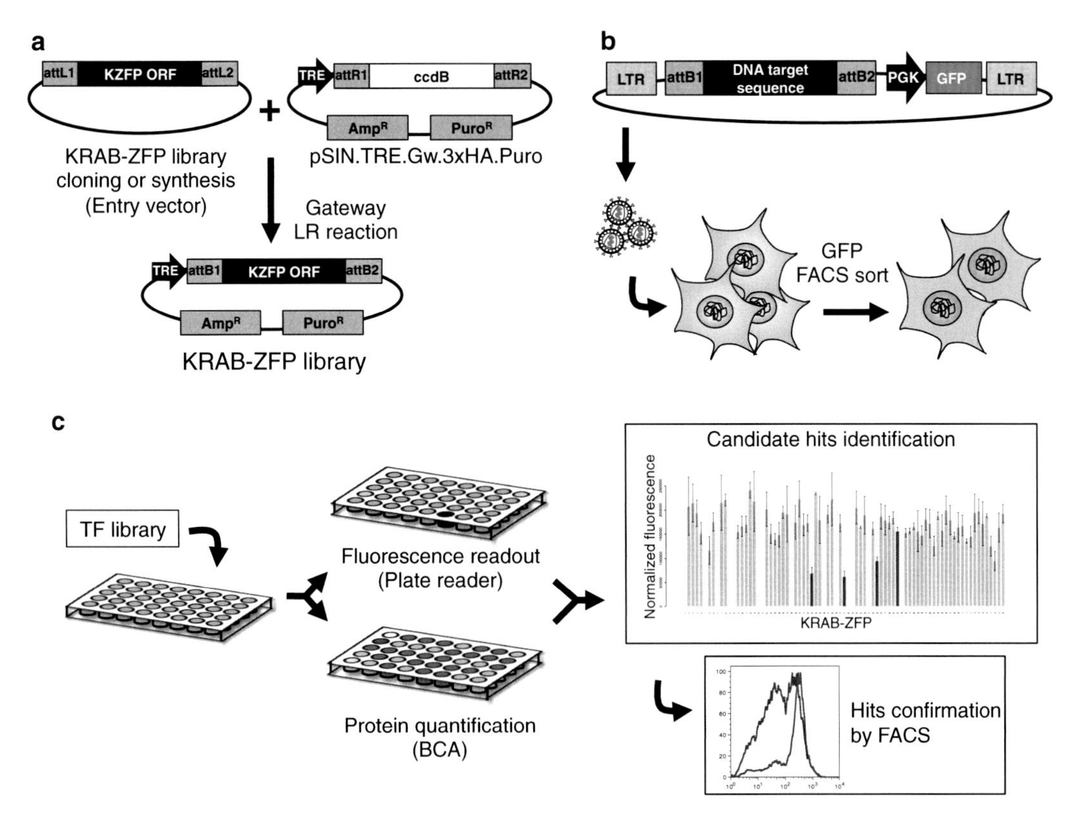

Fig. 1 Screen overview. (**a**) Transcription factor library construction. A library of KRAB-ZFPs is synthesized or PCR-amplified into gateway Entry vectors and used in LR reactions with the pSIN.TRE.Gw.3xHA.Puro vector to generate a final KRAB-ZFP library. (**b**) DNA target sequence cloning and establishment of stable cell lines. The DNA target sequence of interest, in this case a retrotransposon sequence, is cloned into the pRRL.R1R2.PGK. GFP reporter vector using pENTR/D-TOPO cloning. The plasmids are used for lentiviral vector production, which is used to transduce 293T cell. The stable cell line is established after sorting the GFP-positive cells. (**c**) Screening assay. The established cell lines are transfected against each KRAB-ZFP in triplicates in 96-well plates. After 6 days of induction of expression and selection, GFP fluorescence is assessed with a plate reader and protein amount is quantified with BCA. The normalized fluorescence is calculated and candidate hits are identified. As an example, in the graph three candidate hits and the control transfection are highlighted. The candidate hits are tested by flow cytometry (FACS) and the hits are identified when there is specific repression of the DNA target sequence

intensity by plate reader, coupled with protein quantification, followed by the confirmation of the hits by flow cytometry.

We found this screening method, which uses the functional interaction of KRAB-ZFPs with the repressor complex as readout, as capable of identifying a high percentage of DNA target-KRAB-ZFP pairs, with negligible rates of false positivity. Furthermore, such DNA-centered approach can be used to understand other aspects of the regulatory roles of retroelements. Finally, it can be adjusted to different species and to other types of repressors or other modulators of gene expression, and hence help unravel the roles of many *cis*-acting regulatory sequences and transcription factors. In this chapter, we describe in detail each step of this large-scale screen, here aimed at identifying KRAB-ZFPs binding to specific DNA sequences.

2 Materials

2.1 Transcription Factor Library Construction

1. Library of transcription factors in an Entry plasmid (from synthesis or PCR amplification).
2. Gateway LR Clonase II Enzyme mix (Life Technologies).
3. pSIN.TRE.Gw.3xHA.Puro destination vector, 100 ng/μL.
4. *E. coli* HB101 competent cells.
5. LB medium and LB agar plates.
6. Ampicillin sodium salt.
7. GoTaq Green Master Mix 2× (Promega).
8. CMV1F primer: 5′-GGAGGCCTATATAAGCAGAGCTCGT-3′.
9. PGK3R primer: 5′-GCTGCCTTGGAAAAGGCGCAACC-3′.
10. Agarose, LE, analytical grade.
11. Glycerol 60 % (v/v) in ultrapure water: Autoclave it before use, and store it at room temperature.
12. Nucleospin 96 Plasmid Kit (Macherey-Nagel).
13. Nanodrop 8000 Spectrophotometer.
14. 50 μg/mL Proteinase K solution.

2.2 DNA Target Sequence Cloning

1. dNTP set, 100 mM.
2. Pfu DNA polymerase.
3. pENTR/D-TOPO Cloning Kit (Life Technologies).
4. One Shot TOP10 competent bacteria.
5. Kanamycin sulfate.
6. LB medium and LB agar plates.
7. Restriction enzymes.
8. QIAprep Spin Miniprep Kit or similar.
9. M13F primer: 5′-TGTAAAACGACGGCCAG-3′.
10. M13R primer: 5′-CAGGAAACAGCTATGAC-3′.
11. Gateway LR Clonase II Enzyme mix (Life Technologies).
12. pRRL.R1R2.PGK.GFP vector, 100 ng/μL.

2.3 Establishment of Stable Cell Lines Containing the DNA Target Sequence

1. Lentiviral vectors containing the sequence of interest (for lentiviral vector production protocol, *see* [18]).
2. HEK 293T cells.
3. 10 cm round tissue culture dishes.

4. Dulbecco's modified Eagle medium, DMEM.
5. Fetal calf serum (FCS).
6. Penicillin-streptomycin (PS) solution, 100×.
7. Dulbecco's PBS.
8. 0.05 % Trypsin-EDTA.
9. Flow cytometry cell sorter.

2.4 Screening Assay

1. DMEM.
2. Fetal calf serum (FCS).
3. Penicillin-streptomycin (PS) solution, 100×.
4. Dulbecco's PBS.
5. 0.05 % Trypsin-EDTA.
6. 96-Well tissue culture plate, flat bottom.
7. Fugene 6 Transfection Reagent (Promega).
8. Opti-MEM Reduced Serum Media (Life Technologies).
9. 96-Well conical bottom plates.
10. Sciclone ALH 3000 (Caliper Life Sciences).
11. Multidrop Combi dispenser and standard tube dispensing cassette.
12. Doxycycline Hyclate (Sigma-Aldrich).
13. Protease Inhibitor Cocktail Tablets, cOmplete, Mini, EDTA-free (Roche).
14. Puromycin dihydrochloride, 10 mg/mL.
15. RIPA buffer: 50 mM Tris-HCl pH 8.0, 150 mM NaCl, 1 % (v/v) NP40, 0.5 % (w/v) sodium deoxycholate, 0.1 % (w/v) SDS. Add protease inhibitors freshly before use.
16. Black 96-well polystyrene plate, non-treated, flat bottom.
17. Infinite F500 Plate Reader (Tecan).
18. BCA Protein Assay Reagent (Thermoscientific).
19. Flow cytometer.

3 Methods

3.1 Transcription Factor Library Construction

1. For the construction of the transcription factor (TF) library, a collection of KRAB-ZFPs is obtained by PCR amplification or cDNA synthesis and is then placed in the destination gateway-compatible vector (pSIN.TRE.Gw.3xHA.Puro). For the purposes of this protocol, cDNA synthesis should be done directly in a gateway-compatible Entry vector (L1-L2). When performing the synthesis, suppress the stop codon from the

cDNA, in order to obtain HA-fused proteins in the final vector. For PCR amplification of the open reading frame (ORF) of the TF of interest up to an Entry clone, please refer to [19] (*see* **Note 1**).

2. Once you have the purified Entry plasmids, proceed to the LR reaction. Mix together 100 ng of the Entry vector and 100 ng of the pTRE destination vector, and complete to a final volume of 4 μL with TE buffer. Add 1 μL of LR clonase and incubate from 1 h to overnight (ON) (for large fragments or when performing several reactions at once, we recommend the latter). Add 1 μL of proteinase K, incubate for 10 min at 37 °C, and transform everything into 50 μL of competent bacteria (we typically use HB101). Plate the transformed bacteria into LB agar plates containing 100 μg/mL ampicillin and incubate the plates overnight at 37 °C (*see* **Note 2**).
3. Check the cloned fragments by colony PCR. Pick 3–5 colonies from each plate. Prepare a PCR reaction containing 10 μL of GoTaq Green Master Mix 2× (Promega), 1 μL of 10 μM CMV1F primer, and 1 μL of 10 μM PGK3R primer, and complete to 20 μL final volume with nuclease-free water. Pick a colony by touching it with a sterile tip, then touch a new LB agar plate containing 100 μg/mL ampicillin, and finally transfer the tip to the tube containing the PCR mixture. Mix well and remove the tip after 5 min. Do the PCR reaction in a thermal cycler with a heated lid using the following program:

1	95 °C	2 min
2	95 °C	1 min
3	55 °C	1 min
4	72 °C	4 min
5	72 °C	2 min
6	4 °C	hold

* cycle between steps 2–4 and perform from 25 to 30 cycles.

Evaluate the PCR products on an agarose gel and select the colonies with the correct molecular weight.

4. Culture a positive bacterial colony per TF in 2 mL of LB media supplemented with 50 μg/mL ampicillin at 37 °C for 16–18 h (*see* **Note 3**). Use this culture to perform a glycerol stock for long-term storage and the remaining for plasmid DNA purification. For the glycerol stock, mix together 100 μL of bacterial culture and 100 μL of sterile glycerol 60 %. Store the glycerol stock at −80 °C (*see* **Note 4**). Spin the remaining culture at 3600 × *g* for 15 min and use it to perform DNA extraction with the Nucleospin 96 Plasmid Kit (Macherey-Nagel). Quantify

the purified DNA (we typically use 8-channel Nanodrop) and dilute them to 20 ng/μL. Arrange the DNA stock in 96-well plates and store them at −20 °C until the day of the screen.

3.2 DNA Target Sequence Cloning

1. For the DNA sequence cloning, the retrotransposon is cloned using the pENTR/D-TOPO Cloning Kit (Life Technologies) and it is further transferred to the gateway-compatible destination vector (pRRL.R1R2.PGK.GFP) by an LR reaction. Design specific primers for your sequence of interest (we had good results with Primer 3 software in the past [20]) and add the sequence CACC on the 5′ end of the forward primers (*see* **Note 5**).
2. Use these primers to set up a PCR reaction containing 1 μL of dNTP mix (10 mM of each dNTP), 1 μL of each primer (10 μM), 3 μL of 10× polymerase buffer, 0.5 μL Pfu DNA polymerase, and the source of cDNA (typically we use 50 ng of genomic DNA), and complete to a final volume of 30 μL with nuclease-free water. Do the PCR reaction(s) in a thermal cycler with a heated lid using the following program:

1	95 °C	2 min
2	95 °C	1 min
3	Tm − 5 °C	1 min
4	72 °C	2 min/kb of ORF
5	72 °C	2 min
6	4 °C	hold

* cycle between steps 2–4 and perform from 25 to 30 cycles.

Evaluate the PCR products on an agarose gel and proceed to the TOPO reaction (*see* **Note 6**).

3. Set up a TOPO reaction containing 2 μL of the PCR product, 0.5 μL of salt solution, and 0.5 μL of pENTR/D-TOPO vector. Incubate for 5 min at RT and transform the entire reaction into 25 μL of One Shot TOP10 competent bacteria. Plate the transformed bacteria onto LB agar plate containing 50 μg/mL of kanamycin and incubate the plate ON at 37 °C.
4. Check the size and identity of the plasmids by restriction enzyme digestion (*see* **Note 7**). Pick a few colonies and start bacterial cultures in 2 mL of LB supplemented with 50 μg/mL of kanamycin (we typically screen 4–5 colonies per TOPO reaction). Perform DNA purification of the plasmids (we use the QIAprep Spin Miniprep Kit, Qiagen) and screen them using restriction enzyme digestion. Select the appropriate enzymes to verify your sequence (*see* **Note 8**) and assemble the following reaction: 2 μL of enzyme buffer, 2 μL of BSA (1 mg/

mL), 3 μL of purified plasmid, and 0.5 μL of each restriction enzyme and complete to 20 μL with distilled water. Incubate for the time and temperature specified by the enzyme manufacturer and analyze the resulting fragments on agarose gel. Select at least one positive plasmid from the restriction enzyme screen and verify the cloned sequence by Sanger sequencing using M13 primers.

5. Select the correct clones and use them for the LR reaction. Assemble an LR reaction as described in **step 2** of Subheading 3.1, using the pRRL.R1R2.PGK.GFP vector as destination plasmid. Select 3–5 colonies to screen by restriction enzyme digestion as described above. Choose one positive plasmid for each sequence and use it for the establishment of the stable cell lines.

3.3 Establishment of Stable Cell Lines Containing the DNA Target Sequence

1. We have established this screening assay using murine KRAB-ZFP proteins, and for that purpose we have chosen human 293T cells. Firstly, these cells are easily transfectable, and secondly, they are functional for KAP1-mediated repression of mouse KRABZFPs [21]. The advantage of this system is that, since KRAB-ZFPs have evolved to become species specific, human 293T cells can be used to assess the function of mouse KRAB-ZFPs without interference from endogenous human factors. For the establishment of stable cell lines, 293T cells are transduced at a low multiplicity of infection (MOI) with the DNA target pRRL.R1R2.PGK.GFP lentiviral vector in order to achieve approximately one provector copy per cell. For lentiviral vector production *see* [18].
2. Plate 293T cells at a confluence of 1×10^5 cells per well into a 12-well dish. After 6 h, transduce them with lentiviral vectors containing the DNA target pRRL.R1R2.PGK.GFP plasmid at MOIs of 0.05 and 0.1 (titers calculated in 293T) (*see* **Note 9**). After 72 h, verify the GFP expression of transduced cells by flow cytometry. Select the samples with a percentage of GFP-positive cells between 5 and 20 % and expand those cells into a 10 cm culture dish.
3. Perform flow cytometry sorting of the GFP-positive cells. For the sorting, we typically use the FACSAria II machine (BD Biosciences). Put the GFP-positive sorted cells back in culture, expand them, and keep a stock in the liquid nitrogen (*see* **Note 10**).

3.4 Screening Assay

1. The screen is done by transfecting the cell line—containing the *cis*-acting sequence of interest—against the library of TFs. Before starting, make sure that you have the cell lines in culture and enough plasmid DNA for the transfections. Also make sure that you have the control cell lines and control TFs of

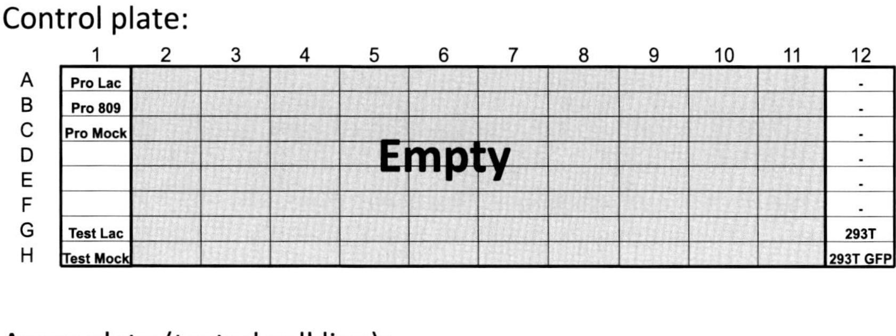

Assay plate (tested cell line):

1 2 3 4 5 6 7 8 9 10 11 12

A B C D E F G H

Empty

Lac

Mock

Tested TFs

Lac

Mock

Empty

Fig. 2 Control and assay plate design with overview of the cell lines and plasmids to be used in transfections. The control plate is done with the control cell line containing the PBS^{Pro} sequence. The PBS^{Pro} line is transfected with its known interactor, ZFP809, with a control plasmid (in this case, expressing the protein LacZ), or with mock transfection (control for puromycin selection). A control of the tested cell line (to be transfected with the control LacZ plasmid and mock transfection) can be included. On column 12, a negative 293T control and a transfection control with 293T cells transfected with an empty pRRL.R1R2.PGK.GFP are performed to check for transfection efficiency. Other controls can be added in the empty wells. The assay plates contain the tested KRAB-ZFPs that will be used to transfect the tested cell line, as well as the transfection and mock control (which should be present at least twice)

interest. Figure 2 depicts an example of control and assay plates (*see* **Note 11**).

Each assay plate should have a negative control (we typically use a plasmid with LacZ in place of the TF ORF) and a mock transfection (control for puromycin) that are present at least twice in each plate. Also, in the control plate, as positive control for the repression assay we use KRAB-ZFP809 and its previously identified DNA target, the PBS^{Pro} sequence (TGGGGGCTCGTCCGGGATCGGGAGACCCC) [22]. A 293T cell line containing the PBS sequence upstream of the PGK-GFP cassette is generated according to Subheadings 3.2 and 3.3. Finally, a GFP-negative control (293T cells) and a transfection control (293T cells transfected with empty pRRL.R1R2.PGK.GFP plasmid) are also added to the control plate.

2. Harvest the cell line of interest and the control cell lines (*see* **Note 12**). Rinse the cells with PBS, aspirate the PBS, and replace it with enough trypsin to cover the cell layer. After 1–3 min, harvest the cells with a 10× excess of DMEM supplemented with FCS 10 % and PS. Count the cells and put them in

suspension at 5×10^4 cells/mL in DMEM supplemented with FCS and PS (calculate the final volume of cells needed according to the number of transfections).

3. The library of TFs should be arranged in 96-well plates and each transfection is performed in triplicate. Each individual transfection is done with the following volumes: 7 μL of TF DNA plasmid stock (100–150 μg of DNA), 0.6 μL of Fugene 6 Transfection Reagent, and 2.4 μL Opti-MEM (Life Technologies) (*see* **Note 13**). First prepare a general Fugene and Opti-MEM mix for all the plates. Then prepare a 3.5× DNA mix for each 96-well TF DNA stock plate to be tested: in a 96-well V-bottom plate pipet 24.5 μL of each TF plasmid, and add 10.5 μL of the Fugene-Opti-MEM mix to each well. Incubate for 5 min, mix gently, and pipet 10 μL of the DNA mix to each 96-well culture plate triplicate. We typically perform this screen using the Sciclone ALH 3000 (Caliper Life Sciences) liquid-handling robot, but it can also be done with multichannel pipets.
4. Add 100 μL of the cell suspension directly to the 96-well culture plates containing the TF plasmid mix (*see* **Note 14**). To add the cells (and liquids throughout the screen) we typically use the Multidrop Combi dispenser (Thermoscientific) equipped with a standard tube dispensing cassette (*see* **Note 15**).
5. The next day (16–20 h after transfection), add a final concentration of 5 μg/mL of doxycycline to induce TF expression from the TRE promoter. Assemble a mix of DMEM (supplemented with FCS and PS) and 10 μg/mL doxycycline, and add 100 μL to each well. If using the Multidrop dispenser, use medium speed.
6. The following day, add a final concentration of 1 μg/mL of puromycin to each well to select for the presence of the TRE vector. Dilute puromycin in PBS to a concentration of 21 μg/mL and add 10 μL to each well.
7. On day 6 after transfection, harvest the cells for the readout. For the harvesting, we recommend to process one 96-well plate at a time. Using a multichannel pipet, aspirate the media from the wells, rinse the cells with 50 μL of PBS, aspirate the PBS, and add 120 μL of RIPA buffer (supplemented with protease inhibitors) to all the wells. Shake the plate for 15 s (using the Multidrop) and put it on ice for at least 10 min. Keep the plates on ice for the following procedures.
8. Once all the plates are processed mix the cell lysate a couple of times and transfer 90 μL of each well to a new black 96-well flat-bottom plate (*see* **Note 16**). Store the remaining cell lysates at −20 °C for further BCA analysis. Measure the fluorescence of the samples on a plate reader equipped with fluorescence

readout, with excitation set at 485 nm and emission at 520 nm (we typically use the Tecan Infinite F500) (*see* **Note 17**). Calculate the fluorescence value by subtracting the value of a well containing only buffer from all the other wells.

9. Thaw the frozen lysates and perform protein quantification by BCA in 96-well plates. Prepare the BSA standards through serial dilution in RIPA buffer to obtain the following concentrations: 2, 1, 0.500, 0.250, 0.125, 0.062, and 0.031 mg/mL. Place 10 μL of each standard and 10 μL of RIPA buffer supplemented with protease inhibitor (blank) in the first column of the 96-well plate and pipet 10 μL of the lysates into the remaining wells (exclude column 12 of the sample plate) (*see* **Note 18**). Prepare a master mix of BCA reagent mixture containing 200 μL of BCA reagent A and 4 μL of BCA reagent B per well. Add 200 μL of BCA reagent mix to each well and incubate the plates for 30 min at 37 °C in the dark. Immediately measure the absorbance at 570 nm and use the standards to calculate the protein concentration of the samples.
10. Calculate the normalized fluorescence by dividing the fluorescence by the protein concentration. Calculate the mean and standard deviation of the triplicates and plot them for visualization. We typically analyze the results using R. The candidate hits can be identified by first selecting the ten TFs with the lowest normalized fluorescence values per plate, and then we select only the ones that are among the ten lowest on all three replicates of the plate (*see* **Note 19**).
11. After selecting the candidates, they need to be tested by flow cytometry to identify the hits. Transfect the 293T cell line containing the retrotransposon upstream of the PGK-GFP cassette with the candidate TFs. Plate the 293T cells in 24-well plates at 1.5×10^4 cells/well. From 3 to 10 h later, transfect the cells with the TF plasmid from the stock. Mix 0.38 μL Fugene 6, 14.62 μL Opti-MEM, and 10 μL DNA from the stock (150–200 ng), per transfection, per well (*see* **Note 20**). Incubate the mix for 5 min at RT and add 25 μL to the cells. Transfect each TF in duplicate and also perform a transfection control in an unrelated 293T pRRL.R1R2.PGK.GFP cell line (e.g., the PBS^{Pro} cell line), to control for unspecific effects. Include LacZ as a negative transfection control for the test cell lines. Also, as a positive repression control, transfect the PBS^{Pro} cell line with ZFP809. On the next day (16–20 h after transfection), add doxycycline to a final concentration of 5 μg/mL to induce TF expression. The following day, add puromycin to a final concentration of 1 μg/mL. On day 6 after transfection, harvest the cells, resuspend them in DMEM, spin them, remove the media, and resuspend them in 300 μL of PBS supplemented with 2 % FCS. Analyze them by flow cytometry to evaluate

GFP expression. Calculate the median fluorescence intensity of the different TFs and compare them with the control transfection (LacZ) and with the control cell line. You have a hit when a TF specifically represses the DNA sequence of interest.

4 Notes

1. We recommend verification of the cloned/synthesized cDNA at the Entry plasmid level by DNA sequencing to avoid perpetuating errors or mutations to the next steps.
2. For a large number of TFs, LR reactions can be performed in 96-well plates and transformed into 96-well plates containing competent bacteria. In that case, instead of plating them onto agar plates, transformed bacteria are grown in 96-well 1 mL culture plates in LB medium supplemented with ampicillin (50 μg/mL). Hence, there is no selection of single clones and the positively transformed population is directly used for the bacterial culture, glycerol stock, and PCR check.
3. When dealing with repetitive sequences, as in the case of KRAB-ZFPs (and also for the DNA target sequence when cloning retrotransposons), we recommend performing the bacterial culture at 30 °C for 24 h instead of 37 °C ON. The lower temperature decreases the potential plasmid recombination.
4. Manipulate the glycerol stock very carefully, avoiding unnecessary freezing and thawing. Furthermore, besides the glycerol stock, we advise keeping a stock of purified DNA.
5. Make sure that the designed primers are unique. We typically use the UCSC genome browser PCR tool (http://genome.ucsc.edu/cgi-bin/hgPcr?command=start), NCBI primer blast (http://www.ncbi.nlm.nih.gov/tools/primer-blast/), and SIB Tag Scan (http://ccg.vital-it.ch/tagger/tagscan.html). Furthermore, verify that the reverse primer does not contain the complementary sequence (CACC) to the TOPO overhang at its 5′ end. Furthermore, when cloning several *cis*-acting sequences at once we recommend the insertion of a unique site for restriction enzyme on the 5′ end (remember to keep the CACC on the outer 5′) of at least one of the primers. This facilitates the bacterial colony screening.
6. The PCR reaction should be optimized to obtain as product a clear single band of the correct size. If this is not possible, the band of interest can be cut from the gel and purified using the MinElute Gel Extraction kit (Qiagen).
7. When cloning many target sequences, colony PCR screening can be used instead of restriction enzyme digestion. We prefer

the latter because it is more effective at detecting undesired sequences.

8. *Ssp*I can be used as a single cutter in the plasmid and another single cutter for the insert should be selected. If unique restriction enzyme sites were included on the primers sequences, they can be used at this step.
9. We have tested different MOIs, and an MOI of 0.1 or 0.05 usually yields a percentage of GFP-positive cells between 5 and 20 % for 293Ts, but other MOIs can be tested. This percentage is predicted to yield about one vector copy number per cell and, according to our copy number tests, that is normally true. If using cells other than 293Ts, the MOI needs to be adjusted to yield between 5 and 20 % of GFP-positive cells.
10. The established cell lines can be checked for vector copy number by qPCR in order to determine if they are in a range of 1 copy per cell.
11. We recommend separate assembling of the control and assay plates, for easier reuse and changes of the plates. The control plate can also be adapted to each screen and can include other positive and negative controls.
12. We recommend splitting the 293T cell lines the day before the screen, from a confluent plate to a new one at 1:3. We have observed that this can improve the transfection efficiency.
13. We have obtained good transfection efficiency with these amounts of reagents, but it can vary greatly according to the DNA quality. If necessary, different amounts and ratios of Fugene and DNA can be tested.
14. The cells have to be added to the plates containing the DNA-Fugene mix at most 30 min to 1 h after the DNA mix has been prepared. Waiting for longer periods can decrease the transfection efficiency.
15. When using the dispenser, make sure to keep it sterile by placing it inside a laminar flow and by decontaminating it with ethanol before use. Unless otherwise stated, use maximum speed for liquid dispensing. Also, in the case of the control cell lines that are plated only into few wells, manual handheld dispenser pipets can be used, instead of the dispenser.
16. Be careful not to take up any bubbles when transferring the lysate to the new plate. As the readout is done from the upper part of the plate, the presence of any bubbles can compromise the readout.
17. For the readout it can be necessary to adjust measurement parameters according to the cell line. Before each readout we typically optimize the gain and the Z-position.
18. Column 12 is excluded for practical reasons and it is dispensable at this step.

19. The candidate hits can also be identified by significant differences using paired *t*-test, when compared to LacZ control transfections. However, this method has proven to be too stringent in some cases and some true biological candidates are missed. For that reason, we have opted to use a less stringent criterion (ten lowest candidates). We have observed that the latter decreases the chances of identifying false negatives that could be discarded as candidates due to the noise. On the other hand, this method leads to the increase of false positives as candidate hits, which are included in the candidate test by flow cytometry. However, it should be noted that after the flow cytometry test we have observed no false positives, and all candidates we have identified so far are be *bona fide* interactors.
20. We typically use these amounts of DNA and Fugene for the transfection in 24-well plates in order to save reagents. However, if desired, an alternative is to use 2.5 times the amounts used in the 96-well plate transfections mentioned in **step 3** of Subheading 3.4.

Acknowledgements

We thank P. Turelli and C. Delattre-Gubelman for advice, C. Raclot and S. E. Offner for technical support, M. Chambon and J. B. Chapalay from the EPFL Biomolecular Screening Facility for advice and help with the robotics, and the EPFL Flow Cytometry Core Facility for FACS sorting. This work was supported by funds from the Swiss National Science Foundation and from the European Research Council.

References

1. Rowe HM, Trono D (2011) Dynamic control of endogenous retroviruses during development. Virology 411(2):273–287. doi:10.1016/j.virol.2010.12.007
2. Leung DC, Lorincz MC (2012) Silencing of endogenous retroviruses: when and why do histone marks predominate? Trends Biochem Sci 37(4):127–133. doi:10.1016/j.tibs.2011.11.006
3. Emerson RO, Thomas JH (2009) Adaptive evolution in zinc finger transcription factors. PLoS Genet 5(1):e1000325
4. Vaquerizas JM, Kummerfeld SK, Teichmann SA, Luscombe NM (2009) A census of human transcription factors: function, expression and evolution. Nat Rev Genet 10(4):252–263
5. Friedman JR, Fredericks WJ, Jensen DE, Speicher DW, Huang XP, Neilson EG, Rauscher Iii FJ (1996) KAP-1, a novel corepressor for the highly conserved KRAB repression domain. Genes Dev 10(16):2067–2078
6. Schultz DC, Friedman JR, Rauscher Iii FJ (2001) Targeting histone deacetylase complexes via KRAB-zinc finger proteins: the PHD and bromodomains of KAP-1 form a cooperative unit that recruits a novel isoform of the Mi-2α subunit of NuRD. Genes Dev 15(4):428–443
7. Schultz DC, Ayyanathan K, Negorev D, Maul GG, Rauscher Iii FJ (2002) SETDB1: a novel KAP-1-associated histone H3, lysine 9-specific methyltransferase that contributes to HP1-mediated silencing of euchromatic genes by KRAB zinc-finger proteins. Genes Dev 16(8):919–932
8. Nielsen AL, Ortiz JA, You J, Oulad-Abdelghani M, Khechumian R, Gansmuller A, Chambon P, Losson R (1999) Interaction with members of

the heterochromatin protein 1 (HP1) family and histone deacetylation are differentially involved in transcriptional silencing by members of the TIF1 family. EMBO J 18(22):6385–6395

9. Sripathy SP, Stevens J, Schultz DC (2006) The KAP1 corepressor functions to coordinate the assembly of de novo HP1-demarcated microenvironments of heterochromatin required for KRAB zinc finger protein-mediated transcriptional repression. Mol Cell Biol 26(22):8623–8638
10. Rowe HM, Jakobsson J, Mesnard D, Rougemont J, Reynard S, Aktas T, Maillard PV, Layard-Liesching H, Verp S, Marquis J, Spitz F, Constam DB, Trono D (2010) KAP1 controls endogenous retroviruses in embryonic stem cells. Nature 463(7278):237–240. doi:10.1038/nature08674
11. Turelli P, Castro-Diaz N, Marzetta F, Kapopoulou A, Raclot C, Duc J, Tieng V, Quenneville S, Trono D (2014) Interplay of TRIM28 and DNA methylation in controlling human endogenous retroelements. Genome Res 24(8):1260–1270. doi:10.1101/gr.172833.114
12. Matsui T, Leung D, Miyashita H, Maksakova IA, Miyachi H, Kimura H, Tachibana M, Lorincz MC, Shinkai Y (2010) Proviral silencing in embryonic stem cells requires the histone methyltransferase ESET. Nature 464(7290):927–931. doi:10.1038/nature08858
13. Liu S, Brind'Amour J, Karimi MM, Shirane K, Bogutz A, Lefebvre L, Sasaki H, Shinkai Y, Lorincz MC (2014) Setdb1 is required for germline development and silencing of H3K9me3-marked endogenous retroviruses in primordial germ cells. Genes Dev 28(18):2041–2055. doi:10.1101/gad.244848.114
14. Castro-Diaz N, Ecco G, Coluccio A, Kapopoulou A, Yazdanpanah B, Friedli M, Duc J, Jang SM, Turelli P, Trono D (2014) Evolutionally dynamic L1 regulation in embryonic stem cells. Genes Dev 28(13):1397–1409. doi:10.1101/gad.241661.114
15. Jacobs FMJ, Greenberg D, Nguyen N, Haeussler M, Ewing AD, Katzman S, Paten B, Salama SR, Haussler D (2014) An evolutionary arms race between KRAB zinc-finger genes ZNF91/93 and SVA/L1 retrotransposons. Nature 516(7530):242–245. doi:10.1038/nature13760
16. Simicevic J, Deplancke B (2010) DNA-centered approaches to characterize regulatory protein-DNA interaction complexes. Mol Biosyst 6(3):462–468. doi:10.1039/b916137f
17. Dey B, Thukral S, Krishnan S, Chakrobarty M, Gupta S, Manghani C, Rani V (2012) DNA-protein interactions: methods for detection and analysis. Mol Cell Biochem 365(1-2):279–299. doi:10.1007/s11010-012-1269-z
18. Barde I, Salmon P, Trono D (2010) Production and titration of lentiviral vectors. In: Jacqueline N Crawley et al. (ed) Current protocols in neuroscience, Chapter 4: Unit 4.21. doi:10.1002/0471142301.ns0421s53
19. Hens K, Feuz JD, Deplancke B (2012) A high-throughput gateway-compatible yeast one-hybrid screen to detect protein-DNA interactions. Methods Mol Biol 786:335
20. Untergasser A, Nijveen H, Rao X, Bisseling T, Geurts R, Leunissen JAM (2007) Primer3Plus, an enhanced web interface to Primer3. Nucleic Acids Res 35(Suppl 2):W71–W74
21. Rowe HM, Friedli M, Offner S, Verp S, Mesnard D, Marquis J, Aktas T, Trono D (2013) De novo DNA methylation of endogenous retroviruses is shaped by KRAB-ZFPs/KAP1 and ESET. Development 140(3):519–529. doi:10.1242/dev.087585
22. Wolf D, Goff SP (2009) Embryonic stem cells use ZFP809 to silence retroviral DNAs. Nature 458(7242):1201–1204

Chapter 26

Reprogramming of Human Fibroblasts to Induced Pluripotent Stem Cells with *Sleeping Beauty* Transposon-Based Stable Gene Delivery

Attila Sebe and Zoltán Ivics

Abstract

Human induced pluripotent stem (iPS) cells are a source of patient-specific pluripotent stem cells and resemble human embryonic stem (ES) cells in gene expression profiles, morphology, pluripotency, and in vitro differentiation potential. iPS cells are applied in disease modeling, drug screenings, toxicology screenings, and autologous cell therapy. In this protocol, we describe how to derive human iPS cells from fibroblasts by *Sleeping Beauty* (SB) transposon-mediated gene transfer of reprogramming factors. First, the components of the non-viral *Sleeping Beauty* transposon system, namely a transposon vector encoding reprogramming transcription factors and a helper plasmid expressing the SB transposase, are electroporated into human fibroblasts. The reprogramming cassette undergoes transposition from the transfected plasmids into the fibroblast genome, thereby resulting in stable delivery of the reprogramming factors. Reprogramming by using this protocol takes ~4 weeks, after which the iPS cells are isolated and clonally propagated.

Key words Reprogramming, Transposon, Genomic integration, Transgene, Fibroblast

1 Introduction

Human pluripotent stem cells represent valuable tools for biomedical research and hold a major promise for therapeutic applications. Because of their potential to give rise to almost all cell types of the body, human embryonic stem (ES) cells generated considerable enthusiasm among scientists and patients since their first isolation in 1998. However, the therapeutic hope represented by ES cells has been hampered by several concerns, such as ethical issues with regard to derivation of these cells from human embryos, and immune rejection once transplanted into incompatible recipients.

The field literally exploded in 2007, when the first human induced pluripotent stem (iPS) cells were generated following the pioneering work of Nobel Laureate Shinya Yamanaka [1]. The iPS cells represent a new, powerful alternative, which overcomes the

Jose L. Garcia-Pérez (ed.), *Transposons and Retrotransposons: Methods and Protocols*, Methods in Molecular Biology, vol. 1400, DOI 10.1007/978-1-4939-3372-3_26, © Springer Science+Business Media New York 2016

limitations of ES cells and provides promising tools for cell and gene therapy applications. Generation of iPS cells has been successfully achieved by simultaneous overexpression of defined reprogramming transcription factors (*Oct4*, *Sox2*, *Klf4*, and *c-Myc*). iPS cells resemble ES cells in gene expression profiles, morphology, pluripotency, and in vitro differentiation potential. In contrast to ES cells, iPS cells can be obtained from autologous, adult somatic cells. iPS cells can be genetically modified and can be directed to differentiate into endodermal, mesodermal, and ectodermal cell types. iPS cells present a promise for vast application fields, including modeling of monogenic and complex, multigenic diseases, drug development and screening, toxicology screenings, and autologous cell therapy, and are a unique tool for dissecting the molecular events of cellular differentiation [2–5]. Potential clinical applications such as cellular replacement/transplantation therapies have already been addressed soon after iPS cells were first described: iPS cells derived from fibroblasts of sickle cell anemia mice were genetically corrected by replacing the mutant β-globin allele with a wild-type allele by means of homologous recombination. This provided a source of iPS cells able to differentiate into disease-free hematopoietic precursors that cured the afflicted mice following transplantation [3]. Extensive research and development ultimately led to the first clinical trial where iPS cell-derived retinal pigment epithelium cells were implanted into an eye of a patient suffering from age-related macular degeneration, a common eye condition of the elderly that can lead to blindness [6].

One field of iPS-related research that presents constant innovations is the method of reprogramming factor delivery. Reprogramming to an iPS cell state can be achieved through either integrative (retroviruses, lentiviruses, transposons) or non-integrative (adenoviruses, Sendai virus, episomal vectors, mRNA) methods. The early studies applied retroviral and later lentiviral vectors for stable genomic insertion of the reprogramming genes [7–9]. However, the potential for insertional mutagenesis and deregulation of gene expression caused by integrating retroviral gene transfer vectors limits the value of the resulting iPS cells for clinical applications [10]. These concerns led to the development of alternative approaches to reprogramming, including elimination of chromosomally integrated reprogramming genes from iPS cells by using Cre/lox or Flp technology [11–13], or the use of non-integrating gene transfer systems to deliver the reprogramming genes, such as repeated transient plasmid transfections [14], adenoviral gene delivery [15], Sendai viral vectors [16], Epstein-Barr virus-derived oriP/EBNA1 episomal vector system [17], delivery of synthetic mRNAs encoding the reprogramming factors [18], and transfection of miRNAs [19]. These reports provide proof of concept for the generation of iPS cells without transgene integration, but at a lower efficiency.

DNA transposons are discrete pieces of DNA with the ability to change their positions within the genome via a cut-and-paste mechanism called transposition. These mobile genetic elements can be harnessed as gene delivery vector systems that can be used as tools for versatile applications. In contrast to viral vectors, transposon vectors can be maintained and propagated as plasmid DNA, thereby providing simplicity and safety to the user. The *Sleeping Beauty* (SB) transposon gene delivery system (Fig. 1a) has a broad utility in genetic applications [20], and consists of a plasmid construct containing the terminal inverted repeats (TIRs) of the transposon that flank a gene of interest (GOI), and a second plasmid that provides transient expression of the transposase enzyme to enable excision of the GOI from the plasmid and its permanent integration into the chromosomes of the cells [21]. Transposition efficiency using the hyperactive SB100X transposase [22] yields stable gene transfer efficiencies comparable to stable transduction efficiencies of integrating viral vectors, enabling high percentage of transgene integration and expression.

We previously established a method of reprogramming human fibroblasts using the SB100X transposase in combination with an SB transposon construct containing a CAG promoter-driven single-expression cassette of Oct4, Sox2, Klf4, and c-Myc with or without Lin28 (pT2-OSKML and pT2-OSKM, respectively), in which the individual genes are linked by 2A self-cleaving peptide sequences, cloned between the transposon TIRs as presented in

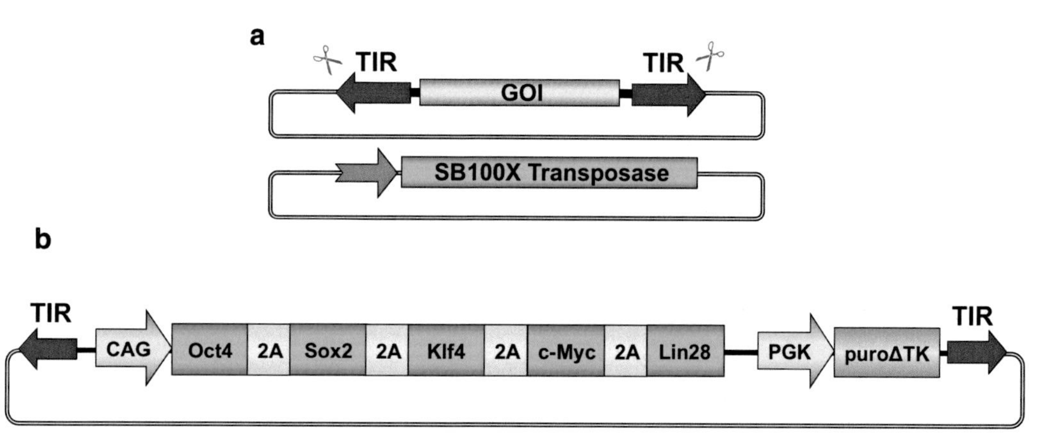

Fig. 1 Components of the *Sleeping Beauty* transposon vector system required for reprogramming. (**a**) The *Sleeping Beauty* transposon system consists of a plasmid vector harboring the transposon containing a gene of interest (GOI) flanked by the transposon terminal inverted repeats (TIRs) and a second plasmid that expresses the SB100X transposase from a suitable promoter (*orange arrow*), such as the CMV promoter. (**b**) The pT2-OSKML *Sleeping Beauty* transposon carrying a reprogramming cassette that contains a CAG promoter-driven fusion gene encoding the reprogramming factors Oct4, Sox2, Klf4, c-Myc, and Lin28, in which the individual cDNAs are linked by 2A self-cleaving peptide sequences, cloned between the transposon TIRs (the construct also contains a separate puroΔTK selection marker expression cassette)

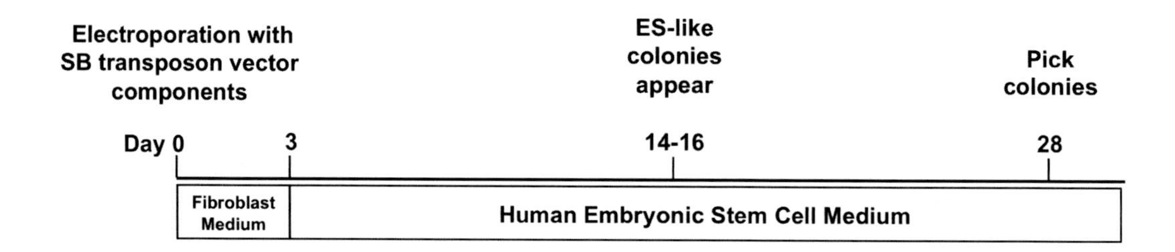

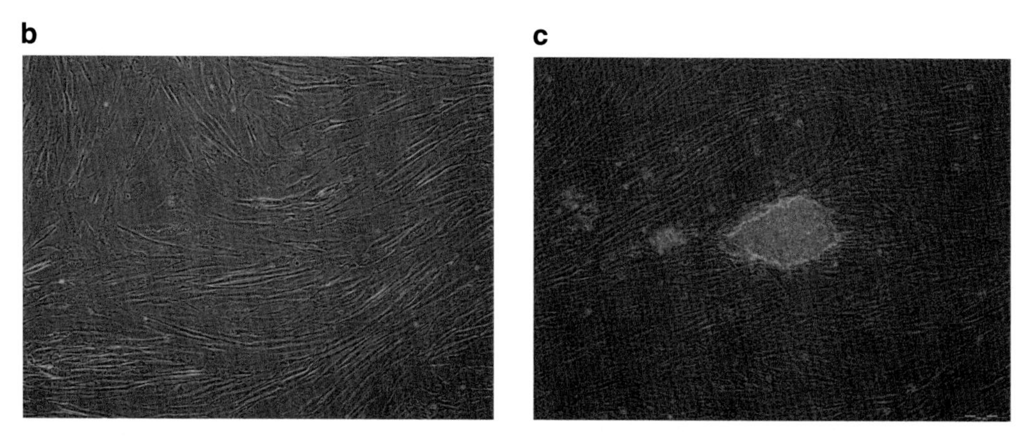

Fig. 2 Human iPS cell reprogramming with *Sleeping Beauty* transposon vectors. (**a**) Timeline of the reprogramming protocol. (**b**) Confluent HFF-1 cells grow in monolayers showing a typical fibroblast morphology. (**c**) iPS colony formed during the reprogramming experiment

Fig. 1b (the construct also contains a separate puroΔTK selection marker expression cassette). pT2-OSKML is more efficient in reprogramming than pT2-OSKM in both mouse and human fibroblasts [23]. For reprogramming we applied a protocol depicted in Fig. 2a. We have shown that this approach of generating iPS cells is a viable and efficient alternative to other integrating reprogramming methods. SB transposon-based reprogramming efficiency (~0.02 % of transfected cells) is similar to values obtained with other integrating gene transfer methods [23].

2 Materials

2.1 Common Reagents

1. Fibroblasts to be reprogrammed, such as human foreskin fibroblasts (obtained from ATCC or alternatives).
2. Dulbecco's modified Eagle medium (DMEM).

3. Dulbecco's modified Eagle medium/nutrient mixture F-12 (DMEM/F-12).
4. Fetal bovine serum (FBS).
5. 0.05 % trypsin (diluted in PBS from 0.25 % trypsin-EDTA solution).
6. Knockout-DMEM (Life Technologies).
7. Knockout Serum Replacement (Life Technologies).
8. L-Glutamine.
9. MEM non-essential amino acid solution (NEAA).
10. 2-Mercaptoethanol.
11. Recombinant basic fibroblast growth factor, human (bFGF, Life Technologies).
12. mTeSR1 (Stem Cell Technologies).
13. Corning Matrigel hESC-Qualified Matrix (Corning).
14. mFreSR Defined Cryopreservation Medium (Stem Cell Technologies).
15. Electroporator (for example: Lonza 4D Nucleofector System/X Unit) with suitable electroporation kit (P2 Primary Cell 4D-Nucleofector® X Kit L; cat no: V4XP-2012).
16. 1× Phosphate-buffered saline (PBS).
17. DNA plasmid pT2-OSKML [23].
18. SB100X transposase expression plasmid, such as pCMV(CAT) T7-SB100X, available from Addgene (http://www.addgene.org/34879/).

2.2 Media Preparation

(a) Fibroblast growth medium (FGM): DMEM + 10 % FBS.

(b) Human embryonic stem cell medium (hESCM): 80 % Knockout DMEM, 20 % knockout serum replacement, 1 % NEAA, 1 mM L-glutamine, 0.1 mM 2-mercaptoethanol, 4 ng/ml bFGF.

(c) Stop medium (SM): Knockout-DMEM + 15 % knockout serum replacement.

3 Methods

1. Coat wells of a 6-well plate with Matrigel according to the instructions of the manufacturer. Aspirate the medium from Matrigel-coated wells (*see* **Note 1**), and add 2 ml FGM pre-warmed to 37 °C per well.

2. Wash the fibroblast culture (*see* **Note 2**) once with 1× PBS and add the required amount of 0.05 % trypsin-EDTA solution pre-warmed to 37 °C (for example, 4 ml for a 75 cm^2 cell culture flask). Incubate for 3 min at 37 °C.
3. After incubation, add 6 ml of FGM, suspend the cells by gently pipetting, and transfer the cell suspension to a 15 ml conical tube.
4. Centrifuge cells at 200 × *g* for 5 min.
5. Discard the supernatant and resuspend the cells in 10 ml of FGM. Count cell number.
6. Pipet 5–8 × 10^5 cells into a fresh tube and centrifuge again at 200 × *g* for 5 min (upscale depending on the number of conditions in your experiment). Discard supernatant.
7. Premix in one Eppendorf tube 2 μg reprogramming transposon and 0.4 μg SB100X plasmids. Resuspend cells in 100 μl electroporation buffer, add the cell suspension to the tube containing the plasmids, mix, and transfer the cell suspension/plasmid DNA mix into an electroporation cuvette.
8. Proceed with electroporation according to the instructions of the manufacturer using the DT130 program.
9. After electroporation seed the cells into 2 ml FGM prewarmed to 37 °C in a well of the Matrigel-coated plate from **step 1**. Change the medium with 2 ml of fresh FGM 24 h later. Incubate cells at 37 °C and 5 % CO_2 in a humidified incubator throughout the whole reprogramming experiment.
10. Maintain cells in 2 ml of FGM for 48 h. Then change medium to hESCM (2 ml per well). Change medium daily. The fibroblast cells will divide until they reach confluence and will form a monolayer (Fig. 2b).
11. iPS colonies will appear towards the end of the second week, beginning in the third week of reprogramming (Fig. 2a).
12. After 4 weeks iPS colonies ready to be picked are visible by eye (Fig. 2c). Pick iPS colonies manually by cutting colonies from the neighboring fibroblasts using a syringe needle. Aspirate the colony by a 100 μl pipette in the lowest possible volume of medium. Transfer the colony into 20 μl of 0.05 % trypsin prewarmed to 37 °C in an Eppendorf tube; incubate for 3 min at 37 °C.
13. Add 20 μl of SM pre-warmed to 37 °C, and resuspend (by pipetting) the colony by dissociating it into several smaller pieces.

14. Prepare a 24-well Matrigel-coated plate as described in **step 1**, and add 500 μl mTeSR1 medium per well (*see* **Note 3**). Transfer the cell suspension into the well, and change antibiotic-free mTeSR1 medium daily. Incubate iPS cells at 37 °C and 5 % CO_2 in a humidified incubator.
15. Dissociate and transfer the cells into a well of a 6-well Matrigel-coated plate before reaching confluence (*see* **Note 4**). For cell dissociation, remove the medium, wash the cells once with 2 ml PBS pre-warmed to 37 °C, add 0.5 ml 0.05 % trypsin pre-warmed to 37 °C to the well, incubate for 3 min at 37 °C, and inactivate trypsin by adding 1 ml SM pre-warmed to 37 °C (*see* **Note 5**). Expand cells in 2 ml antibiotic-free mTeSR1 medium per well.
16. Archive aliquots of the cells for later use (*see* **Note 6**), and characterize pluripotency (*see* **Note 7**).

4 Notes

1. Incubate the Matrigel-coated plates at room temperature (15–25 °C) for at least 1 h before use.
2. Use fibroblasts with low (up to 4–5) passage numbers.
3. Always expand and grow iPS cells in antibiotic-free mTeSR1 medium.
4. Alternatively, when transferring cells from the 24-well plate, cell expansion can be carried out on feeder cells as well. In this case grow the iPS cells in antibiotic-free hESCM, and change medium daily.
5. Alternatively, other solutions can also be used for iPS cell dissociation: collagenase (Stem Cell Technologies, cat. no. 07902), accutase (Stem Cell Technologies, cat. no. 07920), or dispase (Stem Cell Technologies, cat. no. 07923).
6. Freeze cells in mFreSR medium, one well of a 6-well plate cell culture in 1 ml of freezing medium.
7. iPS cells can be characterized for pluripotent stem cell marker expression and, after subjecting cells to differentiation assays, for markers of the three germ layers. In Fig. 3 iPS cells obtained by SB-mediated reprogramming are shown expressing the pluripotent stem cell marker Nanog, as well as Oct4, which is expressed from the reprogramming cassette too.

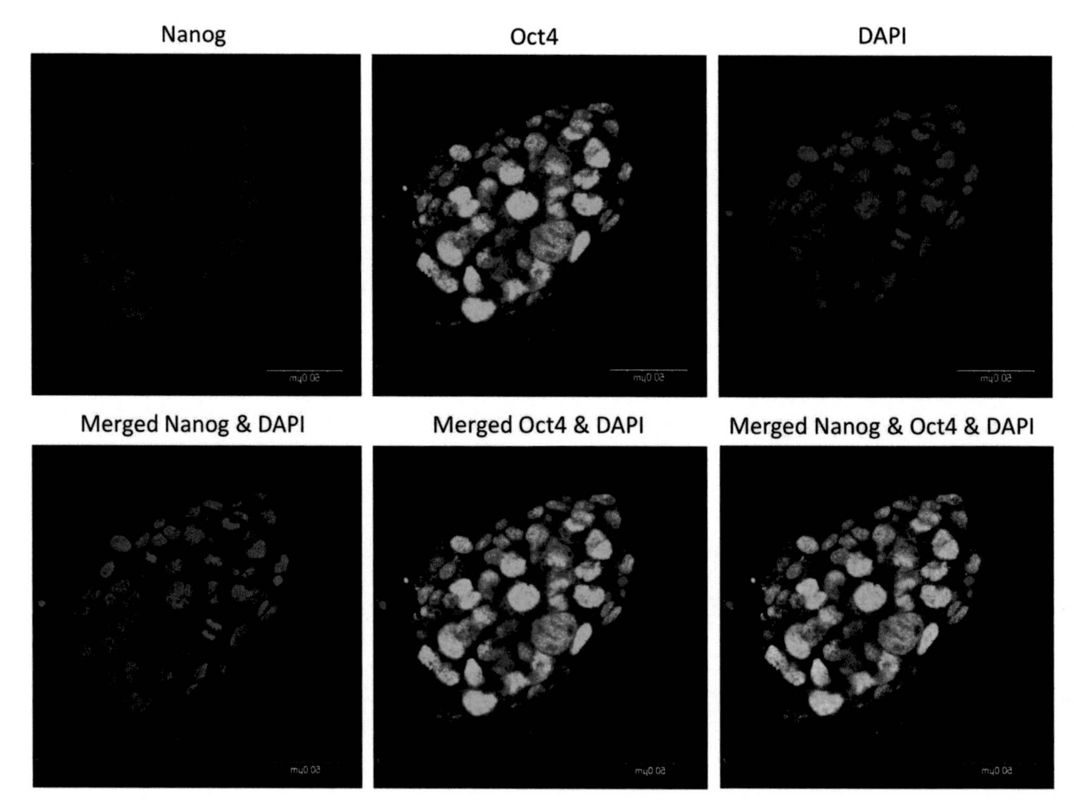

Fig. 3 Immunocytochemical characterization of human iPS cells generated by *Sleeping Beauty* transposon-mediated reprogramming. A typical iPS clump was examined by confocal microscopy after staining with antibodies against the pluriopotency markers Nanog and Oct4

References

1. Takahashi K, Tanabe K, Ohnuki M, Narita M, Ichisaka T, Tomoda K, Yamanaka S (2007) Induction of pluripotent stem cells from adult human fibroblasts by defined factors. Cell 131:861–872
2. Raya A, Rodriguez-Piza I, Guenechea G, Vassena R, Navarro S, Barrero MJ, Consiglio A, Castella M, Rio P, Sleep E, González F, Tiscornia G, Garreta E, Aasen T, Veiga A, Verma IM, Surrallés J, Bueren J, Izpisúa Belmonte JC (2009) Disease-corrected haematopoietic progenitors from Fanconi anaemia induced pluripotent stem cells. Nature 460:53–59
3. Hanna J, Wernig M, Markoulaki S, Sun CW, Meissner A, Cassady JP, Beard C, Brambrink T, Wu LC, Townes TM, Jaenisch R (2007) Treatment of sickle cell anemia mouse model with iPS cells generated from autologous skin. Science 318:1920–1923
4. Xu D, Alipio Z, Fink LM, Adcock DM, Yang J, Ward DC, Ma Y (2009) Phenotypic correction of murine hemophilia A using an iPS cell-based therapy. Proc Natl Acad Sci U S A 106:808–813
5. Zou J, Maeder ML, Mali P, Pruett-Miller SM, Thibodeau-Beganny S, Chou BK, Chen G, Ye Z, Park IH, Daley GQ, Porteus MH, Joung JK, Cheng L (2009) Gene targeting of a disease-related gene in human induced pluripotent stem and embryonic stem cells. Cell Stem Cell 5:97–110
6. Song P, Inagaki Y, Sugawara Y, Kokudo N (2013) Perspectives on human clinical trials of therapies using iPS cells in Japan: reaching the forefront of stem-cell therapies. Biosci Trends 7:157–158.
7. Takahashi K, Yamanaka S (2006) Induction of pluripotent stem cells from mouse embryonic and adult fibroblast cultures by defined factors. Cell 126:663–676
8. Aoi T, Yae K, Nakagawa M, Ichisaka T, Okita K, Takahashi K, Chiba T, Yamanaka S (2008) Generation of pluripotent stem cells from

adult mouse liver and stomach cells. Science 321:699–702

9. Sommer CA, Stadtfeld M, Murphy GJ, Hochedlinger K, Kotton DN, Mostoslavsky G (2009) Induced pluripotent stem cell generation using a single lentiviral stem cell cassette. Stem Cells 27:543–549
10. Okita K, Ichisaka T, Yamanaka S (2007) Generation of germline-competent induced pluripotent stem cells. Nature 448:313–317
11. Kaji K, Norrby K, Paca A, Mileikovsky M, Mohseni P, Woltjen K (2009) Virus-free induction of pluripotency and subsequent excision of reprogramming factors. Nature 458:771–775
12. Soldner F, Hockemeyer D, Beard C, Gao Q, Bell GW, Cook EG, Hargus G, Blak A, Cooper O, Mitalipova M, Isacson O, Jaenisch R (2009) Parkinson's disease patient-derived induced pluripotent stem cells free of viral reprogramming factors. Cell 136:964–977
13. Voelkel C, Galla M, Maetzig T, Warlich E, Kuehle J, Zychlinski D, Bode J, Cantz T, Schambach A, Baum C (2010) Protein transduction from retroviral Gag precursors. Proc Natl Acad Sci U S A 107:7805–7810
14. Okita K, Nakagawa M, Hyenjong H, Ichisaka T, Yamanaka S (2008) Generation of mouse induced pluripotent stem cells without viral vectors. Science 322:949–953
15. Stadtfeld M, Nagaya M, Utikal J, Weir G, Hochedlinger K (2008) Induced pluripotent stem cells generated without viral integration. Science 322:945–949
16. Fusaki N, Ban H, Nishiyama A, Saeki K, Hasegawa M (2009) Efficient induction of transgene-free human pluripotent stem cells using a vector based on Sendai virus, an RNA virus that does not integrate into the host genome. Proc Jpn Acad Ser B Phys Biol Sci 85:348–362
17. Yu J, Hu K, Smuga-Otto K, Tian S, Stewart R, Slukvin II, Thomson JA (2009) Human induced pluripotent stem cells free of vector and transgene sequences. Science 324:797–801
18. Warren L, Manos PD, Ahfeldt T, Loh YH, Li H, Lau F, Ebina W, Mandal PK, Smith ZD, Meissner A, Daley GQ, Brack AS, Collins JJ, Cowan C, Schlaeger TM, Rossi DJ (2010) Highly efficient reprogramming to pluripotency and directed differentiation of human cells with synthetic modified mRNA. Cell Stem Cell 7:618–630
19. Anokye-Danso F, Trivedi CM, Juhr D, Gupta M, Cui Z, Tian Y, Zhang Y, Yang W, Gruber PJ, Epstein JA, Morrisey EE (2011) Highly efficient miRNA-mediated reprogramming of mouse and human somatic cells to pluripotency. Cell Stem Cell 8:376–388
20. Ivics Z, Li MA, Mátés L, Boeke JD, Nagy A, Bradley A, Izsvák Z (2009) Transposon-mediated genome manipulation in vertebrates. Nat Methods 6:415–422
21. Ivics Z, Hackett PB, Plasterk RH, Izsvák Z (1997) Molecular reconstruction of Sleeping Beauty, a Tc1-like transposon from fish, and its transposition in human cells. Cell 91:501–510
22. Mátés L, Chuah MK, Belay E, Jerchow B, Manoj N, Acosta-Sanchez A, Grzela DP, Schmitt A, Becker K, Matrai J, Ma L, Samara-Kuko E, Gysemans C, Pryputniewicz D, Miskey C, Fletcher B, Vandendriessche T, Ivics Z, Izsvák Z (2009) Molecular evolution of a novel hyperactive Sleeping Beauty transposase enables robust stable gene transfer in vertebrates. Nat Genet 41:753–761
23. Grabundzija I, Wang J, Sebe A, Erdei Z, Kajdi R, Devaraj A, Steinemann D, Szuhai K, Stein U, Cantz T, Schambach A, Baum C, Izsvák Z, Sarkadi B, Ivics Z (2013) Sleeping beauty transposon-based system for cellular reprogramming and targeted gene insertion in induced pluripotent stem cells. Nucleic Acids Res 41:1829–1847

Index

Jose L. Garcia-Pérez (ed.), *Transposons and Retrotransposons: Methods and Protocols*, Methods in Molecular Biology, vol. 1400, DOI 10.1007/978-1-4939-3372-3, © Springer Science+Business Media New York 2016

D

E

F

G

H

S

T

Printed in the United States
By Bookmasters